The Proceedings of the 12th International Congress on Mathematical Education

W0112035

Sung Je Cho
Editor

The Proceedings of the 12th International Congress on Mathematical Education

Intellectual and Attitudinal Challenges

8 July – 15 July, 2012, COEX, Seoul, Korea

 Springer Open

Editor
Sung Je Cho
Seoul National University
Seoul
Korea

ISBN 978-3-319-10685-4   ISBN 978-3-319-12688-3   (eBook)
DOI 10.1007/978-3-319-12688-3

Library of Congress Control Number: 2014960270

Springer Cham Heidelberg New York Dordrecht London
© The Editor(s) (if applicable) and the Author(s) 2015. The book is published with open access at
SpringerLink.com.
Open Access This book is distributed under the terms of the Creative Commons Attribution
Noncommercial License which permits any noncommercial use, distribution, and reproduction in any
medium, provided the original author(s) and source are credited.
All commercial rights are reserved by the Publisher, whether the whole or part of the material is
concerned, specifically the rights of translation, reprinting, reuse of illustrations, recitation, broadcasting,
reproduction on microfilms or in any other physical way, and transmission or information storage and
retrieval, electronic adaptation, computer software, or by similar or dissimilar methodology now known
or hereafter developed.
The use of general descriptive names, registered names, trademarks, service marks, etc. in this
publication does not imply, even in the absence of a specific statement, that such names are exempt from
the relevant protective laws and regulations and therefore free for general use.
The publisher, the authors and the editors are safe to assume that the advice and information in this
book are believed to be true and accurate at the date of publication. Neither the publisher nor the
authors or the editors give a warranty, express or implied, with respect to the material contained herein or
for any errors or omissions that may have been made.

Printed on acid-free paper

Springer International Publishing AG Switzerland is part of Springer Science+Business Media
(www.springer.com)

Editor

Sung Je Cho, Seoul National University

Editors

Bill Barton, The University of Auckland
Gail Burrill, Michigan State University
Bernard R. Hodgson, Université Laval
Gabriele Kaiser, University of Hamburg
Oh Nam Kwon, Seoul National University
Hee-Chan Lew, Korea National University of Education

Editorial Board

Michèle Artigue, Université Paris Diderot—Paris 7
Évelyne Barbin, University of Nantes
Morten Blomhøj, IMFUFA Roskilde University
Jaime Carvalho e Silva, Universidade de Coimbra
Johann Engelbrecht, University of Pretoria
Mercy Kazima, University of Malawi
Masataka Koyama, Hiroshima University
Frederick Leung, The University of Hong Kong
Shiqi Li, East China Normal University
Cheryl E. Praeger, The University of Western Australia
Hyun Yong Shin, Korea National University of Education
K. (Ravi) Subramaniam, Homi Bhabha Centre for Science Education
Yuriko Yamamoto Baldin, Universidade Federal de São Carlos

Preface

This book is the result of the 12th International Congress on Mathematical Education (ICME-12), which was held at Seoul, Korea from July 8, 2012 to July 15, 2012.

The International Program Committee (IPC-12) of ICME-12 took on the task of acting as the editorial board to publish the Proceedings of the ICME-12 and Selected Regular Lectures of the ICME-12 in two separate volumes. All conference materials and volumes are accessible through the Open Access Program by Springer, the ICME-12 publisher.

The Proceedings volume of ICME-12 contains the Opening and Award Ceremonies, four Plenary Lectures and three Plenary Panels, four ICMI Awardees Lectures, three Survey Teams reports, five National Presentations, abstracts of 55 Regular Lectures, reports of 37 Topic Study Groups, reports of 17 discussion Groups, Closing Ceremony, and lists of participants. The Selected Regular Lectures volume of ICME-12 contains the full versions of lectures.

The ICME-12 would not have been possible without the contribution from its members and strategic partners. For the first time, all of the Korean mathematical societies united to bid and host the ICME-12. The successes of ICME-12 is closely tied to the tireless efforts of all.

A considerable amount of the ICME-12 budget was funded through private donations by mathematically minded individuals and businesses. ChunJae Education Inc. was one of the largest contributor of funds and services. Printing of the ICME-12 Program Booklets and Abstracts were paid for by ChunJae Education Inc.

The Korean Ministry of Education helped to secure the balance of the budget and assisted in the operation of ICME-12. The City of Seoul, Korea Foundation for the Advancement of Science and Creativity, and Korea Tourism Organization were significant funding bodies as well.

The dedicated members of the Local Organizing Committee, skilled professional conference organizers at MCI, and staff at the COEX (Convention and Exhibition) were integral in the successful planning and execution of ICME-12. The dedication shown by the Local Organizing Committee for the conference was second to none and well beyond expectation.

Finally, the Editor would like to express his sincere thanks to all the members of IPC-12, Korean government agencies, private donors, lecturers, members of Survey Teams, and organizers of Topics Study Groups and Discussion Groups. Gratitude also is extended to the more than 3,000 worldwide attendees who contributed to the success of the ICME-12 by sharing their expertise via paper presentations or participating in discussions. Without Prof. Hee-chan Lew's work and devotion, this extensive volume could not have been completed. The Editor would like to express his heartfelt thanks to him. The Editor believes that the world mathematical society is closer than before and leading towards more productive and friendly mathematics classrooms around the world.

Seoul, Korea   Sung Je Cho
                                                                                                  Editor

Contents

Part V   Survey Teams

Part VI   National Presentations

Part VII   Regular Lectures

Part VIII   Topic Study Groups

Part IX   Discussion Groups

Congratulatory Remarks: President of Korea

Good morning, distinguished ladies and gentlemen. I offer my heartful congratulations to the opening of the 12th International Congress on Mathematical Education (ICME).

I am very pleased to welcome President William Barton of the International Commission on Mathematical Instruction (ICMI), President Ingrid Daubechies of the International Mathematical Union (IMU) and all the mathematical educators here.

As the largest festivities in the realm of mathematical instructions, the ICME has made great contributions to mathematical instruction and popularization of math over the past half century.

Considering the significance of the ICME, I sent a letter to the ICMI requesting to hold the Congress in Korea 9 years ago when I was the Mayor of Seoul.

Today, I am glad to see my wish has come true. Let me take this opportunity to thank all the people who have devoted in preparing for this meeting.

Distinguished educator,

To me, mathematics is a magnificent journey of human reason in search of a clue for the mystery surrounding the universe.

For the past long history, mathematics has been the engine of civilization, striving to liberate humanity from famine, poverty and ignorance.

I hope the rational and creative thinking of young people will be enhanced through math. In this way, they will make the future of humanity better.

I expect all the participants will be able to freely exchange diverse and expert views for the advancement of mathematics at this very meaningful meeting.

As a post script, I would like to remind you that the International Expo is now being held in the southern port city of Yeosu. Please tour the Expo by all means and carry home fond memories of the beautiful natural and cultural landmarks of Korea's South Coast.

Korea has hosted many international events, buy I personally attach extraordinary importance to this congress because the ICME can be described as a "math education forum of Olympic proportions". Once again, welcome to Korea.

Thank you very much.

Lee Myung-bak
10th President of the Republic of Korea
Seoul, Republic of Korea

Part I
Opening Ceremonies

Opening Address: President of IMU

Ingrid Daubechies

It is a great pleasure for me to have the opportunity to address you, during this opening ceremony for the 12th International Congress on Mathematical Education, in my capacity as President of the International Mathematical Union, or IMU.

Officially, that is, with respect to the International Council for Science (or ICSU), which itself reports to UNESCO, IMU is the umbrella organization concerned with matters of global interest to mathematicians worldwide. The International Commission on Mathematical Instruction (or ICMI for brevity), which organizes the quadrennial ICME meetings, is the most important sub-organization of the IMU. In fact, and as ICMI President Bill Barton likes to remind me good-humoredly, ICMI is older than the IMU itself, since it was created in 1908—IMU was created only in 1920, and even then it was an earlier incarnation that stopped functioning in the 1930s; in its present version, it was reborn in 1951.

An extremely important charge for the IMU is to organize the prestigious quadrennial International Congresses of Mathematicians, or ICMs, the first one of which dates back to 1893; it is probably no exaggeration to state that the IMU was first started to ensure a regular and orderly organization of the ICMs. This is similar to the role ICMI plays with respect to the ICME congresses, which are all held under ICMI's auspices and principles. Once the ICME series hit its quadrennial rhythm, it became customary to hold the ICMs and ICMEs in interleaved even-numbered years, keeping stride nicely with the World Cup in Soccer/Football and the Olympic Games, which one could view as a "warm-up" for our more serious pursuits. The next ICM will thus take place in 2014, coincidentally in this very same city, in this very same Conference Center.

Over the years, IMU has come to stand for much more than just the umbrella organization ensuring continuity for the ICMs. In the past few decades, IMU has become more concerned with assisting developing countries build up their own

I. Daubechies (✉)
Duke University, Durham, USA
e-mail: ingrid@math.duke.edu

© The Author(s) 2015
S.J. Cho (ed.), *The Proceedings of the 12th International Congress on Mathematical Education*, DOI 10.1007/978-3-319-12688-3_1

strong mathematics communities. IMU is also solidly and seriously invested in helping develop and sustain excellent mathematics education everywhere, and at all levels—although the work of my colleagues on the Executive Committee of the IMU, as well as my own, is anchored in mathematics research, we all realize fully the importance of teaching mathematical insights, understanding and skills in the best possible way, and we are committed to help ICMI as much as we can in pursuing this goal. These are not empty words—we are acting on our beliefs! The following are just two examples. In setting up the new stable central Secretariat for the IMU, it was viewed as an essential and core part of its charge that it provides a stable administrative support and archival role for ICMI as well. On a different note, IMU is also actively helping ICMI in finding and providing funding for the very important CANP workshops, which build networking for mathematics educators in less developed regions in the world.

I am personally thrilled by this tighter connection between mathematical researchers and experts on, or researchers in, mathematical education. Whether we decide to contribute to mathematical research, or whether we decide to invest our creative energy in mathematics education—you and I, ICME or ICM participants, we are ALL mathematicians, united in our love for mathematics. It was a proud moment in my life when my son announced his decision to become a high school teacher in mathematics; he now teaches in one of the inner city schools in Chicago, and works hard to ignite and keep alive an interest in mathematics among his students, bringing to this the energy and drive that he could easily have taken to graduate school. I respect and value the commitment and engagement of teachers like him, and I encourage all professional research mathematicians to do likewise.

Dear ICME-12 Participants, fellow mathematicians, focused on bringing the best possible mathematics education to future generations, I salute you!

And I wish you a wonderful Congress.

Open Access This chapter is distributed under the terms of the Creative Commons Attribution Noncommercial License, which permits any noncommercial use, distribution, and reproduction in any medium, provided the original author(s) and source are credited.

Opening Address: President of ICMI

Bill Barton

Honourable Mr Lee, Minister of Education

Professor Sung Je Cho, Convenor of this wonderful conference

ICMI Colleagues and friends

Our moment has arrived. Isn't this wonderful!

I am delighted to be here, to open the 12th International Congress of Mathematics Education—to be honest, it is a moment I have been looking forward to for more than 4 years. Our community is very fortunate to have attracted a conference bid from Korea, and our Korean friends are already proving to us that we made a very good decision to accept their bid.

These few minutes are my opportunity to address the wider ICMI community about the things that I believe are important about mathematics education on the international stage. I cannot detail all the many, many activities of ICMI as an organisation: ICMI Studies, Regional conferences, Affiliated organisations, and on and on. I urge everyone in this room to find out who their ICMI country representative is, and ensure that they become part of their national network. You should also subscribe to the ICMI Newsletter (on line) or become a Facebook Friend. We survive as an organisation through your participation.

I wish to mention three topics: our major development project; the Klein Project; and finally some comments on how our community communicates.

Since the last ICME in Mexico, ICMI as an organisation has changed dramatically. We have extended our development activities significantly. It is no longer true that we are primarily an organisation of professionals in mathematics education. Now we spend at least half our efforts and resources on worldwide development activities. A major part of this effort is the Capacity and Networking Project, that we call CANP.

B. Barton (✉)
Former President of International Commission on Mathematical Instruction,
University of Auckland, Auckland, New Zealand

© The Author(s) 2015

S.J. Cho (ed.), *The Proceedings of the 12th International Congress on Mathematical Education*, DOI 10.1007/978-3-319-12688-3_2

The aim of CANP is to support developing regions to form self-sustaining networks of mathematics educators, mathematicians, government officials, and, of course, teachers. What ICMI does is to organise a two-week Workshop in a different region every year. Last year the first was held in Mali, this year the second will be held in Costa Rica, and next year it will be Cambodia. A region of four or five countries is selected, and a Scientific Committee is formed of four people from the international community and four from the region. The Workshop is usually about fifty people representing all the groups in the network. The focus of the Workshop is secondary teacher education, but the aim is really to get key people in the region working together. Funds for each CANP programme are raised separately, we have had significant support from IMU, UNESCO, CIMPA and other organisations.

My second topic is the Klein Project. I invite everyone to turn ON their smartphones or open their computers—please go to the Klein Project Blog <http:// blog.kleinproject.org> ... or at least write this down, and log in at your first opportunity. The Klein Project is a worldwide project to produce writing on contemporary mathematics for secondary school teachers. Note: it is not designed for use in classrooms, but for the pleasure and satisfaction of teachers. In the Klein Blog you will find Klein Vignettes—these are short (4–6 pages) on a contemporary topic, written for secondary school mathematics teachers.

Over the next months you will see the Klein Blog grow—both with new Vignettes, but also as we translate the Vignettes into any and every language. This is a major task for our community, and I seek your help to offer to translate the Vignettes into your languages.

Eventually there will also be a Klein Project book—a small volume aimed at secondary teachers, that they will be able to dip into in the spare moments of their busy teaching lives. A book that will sustain and inspire teachers mathematically.

Please will you have a look, feed back to the project with your reactions, offer to help write more materials, and, most importantly, spread the Blog address amongst your secondary teacher friends and networks—or anyone whom you think would be interested.

I mention the Klein Project not because it is ICMI's only project—it is not, we have several others—but because it illustrates for me an very important point: that ICMI works more closely than ever with IMU, the world body of mathematicians. The Klein Project is a joint project with IMU, and every piece of writing is the result of collaborations between mathematics teachers, educators, and mathematicians.

And lest you think that ICMI is focused only on secondary teachers and mathematicians, let me quickly say: "Look out for the next ICMI Study announcement—it will be on Primary Mathematics". Watch for the announcement in December.

Finally, allow me to note that ICMI is changing in another respect—it is changing in the way the world is changing. New technologies, new modes of communication, new groupings, new social imperatives, new problems to be solved and questions to be answered. ICMI must and does change, and in particular we change in the way we communicate. We have a Facebook page, we have a bank of digitised publications, we have an ever increasing website. In what new ways will we meet and communicate in four years time? We need the new members of our community to lead us in this matter—and I call on you all to embrace the movement forward into new worlds.

But face-to-face communication will, in my opinion, always be highly valued. Being able to Skype my grandchildren or my research colleagues on the other side of the world only makes me want to actually see them and spend time with them so much more.

And this is why we are here. To greet and see and talk to each other. To make new friends and affirm old ones. And we do this with great pleasure at the same time as we work hard to improve the learning of mathematics in classrooms at all levels in every country.

Thus I regard it as one of the greatest honours of my career to declare the 12th International Congress on Mathematics Education Officially open.

Open Access This chapter is distributed under the terms of the Creative Commons Attribution Noncommercial License, which permits any noncommercial use, distribution, and reproduction in any medium, provided the original author(s) and source are credited.

Welcome Address: Chair of IPC

Sung Je Cho

I would like to express my utmost gratitude to His Excellency Lee Myung-bak, the President of the Republic of Korea for preparing a welcoming message for us despite his busy schedule.

Your Excellency Lee Ju Ho, Minister of Education, Science, and Technology, Professor Ingrid Daubechies, the President of IMU, Professor Bill Barton, the President of ICMI, Ladies and Gentlemen, distinguished guests and participants from all around the world, I would like to extend my warmest welcome to you all.

We, the Korean Mathematics Society and Korean Mathematics Education Society, are very proud to host the 12th International Congress on Mathematical Education. Our International Programme Committee has worked tirelessly through two face-to-face meetings and numerous internet discussions. It is needless to say that this Congress would not be possible without the dedicated and coordinated efforts of members of the various committees, presenters and participants. We thank all of you for making this a reality.

Mathematics has been at the heart of human culture, philosophy, technology and advancement since the dawn of civilization. We cannot think of our modern society apart from mathematics because mathematics influences every facet of our daily lives. Due to the far reaching effects of mathematics in our world, mathematics education may be one of the most efficient ways to influence betterment of mankind. For the week starting today, we are gathered here to nurture and cultivate the mathematics educational environment for our future generation so that they may become significant part of the solution and advancement of our society.

S.J. Cho (✉)
International Programme Committee of ICME-12,
Seoul National University, Seoul, Republic of Korea
e-mail: sungjcho@snu.ac.kr

© The Author(s) 2015
S.J. Cho (ed.), *The Proceedings of the 12th International Congress on Mathematical Education*, DOI 10.1007/978-3-319-12688-3_3

It is our sincere hope that this Congress would inspire wider and tighter mathematics education research network as well as inviting and stimulating mathematics classrooms all over the world.

Thank you,

Open Access This chapter is distributed under the terms of the Creative Commons Attribution Noncommercial License, which permits any noncommercial use, distribution, and reproduction in any medium, provided the original author(s) and source are credited.

Congratulatory Remarks: Minister of Education and Science, and Technology

Ju Ho Lee

First of all, congratulations on the opening of the 12th International Congress on Mathematical Education.

I am glad that this important math event is being held in Korea this year.

Also, it is a great pleasure to welcome math education researchers and math teachers from more than 100 countries.

With the aim of transforming Korea into a nation of great science and technology capacity, and a nation of outstanding human talent, the Ministry of Education, Science and Technology of Korea is focusing on three important points in designing and implementing its policies.

The three points are "creativity", "convergence", and "human talent". Creativity enables us to think outside the box, convergence allows us to go beyond the traditional boundaries between disciplines, and finally human talent builds the very foundation that make all these possible.

Without a doubt, these are the most essential elements in today's knowledge-based society.

Math is the very subject that can foster much needed creativity and convergence, and is becoming a core factor in raising national competitiveness.

Math is behind everything.

The ICT revolution would have been impossible without the binary system.

The technology behind the CT scans can be traced back to simultaneous equations.

Open Access This chapter is distributed under the terms of the Creative Commons Attribution Noncommercial License, which permits any noncommercial use, distribution, and reproduction in any medium, provided the original author(s) and source are credited.

An erratum of the original chapter can be found under DOI 10.1007/978-3-319-12688-3_80

J.H. Lee (✉)
Former Minister of Education, Science and Technology, KDI School
of Public Policy and Management, Seoul, Republic of Korea

© The Author(s) 2015
S.J. Cho (ed.), *The Proceedings of the 12th International Congress on Mathematical Education*, DOI 10.1007/978-3-319-12688-3_4

ICMI Awards Report

Carolyn Kieran

A wonderful part of the opening session of the ICME congresses is the ICMI Awards ceremony. The 2012 ceremony, which was presided over by Prof. Carolyn Kieran, the chair of the ICMI Awards Committee, was no exception. Congress participants shared in congratulating the recipients of the 2009 and 2011 competitions for the Klein and Freudenthal awards. The Korean Minister of Education, Science, and Technology, the Honorable Mr. Ju-Ho Lee, did us the honor of presenting each award.

In 2000, the International Commission on Mathematical Instruction decided to create two prizes given in recognition of outstanding achievement in mathematics education research:

- the Felix Klein Award, which honours lifetime achievement in our field, and
- the Hans Freudenthal Award, which honours a major cumulative programme of research.

Each award consists of a medal and a certificate, accompanied by a citation. The two awards are given in odd-numbered years. A six-person Awards Committee is responsible for selecting the awardees and for producing the citations explaining the merits of the awardees. The members, of whom only the Chair is known, are appointed by the President of ICMI and serve on the Committee for 8 years.

Scientific and scholarly quality is of course the fundamental characteristic involved in reviewing the candidates' work and merits. The first Committee, which was appointed in 2002, agreed on four aspects of quality, four criteria of evaluation: impact, sustainability, depth, and novelty. These criteria have been maintained throughout the Committee's work. Nevertheless, the field is influenced by social and cultural conditions, traditions, values, norms, and priorities. So, there are, inevitably, delicate balances to be struck between different dimensions, different traditions, different cultural and ethnic regions, and—indeed—different schools of

C. Kieran (✉)
Université du Québec à Montréal, Montréal, QC, Canada

© The Author(s) 2015   13
S.J. Cho (ed.), *The Proceedings of the 12th International Congress on Mathematical Education*, DOI 10.1007/978-3-319-12688-3_5

thought. Past Klein awardees have been Guy Brousseau (2003), Ubiritan D'Ambrosio (2005), and Jeremy Kilpatrick (2007). Past Freudenthal awardees have been Celia Hoyles (2003), Paul Cobb (2005), and Anna Sfard (2007).

At the 2012 ICMI Awards ceremony, the following four individuals were honored for their contributions to the field.

- *The Felix Klein Medal for 2009*: awarded to IAS Distinguished Professor and Professor Emerita Gilah C. Leder, La Trobe University, Bundoora, Victoria, Australia.
- *The Hans Freudenthal Medal for 2009*: awarded to Professor Yves Chevallard, IUFM d'Aix-Marseille, France.
- *The Felix Klein Medal for 2011*: awarded to the Elizabeth and Edward Connor Professor of Education and Affiliated Professor of Mathematics, Alan H. Schoenfeld, University of California at Berkeley, USA.
- *The Hans Freudenthal Medal for 2011*: awarded to Professor Luis Radford, Université Laurentienne, Sudbury, Canada.

Gilah Leder's citation, which was read by ICMI President Bill Barton, acknowledged her more than thirty years of sustained, consistent, and outstanding lifetime achievement in mathematics education research and development. Her particular emphasis on gender success and equity in mathematics education, but also more broadly her work on assessment, student affect, attitudes, beliefs, and self-concepts in relation to mathematics education from school to university, as well as her research methodology, and teacher education, have contributed to shaping these areas and have made a seminal impact on all subsequent research.

Yves Chevallard's citation, which was read by ICMI Vice-President Mina Teicher, recognized his foundational development of an original, fruitful, and influential research programme in mathematics education. The early years of the programme focused on the notion of didactical transposition of mathematical knowledge from outside school to inside the mathematics classroom, a transposition that also transforms the very nature of mathematical knowledge. The theoretical frame was further developed and gave rise to the anthropological theory of didactics (ATD), which offers a tool for modelling and analysing a diversity of human activities in relation to mathematics.

Alan Schoenfeld's citation, which was read by ICMI Past-President Michèle Artigue, recognized his more than thirty years of scholarly work that has shaped research and theory development in mathematical learning and teaching. His fundamental theoretical and applied work that connects research and practice in assessment, mathematical curriculum, diversity in mathematics education, research methodology, and teacher education has made a seminal impact on subsequent research. Another significant component of his achievements has been the mentoring he has provided to graduate students and scholars, nurturing a generation of new scholars.

Luis Radford's citation, which was read by ICMI Vice-President Angel Ruiz, acknowledged the outstanding contribution of the theoretically well-conceived and highly coherent research programme that he initiated and brought to fruition over the past two decades. His development of a semiotic-cultural theory of learning, rooted in his interest in the history of mathematics, has drawn on epistemology, semiotics, anthropology, psychology, and philosophy, and has been anchored in detailed observations of students' algebraic activity in class. His research, which has been documented in a vast number of scientific articles and in invited keynote presentations, has had a significant impact on the community.

The image of the four awardees standing on the stage together, receiving their medals and accompanying certificates from the Minister of Education—as well as the beautiful bouquets of flowers presented by young Koreans in traditional dress— is one that will stay with us for quite some time.

Open Access This chapter is distributed under the terms of the Creative Commons Attribution Noncommercial License, which permits any noncommercial use, distribution, and reproduction in any medium, provided the original author(s) and source are credited.

Part II
Plenary Lectures

Part II
Plenary lectures

The Butterfly Effect

Étienne Ghys

Abstract It is very unusual for a mathematical idea to disseminate into the society at large. An interesting example is chaos theory, popularized by Lorenz's butterfly effect: "does the flap of a butterfly's wings in Brazil set off a tornado in Texas?" A tiny cause can generate big consequences! Can one adequately summarize chaos theory in such a simple minded way? Are mathematicians responsible for the inadequate transmission of their theories outside of their own community? What is the precise message that Lorenz wanted to convey? Some of the main characters of the history of chaos were indeed concerned with the problem of communicating their ideas to other scientists or non-scientists. I'll try to discuss their successes and failures. The education of future mathematicians should include specific training to teach them how to explain mathematics outside their community. This is more and more necessary due to the increasing complexity of mathematics. A necessity and a challenge!

Introduction

In 1972, the meteorologist Edward Lorenz gave a talk at the 139th meeting of the *American Association for the Advancement of Science* entitled "Does the flap of a butterfly's wings in Brazil set off a tornado in Texas?". Forty years later, a *google* search "butterfly effect" generates ten million answers. Surprisingly most answers are not related to mathematics or physics and one can find the most improbable websites related to movies, music, popular books, video games, religion, philosophy and even Marxism! It is very unusual that a mathematical idea can disseminate into the general society. One could mention Thom's catastrophe theory in the 1970s, or Mandelbrot's fractals in the 1980s, but these theories remained confined to the scientifically oriented population. On the contrary, *chaos theory*, often

É. Ghys (✉)
CNRS-UMPA ENS Lyon, Lyon, France
e-mail: etienne.ghys@ens-lyon.fr

© The Author(s) 2015

S.J. Cho (ed.), *The Proceedings of the 12th International Congress on Mathematical Education*, DOI 10.1007/978-3-319-12688-3_6

19

presented through the butterfly effect, did penetrate the nonscientific population at a very large scale. Unfortunately, this wide diffusion was accompanied with an oversimplification of the main original ideas and one has to admit that the transmission procedure from scientists to nonscientists was a failure. As an example, the successful book *The butterfly effect* by Andy Andrews "reveals the secret of how you can live a life of permanent purpose" and "shows how your everyday actions can make a difference for generations to come" which is not exactly the message of the founding fathers of chaos theory! In Spielberg's movie *Jurassic Park*, Jeff Goldblum introduces himself as a "chaotician" and tries (unsuccessfully) to explain the butterfly effect and unpredictability to the charming Laura Dern; the message is scientifically more accurate but misses the main point. If chaos theory only claimed that the future is unpredictable, would it deserve the name "theory"? After all, it is well known that "Prediction is very difficult, especially the future!".[1] A scientific theory cannot be limited to negative statements and one would be disappointed if Lorenz's message only contained this well known fact.

The purpose of this talk is twofold. On the one hand, I would like to give a very elementary presentation of chaos theory, as a mathematical theory, and to give some general overview on the current research activity in this domain with an emphasis on the role of the so-called *physical measures*. On the other hand, I would like to analyze the historical process of the development of the theory, its successes and failures, focusing in particular on the transmission of ideas between mathematics and physics, or from Science to the general public. This case study might give us some hints to improve the communication of mathematical ideas outside mathematics or scientific circles. The gap between mathematicians and the general population has never been so wide. This may be due to the increasing complexity of mathematics or to the decreasing interest of the population for Science. I believe that the mathematical community has the responsibility of building bridges.

A Brief History of Chaos from Newton to Lorenz

Determinism

One of the main pillars of Science is *determinism*: the possibility of prediction. This is of course not due to a single person but one should probably emphasize the fundamental role of Newton. As he was laying the foundations of differential calculus and unraveling the laws of mechanics, he was offering by the same token a tool enabling predictions. Given a mechanical system, be it the solar system or the collection of molecules in my room, one can write down a differential equation governing the motion. If one knows the present position and velocity of the system, one should

---

[1] See www.peterpatau.com/2006/12/bohr-leads-berra-but-yogi-closing-gap.html for an interesting discussion of the origin of this quotation.

simply solve a *differential equation* in order to determine the future. Of course, solving a differential equation is not always a simple matter but this implies at least the *principle* of determinism: the present situation determines the future. Laplace summarized this wonderfully in his "Essai philosophique sur les probabilités" (Laplace, 1814):

> We ought then to consider the present state of the universe as the effect of its previous state and as the cause of that which is to follow. An intelligence that, at a given instant, could comprehend all the forces by which nature is animated and the respective situation of the beings that make it up, if moreover it were vast enough to submit these data to analysis, would encompass in the same formula the movements of the greatest bodies of the universe and those of the lightest atoms. For such an intelligence nothing would be uncertain, and the future, like the past, would be open to its eyes.

The fact that this quotation comes from a book on *probability theory* shows that Laplace's view on determinism was far from naïve (Kahane 2008). We lack the "vast intelligence" and we are forced to use probabilities in order to understand dynamical systems.

Sensitivity to Initial Conditions

In his little book "Matter and Motion", Maxwell insists on the sensitivity to initial conditions in physical phenomena (Maxwell, 1876):

> There is a maxim which is often quoted, that 'The same causes will always produce the same effects.' To make this maxim intelligible we must define what we mean by the same causes and the same effects, since it is manifest that no event ever happens more that once, so that the causes and effects cannot be the same in *all* respects. [...]
> There is another maxim which must not be confounded with that quoted at the beginning of this article, which asserts 'That like causes produce like effects'. This is only true when small variations in the initial circumstances produce only small variations in the final state of the system. In a great many physical phenomena this condition is satisfied; but there are other cases in which a small initial variation may produce a great change in the final state of the system, as when the displacement of the 'points' causes a railway train to run into another instead of keeping its proper course.

Notice that Maxwell seems to believe that "in great many cases" there is no sensitivity to initial conditions. The question of the frequency of chaos in nature is still at the heart of current research. Note also that Maxwell did not really describe what we would call chaos today. Indeed, if one drops a rock from the top of a mountain, it is clear that the valley where it will end its course can be sensitive to a small variation of the initial position but it is equally clear that the motion cannot be called "chaotic" in any sense of the word: the rock simply goes downwards and eventually stops.

Fear for Chaos

It is usually asserted that chaos was "discovered" by Poincaré in his famous memoir on the 3-body problem (Poincaré 1890). His role is without doubt very important, but maybe not as much as is often claimed. He was not the first to discover sensitivity to initial conditions. However, he certainly realized that some mechanical motions are very intricate, in a way that Maxwell had not imagined. Nevertheless chaos theory cannot be limited to the statement that the dynamics is complicated: any reasonable theory must provide methods allowing some kind of understanding. The following famous quotation of Poincaré illustrates his despair when confronted by the complication of dynamics (Poincaré 1890):

> When we try to represent the figure formed by these two curves and their infinitely many intersections, each corresponding to a doubly asymptotic solution, these intersections form a type of trellis, tissue, or grid with infinitely fine mesh. Neither of the two curves must ever cut across itself again, but it must bend back upon itself in a very complex manner in order to cut across all of the meshes in the grid an infinite number of times. The complexity of this figure is striking, and I shall not even try to draw it. Nothing is more suitable for providing us with an idea of the complex nature of the three-body problem, and of all the problems of dynamics in general [...].

One should mention that ten years earlier Poincaré had written a fundamental memoir "Sur les courbes définies par des équations différentielles" laying the foundations of the qualitative theory of dynamical systems (Poincaré 1881). In this paper, he had analyzed in great detail the behavior of the trajectories of a vector field in the plane, i.e. of the solutions of an ordinary differential equation in dimension 2. One of his main results—the Poincaré-Bendixson theorem—implied that such trajectories are very well behaved and converge to an equilibrium point or to a periodic trajectory (or to a so-called "graphic"): nothing chaotic in dimension 2! In his 1890 paper, he was dealing with differential equations in dimension 3 and he must have been puzzled—and scared—when he realized the complexity of the picture.

Taming Chaos

Hadamard wrote a fundamental paper on the dynamical behavior of geodesics on negatively curved surfaces (Hadamard, 1898). He first observes that "a tiny change of direction of a geodesic [...] is sufficient to cause any variation of the final shape of the curve" but he goes much further and creates the main concepts of the so-called "symbolic dynamics". This enables him to prove positive statements, giving a fairly precise description of the behavior of geodesics. Of course, Hadamard is perfectly aware of the fact that geodesics on a surface define a very primitive mechanical system and that it is not clear at all that natural phenomena could have a similar behavior. He concludes his paper in a cautious way:

Will the circumstances we have just described occur in other problems of mechanics? In particular, will they appear in the motion of celestial bodies? We are unable to make such an assertion. However, it is likely that the results obtained for these difficult cases will be analogous to the preceding ones, at least in their degree of complexity. [...]
Certainly, if a system moves under the action of given forces and its initial conditions have given values *in the mathematical sense*, its future motion and behavior are exactly known. But, in astronomical problems, the situation is quite different: the constants defining the motion are only *physically* known, that is with some errors; their sizes get reduced along the progresses of our observing devices, but these errors can never completely vanish.

So far, the idea that some physical systems could be complicated and sensitive to small variations of the initial conditions—making predictions *impossible in practice* —remained hidden in very confidential mathematical papers known to a very small number of scientists. One should keep in mind that by the turn of the century, physics was triumphant and the general opinion was that Science would eventually explain everything. The revolutionary idea that there is a strong conceptual limitation to predictability was simply unacceptable to most scientists.

Popularization

However, at least two scientists realized that this idea is relevant in Science and tried—unsuccessfully—to advertize it outside mathematics and physics, in "popular books".

In his widely circulated book *Science and Method*, Poincaré expresses the dependence to initial conditions in a very clear way. The formulation is very close to the butterfly slogan and even includes a devastating cyclone (Poincaré 1908):

Why have meteorologists such difficulty in predicting the weather with any certainty? Why is it that showers and even storms seem to come by chance, so that many people think it quite natural to pray for rain or fine weather, though they would consider it ridiculous to ask for an eclipse by prayer? We see that great disturbances are generally produced in regions where the atmosphere is in unstable equilibrium. The meteorologists see very well that the equilibrium is unstable, that a cyclone will be formed somewhere, but exactly where they are not in a position to say; a tenth of a degree more or less at any given point, and the cyclone will burst here and not there, and extend its ravages over districts it would otherwise have spared. If they had been aware of this tenth of a degree they could have known it beforehand, but the observations were neither sufficiently comprehensive nor sufficiently precise, and that is the reason why it all seems due to the intervention of chance.

In 1908 Poincaré was less scared by chaos than in 1890. He was no longer considering chaos as an obstacle to a global understanding of the dynamics, at least from the probabilistic viewpoint. Reading Poincaré's papers of this period, with today's understanding of the theory, one realizes that he had indeed discovered the role of what is called today *physical measures* (to be discussed later) which are at the heart of the current approach. Unfortunately, none of his contemporaries could grasp the idea—or maybe he did not formulate it in a suitable way—and one had to wait for seventy years before the idea could be re-discovered!

You are asking me to predict future phenomena. If, quite unluckily, I happened to know the laws of these phenomena, I could achieve this goal only at the price of inextricable computations, and should renounce to answer you; but since I am lucky enough to ignore these laws, I will answer you straight away. And the most astonishing is that my answer will be correct.

Another attempt to advertize these ideas outside mathematics and physics was made by Duhem (1906) in his book *The aim and structure of physical theory*. His purpose was to popularize Hadamard's paper and he used simple words and very efficient "slogans":

Imagine the forehead of a bull, with the protuberances from which the horns and ears start, and with the collars hollowed out between these protuberances; but elongate these horns and ears without limit so that they extend to infinity; then you will have one of the surfaces we wish to study. On such a surface geodesics may show many different aspects. There are, first of all, geodesics which close on themselves. There are some also which are never infinitely distant from their starting point even though they never exactly pass through it again; some turn continually around the right horn, others around the left horn, or right ear, or left ear; others, more complicated, alternate, in accordance with certain rules, the turns they describe around one horn with the turns they describe around the other horn, or around one of the ears. Finally, on the forehead of our bull with his unlimited horns and ears there will be geodesics going to infinity, some mounting the right horn, others mounting the left horn, and still others following the right or left ear. [...] If, therefore, a material point is thrown on the surface studied starting from a geometrically given position with a geometrically given velocity, mathematical deduction can determine the trajectory of this point and tell whether this path goes to infinity or not. But, for the physicist, this deduction is forever unutilizable. When, indeed, the data are no longer known geometrically, but are determined by physical procedures as precise as we may suppose, the question put remains and will always remain unanswered.

Unfortunately the time was not ripe. Scientists were not ready for the message... Poincaré and Duhem were not heard. The theory went into a coma. Not completely though, since Birkhoff continued the work of Poincaré in a strictly mathematical way, with no attempts to develop a school, and with no applications to natural sciences. One should mention that Poincaré's work had also some posterity in the Soviet Union but this was more related to the 1881 "non chaotic" theory of limit cycles (Aubin and Dahan Dalmedico 2002).

Later I will describe Lorenz's fundamental article which bears the technical title "Deterministic non periodic flow", and was largely unnoticed by mathematicians for about ten years (Lorenz, 1963). Lorenz gave a lecture entitled "Predictability: does the flap of a butterfly's wings in Brazil set off a tornado in Texas?" which was the starting point of the famous butterfly effect (Lorenz, 1972).

If a single flap of a butterfly's wing can be instrumental in generating a tornado, so all the previous and subsequent flaps of its wings, as can the flaps of the wings of the millions of other butterflies, not to mention the activities of innumerable more powerful creatures, including our own species.
If a flap of a butterfly's wing can be instrumental in generating a tornado, it can equally well be instrumental in preventing a tornado.

This is not really different from Poincaré's "a tenth of a degree more or less at any given point, and the cyclone will burst here and not there". However, meanwhile, physics (and mathematics) had gone through several revolutions and non-predictability had become an acceptable idea. More importantly, the world had also gone through several (more important) revolutions. The message "each one of us can change the world[2]" was received as a sign of individual freedom. This is probably the explanation of the success of the butterfly effect in popular culture. It would be interesting to describe how Lorenz's talk reached the general population. One should certainly mention the best seller *Chaos: making a new science* (Gleick 1987) (which was a finalist for the Pulitzer Prize). One should not minimize the importance of such books. One should also emphasize that Lorenz himself published a wonderful popular book The essence of chaos in 1993. Note that the two main characters of the theory, Poincaré and Lorenz, wrote popular books to make their researches accessible to a wide audience.

Lorenz's 1963 Paper

Lorenz's article is wonderful (Lorenz 1963). At first unnoticed, it eventually became one of the most cited papers in scientific literature (more than 6,000 citations since 1963 and about 400 each year in recent years). For a few years, Lorenz had been studying simplified models describing the motion of the atmosphere in terms of ordinary differential equations depending on a small number of variables. For instance, in 1960 he had described a system that can be explicitly solved using elliptic functions: solutions were *quasiperiodic* in time (Lorenz 1960). His article (Lorenz 1962) analyzes a differential equation in a space of dimension 12, in which he numerically detects a sensitive dependence to initial conditions. His 1963 paper lead him to fame.

> In this study we shall work with systems of deterministic equations which are idealizations of hydrodynamical systems.

After all, the atmosphere is made of finitely many particles, so one indeed needs to solve an ordinary differential equation in a huge dimensional space. Of course, such equations are intractable, and one must treat them as partial differential equations. In turn, the latter must be discretized on a finite grid, leading to new ordinary differential equations depending on fewer variables, and probably more useful than the original ones.

The bibliography in Lorenz's article includes one article of Poincaré, but not the right one! He cites the early 1881 "non chaotic" memoir dealing with 2 dimensional dynamics. Lorenz seems indeed to have overlooked the Poincaré's papers that we have discussed above. Another bibliographic reference is a book by Birkhoff (1927)

---

[2] Subtitle of a book by Bill Clinton (2007).

on dynamical systems. Again, this is not "the right" reference since the "significant" papers on chaos by Birkhoff were published later. On the occasion of the 1991 Kyoto prize, Lorenz gave a lecture entitled "A scientist by choice" in which he discusses his relationship with mathematics (Lorenz 1991). In 1938 he was a graduate student in Harvard and was working under the guidance of... Birkhoff "on a problem in mathematical physics". However he seems unaware of the fact that Birkhoff was indeed the best follower of Poincaré. A missed opportunity? On the other hand, Lorenz mentions that Birkhoff "was noted for having formulated a theory of aesthetics".

Lorenz considers the phenomenon of *convection*. A thin layer of a viscous fluid is placed between two horizontal planes, set at two different temperatures, and one wants to describe the resulting motion. The higher parts of the fluid are colder, therefore denser; they have thus a tendency to go down due to gravity, and are then heated when they reach the lower regions. The resulting circulation of the fluid is complex. Physicists are very familiar with the Bénard and Rayleigh experiments. Assuming the solutions are periodic in space, expanding in Fourier series and truncating these series to keep only a small number of terms, Salzman had just obtained an ordinary differential equation describing the evolution. Drastically simplifying this equation, Lorenz obtained "his" differential equation:

$$\frac{dx}{dt} = \sigma(x+y); \; \frac{dy}{dt} = -xz + rz - y; \; \frac{dz}{dt} = xy - bz.$$

Here x represents the intensity of the convection, y represents the temperature difference between the ascending and descending currents, and z is proportional to the "distortion" of the vertical temperature profile from linearity, a positive value indicating that the strongest gradients occur near the boundaries. Obviously, one should not seek in this equation a faithful representation of the physical phenomenon. The constant σ is the *Prandtl number*. Guided by physical considerations, Lorenz was lead to choose the numerical values $r = 28$, $\sigma = 10$, $b = 8/3$. It was a good choice, and these values remain traditional today. He could then numerically solve these equations, and observe a few trajectories. The electronic computer Royal McBee LGP-30 was rather primitive: according to Lorenz, it computed (only!) 1,000 times faster than by hand. The anecdote is well known (Lorenz 1991):

> I started the computer again and went out for a cup of coffee. When I returned about an hour later, after the computer had generated about two months of data, I found that the new solution did not agree with the original one. [...] I realized that if the real atmosphere behaved in the same manner as the model, long-range weather prediction would be impossible, since most real weather elements were certainly not measured accurately to three decimal places.

Let us introduce some basic terminology and notation. For simplicity we shall only deal with ordinary differential equations in $\mathbb{R}^n$ of the form $\frac{dx}{dt} = X(x)$ where x is now a point in $\mathbb{R}^n$ and X is a vector field in $\mathbb{R}^n$. We shall assume that X is transversal to some large sphere, say $\|x\| = R$, pointing inwards, which means that the scalar product $x.X(x)$ is negative on this sphere. Denote by B the ball $\|x\| \leq R$. For any point x in B, there is a unique solution of the differential equation with initial condition x and defined for all $t \geq 0$. Denote this solution by $\phi^t(x)$. The purpose of the theory of *dynamical systems* is to understand the asymptotic behavior of these trajectories when t tends to infinity. With this terminology, one says that X is *sensitive to initial conditions* if there exists some $\delta > 0$ such that for every $\epsilon > 0$ one can find two points x, x' in B with $\|x - x'\| < \epsilon$ and some time $t > 0$ such that $\|\phi^t(x) - \phi^t(x')\| < \delta$.

Lorenz's observations go much further than the fact that "his" differential equation is sensitive to initial conditions. He notices that these unstable trajectories seem to accumulate on a complicated compact set, which is itself *insensitive* to initial conditions and he describes this limit set in a remarkably precise way. There exists some compact set K in the ball such that for almost every initial condition x, the trajectory of x accumulates precisely on K. This attracting set K (now called the *Lorenz attractor*) approximately resembles a surface presenting a "double" line along which two leaves merge.

> Thus within the limits of accuracy of the printed values, the trajectory is confined to a pair of surfaces which appear to merge in the lower portion. [...] It would seem, then, that the two surfaces merely appear to merge, and remain distinct surfaces. [...] Continuing this process for another circuit, we see that there are really eight surfaces, etc., and we finally conclude that there is an infinite complex of surfaces, each extremely close to one or the other of the two merging surfaces.

Lorenz (1963)

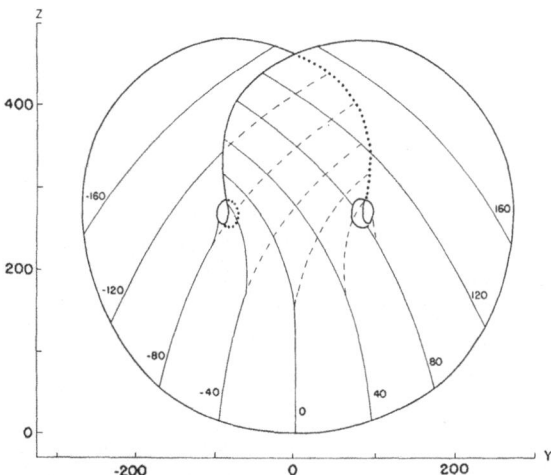

Starting from an initial condition, the trajectory rapidly approaches this "two dimensional object" and then travels "on" this "surface". The trajectory turns around the two holes, left or right, in a seemingly random way. Notice the analogy with Hadamard's geodesics turning around the horns of a bull. Besides, Lorenz studies how trajectories come back to the "branching line" where the two surfaces merge, which can be parameterized by some interval [0,1]. Obviously, this interval is not very well defined, since the two merging surfaces do not really come in contact, although they coincide "within the limits of accuracy of the printed values". Starting from a point on this "interval", one can follow the future trajectory and observe its first return onto the interval. This defines a two to one map from the interval to itself. Indeed, in order to go back in time and track the past trajectory of a point in [0,1], one should be able to select one of the two surfaces attached to the interval. On the figure the two different past trajectories seem to emanate from the "same point" of the interval. Of course, if there are two past trajectories starting from "one" point, there should be four, then eight, etc., which is what Lorenz expresses in the above quotation. Numerically, the first return map is featured on the left part of Figure, extracted from the original paper.

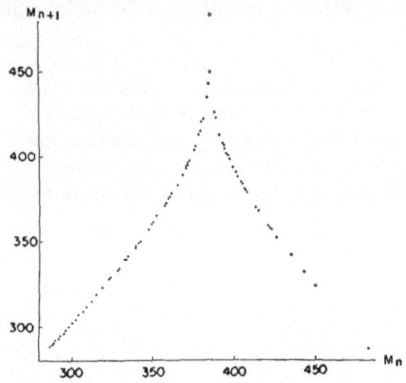 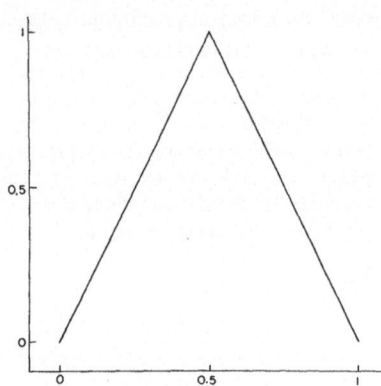

Working by analogy, Lorenz compares this map to the (much simpler) following one: $f(x) = 2x$ if $0 \leq x \leq \frac{1}{2}$ and $f(x) = 2 - 2x$ if $\frac{1}{2} \leq x \leq 1$ (right part of the Figure). Nowadays the chaotic behavior of this "tent map" is well known, but this was much less classical in 1963. In particular, the periodic points of f are exactly the rational numbers with odd denominators, which are dense in [0,1]. Lorenz does not hesitate to claim that the same property applies to the iterations of the "true" return map. The periodic trajectories of the Lorenz attractor are "therefore" dense in K. What an intuition! Finally, he concludes with a lucid question on the relevance of his model for the atmosphere.

> There remains the question as to whether our results really apply to the atmosphere. One does not usually regard the atmosphere as either deterministic or finite, and the lack of periodicity is not a mathematical certainty, since the atmosphere has not been observed forever.

To summarize, this remarkable article contains the first example of a physically relevant dynamical system presenting all the characteristics of chaos. *Individual trajectories are unstable* but their asymptotic behavior seems to be *insensitive to* initial conditions: they converge to the *same* attractor. None of the above assertions are justified, at least in the mathematical sense. How frustrating!

Surprisingly, an important question is not addressed in Lorenz's article. The observed behavior happens to be *robust*: if one slightly perturbs the differential equation, for instance by modifying the values of the parameters, or by adding small terms, then the new differential equation will feature the same type of attractor with the general aspect of a branched surface. This property would be rigorously established much later by Guckhenheimer and Williams.

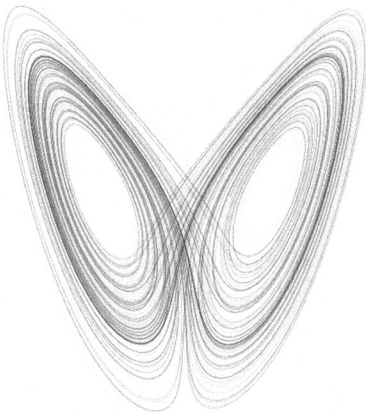

The Lorenz attractor looks like a butterfly

Meanwhile, Mathematicians...

Lack of Communication Between Mathematicians and Physicists?

Mathematicians did not notice Lorenz's paper for more than ten years. The mathematical activity in dynamical systems during this period followed an independent and parallel path, under the lead of Smale. How can one understand this lack of communication between Lorenz—the MIT meteorologist—and Smale—the Berkeley mathematician? Obviously, during the 1960s the scientific community had already reached such a size that it was impossible for a single person to master mathematics and physics; the time of Poincaré was over. No bridge between different sciences was available. Mathematicians had no access to the *Journal of Atmospheric Sciences.*[3]

---

[3] In order to find an excuse for not having noticed Lorenz paper, a famous mathematician told me that Lorenz had published in "some obscure journal"!.

Smale's Axiom A

In 1959 Smale had obtained remarkable results in topology, around the Poincaré conjecture in higher dimension. The main tool was Morse theory describing the gradient of a (generic) function. The dynamics of such a gradient is far from chaotic: trajectories go uphill and converge to some equilibrium point. Smale initiated a grandiose program aiming at a qualitative description of the trajectories of a *generic* vector field (on compact manifolds). His first attempt was amazingly naïve (Smale 1960). He conjectured that a generic vector field has a finite number of equilibrium points, a finite number of periodic trajectories, and that every trajectory converges in the future (and in the past) towards an equilibrium or a periodic trajectory. He was therefore proposing that chaos does not exist! Poincaré, Hadamard or Birkhoff had already published counterexamples many years earlier! Looking back at this period, Smale wrote (1998a, b):

> · It is astounding how important scientific ideas can get lost, even when they are aired by leading scientific mathematicians of the preceding decades.

Smale realized soon *by himself* [4] that the dynamics of a generic vector field is likely to be much more complicated than he had expected. He constructed a counterexample to his own conjecture (Smale 1961). The famous *horseshoe* is a simple example of a dynamical system admitting *an infinite number of periodic trajectories in a stable way*.

In order to describe this example, I should explain a classical construction (due to Poincaré). Suppose we start with a vector field X (in a ball in $\mathbb{R}^n$, as above). It may happen that one can find some $n - 1$ dimensional disc D, which is transverse to X and which is such that the trajectory of every point x in D intersects D infinitely often. In such a situation, one can define a map $F : D \to D$ which associates to each point x in D the next intersection of its trajectory with D. For obvious reasons, this map is called the *first return map*. Clearly the description of the dynamics of X reduces to the description of the iterates of F. Conversely, in many cases, one can construct a vector field from a map F. It is often easier to draw pictures in D since it is one dimension lower than B. In Smale's example, D has dimension 2 and corresponds to a vector field in dimension 3, like in Lorenz's example. The map F is called a *horseshoe map* since the image $F(C)$ of a square C does look like a horseshoe as in the picture.

---

[4] As if obeying Goethe's dictum "Was du ererbt von deinen Vätern hast, erwirb es, um es zu besitzen" ("That which you have inherited from your fathers, earn it in order to possess it.").

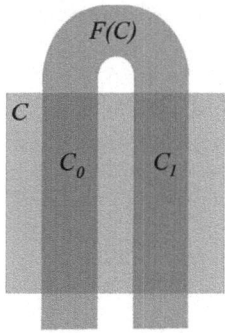

The infinite intersection $\cap_{-\infty}^{+\infty} F^i(C)$ is a nonempty compact set $K \subset D$, and the restriction of F to K is a homeomorphism. The intersection $C \cap F(C)$ consists of two connected components C_0 and C_1. Smale shows that one can choose F in such a way that for every bi-infinite sequence a_i(with $a_i = 0$ or 1), there exists a unique point x in K such that $F^i(x) \in C_i$ for every i. In particular, periodic points of F correspond to periodic sequences a_i; they are dense in K.

More importantly, Smale shows that his example is *structurally stable*. Let us come back to a vector field X defined in some ball in $\mathbb{R}^n$ and transversal to the boundary. One says that X is *structurally stable* if every vector field X' which is close enough to X (say in the C^1 topology) is topologically conjugate to X: there is a homeomorphism h of B sending trajectories of X to trajectories of X'. Andronov and Pontryagin (1937) had introduced this concept in 1937 but in a very simple context, certainly not in the presence of an infinite number of periodic trajectories. The proof that the horseshoe map defines a structurally stable vector field is rather elementary. It is based on the fact that a map F' from D to itself close enough to F is also described by the same infinite sequences a_i.

Smale published this result in the proceedings of a workshop organized in the Soviet Union in 1961. Anosov tells us about this "revolution" in Anosov (2006).

> The world turned upside down for me, and a new life began, having read Smale's announcement of 'a structurally stable homeomorphism with an infinite number of periodic points', while standing in line to register for a conference in Kiev in 1961. The article is written in a lively, witty, and often jocular style and is full of captivating observations. [...] [Smale] felt like a god who is to create a universe in which certain phenomena would occur.

Afterwards the theory progressed at a fast pace. Smale quickly generalized the horseshoe; see for instance (Smale 1966). Anosov proved in 1962 that the geodesic flow on a manifold of negative curvature is structurally stable (Anosov 1962)[5]. For this purpose, he created the concept of what is known today as *Anosov flows*. Starting from the known examples of structurally stable systems, Smale cooked up in 1965 the fundamental concept of dynamical systems satisfying the *Axiom A* and conjectured that these systems are *generic and structurally stable*. Smale's (1967)

---

[5] Surprisingly, he does not seem to be aware of Hadamard's work. It would not be difficult to deduce Anosov's theorem from Hadamard's paper.

article "Differential dynamical systems" represents an important step for the theory of dynamical systems (Smale 1967), a "masterpiece of mathematical literature" according to Ruelle. But, already in 1966, Abraham and Smale found a counterexample to this second conjecture of Smale: Axiom A systems are indeed structurally stable but they are not generic (Smale 1966, Abraham and Smale 1968).

Lorenz's Equation Enters the Scene

Lorenz's equation pops up in mathematics in the middle of the 1970s. According to Guckenheimer, Yorke mentioned to Smale and his students the existence of Lorenz's equation, which did not fit well with their approach. The well-known 1971 paper by Ruelle and Takens (1971) still proposed Axiom A systems as models for turbulence, but in 1975 Ruelle observed that "Lorenz's work was unfortunately overlooked" (Ruelle 1976a). Guckenheimer and Lanford were among the first people to have shown some interest in this equation (from a mathematical point of view) (Guckenheimer 1976; Lanford 1977). Mathematicians quickly adopted this new object which turned out to be a natural counterexample to Smale's conjecture on the genericity of Axiom A systems. It is impossible to give an exhaustive account of all their work. By 1982 an entire book was devoted to the Lorenz's equation, although it mostly consisted of a list of open problems for mathematicians (Sparrow 1982).

Bowen's review article is interesting at several levels (Bowen, 1978). Smale's theory of Axiom A systems had become solid and, although difficult open questions remained, one had a rather good understanding of their dynamics. A few "dark swans" had appeared in the landscape, like Lorenz's examples, destroying the naïve belief in the genericity of Axiom A systems. However mathematicians were trying to weaken the definition of Axiom A in order to leave space to the newcomer Lorenz. Nowadays, Axiom A systems seem to occupy a much smaller place than one thought at the end of the 1970s. The Axiom A paradigm had to abandon its dominant position... According to (Anosov 2006):

> Thus the grandiose hopes of the 1960s were not confirmed, just as the earlier naive conjectures were not confirmed.

For a more detailed description of the "hyperbolic history" one can also read the introduction of (Hasselblatt 2002), or (Ghys 2010). See also "What is... a horseshoe" by one of the main actors of the field (Shub 2005).

Lorenz's Butterfly as Seen by Mathematicians

In order to understand Lorenz's butterfly from a mathematical point of view, Guckhenheimer and Williams (1979) introduced a "geometrical model" in 1979. Remember that Lorenz had observed that "his" dynamics seems to be related to the iterates of a map f from an interval to itself, even though this interval and this map were only defined "within the limits of accuracy of the printed values". The main idea of Guckenheimer and Williams is to *start* from a map f of the interval and to construct some vector field in 3-space whose behavior "looks like" the observed behavior of the original Lorenz equation. The question of knowing if the constructed vector field, called the *geometric Lorenz model*, is actually related to the original Lorenz equation was not considered as important. After all, the original Lorenz equation was a crude approximation of a physical problem and it was unclear whether it was connected with reality, and moreover mathematicians in this group were not really concerned with reality!

The following figure is reprinted from[6] (Guckenheimer and Williams 1979)

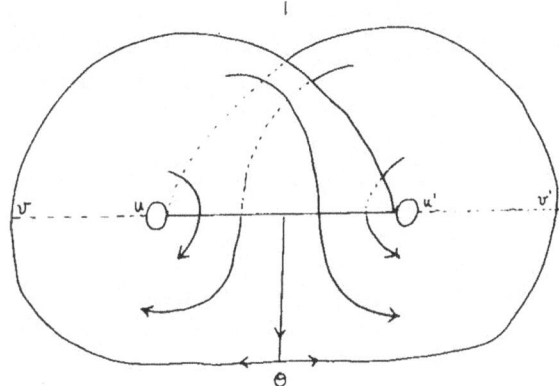

This is a *branched surface* Σ embedded in space. One can define some dynamical system f^t ($t \geq 0$) on Σ whose trajectories are sketched on the figure: a point in Σ has a future but has no past because of the two leaves which merge along an interval. The first return map on this interval is the given map f from the interval to itself. The dynamics of f^t is easy to understand: the trajectories turn on the surface, either on the left or on the right wing, according to the location of the iterates of the original map f. So far, this construction does not yield a vector field. Guckhenheimer and Williams construct a vector field $X(f)$ in some ball B in $\mathbb{R}^3$, transversal to the boundary sphere, whose dynamics mimics f^t. More precisely, denote by $\phi^t(x)$ the trajectories of $X(f)$ and by Λ the intersection $\cap_{t \geq 0}\phi^t(B)$, so that for every point x in B, the accumulation points of the trajectory $\phi^t(x)$ are contained in Λ. The vector field $X(f)$ is such that Λ is very close to Σ and that the trajectories $\phi^t(x)$ shadow f^t. In other words, for every point x in Λ, there is a point x' in Σ such that $\phi^t(x)$ and $f^t(x')$ stay at a very small distance for all positive times $t \geq 0$. This vector field $X(f)$ is not unique but is well defined *up to topological equivalence*, i.e. up to some homeomorphism sending trajectories to trajectories. This justifies Lorenz's intuition,

---

[6] Incidentally, this figure shows that the quality of an article does not depend on that of its illustrations.

according to which the attractor Λ behaves like a branched surface. Moreover, every vector field in B which is close to $X(f)$ is topologically conjugate to some $X(f')$ for some map f' of the interval which is close to f. Furthermore, they construct explicitly a two-parameter family of maps $f_{(a,b)}$ which represent all possible topological equivalence classes. In summary, *up to topological equivalence, the vector fields in the neighborhood of $X(f)$ depend on two parameters and are Lorenz like*. This is the robustness property mentioned above.

Hence, the open set in the space of vector fields of the form $X(f)$ does not contain *any* structurally stable vector field. If Smale had known Lorenz's example earlier, he would have saved time! Lorenz's equation does not satisfy Axiom A and cannot be approximated by an Axiom A system. Therefore any theory describing generic dynamical systems should incorporate Lorenz's equation.

As we have mentioned, the geometric models for the Lorenz attractor have been *inspired* by the original Lorenz equation, but it wasn't clear whether the Lorenz equation indeed behaves like a geometric model. Smale chose this question as one of the "mathematical problems for the next century" in 1998. The problem was positively solved in Tucker (2002). For a brief description of the method used by Tucker, see for instance (Viana 2000).

The Concept of Physical SRB Measures

Poincaré

The main method to tackle the sensitivity to initial conditions uses *probabilities*. This is not a new idea. As mentioned earlier, Laplace realized that solving differential equations requires a "vast intelligence" that we don't have... and suggested developing probability theory in order to get some meaningful information. In his "Science and method", Poincaré gives a much more precise statement. Here is an extract of the chapter on "chance":

> When small differences in the causes produce great differences in the effects, why are the effects distributed according to the laws of chance? Suppose a difference of an inch in the cause produces a difference of a mile in the effect. If I am to win in case the integer part of the effect is an even number of miles, my probability of winning will be ½. Why is this? Because, in order that it should be so, the integer part of the cause must be an even number of inches. Now, according to all appearance, the probability that the cause will vary between certain limits is proportional to the distance of those limits, provided that distance is very small.

This chapter contains much more information about Poincaré's visionary idea and one can even read some proofs between the lines... In modern terminology, Poincaré considers a vector field X in a ball B in $\mathbb{R}^n$, as before. Instead of considering a single point x and trying to describe the limiting behavior of $\phi^t(x)$, he suggests choosing some probability distribution μ in the ball B and to study its evolution $\phi_\star^t \mu$ under the dynamics. He then gives some arguments showing that if μ has a *continuous* density, and if there is "a strong sensitivity to initial conditions", the family of measures $\phi_\star^t \mu$ should converge to some limit v which is *independent of the initial distribution μ*.[7] Even though individual trajectories are sensitive to initial conditions, the asymptotic *distribution* of trajectories is independent of the initial distribution, assuming that this initial distribution has a continuous density. Amazingly, none of his contemporaries realized that this was a fundamental contribution. This may be due to the fact that Poincaré did not write this idea in a formalized mathematical paper but in a popular book. One would have to wait for about seventy years before this idea could surface again.

Lorenz

We have seen that the 1972 conference of Lorenz on the butterfly emphasized the sensitivity to initial conditions and that this idea eventually reached the general public. However, this conference went much further:

> More generally, I am proposing that over the years minuscule disturbances neither increase nor decrease the frequency of occurrence of various weather events such as tornados; the most they may do is to modify the sequence in which these events occur.

This is the real message that Lorenz wanted to convey: the *statistical* description of a dynamical system could be *insensitive* to initial conditions. Unfortunately, this idea is more complicated to explain and did not become as famous as the "easy" idea of sensitivity to initial conditions.

Sinai, Ruelle, Bowen

Mathematicians also (re)discovered this idea in the early 1970s, gave precise definitions and proved theorems. A probability measure v in the ball B, invariant by ϕ^t, is an *SRB measure* (for Sinai-Ruelle-Bowen), also called a *physical measure*, if, for each *continuous* function $u : B \to \mathbb{R}$, the set of points x such that

$$\lim_{T \to \infty} \frac{1}{T} \int_0^T u(\phi^t(x))dt = \int_B u \, dv$$

---

[7] I may be exaggerating because of my excessive worship of Poincaré, but it seems to me that, in modern terminology, Poincaré explains that the limiting probability v is absolutely continuous on instable manifolds and may not be continuous on stable manifolds.

has *nonzero Lebesgue measure.* This set of points is called the *basin* of v and denoted by $B(v)$. Sinai, Ruelle and Bowen (Sinai 1972; Ruelle 1976b; Bowen 1978) proved that this concept is indeed relevant in the case of Axiom A dynamics. If X is such a vector field in some ball B, there is a finite number of SRB measures $v_1,...,v_k$ such that the corresponding basins $B(v_1),...,B(v_k)$ cover B, up to a Lebesgue negligible set. Of course, the proof of this important theorem is far from easy but its general structure follows the lines sketched in Poincaré paper...

In summary, the existence of SRB measures is the right answer to the "malediction" of the sensitivity to initial conditions. In the words of Lorenz, "the frequency of occurrence of various weather events such as tornados" could be insensitive to initial conditions. If for example the ball B represents the phase space of the atmosphere and $u : B \to \mathbb{R}$ denotes the temperature at a specific point on the Earth, the average $\frac{1}{T}\int_0^T u(\phi^t(x))dt$ simply represents the average temperature in the time interval $[0,T]$. If there is an SRB measure, this average converges to $\int u \, dv$, *independently* of the initial position x (at least in the basin of v). The task of the forecaster changed radically: instead of guessing the position of $\phi^t(x)$ for a large t, he or she tries to estimate an SRB measure. *This is a positive statement about chaos as it gives a new way of understanding the word "prevision". It is unfortunate that such an important idea did not reach the general population.* Poor Brazilian butterflies! They are now unable to change the fate of the world!

The quest for the weakest conditions that guarantee the existence of SRB measures is summarized in the book (Bonatti et al. 2005). This question is fundamental since, as we will see, one hopes that "almost all" dynamical systems admit SRB measures.

The geometric Lorenz models are not Axiom A systems, hence are not covered by the works of Sinai, Ruelle and Bowen. However, it turns out that *the Lorenz attractor supports a unique SRB measure* (Bunimovich 1983; Pesin 1992). Lorenz was right!

Palis

The history of dynamical systems seems to be a long sequence of hopes... quickly abandoned. A non chaotic world, replaced by a world consisting of Axiom A systems, in turn destroyed by an abundance of examples like Lorenz's model. Yet, mathematicians are usually optimists, and they do not hesitate to remodel the world according to their present dreams, hoping that their view will not become obsolete too soon. Palis (1995, 2005, 2008) proposed such a vision in a series of three articles. He formulated a set of conjectures describing the dynamics of "almost all" vector fields. These conjectures are necessarily technical, and it would not be useful to describe them in detail here. I will only sketch their general spirit.

The first difficulty—which is not specific to this domain—is to give a meaning to "almost all" dynamics. The initial idea from the 1960s was to describe an *open dense* set in the space of dynamical systems, or at least, a countable intersection of open dense sets, in order to use *Baire genericity*. Yet, this notion has proved to be too strict. Palis uses a concept of "prevalence" whose definition is technical but which is close in spirit to the concept of "full Lebesgue measure". Palis *finiteness conjecture* asserts that in the space of vector fields on a given ball *B*, *the existence of a finite number of SRB measures whose basins cover almost all the ball is a prevalent property.*

Currently, the Lorenz attractor serves as a model displaying phenomena that are believed be characteristic of "typical chaos", at least in the framework of mathematical chaos. Even so, the relevance of the Lorenz model to describe meteorological phenomena remains largely open (Robert 2001).

Communicating Mathematical Ideas?

In Poincaré's time, the total number of research mathematicians in the world was probably of the order of 500. Even in such a small world, even with the expository talent of Poincaré as a writer, we have seen that some important ideas could not reach the scientific community. The transmission of ideas in the theory of chaos, from Poincaré to Palis has not been efficient. In the 1960s we have seen that the Lorenz equation took ten years to cross America from the east coast to the west coast, and from physics to mathematics. Of course, the number of scientists had increased a lot. In our 21st century, the size of the mathematical community is even bigger ($\sim 50,000$ research mathematicians?) and the physical community is much bigger. Nowadays, the risk is not only that a good idea could take ten years to go from physics to mathematics: there could be tiny subdomains of mathematics that do not communicate at all. Indeed, very specialized parts of mathematics that look tiny for outsiders turn out to be of a respectable size, say of the order of 500, and can transform into "scientific bubbles". As Lovász (2006) writes in his "Trends in Mathematics: How they could Change Education?":

> A larger structure is never just a scaled-up version of the smaller. In larger and more complex animals an increasingly large fraction of the body is devoted to 'overhead': the transportation of material and the coordination of the function of various parts. In larger and more complex societies an increasingly large fraction of the resources is devoted to non-productive activities like transportation information processing, education or recreation. We have to realize and accept that a larger and larger part of our mathematical activity will be devoted to communication.

Of course, this comment does not only apply to mathematics but to Science in general and to the society at large. Nowadays, very few university curricula include courses on communication aimed at mathematicians. We need to train mediators who can transport information at all levels. Some will be able to connect two different areas of mathematics, some will link mathematics and other sciences, and some others will be able to communicate with the general public. It is important that we consider this kind of activity as a genuine part of scientific research and that it could attract our most talented students, at an early stage of their career. We should not only rely on journalists for this task and we should prepare some of our colleagues for this noble purpose. We have to work together and to improve mathematical communication. We should never forget that a mathematical giant like Poincaré took very seriously his popular essays and books, written for many different audiences.

Open Access This chapter is distributed under the terms of the Creative Commons Attribution Noncommercial License, which permits any noncommercial use, distribution, and reproduction in any medium, provided the original author(s) and source are credited.

References

Abraham, R. & Smale, S. (1968). Nongenericity of Ω-stability. *Global Analysis* (Proc. Sympos. Pure Math., Vol. XIV, Berkeley, Calif., 1968), *5-8*. Amer. Math. Soc., Providence.

Andronov, A., & Pontrjagin, L. (1937). Systèmes grossiers. *Dokl. Akad. Nauk SSSR, 14*.

Anosov, D.V. (1962). Roughness of geodesic flows on compact Riemannian manifolds of negative curvature. *Dokl. Akad. Nauk SSSR, 145*, 707-709.

Anosov, D.V. (2006). Dynamical systems in the 1960s: the hyperbolic revolution. In *Mathematical events of the twentieth century*, 1-17, Springer, Berlin.

Aubin D., & Dahan Dalmedico A. (2002). Writing the history of dynamical systems and chaos: longue durée and revolution, disciplines and cultures. *Historia Math. 29*(3), 273-339.

Birkhoff, G.D. (1927). Dynamical Systems. vol. 9 of the American Mathematical Society Colloquium Publications (Providence, Rhode Island: American Mathematical Society).

Bonatti, C., Diaz, L., & Viana. M. (2005). Dynamics beyond uniform hyperbolicity. A global geometric and probabilistic perspective. Encyclopaedia of Mathematical Sciences, (*102*). Mathematical Physics, III. Springer-Verlag, Berlin.

Bowen, R. (1978). On Axiom A diffeomorphisms. *Regional Conference Series in Mathematics, 35*. American Mathematical Society, Providence, R.I.

Bunimovich, L.A. (1983). Statistical properties of Lorenz attractors. Nonlinear dynamics and turbulence, Pitman, 7192.

Duhem, P. (1906). La théorie physique; son objet, sa structure. English transl. by P.P. Wiener, The aim and structure of physical theory, Princeton University Press, 1954.

Ghys, E. (2010). L'attracteur de Lorenz, paradigme du chaos. Séminaire Poincaré, XIV.

Gleick, J. (1987). Chaos: Making a New Science. Viking Penguin.

Guckenheimer, J. (1976). A strange, strange attractor. In *The Hopf Bifurcation*, Marsden and McCracken, eds. Appl. Math. Sci., Springer-Verlag.

Guckenheimer, J., & Williams, R.F. (1979). Structural stability of Lorenz attractors. *Inst. Hautes Études Sci. Publ. Math., 50*, 59-72.

Hadamard, J. (1898). Les surfaces à courbures opposées et leurs lignes géodésiques. *Journal de mathématiques pures et appliqués*, 5e série *4*, 27-74.

Hasselblatt, B. (2002). Hyperbolic dynamical systems. Handbook of dynamical systems, Vol. 1A, 239-319, North-Holland, Amsterdam.

Kahane, J.-P. (2008). Hasard et déterminisme chez Laplace. *Les Cahiers Rationalistes*, 593.

Lanford, O. (1977). An introduction to the Lorenz system. In *Papers from the Duke Turbulence Conference* (Duke Univ., Durham, N.C., 1976), Paper No. 4, i + 21 pp. Duke Univ. Math. Ser., Vol. III, Duke Univ., Durham, N.C.

Laplace, P.S. (1814). Essai philosophique sur les probabilités. English transl. by A.I. Dale, Philosophical essay on probabilities, Springer, 1995.

Lorenz, E.N. (1960). Maximum simplification of the dynamic equations. *Tellus*, *12*, 243-254.
Lorenz, E.N. (1962). The statistical prediction of solutions of dynamic equations. *Proc. Internat. Sympos. Numerical Weather Prediction*, Tokyo, 629-635.
Lorenz, E.N. (1963). Deterministic non periodic flow. *J. Atmosph. Sci.*, *20*, 130-141.
Lorenz, E.N. (1972). Predictability: does the flap of a butterfly's wings in Brazil set off a tornado in Texas? 139th Annual Meeting of the American Association for the Advancement of Science (29 Dec 1972), in *Essence of Chaos* (1995), Appendix 1, 181.
Lorenz, E.N. (1991). A scientist by choice. Kyoto Award lecture. (available at http://eaps4.mit. edu/research/Lorenz/Miscellaneous/Scientist_by_Choice.pdf)
Lovász, L. (2006). Trends in Mathematics: How they could Change Education? In *The Future of Mathematics Education in Europe*, Lisbon.
Maxwell, J.C. (1876). Matter and Motion. New ed. Dover (1952).
Palis, J. (1995). A global view of dynamics and a conjecture on the denseness of finitude of attractors. Géométrie complexe et systèmes dynamiques (Orsay, 1995). *Astérisque, 261* (2000), xiii-xiv, 335-347.
Palis, J. (2005). A global perspective for non-conservative dynamics. *Ann. Inst. H. Poincaré Anal. Non Linéaire*, *22*(4), 485-507.
Palis, J. (2008). Open questions leading to a global perspective in dynamics. *Nonlinearity*, *21*(4), T37-T43.
Pesin, Y. (1992). Dynamical Systems With Generalized Hyperbolic Attractors: Hyperbolic, Ergodic and Topological Properties, Ergod. Theory and Dyn. Syst., 12:1 (1992) 123-152.
Poincaré, H. (1881). Mémoire sur les courbes définies par une équation différentielle. *Journal de mathématiques pures et appliqués*, *7*, 375-422.
Poincaré, H. (1890). Sur le problème des trois corps et les équations de la dynamique. *Acta Mathematica*, 13, 1-270.
Poincaré, H. (1908). Science et méthode. Flammarion. English transl. by F. Maitland, Science and Method, T. Nelson and Sons, London, 1914.
Robert, R. (2001). L'effet papillon n'existe plus ! *Gaz. Math.*, *90*, 11-25.
Ruelle, D. (1976a). The Lorenz attractor and the problem of turbulence. In *Turbulence and Navier-Stokes equations* (Proc. Conf., Univ. Paris-Sud, Orsay, 1975), 146-158, Lecture Notes in Math., *565*, Springer, Berlin, (1976).
Ruelle, D. (1976b). A measure associated with axiom-A attractors. *Amer. J. Math.*, *98*(3), 619-654.
Ruelle, D., & Takens, F. (1971). On the nature of turbulence. Comm. Math. Phys., *20*, 167-192.
Shub, M. (2005). What is... a horseshoe?. *Notices Amer. Math. Soc.*, *52*(5), 516-517.
Sinai, Ja. G. (1972). Gibbs measures in ergodic theory, (in Russian). *Uspehi Mat. Nauk.*, *27*(4), 21-64.
Smale, S. (1960). On dynamical systems. *Bol. Soc. Mat. Mexicana*, 5, 195-198.
Smale, S. (1961). A structurally stable differentiable homeomorphism with an infinite number of periodic points. Qualitative methods in the theory of non-linear vibrations (Proc. Internat. Sympos. Non-linear Vibrations, Vol. II, 1961) Izdat. Akad. Nauk Ukrain. SSR, Kiev, 365-366.
Smale, S. (1966). Structurally stable systems are not dense. *Amer. J. Math.*, 88, 491-496.
Smale, S. (1967). Differentiable dynamical systems. *Bull. Amer. Math. Soc.*, *73*, 747-817.
Smale, S. (1998). Finding a horseshoe on the beaches of Rio. *Math. Intelligencer, 20* (1), 39-44.
Smale, S. (1998b). Mathematical problems for the next century. *Math. Intelligencer, 20*(2), 7-15.
Sparrow, C. (1982). The Lorenz equations: bifurcations, chaos, and strange attractors. Applied Mathematical Sciences, *41*. Springer-Verlag, New York-Berlin.
Tucker, W. (2002). A rigorous ODE solver and Smale's 14th problem. *Found. Comput. Math., 2* (1), 53-117.
Viana, M. (2000). What's new on Lorenz strange attractors? *Math. Intelligencer, 22*(3), 6-19.
Williams, R.F. (1979). The structure of Lorenz attractors. *Inst. Hautes Études Sci. Publ. Math., 50*, 73-99.

Whither the Mathematics/Didactics Interconnection? Evolution and Challenges of a Kaleidoscopic Relationship as Seen from an ICMI Perspective

Bernard R. Hodgson

Abstract I wish in this lecture to reflect on the links between mathematics and didactics of mathematics, each being considered as a scientific discipline in its own right. Such a discussion extends quite naturally to the professional communities connected to these domains, mathematicians in the first instance and mathematics educators (didacticians) and teachers in the other. The framework I mainly use to support my reflections is that offered by the International Commission on Mathematical Instruction (ICMI), a body established more than a century ago and which has played, and still plays, a crucial role at the interface between mathematics and didactics of mathematics. I also stress the specificity and complementarity of the roles incumbent upon mathematicians and upon didacticians, and discuss possible ways of fostering their collaboration and making it more productive.

Keywords Mathematics · Didactics of mathematics · Mathematicians · Mathematics educators · ICMI

Introduction

I wish in this lecture to reflect on the links between mathematics on the one hand, and the didactics of mathematics on the other, each being considered as a scientific discipline in its own right. From that perspective, mathematics is a domain with a very long history, while didactics of mathematics, or mathematical education as it is predominantly called by Anglophones, is of a much more recent vintage. Such a discussion extends quite naturally to the professional communities connected to these domains, mathematicians in the first instance, and mathematics educators

B.R. Hodgson (✉)
Département de Mathématiques et de Statistique, Université Laval, Québec, Canada
e-mail: Bernard.Hodgson@mat.ulaval.ca

© The Author(s) 2015   41
S.J. Cho (ed.), *The Proceedings of the 12th International Congress
on Mathematical Education*, DOI 10.1007/978-3-319-12688-3_7

(didacticians) and teachers in the other. The general framework I mainly use to support my reflections is that offered by the International Commission on Mathematical Instruction (ICMI), a body established more than a century ago and which has played, and still plays, a crucial role at the interface between mathematics and its teaching, between mathematics and didactics of mathematics.

As shown notably by the history of ICMI, there is a long tradition of eminent mathematicians being professionally involved in educational matters, including with regard to primary or secondary education. But the emergence, during the last decades of the previous century, of didactics of mathematics as an internationally recognized academic discipline has had among its effects an increase of the gap between mathematicians and mathematical educators, culturally and otherwise. Both mathematics and didactics depend for their development on research, founded in each case upon specific paradigms eventually hindering the fluidity of the communication between the two groups. While most professional mathematicians are involved not only in the creation or application of mathematics but also in its teaching, only a small number of them actually pay substantial attention to what recent research in education tells about the difficulties intrinsic to the learning of mathematics at various levels. And the development of didactics of mathematics, as a field both of practice and of research with distinctive concepts and vocabulary, amplifies to a certain extent the opaqueness of its results to the outsider. At the same time some suspicion may have developed within the mathematical education community about the role and importance of mathematicians in education. Such a situation may be reinforced at times by somewhat naive views expressed by some mathematicians in educational debates, as well as by the fact that, in opposition to the early days of didactics of mathematics, a larger proportion of didacticians nowadays, including teacher educators, have had little contact with higher mathematics, say, at the graduate level or even at the advanced undergraduate level.

I mainly base my discussion both upon my 11-year experience as ICMI Secretary-General (1999–2009) and on various elements stemming from activities organised by or under the auspices of ICMI, for instance ICME congresses or ICMI Studies, as well as on episodes from its history. I consider different contexts where mathematics and mathematical education interact and the way these contexts have evolved over the years. In connection with the complexity of educational issues related to both the teaching and the learning of mathematics, I also stress the specificity and complementarity of the roles incumbent upon mathematicians and upon mathematical educators, and examine possible ways of fostering their collaboration and making it more productive, notably in the context of ICMI activities.

Linguistic Prolegomena

Before embarking on my topic per se, it may be helpful to pay attention to some expressions appearing in the title of this lecture, so to make my use of these as clear as possible.

"Whither" or "Wither"

In spite of my patronymic, I share with the majority of the people in this audience the fact that English is not my mother tongue. Besides regretting any inconvenience stemming from my "French English", I need to point to potential problems provoked by the use of a certain vocabulary representing not only a substantial elocutionary challenge for non-native English speakers like me, but that moreover is usually not part of daily discourse. Such is possibly the case with the "whither" in my title. I do not know if many of you had to look into a dictionary for its exact meaning. I definitely did, when I first met this interrogative adverb. If my memory serves me well, my first encounter with this intriguing word—or at least the first time it really caught my attention—was in the title of one of the concluding chapters ("Whither mathematics?") of a thought-provoking book by Kline (1980) about the nature and role of mathematics. I met it again many years later through the plenary lecture "Whither mathematics education?" presented by Anna Sierpinska at ICME-8, in 1996 (Sierpinska 1998). I remember being fascinated by the idea of the likely future of a given matter being concealed in that single word "whither". And this is precisely what I have in mind in this talk about the mathematics/didactics links.

But depending on one's pronunciation of today's *lingua franca*, non-trivial difficulties may arise when using this word. You will have noted the two aitches ("h") in "whither", thus allowing to distinguish (at least visually!) this word from its neighbour "wither", a verb with a totally different meaning. But how is this difference to be communicated orally? I clearly was myself the source of some confusion recently when discussing with a former ICMI officer the topic of the present lecture. Quite obviously I then dropped the first aitch, either inadvertently or by a lack of capacity of rendering it orally in a proper way. "Why are you proposing such a strong title for your talk? was then wondering my colleague. Why do you insist on the possibility that the interconnection between mathematics and didactics may be drying, waning, decaying?" Such is not at all the message I aim at conveying in this lecture, and this is why the initial aitch is so important. As a matter of fact, I am concerned with quite the opposite: how to ensure that this crucial aitch never gets dropped!

Through the Kaleidoscope

Those of you aware of my long-term involvement in the mathematical preparation of primary school teachers will possibly be familiar with my deep interest for the kaleidoscope, a "philosophical toy" invented—and named[1]—in the early 19th

---

[1] The name "kaleidoscope" was coined by Brewster from the Greek words "kalos", *beautiful*, "eidos", *aspect*, and "skopein", *to see*. With a typical poetical flavour, the Chinese name for this instrument, 万花筒 ("wàn huā tǒng"), can be translated literally as *ten thousand flowers cylinder*,

century by the Scottish physicist Sir David Brewster.[2] This instrument, so simple yet so fertile, is in my opinion a wonderful "attention-catcher" eventually leading to scientific thinking, as it fascinates people of all ages through the richness and beauty of the images created by the interplay of mirrors.[3] It is in my opinion an ideal vehicle for putting teachers in contact with geometry, both practical and theoretical. The kaleidoscope has regularly been part of my teaching with primary school teachers for more than three decades (Hodgson 1987), and I still see as an important personal experience for teachers to explore the explosion of images provoked by the actual interaction of physical mirrors, notwithstanding the virtual possibilities offered by the computer (Graf and Hodgson 1990). A thorough theoretical understanding of the mathematical principles underlying the kaleidoscope is a challenge fully appropriate for primary school student teachers, and I am deeply convinced that the mastery of such a mathematical "micro-theory" can have a positive impact on their perception of mathematics and their personal relation to it (Hodgson 2004).

My mention of the kaleidoscope in the context of this talk is more than a mere wink to a mathematical pet subject of mine offering such a fecund pedagogical environment. I use in my title the kaleidoscope as a metaphor in order to suggest the changing nature of the mathematics/didactics relationship, like the stunning, if not unpredictable, alterations provoked on the image generated by a kaleidoscope by even a small shaking of the glass pieces inside the device. The history of ICMI, for instance, vividly illustrates the evolution over the past century of the links between mathematics and didactics, as well as the communities supporting these fields. But more to my point, the complexity and richness of kaleidoscopic rosettes can also serve as an analogy to the potential fruitfulness not only of the connections between mathematics and didactics as scholarly domains, but also of the collaboration between mathematicians and didacticians.

What?—and Who?

I now wish to comment on the mathematics/didactics tandem on which this talk is based. There is possibly no need to expand on the concept of *mathematics* in itself,

---

(Footnote 1 continued)

or more appropriately, *cylinder with myriads of flowers*. In a similar vein, the Korean name, 만화경 ("mân hwa gyong"), can be translated as *ten thousand brightnesses mirrors*, again suggesting the proliferation of a myriad of images. Quite interestingly, the word "myriad", used in English to convey the idea of an extremely large number, originally designated a unit of ten thousand in classical Greek numeration.

[2] Brewster commented about his instrument that "it was impossible not to perceive that it would prove of the highest service in all the ornamental arts, and would, at the same time, become a popular instrument for the purposes of rational amusement." (Brewster 1819, p. 7).

[3] This fascination for the kaleidoscope has possible been rendered no better than by the famous French writer André Gide (1869–1951), 1947 Nobel laureate in literature, in his autobiographical *Si le grain ne meurt* (cf. Graf and Hodgson 1990, p. 42).

except to stress that I am concerned here with mathematics as both a body of knowledge and an academic discipline implemented as a subject-matter in given teaching and learning environments, at different levels of educational systems all around the world.[4] The word *didactics* is slightly more difficult to circumscribe. I have in mind of course didactics *of mathematics*, rather than a kind of general-purpose didactics. I am aware that in English the adjective didactic may come with a pejorative connotation,[5] and that the noun didactics could be interpreted with the somewhat restricted meaning of "the science and art of teaching"[6]—see also Kilpatrick (2003) for similar linguistic comments. Consequently the expression *mathematics education* has become the one typically used among Anglophone circles to designate the scholarly domain that has developed, especially in the second part of the previous century, in relation to the teaching and learning of mathematics.

It is not my intent to enter here into fine discussions about the respective merits or limitations of expressions such as *didactics of mathematics* and *mathematics education*, and to examine their exact scope. Nor do I wish to focus on the specific case of the so-called French school of "didactique des mathématiques"—I refer those interested for instance to the analysis offered by Kilpatrick (2003, 2012). Still I will mostly use here the expression didactics of mathematics (rather than the more frequent mathematics education), partly because of my own linguistic bias, and partly because of a kind of general agreement, especially among some of the European countries, that seems to be emerging about its use, even in English.[7] In doing so, I am in line with the description proposed by Winsløw (2007), where didactics is understood as "the study of the teaching and learning of specific knowledge, usually within a disciplinary domain" (p. 534). In the same paper, Winsløw stresses how in some European contexts. "[d]idactics is regarded as a continuation of the study of the scientific discipline, in much the same way as the study of its history and philosophy" (p. 524).

---

[4] Dossey (1992) offers an overview of various conceptions of mathematics, including in an historical perspective, and discusses "their current and potential impact on the nature and course of mathematics education" (p. 30). See also Kilpatrick (2008, pp. 29–31), for helpful nuances about the question "What is mathematics?" with regard to educational contexts, in particular in connection with the idea of mathematics then becoming a domain of practice.

[5] As is witnessed for instance by the following definition: "in the manner of a teacher, particularly so as to treat someone in a patronizing way", from the *New Oxford American Dictionary* (2nd edition, 2005, electronic version included in the Mac environment).

[6] According to the *Oxford English Dictionary* (online version), this seems to be a typical 19th-century vision. It is in that sense for instance that the word "didactics" is used in the title of one of the sections on the programme of the International Congress of Mathematicians held in Cambridge in 1912—cf. Hobson and Love (1913), Section IV, *Philosophy, History and Didactics*.

[7] It may be of interest to note that as early as 1968, Hans Georg Steiner was using (in English) the expression "didactics of mathematics" to designate the "new discipline" that, he claimed, had to be established to support what he saw as "new possibilities for mathematics teaching and learning" (cf. Steiner 1968, pp. 425–426). He presented this new discipline as "separate from the 'methodology of mathematics teaching'" (p. 426).

Another facet of the mathematics/didactics dichotomy concerns the actors involved in those fields. This is also far from easy to describe, as the context is intrinsically complex and can vary considerably from one country to the other—and even within a single country—, due to economical, social and cultural factors, as well as local traditions. This is why the local educational structures in which these people are to be found (vg, schools, colleges, universities, teacher education institutes, etc., not to speak of research centres and suchlike) come in a variety of forms. That said, I will now try to briefly identify, but without any pretention to exhaustiveness, what may be considered as typical working environments and structural frameworks for the colleagues I have in mind.

One obvious category of actors is that of the *mathematicians*, that is, people whose main interest is with mathematics as a body of knowledge and eventually contributing to its development through research.[8] To borrow from the title of a well-known math book from the time of my graduate studies (Mac Lane 1971), they are "working mathematicians", active in the field. The vast majority of these people, and especially those in the academia, will belong to a mathematics unit (department, etc.) and be involved in some form of teaching, from courses to math majors to large classes of engineers or graduate courses and seminars with a handful of students. Because of such teaching duties, they are undoubtedly "educators", although one could think that for a number of them, educational activities do not represent their main professional concern and would even have a possibly limited impact on the evolution of their career (promotion, etc.). Still there seems to be a growing number of faculty members in mathematics department developing a *bona fide* interest for educational matters, notably at the tertiary level. A crucial issue then becomes how they can find in the community the kind of support needed for their educational endeavour. I shall say a few words about this later.

Among the mathematicians is a subset of specific interest to this talk, and to which I myself belong: those whose teaching is substantially targeted at the mathematical education of teachers, both of primary and of secondary school. I have discussed in (Hodgson 2001) the importance of this specific contribution of mathematicians[9]—a contribution, I maintain, that should be considered as an intrinsic part of the "mission" of a mathematics department.

But mathematicians are of course not the only players involved in the preparation of mathematics schoolteachers. Another group of teacher educators of prime importance will typically be found in faculties of education (or of educational sciences). While many of them would call themselves *mathematics educators*, I prefer to use here the expression *didacticians*, in line with the preceding

---

[8] While I fully adhere with the statement made by IMU president Ingrid Daubechies, in her ICME-12 opening address, that the term "mathematicians" should be construed as including, for instance, participants at an ICME congress, I am using this word, for the purpose of my talk, in a slightly more restrictive (and customary) sense.

[9] "Mathematicians have a major and unique role to play in the education of teachers—they are neither the sole nor the main contributors to this complex process, but their participation is essential." (Hodgson 2001, p. 501).

comments.[10] Besides the graduate supervision of future didacticians or the development of their own research programme, a large portion of the teaching time of didacticians, at the undergraduate level, would mostly be devoted to the education of primary and secondary school teachers. One possible distinction between their contribution to the education of teachers and that of the mathematicians may be the extent to which emphasis is placed on the challenges encountered in the actual teaching and learning of some mathematical topic. This is to be contrasted with the attention mathematicians may give to the mastery of a given mathematical content, both in itself and as a potential piece of mathematics to be taught, as well as its place in the "global mathematical landscape", for instance when seen from an advanced standpoint *à la Klein* (see Klein 1932).

The actual "location" of didacticians inside the academic environment can vary a lot, but they often belong to a faculty of education. A specific case I wish to stress is when didactics of mathematics is attached, as an academic domain, to the same administrative unit (vg, a given university department) to which mathematics belongs[11]—a context that may be seen as related to the comments of Winsløw quoted above. Such a situation is far from being the general rule—and I would not want to push it as an ideal universal model—, but it clearly offers an interesting potential for fostering the links between mathematicians and didacticians, and eventually improving mutual understanding and respect.

More generally, there is an obvious need for a community and a forum where mathematicians and didacticians can meet in connection to issues, general or specific, related to the teaching and learning of mathematics. An interesting context to that effect is that offered by ICMI.

A Glimpse into the History of ICMI

The International Commission on Mathematical Instruction (ICMI) celebrated in 2008 its centennial, an event that stimulated the publication of a number of papers dealing with various aspects of its history. Detailed information about the origins of the Commission and its evolution over the years can be found for instance in Bass (2008b), Furinghetti et al. (2008) and Schubring (2008), three papers appearing in the proceedings of the ICMI centennial symposium. Other papers of a historical nature include Furinghetti (2003) and Schubring (2003), written on the occasion the

---

[10] My reluctance to speak of "mathematics educators" in that context also stems from the fact that in my opinion, expressions such as "mathematics educators" or "teacher educators" should not be construed as belonging exclusively to or denoting specifically either the community of didacticians or that of mathematicians: as stressed earlier, we are *all* educators, but of course with our own specific ways of addressing educational issues.

[11] As a concrete example, I mention that the position in "didactique des mathématiques" created in 1999 at Université Paris Diderot (a scientific university of international research fame) and first occupied by former ICMI president Michèle Artigue is attached to the mathematics department.

centennial of *L'Enseignement Mathématique*—the journal which since the inception of ICMI has been its official organ—, as well as Hodgson (2009). The survey of Howson (1984) was prepared on the occasion of the 75th anniversary of ICMI. Many ICMI-related sections are found in Lehto (1998), a book about the history of the International Mathematical Union (IMU), the organization to which ICMI owes its legal existence.

The beginnings of ICMI can be seen as resting upon the assumption that mathematicians have a role to play in issues related to school mathematics—at least at the secondary level. Its establishment resulted from a resolution adopted at the Fourth International Congress of Mathematicians held in Rome in 1908 and appointing a commission, under the presidency of the eminent German mathematician Felix Klein, with the mandate of instigating "a comparative study of the methods and plans of teaching mathematics at secondary schools" (Lehto 1998, p. 13). This resolution can be seen as addressing concerns present at the turn of the twentieth century in educational debates and provoked by the spreading of mass education combined with a greater sensitivity towards internationalism that stimulated the need for self-reflection, comparison and communication. Still today, the formal definition of ICMI's global mission and framework for action points to the importance of connecting its educational enterprises with the community of mathematicians as represented by IMU. For instance the Terms of reference of ICMI state that "ICMI shall be charged with the conduct of the activities of IMU bearing on mathematical or scientific education". More details are provided below on the recent and current links between ICMI and IMU.

A sharp distinction is manifest between the "old ICMI's tradition" (Furinghetti 2008, p. 49) of publishing national reports and international analyses of school curricula, as done abundantly in its early years,[12] and the activities of ICMI after its rebirth[13] in 1952, at a time when the international mathematical community was being reorganized, as a permanent commission of the then newly established IMU. Furinghetti (2008) stresses how at that latter time "the developments of society and schools were making the mere study and comparison of curricula and programs (...) inadequate to face the complexity of the educational problems" (p. 49). Highlighting the use of the "new expression 'didactical research'" in the title of a short lecture presented at the 1954 International Congress of Mathematicians, she presents this as a sign of an emerging shift about mathematics education, from a "national business" mainly concerned with curricular comparisons to a "personal business" centred on learners and teachers (Furinghetti 2008, pp. 49–50). The 1950s also saw the development of a new community, the *Commission Internationale pour l'Étude et l'Amélioration de l'Enseignement des Mathématiques*

---

[12] Fehr (1920–1921, p. 339) indicated for instance that between 1908 and 1920, ICMI, jointly with eighteen of the countries it gathered, had produced 187 volumes containing 310 reports, for a total of 13,565 pages.

[13] This rebirth followed a hiatus in ICMI activities around the two World Wars. Like most international scientific organizations of that time, ICMI was deeply affected by the ongoing international tensions.

(CIEAEM /International Commission for the Study and Improvement of Mathematics Teaching, ICSIMT), where the importance of reflecting on the students themselves as well as on the teaching processes and classroom interactions was strongly emphasised, in contrast to educational work typical of the time.

Such deep changes were the reflection of the emergence of a new sensitivity with regard to educational issues. As a result, a context arose propitious not only to the development of new approaches to study the teaching and learning of mathematics, but also to the eventual birth of a new academic discipline, gradually accepted and recognized as such, namely didactics of mathematics (i.e., mathematics education in usual parlance). ICMI itself was at times strongly influenced by these changes— Furinghetti et al. (2008) speak of a "Renaissance" of ICMI under the influence of events from the 1950s and 1960s. But ICMI also accompanied the evolution of didactics of mathematics, and at times even fostered it, thus contributing significantly to its acceptance as a *bona fide* academic domain.

This was particularly true during the ICMI presidency of Hans Freudenthal from 1967 to 1970. This particular moment was definitely a turning point in the renewal of ICMI, principally because of two major events that then occurred, essentially at Freudenthal's personal initiative, and that proved to have a considerable long-term impact: the establishment in 1968 of an international research journal in didactics of mathematics (*Educational Studies in Mathematics*, *ESM*), and the launching in 1969 of a new series of international congresses (the International Congress on Mathematical Education, ICME), the twelfth of which we are now celebrating in Seoul.

Bass (2008b) uses the expressions "Klein era" and "Freudenthal era" (from the names of the first and eighth presidents of ICMI) to designate two pivotal segments structuring the life of ICMI up to its 100th anniversary and corresponding more or less to its first two half-centuries: from ICMI beginnings in 1908 up to World War II, and from ICMI rebirth in 1952 to its centennial celebration. Of central interest to my lecture is the distinction Bass introduces about the actors then involved in ICMI circles. While those of the first period were mostly "mathematicians with a substantial, but peripheral interest in education, of whom Felix Klein was by far the most notable example, plus some secondary teachers of high mathematical culture" (Bass 2008b, p. 9), the majority of the players in the Freudenthal era are professional researchers in the teaching and learning of mathematics, i.e., didacticians. Bass also adds that "[i]n this period we see also the first significant examples of research mathematicians becoming professionally engaged with mathematics education even at the scholarly level" (Bass 2008b, p. 10), and suggests Freudenthal as a outstanding example of such a phenomenon—but of course the name of Hyman Bass himself provides an eloquent example of a more recent nature. A thorny question, in that connection, is the extent to which the growing specificity of the main actors of the Freudenthal era may create a widening distance with the "working mathematician" with regard to educational issues.

As discussed in Hodgson (2009), the presidency of Freudenthal resulted in what might be rightly seen as "years of abundance" for ICMI, in the sense that the scope and impact of its actions expanded considerably. Not only were the newly established *ESM* and ICMEs highly successful, but also new elements were gradually added to the mission of ICMI. To name a few, ICMI introduced in the mid-1970s a notion of Affiliated Study Groups, serving specific segments of a community becoming more and more diverse.[14] There was also a regular collaboration between ICMI and UNESCO, contributing in particular to outreach actions of ICMI towards developing countries. And later, in the mid-1980s, the very successful program of ICMI Studies was initiated. Still this deep evolution of ICMI, notably through the influence of Freudenthal himself, did not happen without some tensions with IMU, in particular as it was often the case that IMU faced decisions that were *faits accomplis*, taken without any consultation between the Executive Committees of ICMI and IMU—such had been the case for instance with the launching of the first ICME congress.[15]

Another moment of tension between IMU and ICMI happened in connection with the program of the section on the Teaching and Popularisation of mathematics at the 1998 International Congress of Mathematicians.[16] As a consequence, the first Executive Committee of ICMI on which I served, under the presidency of Hyman Bass, had to deal with an episode of misunderstanding, and even mistrust, between the communities of mathematicians and didacticians as represented by IMU and ICMI. I will come back to this episode later in this lecture and contrast it with the very positive climate of collaboration and mutual respect between these two bodies that now prevails.

This overview of the history of ICMI may help appreciate the origins of didactics of mathematics as an academic domain, as well as its evolution over the years. One can also see the changing profile of both the main actors involved in the reflections about the teaching and learning of mathematics and the communities gathering them, notably via the two main bodies under consideration in the context I am discussing, ICMI and IMU.

---

[14] HPM and PME, the first two Study Groups affiliated to ICMI, both in 1976, are typical of the development of several specific strands in didactics of mathematics that has happened during the last 35 years or so. The affiliation in 1994 of WFNMC, whose action is centered on mathematical competitions, is linked to an interest of a number of mathematicians concerning the identification and nurturing of mathematical talents. In their survey of international organizations in mathematics education, Hodgson et al. (2013) contrast the mere three international bodies established up to the early 1960s (ICMI—1908, CIEAEM—1950 and CIAEM—1961) with the proliferation since the mid-1970s, each new body corresponding to a particular component of the mathematics education landscape. They comment that "[t]he presence of such subcommunities wanting to become institutionalized within the mathematical education world can be interpreted as a sign of the vitality of the field and the diversity of its global community" (p. 935).

[15] The interested reader will find in Lehto (1998) and Hodgson (2009) more information about this episode of tension between IMU and ICMI resulting from Freudenthal's initiatives.

[16] Comments on this episode and its context, notably with respect to the so-called 'Math War' in the USA, can be found in Artigue (2008, p. 189). See also Hodgson (2009, pp. 85–86), and in particular endnote 5, p. 94.

Some Challenges that Mathematicians and Didacticians Are Facing

I commented above on the fact that both mathematicians and didacticians have a specific contribution to bring to educational issues, and in particular to the preparation of mathematical schoolteachers. In a sense they are more or less compelled to collaborate—at least in principle. But that is easier said than done.

One point at stake, in the case of mathematicians, is the extent to which they are willing to fully acknowledge education as part of their real responsibilities. But there are encouraging signs on that account. For instance more and more national societies of mathematicians, most of which are typically centred on research in mathematics, now devote a non-negligible part of their energy and activities to educational issues, very often with a genuine concern. A striking example, to take one close to my personal environment, is given by the American Mathematical Society, definitely an outstanding research-supporting body, but with pertinent and well-focused actions about educational matters. In a similar vein, one could think of the European Mathematical Society, whose Education Committee has launched in 2011 a series of articles in the *Newsletter* of the EMS under the general label *'Solid findings' in mathematics education*. The 'solid findings' papers are designed as "brief syntheses of research on topics of international importance" (Education Committee of the EMS, 2011 p. 47) which aim at presenting to an audience of non-specialists (especially mathematicians and mathematics teachers) what current research may tell us about how to improve the teaching and learning of a given mathematical topic. The message conveyed by such societies is very clear concerning the place that mathematicians may or should occupy with regard to educational matters, and even debates.[17] The message is also clear, consequently, about the responsibilities of a math department in this connection with respect to the inclusion of education as part of its mission. But transferring this into the daily life of the department is far from trivial.

---

[17] In his ICME-10 plenary lecture concerning the educational involvement of mathematicians, Bass (2008a) makes an important *caveat*:

> I choose specifically to focus on the involvement of *research mathematicians*, in part to dispel two common myths. First, it is a common *belief among mathematicians* that attention to education is a kind of pasturage for mathematicians in scientific decline. My examples include scholars of substantial stature in our profession, and in highly productive stages of their mathematical careers. Second, many *educators have questioned* the relevance of contributions made by research mathematicians, whose experience and knowledge is so remote from the concerns and realities of school mathematics education. I will argue that the knowledge, practices, and habits of mind, of research mathematicians are not only relevant to school mathematics education, but that this mathematical sensibility and perspective is essential for maintaining the mathematical balance and integrity of the educational process—in curriculum development, teacher education, assessment, etc. (pp. 42–43).

I dream of a day when it would be normal for a university math department to open a tenure-track position in mathematics but with a very strong educational emphasis, vg with regard to the preparation of schoolteachers or the development of innovative teaching approaches for very large undergraduate classes. Some of this already exists in some places,[18] but at a much too modest level altogether.

But an immediate concern follows: what about promotion to a higher academic rank? Would a significant involvement in education by a mathematician be judged by his peers as a valuable academic activity, on a par, say, with mathematical research or supervising graduate students? Many indicators point to the fact that this may remain for some time a major challenge that university administrations will be facing. But there are signs that mentalities may be changing.[19] Still it would probably be naive to expect a young mathematician recently hired by a math department to devote much time and energy to education matters, unless the position occupied would be very explicit on that account.

In a survey of the ICMI program of actions as seen from a Canadian perspective that I presented at a meeting of the Canadian Mathematics Education Study Group (Hodgson 2011), I suggested as a major challenge for the Canadian community the question of the actual involvement of individual mathematicians—especially the young ones—in educational matters and in activities of a group such as CMESG. The same challenge also exists, at the international level, with regard to the participation of mathematicians in activities of ICMI. What percentage of the people in the present audience, for instance, would consider themselves first and foremost as "working mathematicians"?

That said, past implications of mathematicians in educational matters have not been always optimal, to say the least. The level of rigor typically shown by mathematicians in their own research work is sometimes less perceptible when they come to express opinions about educational matters, sometimes on the basis of extremely naive observations or opinions. Bass and Hodgson (2004) comment for instance that "mathematicians sometimes lack a sufficient knowledge and/or appreciation of the complex nature of the problems in mathematics education" (p. 640). A particularly eloquent episode on that account is probably that of the Math War.[20] In her presidential closing talk at the ICMI Centennial symposium, Artigue (2008) describes not only the role of ICMI at the interface of mathematics and mathematics education, as announced in the title of her paper, but also at the interface of the communities of mathematicians and didacticians. She speaks of the

---

[18] As a concrete example, the mathematics department to which I belong has currently two such positions for mathematicians, one established as early as in the mid-1970s for the mathematical education of primary school teachers, and the other (mid-1990s) for secondary teachers.

[19] I have witnessed, over the past decade or so, a few successful cases of promotion for tenure or for full professorship concerning mathematicians with a career strongly focused on education and belonging to renowned research-oriented math departments.

[20] Bass (2008a) notes about the expression "Math War" that it is "an unfortunate term coined in the U.S. to describe the conflicts between mathematicians and educators over the content, goals, and pedagogy of the curriculum" (p. 42).

tensions that arose in the 1990s between those communities because "the supposed influence of mathematics educators was considered by some mathematicians as an important, if not the major, source of the observed difficulties in mathematics education, leading to such extremes as the so-called Math War in the USA" (p. 189).

Such a perception by mathematicians connects to a comment from Winsløw (2007), when he contrasts the necessary close ties he sees didactics having with the discipline, and the reality of the "[i]nstitutional policies and tradition" that imposes a distance between mathematicians and didactics (p. 533). He adds that "[t]he hesitancy of mathematicians to admit the need or worth of didactics could perhaps also be interpreted as an instance of a more general scepticism, among mathematicians, with respect to educational research." (p. 534)

But another side of the coin is related to the fact that didactics of mathematics has grown over the past decades into a fully-fledged academic domain, so that it has developed its specific paradigms, concepts, vocabulary. An unavoidable and obvious consequence is an increase of the communication gap between mathematicians and didacticians. Issues connected to the teaching and learning of mathematics can no more be approached with mere naive views or ideas—fortunately, one may say! But even mathematicians with a genuine interest in education feel a greater distance, as communication has become less transparent. A body of knowledge has now been developed, which must be grasped to a certain extent by mathematicians wishing to be part of the ongoing reflections.[21] Mathematicians will of course be familiar with this phenomenon internally, from one branch of mathematics to the other, but they may not be sensitive to its importance when it comes to educational contexts, if they have somehow developed the conviction that educational matters could be addressed seriously even through a very rudimentary approach. There is a responsibility for mathematicians here to keep abreast of recent didactical developments. But maybe more to my point, there is a responsibility for didacticians to make their work accessible without imposing unnecessary jargon or constructs. I believe more needs to be done on that account.

I would like to conclude this part of my talk with a comment of a possibly sensitive nature concerning the education of didacticians and the prerequisites they

---

[21] It is of interest to note, in that connection, that without denying the importance for mathematicians of gaining competency with respect to current developments in didactical research, some networks are developing that allow mathematicians to discuss educational issues and develop familiarity with ongoing work in less 'threatening' contexts, so to say. Such is the case for instance of Delta, an informal collaboration network among Southern Hemisphere countries that has developed since the end of the 1990s. In their survey of international organizations in mathematics education, Hodgson et al. (2013) write: "A central idea of Delta is to provide a forum in which mathematicians feel comfortable in discussing issues related to tertiary mathematics teaching and learning without being intimidated by what some may consider educational jargon or constructs. Many participants at the conferences are thus mathematicians wishing to report about a teaching experience or experiment that would normally not classify as bona fide research in mathematics education, but may still be helpful in inspiring those who want to reflect on their teaching" (p. 927).

should meet to be recognized as such. To make my case clear, I have in mind here the mathematical prerequisites. This issue is even more difficult to circumscribe as it does vary considerably from one country to the next.

As a starting vantage point, let me stress that the majority of the didacticians of my generation, if not all, had a substantial education in mathematics before switching to didactics of mathematics. The reason is simply that graduate studies in mathematics education are still, in most places, of a somewhat recent vintage. So it would not be so uncommon for a didactician of my age to have first done a certain amount of studies in mathematics, even at the graduate level. Today, with the development of didactics of mathematics as an autonomous academic field, the situation has changed substantially. While in many countries the road to didactics of mathematics is still intertwined with an important mathematical component, often of an advanced nature, I am aware of contexts where such is not the case, contexts where someone could be called a didactician of mathematics while having a rather limited experience of undergraduate mathematics, if any, even of the level of basic calculus or linear algebra. I must say that I really see problems with such a possibility. I do not wish here, of course, to express any opinion that may be received as offensive or as a personal criticism by any individual. It is more the "system" allowing this to happen that I want to comment on.

A didactician with no personal direct experience of mathematics at a somewhat advanced level will in my opinion lack a global "vision of the mathematical landscape" that I see as crucial, some aspects of it will escape his or her expertise. I am not at all suggesting here that all didacticians of mathematics should have followed loads of graduate math courses or experienced highly specialized mathematics research. But to take a concrete example, a deep understanding of basic number systems is clearly facilitated when these are considered as steps on the road towards the real numbers, the basic context for elementary analysis.

The present context does not allow me here to enter into fine discussions about the mathematical background that I would hope didacticians to have experienced. In a certain way, as may be the case with the mathematical education of teachers, rather than a simple matter of "doing more math", it is a matter of doing more math that may prove to be significant in order to allow the development of a deep intuition of the mathematical objects one is bound to meet in didactical situations.

Paying attention to this aspect is clearly a good way of facilitating communication between mathematicians and didacticians, as well as helping to foster mutual respect and understanding, unquestionably a vital ingredient in my opinion.

ICMI at the Dawn of Its Second Century

In this final section I examine selected actions recently launched by ICMI that may offer ways of fostering the collaboration between mathematicians and didacticians, and making it more productive. I am not proposing these undertakings as representing a kind of "ideal future" for mathematics or for didactics, nor for their interconnection. But these may be considered as pointing to possible models for concrete joint efforts bringing together the two communities discussed in this paper.

A common feature of the three projects that I discuss below is that they have been launched jointly by ICMI and its mother organization IMU. They thus represent meeting grounds for mathematicians and didacticians as they are represented by these two bodies. It is appropriate from that perspective to go back to the time of the beginnings of the term of office of the first ICMI Executive Committee under the presidency of Hyman Bass. I have already alluded earlier in this paper to two previous events that had provoked not only tensions between ICMI and IMU as bodies, but also between the two communities of mathematicians and didacticians: the so-called Math War in the USA and the turmoil resulting from the setting up of the program of the section on Teaching and Popularization of mathematics at the 1998 ICM. To use the words of Artigue (2008) in her description of the resulting context, "tension was at its maximum" (p. 189). She also comments that when the 1999–2002 ICMI Executive started its term of office, the situation had evolved so badly that "[v]oices asking ICMI to take its independence from a mother institution that expressed such mistrust were becoming stronger and stronger" (p. 189). But she finally concludes:

> Retrospectively this crisis was beneficial. It obliged the ICMI EC to deeply reflect about the nature of ICMI and what we wanted ICMI to be. This led us to reaffirm the strength of the epistemological links between mathematics and mathematics education (...). At the same time, we were convinced that making these links productive needed combined efforts from IMU and ICMI; the relationships could not stay as they were. (p. 190)

Conscious and explicit efforts were thus made by the IMU and ICMI Executives to improve the situation. I have described in Hodgson (2008, 2009) some of these efforts, which started with the (re)establishment of regular contacts between the two ECs, and especially between the presidents and secretaries [-general], and eventually resulted in the mounting of joint IMU/ICMI projects. Consequently, "after certain periods of dormancy and at times profound distance" (Hodgson 2008, p. 200), the IMU/ICMI relations were entering a time of welcomed harmony and intense collaboration. Concrete examples of such collaboration are given in Hodgson (2009, p. 87).

It should be mentioned, *en passant*, that a stunning outcome of this reinvigorated relationship, totally unexpected at the time of the 1998 crisis, is the "dramatic and historic change in the governance of ICMI" (Hodgson 2009, p. 87) represented by the fact that since 2008, the election of its Executive occurs at its own General Assembly (such as the one held just prior to this congress), rather than at the IMU GA, as was the case earlier. Such a development is a strong evidence of the maturity not only of the field represented by ICMI, but also of the relationship of ICMI with the organization to which it owes its legal existence.[22] More comments on this quite extraordinary episode can be found in Hodgson (2009).

---

[22] In that connection, the following comment made by IMU President László Lovász in his report to the 2010 IMU General Assembly may be of interest: "The IMU has a Commission, the ICMI, to

I now describe briefly three recent projects organized jointly by ICMI and IMU. I believe these suggest that concrete actions bringing together mathematicians and didacticians may contribute to resolve the issue of the mathematics/didactics interconnection. Additional information on these projects is to be found on the ICMI website.

The "Pipeline" Issue

Already in 2004, IMU approached ICMI, its education commission, expressing concerns in connection with a perceived decline in the numbers and quality of students choosing to pursue mathematics study at the university level and requesting the collaboration of ICMI to better understand this situation. The ensuing discussions pointed to another related phenomenon that needed to be investigated, namely the apparently inadequate supply of mathematically qualified students choosing to become mathematics teachers in the schools. IMU invited ICMI to partner in this undertaking, and take responsibility for its design.

Eventually the project (coined "Pipeline") was connected to, and became an extension of, the work of one of the Survey Teams for ICME-11, on the topic of "Recruitment, entrance and retention of students to university mathematical studies in different countries". It aimed at gathering data about different countries as well as promoting better understanding of the situation internationally. It was decided to focus on eight pilot countries for reasons of manageability (Australia, Finland, France, Korea, New Zealand, Portugal, UK, and USA), and to centre the study around four crucial transition points:

- From school to undergraduate program
- From undergraduate program to teacher education (and to school teaching)
- From undergraduate program to higher degrees in mathematics
- From higher degrees to the workforce

The final report of the Pipeline project was presented in a panel at the last International Congress of Mathematicians held in 2010 in Hyderabad, India. The resulting picture[23] is that there may not be a worldwide crisis in the numbers of mathematically gifted students, but that there is a crisis in some of the pilot countries. The numbers of such students in universities is susceptible to changes in school curricula and examination systems.

(Footnote 22 continued)
deal with math education. The [IMU] General Assembly in 2006 gave a larger degree of autonomy to this Commission, including separate elections for their officials. I would say that this did not loosen the connections between IMU and ICMI, to the contrary, I feel that we have developed an excellent working relationship." (Lovász 2010, p. 13).

[23] From ICMI quadrennial report of activities 2006–2009 submitted to the 2010 IMU General Assembly [cf. *Bulletin of the International Mathematical Union 58* (2010, p. 100)].

ICMI from Klein to Klein

It was at the first meeting of the 2007–2009 ICMI Executive Committee, under the presidency of Michèle Artigue, and in the context of a discussion about worthy projects that would bind the communities of mathematicians and didacticians, that the so-called Klein project was first mentioned. The ICMI EC saw it as a valuable undertaking to revisit the vision of ICMI first President, Felix Klein, in his milestone book *Elementary mathematics from an advanced standpoint*, published a century earlier and based on his lectures to secondary teachers. Klein's aim was on the one hand to help prospective and new teachers connect their university mathematics education with school mathematics and thus overcome the "double discontinuity" which they face when going from secondary school to university, and then back to school as a teacher (cf. Klein 1932, p. 1). But more generally Klein wanted to allow mathematics teachers to better appreciate the recent evolution in mathematics itself and make connections between the school mathematics curricula and research mathematics. This is in line with the view that a fundamental contribution of mathematicians to the reflections on teaching is by providing teachers with access to recent advances in mathematics and to conceptual clarifications (cf. Artigue 2010).

The reflections of the ICMI EC on this project were pursued in conjunction with the IMU EC and a Design Team responsible for the project was jointly appointed in 2008. The Klein project has already provoked a lot of very positive reactions from mathematicians, didacticians and teachers, and it is expected to have a triple output: a book simultaneously published in several languages, a resource DVD for teachers, and a wiki-based web-site continually updated and intended as a vehicle for the people who may wish to contribute to the project in an ongoing way.[24]

Capacity and Networking

The history of ICMI shows a long tradition of outreach initiatives with regard to developing countries. But this prime responsibility of our community has received a renewed attention recently. In her reviews of challenges now facing ICMI, Artigue (2008) stresses the importance, for the successful integration of colleagues from developing countries into the ICMI network, of developing new relationships between "centers and peripheries". She thus points to a necessary evolution from the traditional "North-South" model towards "more balanced views and relationships" (Artigue 2008, p. 195).

The Capacity and Networking Project (CANP) was developed by ICMI with this spirit in mind. It aims at enhancing mathematics education at all levels in developing countries by supporting the educational capacity of those responsible for the

---

[24] More information on the project and its evolution can be found at www.kleinproject.org.

preparation of mathematics teachers, and creating sustained and effective networks of teachers, mathematics educators and mathematicians in a given region. CANP was officially launched in 2011 jointly by IMU and ICMI, in conjunction with UNESCO. A prerequisite for the acceptability of a given proposal is some evidence of existing collaboration between local mathematicians and mathematics educators.

Each CANP program is based on a two-week workshop of about forty participants, half from the host country and half from regional neighbours. It is primarily aimed at mathematics teacher educators, but also includes mathematicians, researchers, policy-makers, and key teachers. Three CANP actions have already taken place or been announced: Mali (2011), Costa Rica (2012) and Cambodia (2013).

Conclusion

This lecture has centred on the specificity and complementarity of the contributions brought by mathematicians and didacticians of mathematics to the reflections on the teaching and learning of mathematics. Another more encompassing approach would be to consider the general framework of the sciences to which research in the didactics of mathematics is connected because of its interdisciplinary nature. The importance of "defining and strengthening the relations to the supporting sciences" is discussed in Blomhøj (2008), where emphasis is placed on the need for mathematics education research "to benefit from new developments in the supporting disciplines" (p. 173). In particular the author stresses that "[o]n a more political level the relationships to the supporting disciplines are very important for the integration of mathematics education research in academia and thereby for the institutionalisation of our research field" (Blomhøj 2008, p. 173). Mathematics appears of course as a fundamental *cas de figure* on that account.

The issue of the mathematics/didactics interconnection is clearly a very vast one and my focus in this talk was to look at it from the vantage point of the International Commission on Mathematical Instruction, through both its history and its current actions. In a survey paper aiming at encouraging mathematicians' participation to the ICME-10 congress, Bass and Hodgson (2004) have raised the question: "So how are mathematics and mathematics education, as domains of knowledge and as communities of practice, now linked, and what could be the most natural and productive kinds of connections?" Their comment was that "ICMI represents one historical, and still evolving, response to those questions at the international level" (p. 640). To borrow from the beautiful title of Artigue (2008), ICMI was, and is still there, at the interface between mathematics and mathematics education.

In his reaction to Kilpatrick's paper (2008) on the development of mathematics education as an academic field, Dorier (2008) mentions the multiple types of cooperation that mathematics education has developed with other academic fields "because the development of research shows that the complexity of the reality of education needs to be tackled from different viewpoints" (p. 45). Emphasizing the importance for mathematics education, amidst this diversity, "to put forward the specificities of its objects, methods, and epistemology" (p. 45) in comparison to other fields connected to educational issues, he notes the following:

> In that sense, the relation [of mathematics education] to mathematics is essential, and the role of ICMI is thus vital in order to maintain and develop in all its variety an academic field specific to mathematics education that maintains a privileged relation with the mathematical community at large. (p. 45)

But seeing as a risk that mathematics education may fail to develop as a fully-fledged autonomous academic domain and be absorbed in related fields, Dorier concludes that "[a] barrier against this possible dilution remains the attachment of mathematics education to mathematics that ICMI can guarantee while encouraging cooperative work with other academic fields connected to education" (p. 45). That describes in a very fitting way the framework I was proposing in this talk to reflect on the links, past and future, between mathematics and didactics and between the main communities that support these domains.

Acknowledgments The author wishes to express his gratitude to the colleagues who brought their support in the preparation of this paper, and in particular to his immediate predecessors as ICMI Secretary-General, A. Geoffrey Howson (1983–1990) and Mogens Niss (1991–1998).

Open Access This chapter is distributed under the terms of the Creative Commons Attribution Noncommercial License, which permits any noncommercial use, distribution, and reproduction in any medium, provided the original author(s) and source are credited.

References

Artigue, M. (2008). ICMI: A century at the interface between mathematics and mathematics education. In M. Menghini, F. Furinghetti, L. Giacardi, & F. Arzarello (Eds.), *The first century of the International Commission on Mathematical Instruction (1908-2008). Reflecting and shaping the world of mathematics education* (pp. 185-198). Rome, Italy: Istituto della Enciclopedia Italiana.

Artigue, M. (2010). Penser les relations entre mathématiciens, enseignement des mathématiques, recherche sur et pour cet enseignement: que nous apportent l'expérience des IREM et celle de la CIEM? [Reflecting on the relations between mathematicians, teaching of mathematics, research on and for this teaching: what do the experience of the IREMs and that of ICMI bring us?] *Les mathématiciens et l'enseignement de leur discipline en France. [Mathematicians and the teaching of their discipline in France.]* Séminaire des IREM et de *Repères IREM* (March 2010). Unpublished manuscript.

Bass, H. (2008a). Mathematics, mathematicians, and mathematics education. In M. Niss (Ed.), *Proceedings of the tenth International Congress on Mathematical Education* (pp. 42-55). Roskilde, Denmark: IMFUFA.

Bass, H. (2008b). Moments in the life of ICMI. In M. Menghini, F. Furinghetti, L. Giacardi, & F. Arzarello (Eds.), *The first century of the International Commission on Mathematical Instruction (1908-2008). Reflecting and shaping the world of mathematics education* (pp. 9-24). Rome, Italy: Istituto della Enciclopedia Italiana.

Bass, H., & Hodgson, B. R. (2004). The International Commission on Mathematical Instruction—What? Why? For whom? *Notices of the American Mathematical Society, 51*(6), 639-644.

Blomhøj, M. (2008). ICMI's challenges and future. In M. Menghini, F. Furinghetti, L. Giacardi, & F. Arzarello (Eds.), *The first century of the International Commission on Mathematical Instruction (1908-2008). Reflecting and shaping the world of mathematics education* (pp. 169-180). Rome, Italy: Istituto della Enciclopedia Italiana.

Brewster, D. (1819). *A treatise on the kaleidoscope*. Edinburgh, UK: Archibald Constable & Co.

Dorier, J.-L. (2008). Reaction to J. Kilpatrick's plenary talk: The development of mathematics education as an academic field. In M. Menghini, F. Furinghetti, L. Giacardi, & F. Arzarello (Eds.), *The first century of the International Commission on Mathematical Instruction (1908-2008). Reflecting and shaping the world of mathematics education* (pp. 40-46). Rome, Italy: Istituto della Enciclopedia Italiana.

Dossey, J. A. (1992). The nature of mathematics: its role and its influence. In: D. A. Grouws (Ed.), *Handbook of research on mathematics teaching and learning* (pp. 39-48). New York, NY: Macmillan.

Education Committee of the European Mathematical Society. (2011). 'Solid findings' in mathematics education. *Newsletter of the European Mathematical Society, 81,* 46-48.

Fehr, H. (1920–1921). La Commission internationale de l'enseignement mathématique de 1908 à 1920: Compte rendu sommaire suivi de la liste complète des travaux publiés par la Commission et les Sous-commissions nationales. *L'Enseignement Mathématique,* s. 1, *21,* 305-339.

Furinghetti, F. (2003). Mathematical instruction in an international perspective: The contribution of the journal *L'Enseignement Mathématique*. In D. Coray, F. Furinghetti, H. Gispert, B. R. Hodgson, & G. Schubring (Eds.), *One hundred years of* L'Enseignement Mathématique*: Moments of mathematics education in the twentieth century* (Monographie 39, pp. 19–46). Geneva, Switzerland: L'Enseignement Mathématique.

Furinghetti, F. (2008). Mathematics education in the ICMI perspective. *International Journal for the History of Mathematics Education, 3*(2), 47–56.

Furinghetti, F., Menghini, M., Arzarello, F., & Giacardi, L. (2008). ICMI Renaissance: The emergence of new issues in mathematics education. In M. Menghini, F. Furinghetti, L. Giacardi, & F. Arzarello (Eds.), *The first century of the International Commission on Mathematical Instruction (1908-2008). Reflecting and shaping the world of mathematics education* (pp. 131-147). Rome, Italy: Istituto della Enciclopedia Italiana.

Graf, K.-D., & Hodgson, B. R. (1990). Popularizing geometrical concepts: the case of the kaleidoscope. *For the Learning of Mathematics, 10*(3), 42-50.

Hobson, E. W., & Love, A. E. H. (1913). *Proceedings of the fifth International Congress of Mathematicians*. Cambridge, UK: Cambridge University Press.

Hodgson, B. R. (1987). La géométrie du kaléidoscope. *Bulletin de l'Association mathématique du Québec, 27*(2), 12-24. Reprinted in: *Plot* (Supplément: Symétrie – dossier pédagogique) *42* (1988), 25-34.

Hodgson, B. R. (2001). The mathematical education of school teachers: role and responsibilities of university mathematicians. In D. Holton (Ed.), *The teaching and learning of mathematics at university level: An ICMI study* (pp. 501-518). (New ICMI Study Series, No. 7) Dordrecht, The Netherlands: Kluwer.

Hodgson, B. R. (2004). The mathematical education of schoolteachers: a baker's dozen of fertile problems. In J.P. Wang & B.Y. Xu (Eds.), *Trends and challenges in mathematics education* (pp. 315-341). Shanghai, China: East China Normal University Press.

Hodgson, B. R. (2008). Some views on ICMI at the dawn of its second century. In M. Menghini, F. Furinghetti, L. Giacardi, & F. Arzarello (Eds.), *The first century of the International Commission on Mathematical Instruction (1908-2008). Reflecting and shaping the world of mathematics education* (pp. 199-203). Rome, Italy: Istituto della Enciclopedia Italiana.

Hodgson, B. R. (2009). ICMI in the post-Freudenthal era: Moments in the history of mathematics education from an international perspective. In K. Bjarnadóttir, F. Furinghetti, & G. Schubring (Eds.), *"Dig where you stand": Proceedings of the conference on On-going research in the history of mathematics education* (pp. 79–96). Reykjavik: University of Iceland, School of Education.

Hodgson, B. R. (2011). Collaboration et échanges internationaux en éducation mathématique dans le cadre de la CIEM: Regards selon une perspective canadienne [ICMI as a space for international collaboration and exchange in mathematics education: Some views from a Canadian perspective]. In P. Liljedahl, S. Oesterle, & D. Allan (Eds.), *Proceedings of the 2010 annual meeting of the Canadian Mathematics Education Study Group/Groupe canadien d'étude en didactique des mathématiques* (pp. 31–50). Burnaby, Canada: CMESG/GCEDM.

Hodgson, B. R., Rogers, L. F., Lerman, S., & Lim-Teo, S. K. (2013). International organizations in mathematics education. In M. A. (Ken) Clements, A. Bishop, C. Keitel, J. Kilpatrick, & F. K. S. Leung (Eds.), *Third international handbook of mathematics education* (pp. 901–947). New York, NY: Springer.

Howson, A. G. (1984). Seventy-five years of the International Commission on Mathematical Instruction. *Educational Studies in Mathematics*, *15*(1), 75-93.

Kilpatrick, J. (2003). Twenty years of French *didactique* viewed from the United States. *For the Learning of Mathematics*, *23*(2), 23-27. [Translation of: Vingt ans de didactique française depuis les USA. In M. Artigue, R. Gras, C. Laborde, & P. Tavignot (Eds.), *Vingt ans de didactique des mathématiques en France: Hommage à Guy Brousseau et Gérard Vergnaud* (pp. 84-96). Grenoble, France: La Pensée Sauvage, 1994.].

Kilpatrick, J. (2008). The development of mathematics education as an academic field. In M. Menghini, F. Furinghetti, L. Giacardi, & F. Arzarello (Eds.), *The first century of the International Commission on Mathematical Instruction (1908-2008). Reflecting and shaping the world of mathematics education* (pp. 25-39). Rome, Italy: Istituto della Enciclopedia Italiana.

Kilpatrick, J. (2012). Lost in translation. *Short Proceedings of the "Colloque hommage à Michèle Artigue"*. Paris: France, Université Paris Diderot – Paris 7. (Accessible at http://www. colloqueartigue2012.fr/).

Klein, F. (1932). *Elementary mathematics from an advanced standpoint: Arithmetic, algebra, analysis*. New York, NY: Macmillan. [Translation of volume 1 of the three-volume third edition of *Elementarmathematik vom höheren Standpunkte aus*. Berlin, Germany: J. Springer, 1924-1928.].

Kline, M. (1980). *Mathematics: The loss of certainty*. New York, NY: Oxford University Press.

Lehto, O. (1998). *Mathematics without borders: A history of the International Mathematical Union*. New York, NY: Springer.

Lovász, L. (2010). Overview on Union activities. (Section 3.1 of the report of the 16th General Assembly of the International Mathematical Union – IMU) *Bulletin of the International Mathematical Union*, *59*, 10-15.

Mac Lane, S. (1971). *Categories for the working mathematician*. New York, NY: Springer.

Schubring, G. (2003). *L'Enseignement Mathématique* and the first international commission (IMUK): The emergence of international communication and cooperation. In D. Coray, F. Furinghetti, H. Gispert, B. R. Hodgson, & G. Schubring (Eds.), *One hundred years of L'Enseignement Mathématique: Moments of mathematics education in the twentieth century* (Monographie 39, pp. 47–65). Geneva, Switzerland: L'Enseignement Mathématique.

Schubring, G. (2008). The origins and early incarnations of ICMI. In M. Menghini, F. Furinghetti, L. Giacardi, & F. Arzarello (Eds.), *The first century of the International Commission on Mathematical Instruction (1908-2008). Reflecting and shaping the world of mathematics education* (pp. 113-130). Rome, Italy: Istituto della Enciclopedia Italiana.

Sierpinska, A. (1998). Whither mathematics education? In C. Alsina, J. M. Alvarez, M. Niss, A. Pérez, L. Rico, & A. Sfard (Eds.), *Proceedings of the 8th International Congress on Mathematical Education* (pp. 21-46). Sevilla, Spain: S.A.E.M. Thales.

Steiner, H. G. (1968). The plans for a Center of didactics of mathematics at the University of Karlsruhe. In N. Teodorescu (Ed.), *Colloque international UNESCO "Modernisation de l'enseignement mathématique dans les pays européens"* (pp. 423-435). Bucharest, Romania: Éditions didactiques et pédagogiques.

Winsløw, C. (2007). Didactics of mathematics: an epistemological approach to mathematics education. *The Curriculum Journal*, *18*(4), 523-536.

Mathematics Education in the National Curriculum—with Some Reflections on Liberal Education

Lee Don-Hee

Abstract Mathematics has been recognized and justified to be placed in the prime core of the formal curriculum for general education. In this paper, however, some reflections are made on the national curriculum together with mathematics education in accordance with the tradition of "liberal education." Liberal education is education for liberal men. The basic education of liberal human being is the discipline of his rational powers and the cultivation of his intellect. It has sustained its meaning and value to be different from the vocational training for the purpose of earning one's living. But John Dewey differently contends that the vocational training may claim a pertinent candidate to the position playing a role in cultivating the human mind, the intellect (or intelligence). For Dewey, important is not the content of teaching but rather the intelligence in its operation.Intelligence is "equipped" with some properties that are functionally related to the properties of the problematic situation, which they take on the character of "method." A kind of mental process, "a methodic process," connecting problematic situations and resolved consequences is what Dewey qualified to be "reflective thinking," where the intelligence keeps itself alive and activating for its full operation. Then, we would have two different, but closely related tasks. One is (i) the self-habituation of methodic activity; and the other is (ii) the nurturing of children in methods. The curricular device is bound to gratify a variety of different needs and motives. No matter how worth studying mathematics may be, it can never be learnt unless the body of learning materials are so organized that students may cope with its degree of difficulty settled for the teaching purpose. Then contents must be appropriately selected and efficiently programmed on the part of learners. Learnability is prior to the academic loftiness at least in educational situations.

L. Don-Hee (✉)
Philosophy of Education, Seoul National University, Seoul, South Korea

© The Author(s) 2015
S.J. Cho (ed.), *The Proceedings of the 12th International Congress on Mathematical Education*, DOI 10.1007/978-3-319-12688-3_8

Why Should We Teach Mathematics?

For the first nine school years of the elementary and secondary education in Korea, mathematics is one of the major subject-matters of which the national curriculum formally consists. In several partial amendments, mathematics has outdone other competing subjects in the official process of allocating the weekly teaching hours. In spite of the fact that mathematics fails to draw students' favor and popularity, it has been recognized and justified to be placed in the prime core of the formal curriculum for general education, both elementary and secondary.

Mathematical study itself has occupied integral part of the human civilization as well as intellectual life. As a matter of fact, mathematics has been taught as the core subject-matter in the history of school curriculum everywhere in any civilized part of the world. Its system of knowledge, together with its language and method, has been shared among the world intellectual communities more than any other discipline, probably more than any other human undertaking.

Now, however, I would like to raise an unexpected question: Why should we teach relevantly mathematics to all the young people at elementary and secondary levels of education? And, does it really deserve attention as a competitive power in curriculum development?

From the standpoint of social utility, among different points of view, there may be four reasons, at least, why we should teach mathematics in the school. First, to raise mathematical specialists; second, to meet needs of mathematical knowledge required for the advanced level of professional services; third, to promote problem-solving abilities, namely those of logical or formal reasoning; and fourth, to help people to be familiar with basic mathematical knowledge necessitated for the ordinary daily life.

It may be realistically the case that there must be those who devote themselves to study the highly advanced and outstanding mathematics in any civilized society; that mathematical knowledge must be applied to a variety of professional services; that mathematics by its own nature shows us how to make our thinking logically valid and how to solve efficiently complicated problems encountered in our daily life; and that even basic rules or ideas of mathematics help us to see the complexity of the world in organized forms by virtue of its symbolic power.

But it seems to be necessitated to recognize that only a limited number of mathematicians and professionals are in need of training at higher levels, some basic parts of which are already embedded in the national curriculum for the upper or even lower secondary education. A greater part of students say that mathematics is too unintelligible for them to learn, and that it gives them toilsome and boring time in the class room situations. You cannot teach students anything if they are not able, and not willing, to learn it properly. And your instructional device cannot work in teaching mathematics if they extremely hate and stubbornly refuse to learn it at their own will.

In order to see why we should teach mathematics in the school, and what kinds of mathematization should be experienced, I, as a student of philosophy of education and, to be sure, a rank outsider, here would like to say something, perhaps what an ordinary consumer of education experiences with reflections on the national curriculum together with mathematics education.

Here, I would like to make some reflections in accordance with the tradition of "liberal education." I believe that the question should be answered in terms of values and implications of liberal education. For it represents, in its very nature of meaning, authentic communications between the human mind and the cultural tradition. But I do not try to make mathematics education fitted into an orthodox admittedly dominant in its tradition, but rather to discuss about how we should understand the idea of liberal education in its consideration with teaching mathematics.

In the Tradition of Liberal Education

Liberal education is education for liberal men. Originally, as Leo Strauss mentioned, a liberal man was a man who possessed a privilege to behave in a manner becoming a free man, as distinguished from a slave (Strauss 1968, p. 10). A slave is also a human being who lives yet for another human being, his master; he has no life of his own. The master, on the other hand, has all his time for himself, that is, for the pursuits becoming him in the world, with its meaning, of his social and intellectual life.

Nowadays, in the democratic society, however, we may say that a liberal man is a man, a rational being, who is to live under his own will, not other's. By education, one becomes, and maintains oneself, a liberal man in the genuine sense. The basic education of liberal human being is the discipline of his rational powers and the cultivation of his intellect. Historically, it is believed that this discipline can be achieved by the liberal arts, basically the communicational arts, namely reading, speaking, writing, listening, reckoning, and reasoning. The three R's (reading, writing, and reckoning), which always signified the formal discipline, are qualified for the essence of liberal or general education.

In the tradition of liberal education, numeracy, together with literacy, has been integral part of human abilities for the societal life civilized more or less so as to engage in liberal education. Plato especially points out that the mathematical studies develop the soul in two ways: In the first place, they provoke reflection and bring out all the contradictions that lie hid in ordinary opinions based on mere sense-knowledge; in the second place, they take him part of the road towards the good which is the goal of all learning and all life (Boyd and King 1975, pp. 34–35).

In his master-work, the Republic, Plato discusses an educational scheme to show how the ideal State might be created out of programs cultivating the mind of the youth. Up to seventeen or eighteen, the children, assumed to be the future rulers, were all to devote themselves to gymnastics and music. After 2 years of physical training, the youth who had proved themselves capable of more advanced studies were to work at the mathematical sciences–arithmetic, geometry, astronomy, and harmonics (the mathematical theory of music) from twenty to thirty. Finally a select group who had shown distinction both of mind and character throughout the whole course of their previous training were to spend 5 years in the study of dialect (or philosophy), the science of the good (ideas), before taking their place in the ranks of the "guardians."

There may be at least two different conceptions of liberal education: one is intellectualistic while the other is pragmatic. Among others I want to mention here Mortimer J. Adler as an intellectualist who stands against a pragmatist John Dewey. The idea of liberal education itself was genetically aristocratic, for the truly free man who can live in a manner becoming a free man is the man of leisure. But liberal education is not simply entitled to a kind of program for the free man in a political sense, but also understood differently so as to mention a certain principle overriding activities cultivating the human intelligence and creativity.

For liberal education, Adler maintains that the human reason may be at first trained in its proper operations by the communicational arts, since man is a social animal as well as a rational one and his intellectual life is lived in a community which can exist only through the communication of men. The intellect cannot be accomplished merely by the three R's, but, in addition, through furnishing it with knowledge and wisdom, acquainting it with truth, and giving it a mastery of ideas. At this point, he suggests that the other basic feature of liberal education appears, namely the great books, that is, the master productions in all fields, philosophy, science, history, and belles-lettres. These constitute the cultural tradition by which the intellects of each generation must first be cultivated.

Mortimer J. Adler says:

> ... If there is philosophical wisdom as well as scientific knowledge, if the former consists on insights and ideas that change little from time to time, and if even the latter has many abiding concepts and a relatively constant method, lf the great works of literature as well as of philosophy touch upon the permanent moral problems of mankind and express the universal convictions of men involved in moral conflict–if these things are so, then the great books of ancient and medieval, as well as modern, times are repository of knowledge and wisdom, a tradition of culture which must initiate each new generation. (Adler 1939)

In Adler's conception, liberal education is a kind of program which provides the youth with communicational arts (reading, writing, speaking, reckoning etc.), and thereafter with the intellectual mediator for the constant intercourse between them and the greatest minds in the cultural tradition. Liberal education is learning for its own sake or for the sake of all those self-rewarding activities which include the political, aesthetic, and speculative. It differentiates itself from vocational training which no one should have to take without compensation, and which is just pre-paratory to work for the sake of earning. (Adler 1951)

Intelligence, Method, and Methodic

Now, we may ask again "what for liberal education?" It is education to cultivate the human intellect, and thus to liberate the human mind. It has sustained its meaning and value to be different from the vocational training which is confined to learning skills for the purpose of earning one's living. Traditionally, it is literate education of a certain kind: some sort of education in letters or through letters, as tools for developing the intellect. It has been conceived to be a kind of program for teaching the youth in subjects, namely liberal arts, and studying the great books reminding oneself of human excellence, of human greatness.

As John Dewey differently contends, however, that there seems to be no relevant reason why we are to confine ourselves to literate education for teaching in so-called liberal arts. Even the vocational training may claim a pertinent candidate to the position playing a role in cultivating the human mind, the intellect (or intelligence). For this qualification, of course, vocational training is also availed of the capacity as efficiently a tool to be utilized as the traditional program in liberal arts. Dewey says as follows:

> ... Instead of trying to split schools into two kinds, one of a trade type for children whom it is assumed are to be employees and one of a liberal type for the children of the well-to-do, it will aim at such a reorganization of existing schools as will give all pupils a genuine respect for useful work, an ability to render service, and a contempt for social parasites whether they are called tramps or leaders of 'society.'...
> ... It will indeed make much of developing motor and skills, but not of a routine or automatic type. It will rather utilize active and manual pursuits as the means of developing constructive, intentive and creative power of mind... the individual may be able to make his own choices and his own adjustments, and be master, so far as in him lies, of his own economic fate... So far as method is concerned, such a conception of industrial education will prize freedom more than docility; initiative more than automatic skills; insight and understanding more than capacity to recite lessons or to execute tasks under the direction of others... (Dewey 1917)

For Dewey, what must be important is not whether the content of teaching consists of letters or non-letters for developing the mind, indeed the mind of the liberal man, but rather whether "the human intelligence" can work properly in its operation. Intelligence can work to solve the problem situation, trifling or serious, that we encounter in our daily life, such as conflict with neighbors, discord within the family, crises of confidence in business and the like. We need a social intelligence to solve the problem situation, such as deep economic depression, state security risk, vicious inflationary spiral, chronic rebellion, and the like. Academically, a variety of disciplines, theoretical or practical, are products of intelligence managing to work out of the problem situation where academics struggle with a systematic body of highly complicated ideas and matters. Mathematics is a structure of resolutions painstaking with forms of mathematical intelligence.

Intelligence does not operate vacuously: It is "equipped" with some properties that are functionally related to the properties of the problematic situation. When these properties are systematically distinguished, formulated and organized so as to apply to the problematic situation, they take on the character of "method." Method then is not outside of or divorced from material. Method may be philosophical, literary, scientific, mathematical, or technological. Dewey writes, "The fact that the material of a science is organized is evidence that it has already been subjected to intelligence; it has been methodized, so to say" (Dewey 1916, p. 165). Method then is a logical description of intelligence in operation. Indeed, intelligence and method are synonymous.

In this consideration, mathematics as a discipline or a subject-matter may be admittedly said to be a sort of human product subjected to intelligence, thus its material content methodized in such a way that it has characteristically differentiate itself in its properties from other modes of human works.

In the educational discourses, we often refer to "intelligence" as the prime human ability among others to be developed in teaching or training programs. "Habit" is also referred to as objective pertinent to educational activities. But "intelligence" or "abilities" mostly includes those which are characteristically cognitive and self-directive, whereas "habits" mostly represents those which mainly pertain to physical and routine actions. This is the reason why the vocational training makes itself mistakenly different in its mode of learning from the traditional conception of liberal education.

A general theory that accounts for habits and intelligence and their various relationships becomes a matter of our concern. The question here, of course, is what kind of action is both habitual and intelligent: And the problem is to distinguish the appropriate kind of situations for the use of the terms, "habitual" and "intelligent," respectively, to be employed.

If method is a logical description of intelligence in operation, and indeed intelligence and method are synonymous, then we may ask: Could methods be habituated? Could they become habitual? These questions have to do neither with the possibility of forming the habit of adopting methods nor with the evolution of a method into habit. Rather, these questions have to do with the possibility of habituating "methodic" activities. But the habit of methodic activity could still be understood as a habit of translating methods into the pursuit of an end. This sense of "methodic habit" implies a habit of reproduction. The intelligence that has served methods is secondary to the intelligence functioning in methodic activities. For the former intelligence is not activating while the latter intelligence is. Furthermore, the powers that methods may execute are not necessarily powers of intelligence, nor are they human powers. What we actually look for is the habit of methodizing or controlling problem-situations, of pursuing methods, and of utilizing methodized patterns in the pursuit of an end, that is, a methodic habit.

A kind of mental process connecting problematic situations and resolved consequences is what Dewey qualified to be the process of "reflective thinking," where the intelligence keeps itself alive and activating for its full operation.

Dewey's conception of reflective thinking is in somewhat temporal terms, different from a methodological account featuring formal properties. Dewey is not providing a formula, but a temporal account of the activity in which the formula does its work. Dewey's theory of reflective thinking should therefore not be understood to rule out the adoption of ready-made methodic formulae—those which have been already methodized. Indeed, he cautioned us that we ignore these at our peril. He argues that things as methodized represent the office of intelligence, in projection, in pursuit, and in the control of new experience. In short, methodic habits are valued habits of intelligence. They are embodied in, or are cases of, intelligence.

Dewey claims that reflective thinking should be an educational aim since it carries with itself qualities significant as educational values (1933, p. 17). In the first place, it emancipates us from merely impulsive and merely routine activity. It enables us to direct our activities with foresight and to plan according to ends-in-view, or purposes. Secondly, it enables us to develop and arrange "artificial" signs to remind us by representing in advance not only "existential" consequences but also ways of securing and avoiding them. Thirdly, since an object to which we react is not a mere thing, but rather a thing having a definite significance or meaning, reflective thinking enriches the object with further meanings as we compare a thing or event as it was before with what it is after. Intelligent mastery over the object is obtained through thinking.

Here we would have two different, but closely related tasks. One is (i) the self-habituation of methodic activity; and the other is (ii) the nurturing of children in methods. The first is based on the assumption that methodic activity, reflective thinking or the problem-solving process, is not only a methodological mechanism for teaching knowledge of substances (or subject-matters) emerging in the system of educational values, but it is also something to teach. This means that methodic activity is not merely a means serving in the pursuit of various educational objectives, but also is itself a candidate for being an educational objective.

But the second task should not be understood as the fact that the educative process is methodic because it is a process of applying the method supposedly common to all disciplines. That is, if all disciplines are cases of method then they should display common properties—common formal properties. The ends of various disciplines differ in form and substance; the means differ in the force of applicability of their theories. But each is an affair of controlled means and ends. To say again, mathematics is a discipline of mathematical method as well as mathematical intelligence.

Methods are symbolic expressions of what is performed in the process of controlled activity. They represent among other matters the material involved. But substantial materials, for example, problems, issues, situations, events, or reports, are not always of a single type in their mode of placement in the means-end relating.

The objective common to all sciences, including mathematics, is assertion making, conclusion drawing, proposition forming, and possibly theory structuring. Each science is a kind of knowledge forging, hypothesis testing, prediction constructing, and so on. Thus, we have physiological and biological knowledge, economic and astronomical facts, geological and biological hypotheses, historical and anthropological reports, and philosophical and mathematical arguments. We sometimes call all these bodies of knowledge–meaning, of course, the fruits of the inquiries of these sciences. All are equally sciences and human achievements according to scientific method. But each is different in the sense that each is proceeding with different problems, materials, concepts, and terminology. Thus it seems unlikely that one can learn a method in a specific type of problematic situation without experience in the conduct of dealing in and with that situation.

The implication that the educative process is a process of nurturing children in methods leads to a theoretical corollary that the object of educational research or inquiry may be found in the universe of methods. This is to claim that if there is a universe of discourse, a block of conceptual equipment which more adequately deals with the tasks of educators, it must be the universe of discourse about methods or methodic activities in which intelligence is to be embedded. Mathematics is also methodic in its nature.

Conditions of Mathematics for Liberal Education in the National Curriculum

Last February, I found an interesting column in one of the issues of the daily newspaper Joong Ang Ilbo that is published in Seoul. It was titled "No Easy and Interesting Math Learning" written by a professor of mathematics, named Yong-Jin Song of In-Ha University. To partly translate into English as follows:

> ... Mathematics for today has become a systematic discipline which has grown up sophisticated by virtue of great geniuses intelligence in the human history, and thus it must be difficult in its nature for ordinary people to learn. You cannot make yourself master of its hard and tough contents without taking a well-planned course of learning. Mathematics is the supreme product of human intellect such that even its fundamental level requests you to undergo a well-organized training which is somewhat intensive to some extent. To be sure, there exists no mathematics that is easy and, at the same time, interesting; but perhaps rather there may exist such a kind of mathematics that is both difficult and interesting. Many a thing is popular and interesting because of its difficulty: playing a game of go, golf, soccer, computer or the like. (Song 2012)

Professor Song, however, does not mention what kind of mathematics to be taught. Of course, he may presuppose the possibility such that its contents be organized in accordance with the condition of learners describable in terms of age, experience, motivation and cultural orientation. Nevertheless, he seems to assume that the mathematics may be enjoyed exclusively by those who are intellectually equivalent to appreciate of its value. It seems to me that he assumes there exist "the (one and only one in kind) mathematics" which schools should teach to all young people.

If he believes, as Karl Mannheim opposes the possibility of a sociology of mathematics, and as Pythagoreans and Platonists believe, that mathematical truths are eternal objects, not culturally relative (Restivo and Collins 1982), then he may be right in the assertion that we should not be concerned with the degree of difficulty in mathematics education.

But, as Oswald Spengler says that there is no mathematic but only mathematics, we define mathematics as methodic products, we discussed earlier, of intelligence in operation for the struggle with problematic situations. Mathematics is a particular mode of experience, distinguishable from other disciplines and arts, and the character of mathematical inquiries vary with cultural toils, and with problematic motives and interests. There may be quite a few different ways of inclusion and exclusion in organizing contents for different orientations.

And if learning values are appreciated, and academic needs are gratified by the experience of mathematical difficulty, the very difficulty of "the mathematics as such," then it confines to a very limited number of people who can intrinsically enjoy the subject-matter, what is called "mathematics," just as a very limited number of few people, the professional or ardent players, enjoy the game, go, golf, and soccer. The difficulty provides no legitimate ground that mathematics may outdo other competing subjects in the competitive process of allocating the weekly teaching hours, and that it can claim to be the prime core of the national curriculum for general education.

We cultivate the human mind (intellect or intelligence, whatever) by the instrumentality of mathematics in association with other teaching-learning programs, that take care of, and improve, the native faculties of the mind. Therefore, mathematics is to deserve a core subject-matter among those worthwhile to teach for liberal education, the finished product of which is a cultured human being. Mathematics, which cultivates and thus liberates the human mind, consists of intrinsic values, that is, those which are good in itself. We do not necessarily enforce it to demonstrate any practical utilities, that is, extrinsic values which are instrumentally good for something other than itself. Even its applied ramifications may be so organized as to materialize their cultivating and liberating powers to the maximum extend. Even in non-academic activities where mathematics is subsidiary, they must be planned to methodically activate the human potentials of creativity and productivity.

Probably, of course, an outstanding group in mathematics can enjoy its intrinsic value at the highly advanced level. And the well-trained professional proficiency in teaching may open up new path into a more sophisticated realm as a benefit to ambitious students. To them mathematics becomes not any more a painstaking burden, but rather an enjoyable game.

The curricular device is bound to gratify a variety of different needs and motives. No matter how worth studying mathematics may be, it can never be learnt unless the body of learning materials are so organized that students may cope with its degree of difficulty settled for the teaching purpose. Then contents must be appropriately selected and efficiently programmed on the part of learners. Learnability is prior to the academic loftiness in educational situations. You cannot enjoy what you are not learnable. The variety may avail with us widely open learning opportunities where many a different mathematical need may be gratified.

In sofar as mathematical education is concerned, we may justifiably say that learning opportunity in its genuine sense be available to the learners, if and only if it is not the case that its course of study is too unequivalent for the students to carry out in the regular school activities. Especially, it is true of the national curriculum system which is assumed to be compulsory to all youngsters.

Open Access This chapter is distributed under the terms of the Creative Commons Attribution Noncommercial License, which permits any noncommercial use, distribution, and reproduction in any medium, provided the original author(s) and source are credited.

References

Adler, M. J. (1939). The crisis in contemporary education, *The Social Frontier*, *5*, 141-144.

Adler, M. J. (1951) Labor, leisure, and liberal education, *Journal of General Education 6*, 43.

Boyd, W., & Edmund J. K. (1975). *The history of western education.* 7th Edition. London: Adams & Charles Black.

Brown, T. (2010). Cultural continuity and consensus in mathematics education, (Special Issue on Critical Mathematics Education) *Philosophy of Mathematics Education Journal* ISSN 1465-2978 (Online) No. 25.

Dewey, J. (1916). *Democracy and education.* New York: The Macmillan Company.

Dewey, J. (1917). Learning to earn, *School and Society*, *5*, 333–334.

Dewey, J. (1933). *How to think.* Chicago: Henry Regnery Company.

Dewey, J. (1934). *Art as experience.* New York: Capricorn Books.

Dewey, J. (1938). *Logic: The theory of inquiry.* New York: Henry Holt and Company.

Restivo, S., & Randal, C. (1982). Mathematics and civilization: a Word version of the original article from *The Centennial Review* 26(3), 271–301.

Song Y. (2012). No easy and interesting math learning, (in Korean) *Joong Ang Ilbo* February 4. (2012. 2. 4.).

Strauss, L. (1989). *Liberalism ancient and modern.* With a new Foreword by Allan Bloom. Ithaca: Cornell University Press.

Quality Teaching of Mathematical Modelling: What Do We Know, What Can We Do?

Werner Blum

Introduction

The topic of this paper is mathematical modelling or—as it is often, more broadly, called—*applications and modelling*. This has been an important topic in mathematics education during the last few decades, beginning in particular with Henry Pollak's survey lecture (Pollak 1979) at ICME-3, Karlsruhe 1976 (my first ICME). By using the term "applications and modelling", both the products and the processes in the interplay between the real world and mathematics are addressed. In this paper, I will try to summarize some important aspects, in particular, concerning the *teaching* of applications and modelling. For obvious reasons, I have to restrict myself and hence omit some important aspects, such as gender issues or the question of how to embed applications and modelling in curricula and lessons. My paper is mainly a *survey*, only occasionally I can go into depth. I will concentrate on the *secondary school* level. I hope it will become clear that we have made considerable *progress* in the field during the last few decades, both theoretically and empirically, although still a lot remains to be done. For those who would like to find more on this topic I would refer to ICMI Study 14 on Modelling and Applications in Mathematics Education (Blum et al. 2007) where one can also find a short history of the field. Further, I would refer to the Proceedings of the ICTMA conference series (the International Conferences on the Teaching of Mathematical Modelling and Applications), held biennially since 1983. One can see how dynamically the filed develops by only looking at the number of papers in these books (see the last two Volumes: Kaiser et al. 2011, and Stillman et al. 2013).

In this paper, I will switch between theoretical aspects (Parts 2 and 4) and empirical aspects (Parts 3, 5–7). Part 8 is on teacher education, and I will start and close with concrete examples (Parts 1 and 9).

W. Blum (✉)
University of Kassel (Germany), Institute of Mathematics, Kassel, Germany
e-mail: blum@mathematik.uni-kassel.de

© The Author(s) 2015
S.J. Cho (ed.), *The Proceedings of the 12th International Congress on Mathematical Education*, DOI 10.1007/978-3-319-12688-3_9

Two Introductory Real World Examples

I work and live in Kassel, the city where, every five years, the "documentas" take place, the world's most important exhibitions for contemporary art (in 2012 with 850.000 visitors). Each documenta leaves some of its exhibits in the city. One of those is Claes Oldenburg's oversized pick-axe from documenta-7, 1982 (see Fig. 1).

The story that Oldenburg invented and Kassel people like to continue to tell is that Hercules, the landmark of Kassel (see Fig. 2), has thrown this pick-axe from his place, in the mountain park Wilhelmshöhe above Kassel, to the Fulda river. I will come back to this story in the final part of my paper.

Fig. 1 Oldenburg's pick-axe in Kassel

Fig. 2 The Kassel Hercules

The first question is: *How tall would a giant have to be for this pick-axe to fit to him? Would it fit to the Kassel Hercules himself?*

Following Pollak's famous characterization of modelling "Here is a situation—think about it" (Pollak 1969), we begin with comparing the pick-axe and a normal person. Using proportionality, we find that this pick-axe is about 13 m long. A normal pick-axe measures about 1 m. So, using again a proportional model, we find that a suitable giant would be about 25 m tall or, better perhaps, something between 20 and 30 m. The Kassel Hercules measures only 9 m, so he seems a bit too small for this pick-axe, unfortunately.

$$x : 1.80 \approx 13 : 1$$
$$x \approx 23.40$$

A second example from documenta in Kassel: During documenta everything is more expensive in downtown Kassel. A Hercules T-shirt, for instance, as a Kassel souvenir, costs 15.99 € downtown, whereas in the shopping mall dez which is not far away, the same T-shirt costs only 12.99 €. The second question is: *Is it worthwhile to drive to dez in order to buy this T-shirt there?*

We will solve this problem in several steps (the same steps that we have applied also in the pick-axe example without noticing it).

Step 1: We construct a mental model of the situation (Fig. 3).
Step 2: We simplify and structure this mental model by assuming that we go by car, that our car consumes 10 l/100 km in the city, that the gas costs 1.599 €/l and that the distance we have to drive from downtown to the mall is 5 km.

Fig. 3 Mental map of the situation

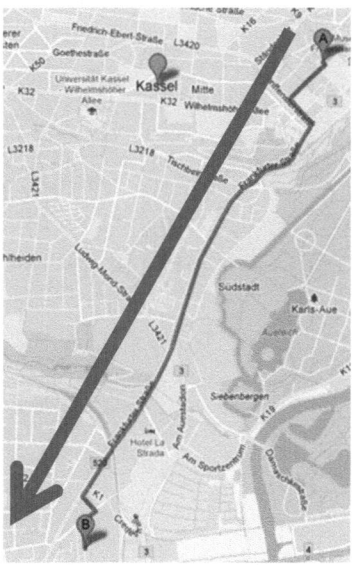

Step 3: We construct a suitable mathematical model by mathematizing these concepts and relations:

> **C-downtown = 15.99 €**
> **C-dez = 12.99 € + 2 · d · a · b, d: distance, a: consumption, b: gas price**
> **C-dez < C-downtown?**

Step 4: We work mathematically by calculating **C-dez ≈ 12.99 € + 1.60 € = 14.59 €** and by comparing: Yes, C-dez < C-downtown!

Step 5: We interpret this mathematical result in the real world: It is indeed by 1.40 € cheaper to drive to the shopping mall!

Step 6: We validate our result: Does it really make sense to drive 10 km in order to save 1.40 €? What about using this time instead to see more of Kassel's beauties? What about the risk of an accident or the air pollution caused by our trip? So perhaps we will refine our model and start again, or we will simply decide against that simple mathematical solution.

Step 7: In the end, we write down the whole solution.

This seven-step-process is one of the many schemas for the modelling process (Fig. 4, see Blum and Leiß 2007a).

Here are a few more such schemas (Fig. 5).

All these schemas have their specific strength and weaknesses, depending on the respective purposes. For cognitive analyses, this seven-step-model seems particularly helpful. It is a blend of models from applied mathematics (Pollak 1979; Burghes 1986), linguistics (Kintsch and Greeno 1985) and cognitive psychology (Staub and Reusser 1995).

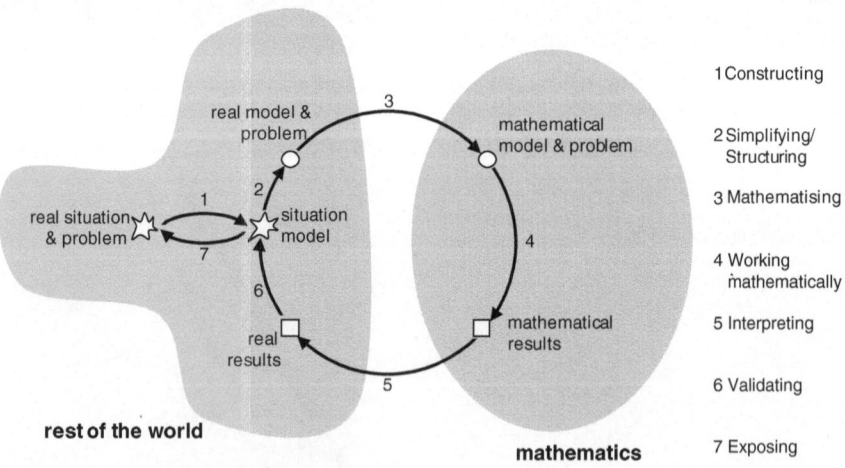

Fig. 4 Seven step modelling schema

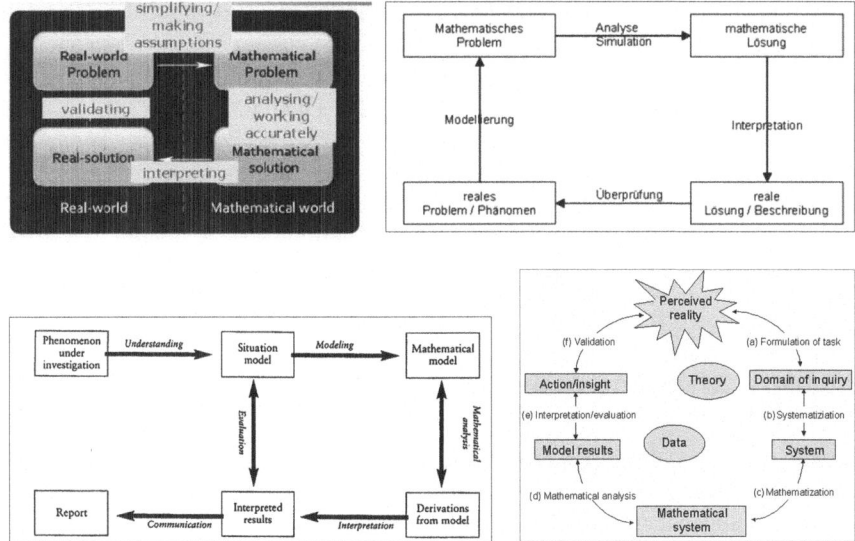

Fig. 5 Modelling schemas

Mathematical Modelling Competency

Here comes some theory. The topic of this paper is the teaching and learning of mathematics in the context of relations between mathematic and the extra-mathematical world. The latter is often called reality or the real world, or better, in the words of Pollak (1979), the "rest of the world", including nature, culture, society, or everyday life. The process of solving real world problems by means of mathematics can, from a cognitive point of view, be described by the schema from Fig. 4. If need be, one has to go round the loop several times. A key concept here is the concept of a model. A *mathematical model* is a deliberately simplified and formalized image of some part of the real world, formally speaking: a triple (D, M, f) consisting of a domain D of the real world, a subset M of the mathematical world and a mapping from D to M (Niss et al. 2007). Among the purposes of models are not only describing and explaining ("descriptive models") but also predicting and even creating parts of the real world ("normative models").

In the language of competencies according to Niss and colleagues (see Niss 2003), the ability to carry out those steps corresponds to certain *competencies* or *sub-competencies* such as understanding a given real world situation or interpreting mathematical results in relation to a situation (Blomhøj and Jensen 2007; Maaß 2006; Kaiser 2007; Turner et al. 2013). Cognitively speaking, an individual's competency is his/her ability to carry out certain actions in a well-aimed way. *Modelling competency* in a comprehensive sense means the ability to construct and to use or apply mathematical models by carrying out appropriate steps as well as to analyse or to compare given models (Blum et al. 2007). It is this comprehensive idea of modelling that will be used in the following.

The Niss competencies are also the conceptual basis for the PISA study and for the heart of PISA, *mathematical literacy* (see, e.g., OECD 2013, p. 23 ff). In large parts, PISA items require some modelling in a broad sense. An important source for the PISA philosophy was Hans Freudenthal's view of "mathematical concepts, structures and ideas as tools to organise the phenomena of the physical, social and mental world" (Freudenthal 1983). It is an open question whether this spirit of PISA will be preserved also in future PISA cycles.

Students' Modelling Activities

Mathematical modelling is a *cognitively demanding* activity since several competencies involved, also non-mathematical ones, extra-mathematical knowledge is required, mathematical knowledge and, in particular for translations, conceptual ideas (in German: "Grundvorstellungen") are necessary (e.g., in the examples in part 1, ideas about proportional functions), and appropriate beliefs and attitude are required, especially for more complex modelling activities.

These cognitive demands are responsible for *empirical difficulty*. Modelling is indeed rather difficult for students (see, for instance, Houston and Neill 2003, or Frejd and Ärlebäck 2011). Figure 6 shows the PISA task "Rock Concert".

The correct solution is C. In the OECD, only 26 % of all 15-year-olds have solved this task correctly, in Finland, one of the top performing countries, only 37 %, and in Korea, another top performing country, even only 21 %. The PISA Mathematics Expert Group has shown that the empirical difficulty of PISA

Fig. 6 PISA task "Rock Concert"

ROCK CONCERT

For a rock concert a rectangular field of size 100 m by 50 m was reserved for the audience. The concert was completely sold out and the field was full with all the fans standing.

Which one of the following is likely to be the best estimate of the total number of people attending the concert?

A   2 000
B   5 000
C   20 000
D   50 000
E   100 000

mathematics tasks can indeed be substantially explained by the competencies needed to solve these tasks (see Turner et al. 2013).

Several studies have shown that each step in the modelling process (see Fig. 4) is a potential cognitive barrier for students, a potential "blockage" or "red flag situation" (Goos 2002; Galbraith and Stillman 2006; Stillman 2011). "The weakest link in their modelling chain will set the limits on what they can do" (Treilibs et al. 1980).

Here are some remarks to *step 1* "Understanding the situation and constructing a situation model". Many students get stuck already here. This is not only or even not primarily a cognitive deficiency. For, many students around the world have learned, as part of the hidden curriculum, that they can survive without the effort of careful reading and understanding given contextual tasks. Instead, they successfully follow a substitute strategy for word problems: *"Ignore the context, just extract all data from the text and calculate something according to a familiar schema"* (see, e.g., Nesher 1980; Baruk 1985; Schoenfeld 1991; Lave 1992; Reusser and Stebler 1997; Verschaffel et al. 2000; Xin et al. 2007; de Bock, Verschaffel et al. 2010). Schoenfeld and Verschaffel speak of the "suspension of sense-making" when playing the "word problem game". This strategy even becomes more popular with age, and in the school context it may indeed make a lot of sense to follow this strategy in order to pass tests and to survive. This is empirically well documented, in very many countries. Here is a well-known example (Verschaffel et al. 2000):

450 soldiers must be bussed to their training site. Each army bus can hold 36 soldiers. How many busses are needed?

Popular answers are "12 busses remainder 18" or "12.5 busses". Another example of a calculation without imagining the situation clearly is:

An orchestra needs 40 min for Beethoven's 6th symphony. How long will it take for Beethoven's 9th symphony?

The popular answer is 60 min. In the PISA task "Rock Concert" (see Fig. 6), the by far most attractive distractor (49 %) was no. 2, the one that follows exactly the substitute strategy: $50 \cdot 100 = 5,000$.

Step 2 "Simplifying and structuring" is a source of difficulties as well. In particular, learners are afraid of making *assumptions* by themselves.

Step 6 "Validating" is mostly not present at all in students' solutions. Here (Fig. 7) is a solution of the pick-axe task.

The answer 254.84 m for the giant's height is, first, ridiculously accurate (rounding off is a rare event in mathematics classrooms) and, second, obviously much too big. However, students normally do not validate their solutions, it seems to be part of the "contract didactique": Checking the correctness and suitability of a solution is exclusively the teacher' responsibility!

I would like to mention a few other important empirical results concerning students' dealing with modelling tasks. Several studies have shown (Matos and Carreira 1997; Leiß 2007; Borromeo Ferri 2011; Schukajlow 2011; Sol et al. 2011): If students are dealing with modelling tasks independently, the process is normally *non-linear* according to one of those ideal-typical loops but rather characterized by jumps forth and back, by omissions or mini-loops. Borromeo Ferri (2007) speaks of

Fig. 7 A student solution of the pick-axe task

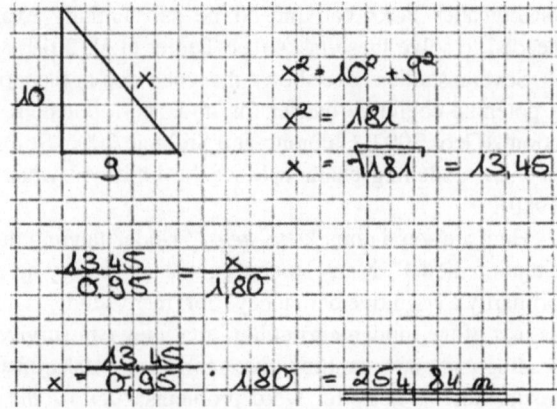

"individual modelling routes" which are determined by individual knowledge and preferences such as individual thinking styles.

Another well-documented observation is that students normally do not have *strategies* available for solving real world problems. More generally, students usually do not reflect upon their activities and, closely related to that, are not able to *transfer* their knowledge and skills from one context or task to a different context or task, even if there are structural similarities. For instance, in one of our projects, grade 9 students dealt in a lesson with the "Filling up" task (see Blum and Leiß 2006) which is quite analogous to the T-Shirt task from part 1. The question is whether it is worthwhile for a certain Mrs. Stone to drive from her hometown Trier across the nearby border of Luxemburg, where the gas is cheaper, in order to fill up her car there. In the following test, the students had to solve very similar tasks, among others whether it is worthwhile to drive to a nearby strawberry field in order to pick the berries for a cake instead of buying them in a supermarket, or whether it is worthwhile to use cloth-diapers instead of disposable ones. For many students, these were totally new challenges, now about strawberries and diapers instead of cars. The PISA study also demonstrates every three years how difficult it is for 15-year-olds to transfer their school knowledge to real world problem situations.

The phenomena just described are, as is well-known, special instances of *situated cognition*, or in the words of Jürgen Baumert: Every learning topic carries with it the "indices" referring to its learning context. This is particularly relevant for learning in the field of relations between the real world and mathematics (DeCorte et al. 1996; Niss 1999). Actually, when we report on empirical results about "modelling competency" we have to write this construct with several indices, especially referring to the mathematical topics and the extra-mathematical contexts involved. The question is even: Is there a "general modelling competency" at all? Much more research is necessary into how and how far the desired transfer can be achieved. I will come back to this aspect in parts 5 and 7.

Aims and Perspectives of Modelling

We come back to theoretical aspects of modelling. Modelling is a cognitively demanding activity, so why should learners have to deal with such activities? Why is it not sufficient to learn pure mathematics in order to achieve the aims of mathematics as a school subject? Mathematics is, as we know, a compulsory subject at school for the following reasons (see, e.g., Niss 1996): Mathematics as

- a powerful tool for better understanding and mastering present or future real world situations,
- a tool to develop general mathematical competencies,
- an important part of culture and society, and a world of its own.

The basis for that are general educational goals such as the ability to take part in social life as an independent and responsible citizen.

On this background, we can distinguish between four groups of *justifications* for the inclusion of applications and modelling in curricula and everyday teaching (see, e.g., Blum and Niss 1991; Blum 2011):

1. *"pragmatic"* justification: In order to understand and master real world situations, suitable applications and modelling examples have to be explicitly treated; we cannot expect any transfer from intra-mathematical activities.
2. *"formative"* justification: Competencies can be advanced also by engaging in modelling activities; in particular, modelling competency can only be advanced in this way, and argumentation competency can be advanced by "reality-related proofs" (Blum 1998).
3. *"cultural"* justification: Relations to the extra-mathematical world are indispensable for an adequate picture of mathematics as a science in a comprehensive sense.
4. *"psychological"* justification: Real world examples may contribute to raise students' interest in mathematics, to motivate or structure mathematical content, to better understand it and to retain it longer.

We can see a certain *duality* here (Niss et al. 2007): Whereas the first aspect deals with mathematics as an aid for the real world, the other three aspects deal with the opposite direction, the real world as an aid for mathematics, in a broad sense. Instead of "justifications for the inclusion of applications and modelling" we could also say "*aims* of the teaching of applications and modelling".

In order to advance those aims, suitable examples are needed. There is a broad spectrum of real world examples, from small dressed-up word problems to authentic modelling problems or projects that require days or weeks. The justifications or aims just mentioned require certain specific *types of examples*:

- *"pragmatic"*: concrete authentic examples (from shopping, newspapers, taxes, traffic flow, wind park planning, air fare calculation, ...);
- *"formative"*: cognitively rich examples, accompanied by meta-cognitive activities;

- *"cultural"*: either authentic examples that show students how strongly mathematics shapes the world (sometimes hidden and invisible, embedded in technology—the famous relevance paradox, see e.g. Niss 1999) or epistemologically rich examples that shed some light on mathematics as a science (including ethno-mathematical examples); in both cases, the role of mathematics and its relations to the real world must be made more conscious;
- *"psychological"*: either interesting examples for motivation or illustration purposes, to make mathematics better marketable for students (these examples might quite well be dressed-up or whimsy problems, it is only a matter of honesty), or mathematically rich examples that serve the purpose to make certain mathematical topics better comprehensible.

So, examples are not good or bad per se, it depends on their *purpose*.

It was Gabriele Kaiser's idea, together with colleagues (see Kaiser et al. 2006), to distinguish between various *perspectives* of modelling. On the basis of what I have just presented, I have conceptualized the notion of "perspective" a bit more formally, as a pair (aim | suitable examples), with a slightly different terminology. So we can distinguish between six perspectives.

- (pragmatic | authentic) → *"applied* modelling" (Burghes, Haines, Kaiser, and others; particularly rooted in the Anglo-Saxon tradition)
- (formative | cognitively rich) → *"educational* modelling" (Burkhardt/Swan, Blomhøj, and others)
- (cultural with an emancipatory intention | authentic) → *"socio-critical* modelling" (Keitel/Jablonka, Skovsmose, Julie, Barbosa, and others)
- (cultural concerning mathematics | epistemologically rich) → *"epistemological* modelling" (d'Ambrosio, Garcia, Bosch, and others; more rooted in the Romanic tradition)
- (psychological with marketing intention | motivating) → *"pedagogical* modelling" (by far the most important aspect in school)
- (psychological | mathematically rich) → *"conceptual* modelling" (Freudenthal, de Lange, Gravemeijer, and others)

For each perspective, there is a certain model of the modelling process that is best suitable for that purpose. For instance, for applied modelling, a four step model "Mathematising → Math. Working → Interpreting → Validating" seems most appropriate. There is no space here to elaborate more on this. In effect, it is more appropriate to conceptualise a "perspective" as a tripel (aim | examples | cycle).

All these perspectives also contribute to the question of *sense-making*. Here, I mean by the "sense" of an activity the subjective meaning of this activity to the individual whereby the individual can understand the purpose of this activity. Each perspective offers to learners a specific aspect of sense:

- "applied": sense through understanding and mastering real world situations
- "educational": sense through realizing own competency growth
- "socio-critical": sense through understanding the role of mathematics
- "epistemological": sense through comprehending mathematics as a science

- "pedagogical": sense through enjoying doing mathematics
- "conceptual": sense through understanding mathematical concepts

It is important to offer various aspects of sense since learners will react differently, also according to their beliefs about and attitudes towards mathematics. The hope is that, by offering various aspects of sense, students' beliefs will become broader, and their attitudes will become more positive.

Teaching Modelling

Back from theory to practice. In the first few parts of this paper, the focus was on learning. It is clear that all aims and purposes can only be reached by *high-quality teaching*. Applications and modelling are important, and learning applications and modelling is demanding. This implies that there have to be particularly big efforts to make applications and modelling *accessible* for learners. In fact, there are such efforts in many countries around the world. However, in everyday mathematics teaching practice in most countries, there is still relatively few modelling. Applications in the classroom still occur mostly in the context of dressed-up word problems. We have been deploring this *gap* between the educational debate and classroom practice for decades. Why do we still have this gap? The main reason is that teaching applications and modelling is demanding, too (Freudenthal 1973; Pollak 1979; DeLange 1987; Burkhardt 2004; Ikeda 2007). Also the teachers have to have various competencies available, mathematical and extra-mathematical knowledge, ideas for tasks and for teaching as well as appropriate beliefs. Instruction becomes more open and assessment becomes more complex. This is the main barrier for applications and modelling.

What can we do to improve the situation? What do we know empirically about effective teaching of applications and modelling according to those various aims and purposes? Generally speaking, the well-known findings on quality mathematics teaching of mathematics hold, of course, also for teaching mathematics in the context of relations to the real world. This seems self-evident but is ignored in classrooms around the world every day a million times.

In the following, I will present ten—in my view—important aspects for a *teaching methodology* for applications and modelling, based on empirical findings.

1. A necessary condition is an *effective and learner-oriented classroom management* (see, e.g., Baumert et al. 2004; Hattie 2009; Timperley 2011; Kunter and Voss 2013): using time effectively, separating learning and assessment recognisably, using students' mistakes constructively as learning opportunities (motto: every wrong answer is the right answer to a different question), or varying methods and media flexibly. For modelling, group work is particularly suitable (Ikeda and Stephens 2001). The group is not only a social but also a cognitive environment (co-constructive group work; see Reusser 2001).

2. Just as necessary is to *activate learners cognitively*, to stimulate students' own activities. "Modelling is not a spectator sport" (Schoenfeld, personal communication), one can expect learning effects at most if students engage actively in modelling. This is not a matter of surface structures such as whole-class teaching versus group work versus individualized teaching, which may be dependent on cultural backgrounds. What only counts is that learners are cognitively active (Schoenfeld 1992). We have to distinguish carefully here between students working independently with teacher support, on the one hand, and, on the other hand, students working on their own, alone. Crucial for teaching is a permanent *balance* between students 'independence and teacher's guidance, according to Aebli's famous "Principle of minimal support" (Aebli 1985). I will come back to this aspect in part 6 of this paper.

3. Learners have to be activated not only cognitively but also *meta-cognitively*. All activities ought to be accompanied by reflections and ought to be reflected in retrospective, with the aim to advance appropriate learning *strategies*. Again this is not a matter of lesson surface structures. I will elaborate more on this aspect in part 7 of this paper.

4. There has to be a broad variety of suitable examples as the substance of mathematics lessons since we cannot expect any mystical transfer from one example or context to another. In particular, there has to be a well-aimed variation of real world contexts as well as of mathematical contexts and topics. As I have said in part 4, different kinds of examples may serve different purposes and authenticity is not always required. However, if contexts are made more authentic, the "suspension of sense-making" (see part 3) can be reduced substantially (Palm 2007; Verschaffel et al. 2010). For instance, if the "Army bus" task (see part 3) is embedded in a credible context where students have to write an order form for a bus company, the number of reasonable solutions increases substantially.

There are a lot of rich teaching/learning environments available for all aims of application and modelling, among many others the following:

- A wealth of materials from the Shell Centre in Nottingham, the UCSMP project, Roskilde University, the Freudenthal institute (RME) and much more (see Blum et al. 2007, part 6).
- Dick Lesh's Model Eliciting Activities (Lesh and Doerr 2003); they are primarily meant as a research tool, but they can be used equally well for teaching purposes, together with his Model Exploration Activities and Model Adaptation Activities.
- "Real objects, contexts and actions" and "local applications" (Alsina 2007); other outdoor activities in the same spirit are "Maths trails" (see, e.g., Shoaf et al. 2004).
- Materials from the modelling weeks in various cities, in Germany, Singapore or Queensland.

5. Teachers ought to encourage *individual solutions* of modelling tasks. In everyday teaching practice, however, teachers tend to favour strongly their own solution, without even noticing it (Leikin and Levav-Waynberg 2007; Borromeo Ferri and Blum 2009), also because of a limited knowledge of the "task space". There are several reasons for encouraging *multiple solutions* (Schoenfeld 1988; Hiebert and Carpenter 1992; Krainer 1993; Neubrand 2006; Rittle-Johnson and Star 2009; Tsamir et al. 2010): These comply with students' individual preferences, support internal differentiation in the classroom, reflect the genuine spirit of mathematics, and enable comparisons between and reflections on different solutions on a meta-level. In the current project MultiMa (Schukajlow and Krug 2013), two independency-oriented teaching units with modelling tasks are compared where in one unit students are explicitly required to produce multiple solutions. It turned out that those students who developed several solutions had higher learning gains.

6. Competencies such as modelling evolve in *long-term* learning processes, beginning already in primary school with "implicit models" (Greer and Verschaffel 2007; Borromeo Ferri and Lesh 2013) and continuing forever. Necessary and not at all out-of-date are permanent integrated repeating and intelligent practising. It is also important to have a permanent balance between focussing on sub-competencies of modelling and focussing on modelling competency as a whole. It is an open research question what such a balance would look like. What would be needed is a competency development model for modelling, theoretically sound and empirically well-founded, or several such models. This is a big deficit in research.

An interesting approach to describe competency development comes from the Danish KOM project (Blomhøj and Jensen 2007; Niss and Højgaard Jensen 2011). The authors distinguish between three dimensions in an individual's possession of a given mathematical competency: the "degree of coverage" of aspects of this competency, the "radius of action" that indicates the spectrum of contexts and situations, and the "technical level" that indicates the conceptual and technical level of the involved mathematical entities.

7. Not only teaching but also *assessment* has to reflect the aims of applications and modelling appropriately. Quality criteria such as variation of methods are relevant here, too (Haines and Crouch 2001; Izard et al. 2003; Houston 2007; Antonius et al. 2007; Vos 2007). One method is, of course, to work with *tests*. As we know, tests have several functions, among others to set norms and to illustrate the aspired aims ("What You Test Is What You Get"), but also and particularly to diagnose students' strengths and weaknesses in order to know better how to help.

An interesting research question is whether and how it is possible to assess modelling sub-competencies and general modelling competency separately. Zöttl et al. (2011) have found that the following model describes their data best (Fig. 8): Some items measure certain sub-competencies and all items measure a general competency.

Fig. 8 Model for modelling
sub-competencies

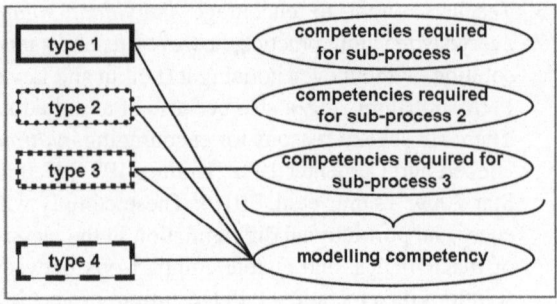

8. It is important to care for a parallel development of competencies and appropriate *beliefs* and *attitudes*. Taking into account the remarkable stability of beliefs and attitudes, this also requires long-term learning processes.

9. There are a lot of case studies that show that *digital technologies* can be used as powerful tools for modelling activities, not only in the intra-mathematical phases (see, e.g., Borba and Villarreal 2005; Henn 2007; Geiger 2011; Greefrath et al. 2011). Computers can be used for experiments, investigations, simulations, visualisations or calculations. Greefrath suggests to extend the modelling cycle by adding a third world: the technological world (Fig. 9).

   What we need here are much more controlled studies into the effects of digital technologies on modelling competency development.

10. The best message comes last. Several case studies have shown that mathematical modelling can in fact be learned by secondary school students supposed there is quality teaching (a.o. Kaiser-Meßmer 1987; Galbraith and Clatworthy 1990; Abrantes 1993; Maaß 2007; Biccard and Wessels 2011; Blum and Leiß 2007b; Schukajlow et al. 2012). Some studies have shown that also students' beliefs about mathematics can be broadened by appropriate quality teaching.

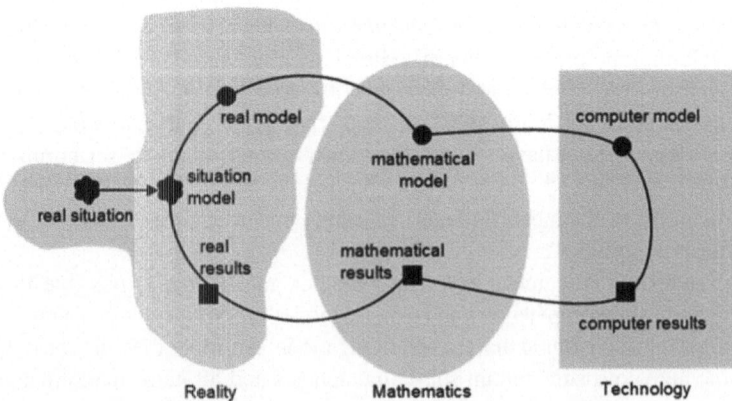

Fig. 9 The extended modelling cycle

However, much more research is needed, especially small-scale studies using a mixture of qualitative and quantitative methods.

In closing part 5, I would like to emphasise that all these efforts will not be sufficient to assign applications and modelling its proper place in curricula and classrooms and to ensure effective and sustainable learning. The implementation of applications and modelling has to take place *systemically*, with all system components collaborating closely: curricula, standards, instruction, assessment and evaluation, and teacher education. I cannot elaborate more on this aspect.

Teacher Support for Modelling Activities

I would like to go more deeply into the second aspect mentioned in part 5: How can the balance between students' independent work and teacher's guidance be put into practice, what does "minimal support" look like? The key concept is *adaptive teacher invention* (see Leiß 2010, for an overview). Such an intervention allows students to continue their work without losing their independence—in the Vygotski terminology: an intervention in the Zone of Proximal Development. Whether an intervention was adaptive or not can, on principle, be only judged afterwards: Is the cognitive barrier really overcome, has the "red flag" vanished? Adaptive interventions can be regarded as a special case of *scaffolding* (Smit et al. 2013). A necessary basis for such a temporary support is a good diagnosis.

In everyday classrooms, teachers tend to strong, content-related interventions, sometimes in order to prevent mistakes or blockages before they occur. According to several studies (see, e.g., Leiß 2007), there are only very few *strategic* interventions, and most interventions seem to be not adaptive. However, especially strategic interventions have the potential of being adaptive (for an impressive example of a successful strategic intervention see Blum and Borromeo Ferri 2009). Here are some examples of strategic interventions:

Read the text carefully! Imagine the situation clearly! Make a sketch! What do you aim at? What is missing? Which data do you need? How far have you got? Does this result make sense for the real situation?

In the DISUM project (see Blum and Leiß 2007b), a ten lesson teaching unit on modelling in 18 grade 9 classes proved to produce significantly higher learning gains in modelling competency in a teaching design oriented towards students' independence with adaptive teacher interventions compared to a design with directive teaching; see Schukajlow et al. (2012) for more details.

Strategies for Learning Modelling

All teacher interventions and support as just discussed will have no long-term effects if they are only applied situationally, transfer cannot be expected. Only accompanying meta-cognitive activities may promise sustainable effects. Students have to be enabled to see the general feature in the concrete step, in the concrete cognitive barrier: How can I help myself in such a difficulty? How can I solve such kind of tasks by myself? For, in assessment situations or in real life contexts, there is no teacher support available.

A promising approach is to teach *learning strategies*, cognitive strategies as well as meta-cognitive strategies such as planning, controlling or regulating. There are a lot of empirical results concerning the effects of using strategies, mostly encouraging, some also disappointing (Tanner and Jones 1993; Schoenfeld 1992, 1994; Matos and Carreira 1997; Stillman and Galbraith 1998; Kramarski et al. 2002; Burkhardt and Pollak 2006; Desoete and Veenman 2006; Stillman 2011; for an overview see Greer and Verschaffel 2007). One of the problems in these empirical studies is: how to measure strategy knowledge, on the one hand, and strategy use, on the other hand, and another problem is how to reliably link students' activities to their strategies.

In particular for novices in modelling there are two strategic instruments that I would like to mention since they turned out to be successful: First, the heuristic worked examples in the KOMMA project, with a three step schema (see Zöttl et al. 2011). Second, the DISUM four step schema ("Understanding task/ Searching mathematics/ Using mathematics/ Explaining result"; see Blum 2011, for more details). This is not meant as a schema that students must follow but as a guiding line, a meta-cognitive aid, particularly in case of difficulties. The problem for students with such strategic devices is: What do these hints mean concretely (for instance in step 2 "Make assumptions": which, how, how many?)? Much more research is needed into the design and use of strategic instruments for modelling.

Teacher Competencies for Modelling

Several empirical studies tell us (recently the comparative study TEDS-M, see Schmidt et al. 2007; Blömeke et al. 2010): The teacher matters most! For quality teaching of applications and modelling, the teacher needs a lot of different competencies. As a theoretical foundation, I would like to use the competence model from the COACTIV project (see Baumert and Kunter 2013). Here, as part of the professional knowledge, five categories are distinguished, especially content knowledge (CK), pedagogical content knowledge (PCK), and pedagogical/psychological knowledge (PK), along the distinction made by Shulman and others. Based on the fundamental assumption about the impact of teaching on learning teacher competencies → quality teaching → student learning,

the COACTIV project has shown, for a representative sample of German secondary mathematics teachers, that subject-related teacher competencies have a strong influence on students' performance (see Baumert et al. 2010). Among the mediators that significantly influence students' performance are classroom management and the cognitive level of tasks set for written class tests. And the TEDS-M study has shown that competencies of beginning teachers vary a lot across different countries, dependent on their learning opportunities. Therefore, *teacher education* is crucial.

What PCK is needed especially for teaching applications and modelling (see Ball et al. 2005, in general and, in particular for modelling, Doerr 2007; Lingefjärd 2013; Kaiser et al. 2010). Borromeo Ferri and Blum (2010) distinguish, in their model, between four dimensions of teachers' PCK for modelling: (1) a *theoretical* dimension (incl. modelling cycles or aims and perspectives of modelling as background knowledge), (2) a *task* dimension (incl. multiple solutions or cognitive analyses of modelling tasks), (3) an *instructional* dimension (incl. interventions, support and feedback), and (4) a *diagnostic* dimension (incl. recognising students' difficulties and mistakes). Also for teachers' learning, no transfer can be expected. Hence, all these elements have to be included as compulsory components in teacher education and professional development. Obviously, in most places where maths teachers are trained, this is not (yet) the case, that means the naïve faith in some mystic transfer is strong here, too. Another myth is that teachers will gain their necessary professional knowledge just by teaching practice. However, in the CO-ACTIV project, there was no correlation between experience and professional knowledge (see Kunter et al. 2013).

One way of providing future teachers with the necessary professional knowledge is to offer specific modelling seminars already at the university, with compulsory own teaching experiences (Borromeo Ferri and Blum 2010). Also the Model Eliciting Activities mentioned in part 5 (see Doerr and Lesh 2011) are very efficient learning environments both for future and for practicing teachers. Nevertheless, a lot has still to be done in research as well, in particular: How will the various teacher competencies play out in teaching practice and how will they influence student learning about applications and modelling?

A Final Real World Example

I would like to come back to the example in part 1, Oldenburg's oversized pick-axe in Kassel. The story that the Kassel Hercules has thrown this pick-axe to the Fulda river is very nice, but we may ask: Is it conceivable?

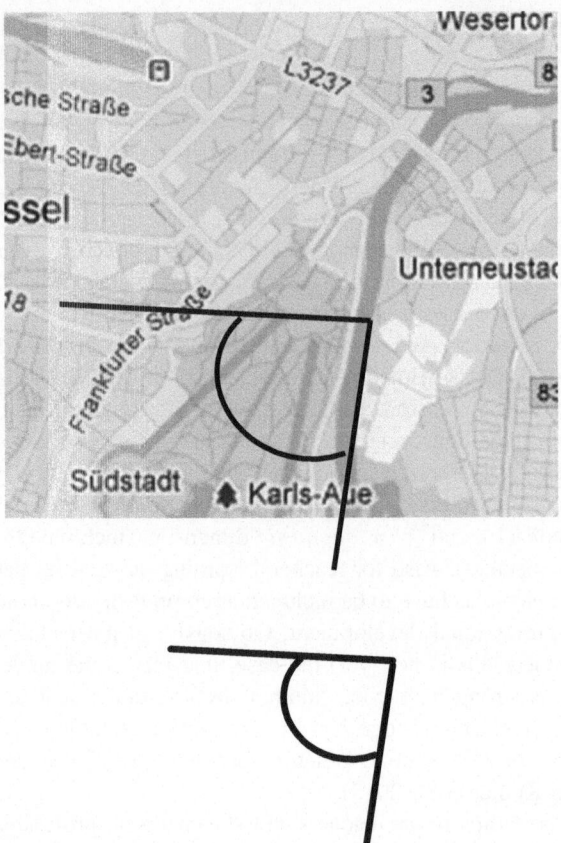

Fig. 10 The angle between axe and river

The first question is: Is the axis correct from the pick-axe to the Hercules? Hercules cannot be seen from the Fulda bank, but we can just measure the angle between the pick-axe and the Fulda River in reality and at the same time the angle between the line Hercules-axe and the river on the map (Fig. 10).

In both cases we find approximately 85°. Since angles are preserved under similarity transformations, this shows that the axis is correct indeed.

The second question is: Can Hercules really throw that far? This depends on a more basic question: Is Hercules able to hold this pick-axe at all? See part 1: Oldenburg's pick-axe is 13 times as long as a normal axe. So, using a cubic model, it weighs more than 2,000 times a normal axe, thus approximately 5 tons. Kassel's Hercules measures 9 m, 5 times a normal man's height, and Hercules is, as one knows from history, much stronger than normal people. The world record in weight-lifting is ¼ ton. Now we can apply two different models. If we assume that the power for weight-lifting only grows proportionally with height, Hercules will be able to hold at most 1.5 tons but not 5 tons, unfortunately. However, if we assume quadratic growth with height, Hercules will be able to hold even 6 tons. I would like to leave this question open: Which model is more appropriate? Personally, I prefer the quadratic model in order not to run down such a nice story about a hero and his pick-axe.

Open Access This chapter is distributed under the terms of the Creative Commons Attribution Noncommercial License, which permits any noncommercial use, distribution, and reproduction in any medium, provided the original author(s) and source are credited.

References

Abrantes, P. (1993). Project work in school mathematics. In: De Lange, J. et al. (Eds), *Innovation in Maths Education by Modelling and Applications*. Chichester: Horwood, 355-364.

Aebli, H. (1985). *Zwölf Grundformen des Lehrens*. Stuttgart: Klett-Cotta.

Alsina, C. (2007). Less chalk, less words, less symbols ... More objects, more context, more actions. In: Blum, W. et al. (Eds), *Modelling and Applications in Mathematics Education*. New York: Springer, 35-44.

Antonius, S. et al. (2007). Classroom activities and the teacher. In: Blum, W. et al. (Eds), *Modelling and Applications in Mathematics Education*. New York: Springer, 295-308.

Ball, D.L., Hill, H.C. & Bass, H. (2005). Knowing mathematics for teaching. In: *American Educator*, 29 (3), 14-46.

Baruk, S. (1985). *L'age du capitaine. De l'erreur en mathematiques*. Paris: Seuil.

Baumert, J., Kunter, M. & Blum, W. et al. (2004). Mathematikunterricht aus Sicht der PISA-Schülerinnen und -Schüler und ihrer Lehrkräfte. In: Prenzel, M. et al. (Eds), *PISA 2003. Der Bildungsstand der Jugendlichen in Deutschland – Ergebnisse des zweiten internationalen Vergleichs*. Waxmann, Münster, 314-354.

Baumert, J. & Kunter, M. (2013). The COACTIV Model of Teachers' Professional Competence. In: Kunter, M., Baumert, J., Blum, W. et al. (Eds), *Cognitive Activation in the Mathematics Classroom and Professional Competence of Teachers – Results from the COACTIV Project*. New York: Springer, 25-48.

Baumert, J., Kunter, M., Blum, W. et al. (2010): Teachers' Mathematical Knowledge, Cognitive Activation in the Classroom, and Student Progress. In: *American Educational Research Journal* 47(1), 133-180.

Biccard, P. & Wessels, D.C.J. (2011). Documenting the Development of Modelling Competencies of Grade 7 Students. In: Kaiser, G. et al. (Eds). *Trends in Teaching and Learning of Mathematical Modelling (ICTMA 14)*. Dordrecht: Springer, 375-383.

Blömeke, S., Kaiser, G. & Lehmann, R. (Eds, 2010). *TEDS-M 2008: Professionelle Kompetenz und Lerngelegenheiten angehender Mathematiklehrkräfte für die Sekundarstufe I im internationalen Vergleich*. Münster: Waxmann.

Blomhøj, M. & Jensen, T.H. (2007). What's all the fuss about competencies? In: Blum, W. et al. (Eds), *Modelling and Applications in Mathematics Education*. New York: Springer, 45-56.

Blum, W. (1998). On the role of "Grundvorstellungen" for reality-related proofs – examples and reflections. In: Galbraith, P. et al. (Eds), *Mathematical Modelling – Teaching and Assessment in a Technology-Rich World*. Chichester: Horwood, 63-74.

Blum, W. (2011). Can Modelling Be Taught and Learnt? Some Answers from Empirical Research. In: Kaiser, G. et al. (Eds), *Trends in Teaching and Learning of Mathematical Modelling (ICTMA 14)*. Dordrecht: Springer, 15-30.

Blum, W. & Borromeo Ferri, R. (2009). Mathematical Modelling: Can it Be Taught and Learnt? In: *Journal of Mathematical Modelling and Application* 1(1), 45-58.

Blum, W. & Leiß, D. (2006). "Filling up" – The problem of independence-preserving teacher interventions in lessons with demanding modelling tasks. In: Bosch, M. (Ed.), *CERME-4 – Proceedings of the Fourth Conference of the European Society for Research in Mathematics Education*. Guixol.

Blum, W. & Leiß, D. (2007a). How do students' and teachers deal with modelling problems? In: Haines, C. et al. (Eds), *Mathematical Modelling: Education, Engineering and Economics*. Chichester: Horwood, 222-231.

Blum, W. & Leiß, D. (2007b). Investigating Quality Mathematics Teaching – the DISUM Project. In: Bergsten, C. & Grevholm, B. (Eds), *Developing and Researching Quality in Mathematics Teaching and Learning, Proceedings of MADIF 5*. Linköping: SMDF, 3-16.

Blum, W. & Niss, M. (1991). Applied mathematical problem solving, modelling, applications, and links to other subjects – state, trends and issues in mathematics instruction. In: *Educational Studies in Mathematics* 22(1), 37-68.

Blum, W., Galbraith, P., Henn, H.-W. & Niss, M. (Eds, 2007). *Modelling and Applications in Mathematics Education*. New York: Springer.

Borba, M.C. & Villarreal, M.E. (2005). *Humans-with-Media and the Reorganization of Mathematical Thinking – Informations and Communication Technologies, Modeling, Experimentation and Visualization*. New York: Springer.

Borromeo Ferri, R. (2007). Modelling problems from a cognitive perspective. In: Haines, C. et al. (Eds), *Mathematical Modelling: Education, Engineering and Economics*. Chichester: Horwood, 260-270.

Borromeo Ferri, R. (2011). *Wege zur Innenwelt des mathematischen Modellierens: Kognitive Analysen zu Modellierungsprozessen im Mathematikunterricht*. Wiesbaden: Vieweg+Teubner.

Borromeo Ferri, R. & Blum, W. (2009). Insight into Teachers' Unconscious Behaviour in Modeling Contexts. In: Lesh, R. et al. (Eds), *Modeling Students' Mathematical Modeling Competencies*. New York: Springer, 423-432.

Borromeo Ferri, R. & Blum, W. (2010). Mathematical Modelling in Teacher Education – Experiences from a Modelling Seminar. In: Durand-Guerrier, V., Soury-Lavergne, S. & Arzarello, F. (Eds), *CERME-6 – Proceedings of the Sixth Congress of the European Society for Research in Mathematics Education*. INRP, Lyon 2010, 2046-2055.

Borromeo Ferri, R. & Lesh, R. (2013). Should Interpretation Systems be Considered to be Models if they only Function Implicitly? In: Stillman, G. et al. (Eds). *Teaching Mathematical Modelling: Connecting to Teaching and Research Practice – the Impact of Globalisation*. New York: Springer.

Burghes, D. (1986). Mathematical modelling – are we heading in the right direction? In: J. Berry et al. (Eds), *Mathematical Modelling Methodology, Models and Micros*. Chichester: Horwood, 11-23.

Burkhardt, H. (2004). Establishing modelling in the curriculum: barriers and levers. In: Henn, H. W. & Blum, W. (Eds), *ICMI Study 14: Applications and Modelling in Mathematics Education Pre-Conference Volume*. University of Dortmund, 53-58.

Burkhardt, H. & Pollak, H.O. (2006). Modelling in mathematics classrooms: reflections on past developments and the future. In: *Zentralblatt für Didaktik der Mathematik* 38(2), 178-195.

DeCorte, E., Greer, B. & Verschaffel, L. (1996). Mathematics teaching and learning. In: Berliner, D.C. & Calfee, R.C. (Eds.), *Handbook of Educational Psychology*. New York: Macmillan, 491-549.

DeLange, J. (1987). *Mathematics, Insight and Meaning*. Utrecht: CD-Press.

Desoete, A. & Veenman, M.V.J. (2006). *Metacognition in mathematics education*. Hauppauge: Nova Science Publishers.

Doerr, H. (2007). What knowledge do teachers need for teaching mathematics through applications and modelling? In: Blum, W. et al. (Eds), *Modelling and Applications in Mathematics Education*. New York: Springer, 69-78.

Doerr, H. & Lesh, R. (2011). Models and Modelling Perspectives on Teaching and Learning Mathematics in the Twenty-First Century. In: Kaiser, G. et al. (Eds). *Trends in Teaching and Learning of Mathematical Modelling (ICTMA 14)*. Dordrecht: Springer, 247-268.

Frejd, P. & Ärlebäck, J. (2011). First Results from a Study Investigating Swedish Upper Secondary Students' Mathematical Modelling Competencies. In: Kaiser, G. et al. (Eds). *Trends in Teaching and Learning of Mathematical Modelling (ICTMA 14)*. Dordrecht: Springer, 407-416.

Freudenthal, H. (1973). *Mathematics as an Educational Task*. Dordrecht: Reidel.

Freudenthal, H. (1983). *Didactical Phenomenology of Mathematical Structures*. Dordrecht: Reidel.

Galbraith, P. & Clathworthy, N. (1990). Beyond standard models – Meeting the challenge of modelling. In: *Educational Studies in Mathematics* 21(2), 137-163.

Galbraith, P. & Stillman, G. (2006). A framework for identifying student blockages during transitions in the modelling process. In: *Zentralblatt für Didaktik der Mathematik* 38(2), 143-162.

Geiger, V. (2011). Factors Affecting Teachers' Adoption of Innovative Practices with Technology and Mathematical Modelling. In: Kaiser, G. et al. (Eds, 2011), *Trends in Teaching and Learning of Mathematical Modelling (ICTMA 14)*. Dordrecht: Springer, 305-314.

Goos, M. (2002). Understanding metacognitive failure. In: *Journal of Mathematical Behavior* 21 (3), 283-302.

Greefrath, G., Siller, H.-S. & Weitendorf, J. (2011). Modelling Considering the Influence of Technology. In: Kaiser, G. et al. (Eds, 2011), *Trends in Teaching and Learning of Mathematical Modelling (ICTMA 14)*. Dordrecht: Springer, 315-329.

Greer, B. & Verschaffel, L. (2007). Modelling competencies – overview. In: Blum, W. et al. (Eds). *Modelling and Applications in Mathematics Education*. New York: Springer, 219-224.

Haines, C. & Crouch, R. (2001). Recognizing constructs within mathematical modelling. In: *Teaching Mathematics and its Applications* 20(3), 129-138.

Hattie, J.A.C. (2009): *Visible Learning. A synthesis of over 800 meta-analyses relating to achievement*. London & New York: Routledge.

Henn, H.-W. (2007). Modelling pedagogy – Overview. In: Blum, W. et al. (Eds), *Modelling and Applications in Mathematics Education*. New York: Springer, 321-324.

Hiebert, J. & Carpenter, T.P. (1992). Learning and teaching with understanding. In: D.A. Grouws (Ed.), *Handbook of research on mathematics teaching and learning*. New York: Macmillan, 65-97.

Houston, K. (2007). Assessing the "phases" of mathematical modelling. In: Blum, W. et al. (Eds), *Modelling and Applications in Mathematics Education*. New York: Springer, 249-256.

Houston, K., & Neill, N. (2003). Assessing modelling skills. In: Lamon, S.J., Parker, W.A. & Houston, S.K. (Eds), *Mathematical modelling: A way of life – ICTMA 11*. Chichester: Horwood, 155-164.

Ikeda, T. (2007). Possibilities for, and obstacles to teaching applications and modelling in the lower secondary levels. In: Blum, W. et al. (Eds), *Modelling and Applications in Mathematics Education*. New York: Springer, 457-462.

Ikeda, T. & Stephens, M. (2001). The effects of students' discussion in mathematical modelling. In: Matos, J.F., Blum, W., Houston, S.K. & Carreira, S.P. (Eds.), *Modelling and Mathematics Education: Applications in Science and Technology*. Chichester: Horwood, 381-390.

Izard, J., Haines, C.R., Crouch, R.M., Houston, S.K. & Neill, N. (2003). Assessing the impact of the teaching of modelling. In: Lamon, S., Parker, W. & Houston, S.K. (Eds), *Mathematical Modelling: A Way of Life*. Chichester: Horwood, 165-178.

Kaiser-Meßmer, G. (1987). Application-oriented mathematics teaching. In: Blum, W. et al. (Eds), *Applications and Modelling in Learning and Teaching Mathematics*. Chichester: Horwood, 66-72.

Kaiser, G. (2007). Modelling and modelling competencies in school. In: Haines, C. et al. (Eds), *Mathematical Modelling: Education, Engineering and Economics*. Chichester: Horwood, 110-119.

Kaiser, G., Blum, W., Borromeo Ferri, R. & Stillman, G. (Eds, 2011). *Trends in Teaching and Learning of Mathematical Modelling (ICTMA 14)*. Dordrecht: Springer.

Kaiser, G., Blomhøj, M. & Sriraman, B. (Eds, 2006). Mathematical modelling and applications: empirical and theoretical perspectives. In: *Zentralblatt für Didaktik der Mathematik* 38(2).

Kaiser, G., Schwarz, B. & Tiedemann, S. (2010). Future Teachers' Professional Knowledge on Modeling. In: Lesh, R., Galbraith, P.L., Haines, C.R. & Hurford, A. (Eds): *Modeling Students' Mathematical Modeling Competencies. ICTMA 13*. New York: Springer, 433-444.

Kintsch, W. & Greeno, J. (1985). Understanding word arithmetic problems. In: *Psychological Review* 92 (1), 109-129.

Krainer, K. (1993). Powerful tasks: A contribution to a high level of acting and reflecting in mathematics instruction. In: *Educational Studies in Mathematics* 24, 65−93.

Kramarski, B., Mevarech, Z.R. & Arami, V. (2002). The effects of metacognitive instruction on solving mathematical authentic tasks. In: *Educational Studies in Mathematics* 49(2), 225-250.

Kunter, M. & Voss, T. (2013). The Model of Instructional Quality in COACTIV: A Multicriteria Analysis. In: Kunter, M., Baumert, J., Blum, W. et al. (Eds), *Cognitive Activation in the Mathematics Classroom and Professional Competence of Teachers – Results from the COACTIV Project*. New York: Springer, 97-124.

Kunter, M., Baumert, J., Blum, W. et al. (Eds, 2013). *Cognitive Activation in the Mathematics Classroom and Professional Competence of Teachers – Results from the COACTIV Project*. New York: Springer.

Lave, J. (1992). Word problems: a microcosm of theories of learning. In: Light, P. & Butterworth, G. (Eds). Context and cognition: *Ways of learning and knowing*. New York: Harvester Wheatsheaf, 74-92.

Leikin, R. & Levav-Waynberg, A. (2007). Exploring mathematics teacher knowledge to explain the gap between theory-based recommendations and school practice in the use of connecting tasks. In: *Educational Studies in Mathematics* 66, 349-371.

Leiß, D. (2007). *Lehrerinterventionen im selbständigkeitsorientierten Prozess der Lösung einer mathematischen Modellierungsaufgabe*. Hildesheim: Franzbecker.

Leiß, D. (2010). Adaptive Lehrerinterventionen beim mathematischen Modellieren – empirische Befunde einer vergleichenden Labor- und Unterrichtsstudie. In: Journal für Mathematik-Didaktik 31 (2), 197-226.

Lesh, R.A. & Doerr, H.M. (2003). *Beyond constructivism: A models and modelling perspective on teaching, learning, and problem solving in mathematics education*. Mahwah: Lawrence Erlbaum.

Lingefjärd, T. (2013). Teaching mathematical modeling in teacher education: Efforts and results. In: Yang, X.-S. (Ed.), *Mathematical Modeling with Multidisciplinary Applications*. Holboken, Wiley, 57-80.

Maaß, K. (2006). What are modelling competencies? In: *Zentralblatt für Didaktik der Mathematik* 38(2), 113-142.

Maaß, K. (2007). Modelling in Class: What do we want the students to learn? In: Haines, C. et al. (Eds), *Mathematical Modelling: Education, Engineering and Economics*. Chichester: Horwood, 63-78.

Matos, J.F. & Carreira, S. (1997). The quest for meaning in students' mathematical modelling activity. In: Houston, S.K. et al. (Eds), *Teaching & Learning Mathematical Modelling*. Chichester: Horwood, 63-75.

Nesher, P. (1980). The stereotyped nature of school word problems. In: *For the Learning of Mathematics* 1(1), 41-48.

Neubrand, M. (2006). Multiple Lösungswege für Aufgaben: Bedeutung für Fach, Lernen, Unterricht und Leistungserfassung. In: Blum, W., Drüke-Noe, C., Hartung, R. & Köller, O. (Eds), *Bildungsstandards Mathematik: konkret. Sekundarstufe I: Aufgabenbeispiele, Unterrichtsanregungen, Fortbildungsideen*. Berlin: Cornelsen, 162-177.

Niss, M. (1996). Goals of mathematics teaching. In: Bishop, A. et al. (Eds), *International Handbook of Mathematical Education*. Dordrecht: Kluwer, 11-47.

Niss, M. (1999). Aspects of the nature and state of research in mathematics education. In: *Educational Studies in Mathematics* 40, 1-24.

Niss, M. (2003). Mathematical Competencies and the Learning of Mathematics: The Danish KOM Project. In: Gagatsis, A. & Papastavridis, S. (Eds), *3rd Mediterranean Conference on Mathematical Education.* Athens: The Hellenic Mathematical Society, 115-124.

Niss, M. & Højgaard Jensen, T. (Eds, 2011). *Competencies and Mathematical Learning.* Roskilde University.

Niss, M., Blum, W. & Galbraith, P. (2007). Introduction. In: W. Blum et al. (Eds), *Modelling and Applications in Mathematics Education.* New York: Springer, 3-32.

OECD (2013). *PISA 2012 Assessment and Analytical Framework: Mathematics, Reading, Science, Problem Solving and Financial Literacy.* Paris: OECD Publishing.

Palm, T. (2007). Features and impact of the authenticity of applied mathematical school tasks. In: Blum, W. et al. (Eds), *Modelling and Applications in Mathematics Education.* New York: Springer, 201-208.

Pollak, H.O. (1969). How can we teach applications of mathematics? In: *Educational Studies in Mathematics* 2, 393-404.

Pollak, H. (1979). The Interaction between Mathematics and Other School Subjects. In: UNESCO (Ed.), *New Trends in Mathematics Teaching IV.* Paris, 232-248.

Reusser, K. (2001). Co-constructivism in educational theory and practice. In: Smelser, N.J., Baltes, P. & Weinert, F.E. (Eds), *International Encyclopedia of the Social and Behavioral Sciences.* Oxford: Pergamon/Elsevier Science, 2058-2062.

Reusser & Stebler (1997). Every word problem has a solution: The suspension of reality and sense-making in the culture of school mathematics. In: *Learning and Instruction* 7, 309-328.

Rittle-Johnson, B. & Star, J.R. (2009). Compared With What? The Effects of Different Comparisons on Conceptual Knowledge and Procedural Flexibility for Equation Solving. In: *Journal of Educational Psychology* 101(3), 529-544.

Schmidt, W.H., Tatto, M.T., Bankov, K., Blömeke, S., Cedillo, T., Cogan, L., et al. (2007). *The preparation gap: Teacher education for middle school mathematics in six countries (MT21 Report).* East Lansing: MSU Center for Research in Mathematics and Science Education.

Schoenfeld, A.H. (1988). When good teaching leads to bad results: The disasters of "well-taught" mathematics courses. In: *Educational Psychologist* 23, 145-166.

Schoenfeld, A.H. (1991). On mathematics as sense-making: An informal attack on the unfortunate divorce of formal and informal mathematics. In: Voss, J.F., Perkins, D.N. & Segal, J.W. (Eds), *Informal Reasoning and Education.* Hillsdale: Erlbaum, 311-343.

Schoenfeld, A.H. (1992). Learning to think mathematically: problem solving, metacognition, and sense-making in mathematics. In: Grouws, D. (Ed.), *Handbook for Research on Mathematics Teaching and Learning.* New York: MacMillan, 334-370.

Schoenfeld, A.H. (1994). *Mathematical Thinking and Problem Solving.* Hillsdale: Erlbaum.

Schukajlow, S. (2011). *Mathematisches Modellieren. Schwierigkeiten und Strategien von Lernenden als Bausteine einer lernprozessorientierten Didaktik der neuen Aufgabenkultur.* Münster: Waxmann.

Schukajlow, S., Leiss, D., Pekrun, R., Blum, W., Müller, M. & Messner, R. (2012). Teaching methods for modelling problems and students' task-specific enjoyment, value, interest and self-efficacy expectations. In: *Educational Studies in Mathematics* 79(2), 215-237.

Schukajlow, S. & Krug, A. (2013). Considering multiple solutions for modelling problems – design and first results from the MultiMa-Project. In: Stillman, G. et al. (Eds), *Teaching Mathematical Modelling: Connecting to Teaching and Research Practice – the Impact of Globalisation.* New York: Springer,.

Shoaf, M.M., Pollak, H. & Schneider, J. (2004). *Math Trails.* Lexington: COMAP.

Smit, J., van Eerde H. A. A. & Bakker, A. (2013). A conceptualisation of whole-class scaffolding. *British Educational Research Journal* 39(5), 817-834.

Sol, M., Giménez, J. & Rosich, N. (2011). Project Modelling Roites in 12- 16-Year-Old Pupils. In: Kaiser, G. et al. (Eds). *Trends in Teaching and Learning of Mathematical Modelling (ICTMA 14).* Dordrecht: Springer, 231-240.

Staub, F.C. & Reusser, K. (1995). The role of presentational structures in understanding and solving mathematical word problems. In: Weaver, C.A., Mannes, S. & Fletcher, C.R. (Eds), *Discourse comprehension. Essays in honor of Walter Kintsch*. Hillsdale: Lawrence Erlbaum, 285-305.

Stillman, G. (2011). Applying Metacognitive Knowledge and Strategies in Applications and Modelling Tasks at Secondary School. In: Kaiser, G. et al. (Eds). *Trends in Teaching and Learning of Mathematical Modelling (ICTMA 14)*. Dordrecht: Springer, 165-180.

Stillman, G. & Galbraith, P. (1998). Applying mathematics with real world connections: Metacognitive characteristic of secondary students. In: *Educational Studies in Mathematics 36* (2), 157-195.

Stillman, G., Kaiser, G., Blum, W. & Brown, J. (Eds, 2013). *Teaching Mathematical Modelling: Connecting to Teaching and Research Practice – the Impact of Globalisation*. New York: Springer.

Timperley, H.S. (2011). *Realizing the Power of Professional Learning*. London: Open University Press.

Treilibs, V., Burkhardt, H. & Low, B. (1980). *Formulation processes in mathematical modelling*. Nottingham: Shell Centre for Mathematical Education.

Tsamir, P., Tirosh, D., Tabach, M. & Levenson, E. (2010). Multiple solution methods and multiple outcomes—is it a task for kindergarten children? In: *Educational Studies in Mathematics 73*, 217-231.

Turner, R., Dossey, J., Blum, W. & Niss, M. (2013). Using Mathematical Competencies to Predict Item Difficulty in PISA: A MEG Study. In: Prenzel, M., Kobarg, M., Schöps, K. & Rönnebeck, S. (Eds.), *Research on PISA – Research Outcomes of the PISA Research Conference 2009*. New York: Springer, 23-37.

Verschaffel, L., Greer, B. & DeCorte, E. (2000). *Making Sense of Word Problems*. Lisse: Swets & Zeitlinger.

Verschaffel, L., van Dooren, W., Greer, B. & Mukhopadhyay, S. (2010). Reconceptualising Word Problems as Exercises in Mathematical Modelling. In: *Journal für Mathematik-Didaktik 31*(1), 9-29.

Vos, P. (2007). Assessment of applied mathematics and modelling: Using a laboratory-like environment. In: Blum, W. et al. (Eds), *Modelling and Applications in Mathematics Education*. New York: Springer, 441-448.

Xin, Z., Lin, C., Zhang, L. & Yan, R. (2007). The performance of Chinese primary school students on realistic arithmetic word problems. In: *Educational Psychology in Practice 23*, 145-159.

Zöttl, L., Ufer, S. & Reiss, K. (2011). Assessing modelling competencies using a multidimensional IRT approach. In: Kaiser, G. et al. (Eds), *Trends in Teaching and Learning of Mathematical Modelling (ICTMA 14)*. Dordrecht: Springer, 427-437.

Part III
Plenary Panels

The TEDS-M: Important Issues, Results and Questions

Konrad Krainer, Feng-Jui Hsieh, Ray Peck and Maria Teresa Tatto

Abstract Until the last decade, international comparative studies in mathematics education focused primarily on the knowledge and beliefs of school students. Recently, the focus has shifted towards research on teachers and teacher education. The Teacher Education and Development Study in Mathematics (TEDS-M) is the first international large-scale study about (initial primary and secondary) mathematics teacher education with 17 countries participating. The importance of large-scale research in mathematics teacher education is mirrored in the decision to organize a Plenary Panel on TEDS-M at the 12th International Congress on Mathematical Education (ICME-12). This paper sketches the background of the study, main program features and major inputs of the Plenary Panel.

Keywords Mathematics teacher education · Content knowledge · Pedagogical content knowledge · Survey research · International and comparative studies

K. Krainer (✉)
Alpen-Adria-Universität Klagenfurt, Klagenfurt, Austria
e-mail: konrad.krainer@aau.at

F.-J. Hsieh
National Taiwan Normal University, Taipei, Taiwan
e-mail: hsiehfj@math.ntnu.edu.tw

R. Peck
Australian Council for Educational Research, Sydney, Australia
e-mail: peck@acer.edu.au

M.T. Tatto
Michigan State University, Michigan, USA
e-mail: mttatto@msu.edu

© The Author(s) 2015
S.J. Cho (ed.), *The Proceedings of the 12th International Congress on Mathematical Education*, DOI 10.1007/978-3-319-12688-3_10

Introduction

Empirical research over recent decades points to the high influence of teachers on students' learning of mathematics. Teachers have been identified as key agents of educational change (Fullan 1993; Krainer 2011). Amongst others, the comprehensive meta-analysis on student learning by Hattie (2003) found that teachers' impact on students' learning is high. Identified factors that contribute to major sources of variation in student performance include the students (50 %) and teachers (30 %) as the most important factors, whereas home, schools, principals, peer effects (altogether 20 %) play a less important role.

Thus intensive research in mathematics teacher education is needed. There is increasing literature about relevant results, however, large-scale findings about the conditions, processes, and effects of mathematics teacher education are rare (Adler et al. 2005). Since Mathematical Content Knowledge (MCK) and Mathematical Pedagogical Content Knowledge (MPCK) play a fundamental role for teachers' effectiveness (Shulman 1986; Baumert et al. 2010), the education of future teachers is a crucial phase in teachers' professional development and a key time for communicating pedagogical innovations, especially because many teachers tend to teach as they have been taught.

The Teacher Education and Development Study in Mathematics (TEDS-M) is the first cross-national data-based study about initial mathematics teacher education with large-scale samples (Tatto et al. 2011, 2012; Loewenberg-Ball et al. 2012). The study collected data from 23,000 future mathematics teachers (primary and lower-secondary) from 17 countries[1] in 2008–2009.

The TEDS-M study drew nationally representative samples and conducted large scale surveys of teacher education institutions, teacher educators, and future teachers to provide substantive information on how institutions organize and prepare future teachers to teach mathematics at the primary and secondary levels. The study also successfully created instruments for measuring the MCK and MPCK of future teachers at the international level in different types of program groups.

TEDS-M was a collaborative effort of worldwide institutions, launched by the International Association for the Evaluation of Educational Achievement (IEA) to address concerns raised by the Third International Mathematics and Science Study (TIMSS). The study is an ambitious attempt to move the study of teacher education and its outcomes in the direction of scientific research with the goal to inform policy. The study was directed by Michigan State University (MSU) in collaboration with the Australian Council for Educational Research (ACER), and National Research Centres in all 17 countries and received important funding from the National Science Foundation (USA), and the IEA.

---

[1] Botswana, Canada (was unable to meet IEA sampling requirements), Chile, Chinese Taipei (Taiwan), Georgia, Germany, Malaysia, Norway, Philippines, Oman, Poland, Russia, Singapore, Spain, Switzerland, Thailand, and USA.

TEDS-M posed questions at three levels: (*a*) *Policies*: What are the teacher education policies of the participating countries that support the mathematics and related knowledge for teaching of their future teachers? (*b*) *Practices*: What learning opportunities in teacher education programs allow future teachers to attain mathematics and related knowledge for teaching? (*c*) *Outcomes*: What is the level and depth of the mathematics and related knowledge for teaching attained by future teachers at the end of their initial teacher education programs? TEDS-M aimed at bringing these three components—policies, practices and outcomes of mathematics teacher education—together. As a result, the findings should be of interest to educational policy makers and researchers, mathematicians and mathematics educators. In the same way that teachers are the key to educational change in schools, mathematicians and mathematics educators are—together with the future teachers themselves—the key drivers of change and innovation in mathematics teacher education.

Comparisons between countries are complex. Outcomes from the study show significant differences in outcome measures between future teachers in different programs in different countries. Since the participating countries have a diverse level of "human development" (formerly "standard of living"), as measured by the Human Development Index (HDI),[2] it is important to take this into account when comparing countries performance in TEDS-M. A study by Blömeke (2011, p. 19) shows a close correlation between the countries' TEDS-M outcome measures and their HDI. However, related to this index, some countries achieved higher than expected in TEDS-M, others lower. The Blömeke study indicates Taiwan, Russia, and Thailand as "overachieving" countries and the USA, Norway, and Chile as "underachieving" countries compared to their level of human development. From the case of Taiwan, we will learn what factors may have a positive influence on the education of future mathematics teachers graduating with high levels of MCK and MPCK. We will also see that Chile and Norway, both performing below their expectations compared to HDI, started reforms as a consequence of their TEDS-M results. Thus, this study offers opportunities to compare with other countries, to look for communalities and differences, as well as for (relative) strengths and weaknesses. However, in order to learn more deeply from other countries and probably to take relevant actions fitting to a country's own context, it is important to look in a more detailed way at program characteristics. TEDS-M is both, a starting point for diverse comparisons among countries, as well as a chance to investigate the quality of teacher education programs and the learning opportunities they offer to future teachers of mathematics.

---

[2] The HDI is a comparative measure of life expectancy, literacy, education, and standards of living for countries worldwide.

The Organization of the Plenary Panel on TEDS-M

The Plenary Panel on TEDS-M at ICME-12 involved four Panel Members: *Feng-Jui Hsieh* (Taiwan), *Konrad Krainer* (Austria, Chair), *Ray Peck* (Australia), and *Maria Teresa Tatto* (USA).

After a short introduction of the Plenary Panel members by *Mi-Kyung Ju* (Korea, Presider), some basic information about TEDS-M and the Plenary Panel by the chair, the other Panel members gave inputs on the following topics:

- Teaching and teacher knowledge: A focus on MCK and MPCK (Ray Peck)
- Teacher education and quality: The performance of Taiwan in an international context (Feng-Jui Hsieh)
- Research in teacher education and TEDS-M: International findings and implications for future policy research (Maria Teresa Tatto)

In order to support the audience in actively following the presentations, each input included a short activity for the whole audience. Given the fact, that in a Plenary Panel with some thousand people it is not easy to have open discussions, the Panel team invited Audience representatives. They are well-known experts with diverse background (mathematics, mathematics education or pedagogy), some having deeper knowledge about TEDS-M: *Deborah Loewenberg Ball* (USA), *Mellony Graven* (South Africa), *Maitree Inprasitha* (Thailand), *Liv Sissel Gronmo* (Norway), *Leonor Varas* (Chile), and *Ildar Safuanov* (Russia).

The Audience representatives were prepared to respond to questions raised by the chair of the Panel each related to the corresponding topic presented by the three panelists.

Teaching and Teacher Knowledge: A Focus on MCK and MPCK

Why Is Teacher Knowledge Important?

Anthony and Walshaw (2009, p. 25) remind us that knowledge helps teachers recognize, and then act upon, the teaching opportunities that come up in the moment. Understanding the 'big ideas' of mathematics, permits teachers to recognize mathematics as a 'coherent and connected system'. This in turn enables them to 'make sense of and manage multiple student viewpoints'. With strong content and pedagogical content knowledge teachers can help students to develop 'mathematically grounded understandings'.

Research into student achievement in mathematics has strongly supported the importance and significance of teacher knowledge. For example, Hill et al. (2005), found that the mathematical knowledge of teachers was significantly related to student achievement gains in both first and third grades after controlling for key student- and teacher-level covariates.

Defining Teacher Knowledge in TEDS-M

Teacher knowledge for teaching mathematics in TEDS-M was narrower than that defined by Shulman (1986). It was limited to the knowledge that could be reasonably demonstrated by future teachers in their final year of their programs on a written 60 min assessment. It was also limited to the knowledge that was considered important and culturally meaningful to the 17 participating countries.

In short, the knowledge for teaching mathematics in TEDS-M was confined to two dimensions—mathematics content knowledge (MCK) and mathematics pedagogical content knowledge (MPCK). MCK is mathematics that teachers know and can do whereas MPCK is knowledge about how to assist students to learn mathematics. MPCK is **not** knowledge that ordinary citizens possess. It is theoretical and experiential knowledge learned from studying and working in mathematics education. The focus of MCK in TEDS-M was on the mathematics that the future teachers would be required to teach plus some content 2 or 3 years beyond that.

Because TEDS-M was an international study, the decision was taken to make use of the TIMSS content frameworks for Year 8 and Advanced (Mullis et al. 2005; Garden et al. 2006). The MPCK framework in TEDS-M was developed by the TEDS-M international team, after a review of the literature and was informed in part by the framework used by the Mathematics Teaching in the 21st Century Project (MT21) (Blömeke et al. 2008; Schmidt et al. 2011) which focused on middle school mathematics teacher preparation in six countries. The final version of the MPCK framework was arrived at following a critical review by international experts in the field.

The TEDS-M MPCK framework consists of three sub-domains.

Mathematical curricular knowledge:

knowing the school mathematics curriculum, establishing appropriate learning goals, identifying key ideas in learning programs, selecting possible pathways and seeing connections within the curriculum, knowing different assessment formats and purposes

Knowledge of planning for mathematics teaching and learning:

selecting appropriate activities, predicting typical students' responses, including misconceptions, planning appropriate methods for representing mathematical ideas, linking didactical methods and instructional designs, identifying different approaches for solving mathematical problems, choosing assessment formats and items

Enacting mathematics for teaching and learning:

explaining or representing mathematical concepts or procedures, generating fruitful questions, diagnosing responses, including misconceptions, analysing or evaluating students' mathematical solutions or arguments, analysing the content of students' questions, responding to unexpected mathematical issues, providing appropriate feedback

Measuring Teacher Knowledge in TEDS-M

The TEDS-M study measured knowledge found 'in the mind', not that 'in the body' as seen and found 'in our practices' (Connelly and Clandinin 1988). So, unlike the study by Huckstep et al. (2003), there was not the opportunity in TEDS-M to observe how the mathematics content knowledge of future primary teachers was enacted in practical teaching during school-based placements.

In TEDS-M, content knowledge was assessed by a combination of simple and complex multiple-choice items, together with short and extended constructed response items. Scoring guides for the constructed-response items were refined using responses from the field trial and for most extended constructed-response items, partial credit could be awarded.

Short activity for the audience
In order to sketch the difference between MCK and MPCK items, selected MCK and MPCK examples covering a range of attributes from the released TEDS-M item pool were presented to the audience including item statistics. The audience was invited to participate in providing "informed answers" to the items presented. Their answers were then contrasted with those obtained in the TEDS-M study by using "percent correct" information.

The total score points for each future teacher were analyzed using item response theory (Wu et al. 2007). This enabled four scales for knowledge for teaching mathematics to be constructed: MCK and MPCK for both primary and secondary. Tables and charts were created showing the distribution of country scale scores by program group.

Six "anchor points" were defined and described, two for each MCK scale and one for each MPCK scale. This enabled the achievement of future teachers in each program group to be described against the anchor points. It is hoped that these will provide useful benchmarks for future work. An example of the primary MPCK anchor point follows.

Primary MPCK Anchor Point

Future primary teachers who scored at this anchor point were generally able to recognize the correctness of a teaching strategy for a particular concrete example, and to evaluate students' work when the content was conventional or typical of primary grades. They were likely to identify the arithmetic elements of single-step story problems that influence their difficulty. Although future primary teachers at the primary MPCK anchor point were likely to be able to interpret some students' work, their responses were often unclear or imprecise. In addition, future teachers at the anchor point were unlikely to use concrete representations to support students' learning or to recognize how a student's thinking is related to a particular algebraic representation. They generally were unlikely to understand some measurement or probability concepts needed to reword or design a task. These future teachers also were unlikely to know why a particular teaching strategy made sense, if it would always work, or whether a strategy could be generalized to a larger class of problems. They were unlikely to be aware of common misconceptions or to conceive useful representations of numerical concepts.

For the 15 countries whose data could be analyzed, nine of the 21 program types across the four defined program groups had the majority of their future teachers at or above this anchor point on the MPCK scale. In some cases, items worth two score points (partial credit items) were able to measure levels of knowledge above and below anchor points. An example of this is item MFC410[3] shown in Fig. 1.

Future teachers at the primary MPCK anchor point were able to achieve partial credit (1 out of a maximum of two score points) with a probability of at least 0.7 on this

MFC410

Imagine that two <primary> students in the same class have created the following representations to show the number of teeth lost by their classmates.

[Mary] drew pictures of her classmates on cards to make this graph.

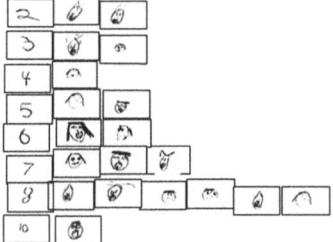

[Sally] cut out pictures of teeth to make this graph.

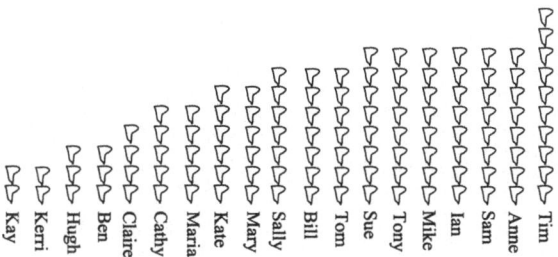

From a data presentation point of view, how are the representations alike and how are they different?

| Alike: |
| Different: |

Fig. 1 Item MFC410, primary MPCK—sub-domain Enacting, data, two score points

[3] Alejandra Sorto, formerly of Michigan State University, is acknowledged for this item.

Table 1 Scoring guide for MFC410

Code	Response	Item: MFC410
	Correct response	
20	Responses that indicate how the representations are **alike** AND how they are **different** 'Alike' examples • They both show the same data/same number of teeth lost • They are both pictorial representations • They are both forms of bar graphs • They are both skewed in the same direction 'Different' examples • Mary has grouped the data/done a frequency tally whereas Sally has not • In Mary's graph each bar or column represents the number of teeth lost, whereas in Sally's graph each column or stack represents a student • Mary's graph is categorized by the number of teeth lost whereas Sally's is person by person	
	Partially correct response	
10	**The 'alike' description is acceptable** but the 'different' description is not acceptable, trivial or is missing 'Alike' example • They both show the same number of teeth lost 'Different' example • Mary's is easier to comprehend than Sally's	
11	**The 'different' description is acceptable** but the 'alike' description is not acceptable, trivial or is missing 'Alike' example • They both made graphs about teeth (Trivial) 'Different' example • Sally made a column for each student whereas Mary made a column for each number of teeth lost	
	Incorrect response	
70	Responses that are insufficient or trivial 'Alike' examples • They are both graphs • Both graphs are about teeth 'Different' examples • Mary used numbers, Sally didn't • Mary's is hard to read, Sally's is easier	
79	Other incorrect (including crossed out, erased, stray marks, illegible, or off task)	
99	**Non-response** (blank)	

item. Only future teachers well above the anchor point were able to achieve full credit on this item. Twenty-nine percent (29 %) of the international sample of future teachers achieved full credit on this item and another 37 % were able to achieve partial credit.

The following Table 1 shows the scoring guide for MFC410. On this item, for the international sample, 29 % were awarded full credit, 37 % partial credit, and 23 % no credit. Eleven percent (11 %) of the international sample of future teachers chose not to respond. The future teachers who achieved partial credit found it harder to say how the representations were different (6 %) than how they were alike (31 %).

This work is described in more detail in recent TEDS-M publications (Senk et al. 2012; Tatto et al. 2012; Tatto 2013).

Views from Audience Representatives

The chair asked two Audience representatives to respond to two questions: "Is what TEDS-M measured valued by the mathematics education community (with a particular focus on the MPCK items)? How well has TEDS-M contributed knowledge to the field?"

Maitree Inprasitha (Thailand) stressed that before TEDS-M, most education faculties in Thailand provided only mathematics content courses (MCK) to future teachers. Now education faculties have started incorporating the idea of MPCK into teacher preparation curriculum. More recently, the Khon Kaen University received a grant to create a network among education faculties in order to redefine courses for future teachers who are majoring in mathematics education. Through this network, mathematics education faculty staff attend seminars and workshops hosted by the education faculty of Khon Kaen university.

Ildar Safuanov (Russia) indicated extensive research arising from TEDS-M in his country. Although Russia has strong MCK and MPCK results, research looks for fields where future teachers have difficulties (e.g., in constructing different interpretations of theoretical contents) in order to achieve improvements. Research also shows that there is a relationship between the quality of education of future teachers and their attitudes to teaching mathematics (e.g., related to an orientation on conceptual models and cognitive-constructivist approaches to teaching mathematics).

Teacher Education and Quality: The Performance of Taiwan in an International Context

Becoming a Teacher in Taiwan

Teaching in Taiwan is attractive in terms of income, working hours, career development opportunities, and job security. As a result, candidates face rigorous evaluation and serious competition throughout the process of becoming a teacher.

Future teachers must obtain a bachelor's degree, complete the initial teacher education curriculum, and finish a practicum before they are evaluated in the yearly-held, national-common teacher qualification assessment. The average passing rates of the qualification assessments for the years of 2007–2010 was 67.4 %.

To get a tenure teaching position, qualified teachers must also undergo a public, competitive, on-site-screening process administered by the school district or individual schools. The screenings are not held only for future teachers, but for all the practicing teachers who want to change schools. The average pass rates of the screenings across the country for the recent years 2007–2010 at the primary, lower secondary and upper secondary levels were 3.5, 11.9, and 6.5 %,[4] respectively (Hsieh et al. 2012a). Regarding the future teachers, the average rates of employment for tenure teaching positions for 2007–2010 were lower than 3.4 % for the primary level and 20.2 % for the secondary level.

What Taiwan Learned from TEDS-M on Teaching Knowledge

As a participating country in TEDS-M, Taiwan intended to examine how future teachers performed and what the weaknesses and strengths of teachers were on teaching knowledge as compared to other countries. The results of MCK and MPCK achievement for future teachers, especially at the primary level, challenged the expectations of Taiwanese scholars in two areas. First, Taiwan ranked number one in performance. Second, Taiwan's percentages of correct answers for some primary items with low-level of difficulty were low.

In Taiwan, future teachers are expected to be knowledgeable and to master the concepts and skills on the field they intend to teach. It is expected that at least 80 % (if not 100 %) of future teachers should provide correct answers for any item at their teaching level. However, Taiwan's data showed that, in the lower secondary-level study, 30 % of MCK and 33 % of MPCK items did not meet the desired 80 % threshold. For the primary-level study, 36 % of MCK and 83 % of MPCK items did not achieve the 80 % threshold. For the type of *thought-oriented mathematical competence* primary-level items,[5] a high rate of 70 % of items did not reach the 80 % threshold. These results are a strong warning for the Taiwanese teacher education system.

---

[4] People may attend many screenings, so the actual rates of people who pass the screenings should be higher than these data.

[5] This is a type of MCK that contrasts with another MCK-type: content-oriented mathematical competence. For more information concerning this section, see the relevant article by Hsieh et al. (2012b).

Why Taiwan Performed Well

The Taiwan TEDS-M team was interested in analyzing how Taiwanese future teachers performed for MCK items with respect to different curricular levels. For this analysis, TEDS-M knowledge items were classified according to four curricular levels: primary, lower secondary, upper secondary or tertiary. The results showed that, in comparison to all participating countries, Taiwan demonstrated a unique pattern in the lower secondary-level study. As shown in Fig. 2, the pattern exhibited in Taiwan was high achievement with respect to the percentage of correct answers for items from primary, lower secondary, and upper secondary levels, but a sharp decline in percent correct on the tertiary level items. Singapore, which demonstrated performance similar to Taiwan for TIMSS, showed MCK achievement patterns different from those for Taiwan. Singapore, Germany, and Switzerland did not show achievement on primary-level MCK items as high as Taiwan but did show a sharp decline from primary to upper secondary levels. For all other countries (except for Taiwan, Singapore, Germany, and Switzerland), MCK achievement remained approximately the same from secondary to tertiary levels. Since Taiwanese lower secondary-level teacher education programs emphasize mostly tertiary-level mathematics (but do not cover primary-level mathematics), these data show that one of the reasons Taiwan performed better in MCK is that it recruits high-achieving students for secondary teacher education programs.

This idea also explains why Taiwan performed well in MPCK for the lower secondary-level study. Mathematical concepts applied for almost all MPCK items appear in the lower secondary-level, a level in which Taiwan excelled.

For the primary-level study, future Taiwanese teachers achieved high results for primary-level MCK items, lower secondary- and upper secondary-level items (see Fig. 3).

This result may demonstrate that Taiwan recruits high-achieving students for primary teacher education programs. However, a question remains as to why

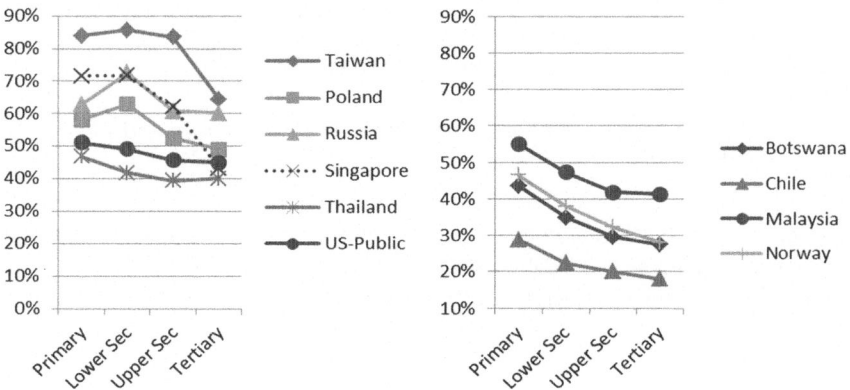

Fig. 2 Percentage of correct answers for MCK items across different levels in the lower secondary-level study for certain countries

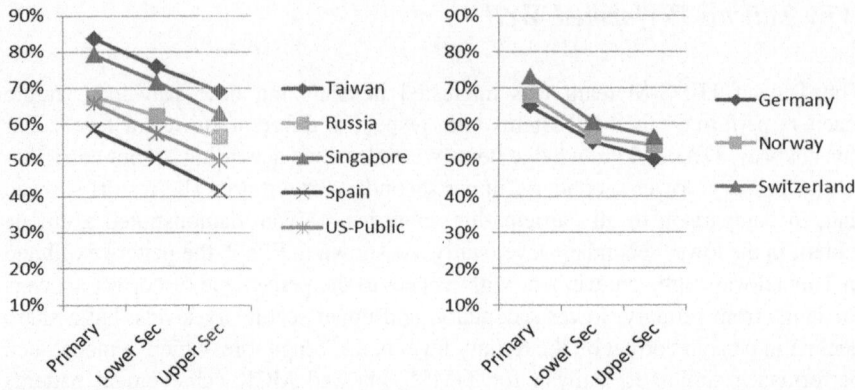

Fig. 3 Percentage of correct answers for MCK items across different levels in the primary-level study for certain countries

Singapore, which performed worse on the MCK than Taiwan, achieved results similar to Taiwan for the MPCK test. Further research is required to examine factors influencing relevant knowledge achievements.

Other TEDS-M data show that Taiwan may have demonstrated superior performance, especially in the lower secondary-level, because of the following reasons: Taiwan teaches more topics in both school- and tertiary-level mathematics than other countries, and future Taiwanese teachers have increased opportunities to perform challenging problems (thought-oriented). This finding is consistent with findings from analyses of relationships between Opportunity to Learn (OTL), MCK, and MPCK (Hsieh et al. 2012a).

Short activities related to single-item performance
The following questions were posed to the audience.
Example 1: (a lower secondary-level item)

Let $A = \begin{bmatrix} p & q \\ r & s \end{bmatrix}$ and $B = \begin{bmatrix} t & u \\ v & w \end{bmatrix}$. Then $A \otimes B$ is defined to be $\begin{bmatrix} pt & qu \\ rv & sw \end{bmatrix}$.

Is it true that if $A \otimes B = O$, then either $A = O$ or $B = O$ (where O represents the zero matrix)? Justify your answer.

The operation defined in Example 1, MFC814,[6] a tertiary-level MCK item, is not taught in relevant courses. To correctly answer this problem, a test-taker must observe the relationships between mathematical objects, devise formal or informal mathematical arguments, and transform heuristic arguments into valid proofs.

---

[6] The *Knowledge of Algebra for Teaching* (KAT) project, Michigan State University, is acknowledged for item MFC814.

Question 1: What percentage of future mathematics teachers at the secondary level can answer this item correctly in your country?

Example 2: (a lower secondary-level item)

A mathematics teacher wants to show some students how to prove the quadratic formula.

Determine whether each of the following types of knowledge is needed in order to understand a proof of this result.

Check one box in each row.

		Needed	Not needed
A.	How to solve linear equations.	$\square_1$	$\square_2$
B.	How to solve equations of the form $x^2 = k$, where $k > 0$.	$\square_1$	$\square_2$
C.	How to complete the square of a trinomial.	$\square_1$	$\square_2$
D.	How to add and subtract complex numbers.	$\square_1$	$\square_2$

Example 2, MFC712, is an MPCK item concerning a formal approach for teaching the quadratic formula. Option C is considered necessary to understanding a proof of the formula.

Question 2: In which of the following situations will future teachers in your country know that option C is necessary? Answer Yes or No to each.

- If they know how to prove the quadratic formula and attempt to prove it when answering this item.
- If they know the pre-requisites for learning how to prove the quadratic formula.
- If they have watched a teacher teaching approaches for proving the quadratic formula.
- If they have had experience teaching how to prove the quadratic formula.
- If they have been taught by faculty in their teacher education programs how to demonstrate the quadratic formula.

Example 3: (a primary-level item)

How many decimal numbers are there between 0.20 and 0.30?

Check one box.

A.	9	$\square_1$
B.	10	$\square_2$
C.	99	$\square_3$
D.	An infinite number	$\square_4$

A special feature of Example 3, MFC304,[7] is that 0.2 is expressed as 0.20. Question 3: At what grade do teachers teach the addition of decimals with three digits in your country?

---

[7] Item MFC304 is one of a pool of items developed for TEDS-M by Doug Clarke, Peter Sullivan, Kaye Stacey, Ann Roche, and Ray Peck, Melbourne, Australia.

Discussion of other MCK and MPCK items can be found in an article by Hsieh et al. (2012b).

Views from Audience Representatives

After a short exchange with the audience, the chair asked two Audience representatives to "describe any interventions that have been undertaken in Chile and Norway as a consequence of disappointing TEDS-M results."

Liv Sissel Gronmo (Norway) stressed that although there was disappointment with the results, there have been few interventions so far. In particular, concerning the problem that future teachers do not have the necessary competence in mathematics, no measures have been taken so far. On the contrary, a recent change in teacher education has expanded the amount of general pedagogy which seems to be a step in the wrong direction.

María Leonor Varas (Chile) reported that TEDS-M results had—after a first shock—a distinguishable impact in Chile at different levels. For example, it accelerated decisions and deepened interventions that were in the process of implementation (e.g., outcome standards for teacher preparation programs and entrance examinations for teachers). It also led to an increased engagement of mathematicians in teacher preparation in collaboration with mathematics educators (e.g., jointly developing standards for teacher preparation as well as preparing books and materials to support its implementation).

Research in Teacher Education and TEDS-M: International Findings and Implications for Future Policy Research

Research has begun to advance our understanding of the knowledge considered most important for school mathematics teaching (e.g., Baumert et al. 2010; Hill et al. 2007; Schmidt et al. 2011; Tatto 2008; Tatto et al. 2010). For more than a decade, recommendations from relevant societies and expert groups have emphasized that future teachers of school mathematics need to develop a deep understanding of the mathematics they will teach (Conference Board of Mathematical Sciences 2001), and that to be successful "... mathematics teachers need preparation that covers knowledge of mathematics, of how students learn mathematics and of mathematical pedagogy" (National Research Council 2010, p. 123; Education Committee of the EMS 2012). Importantly for our discussion today are calls to collect "... quantitative and qualitative data about the programs of study in mathematics offered and required at teacher preparation institutions ... to improve understanding of what sorts of preparation approaches are most effective at developing effective teachers" (National Research Council 2010, p. 124). In this session, we will present some of the challenges involved in doing research in teacher education, the main findings that are emerging from the study, and plans for future research including a new study of novice mathematics teachers.

To recap, the overall goal of TEDS-M was to study in a group of countries how primary and secondary school mathematics teachers learn to teach subject matter content effectively to a wide variety of students as a result of their preparation programs. This comparative approach to exploring teacher education and its influence cross-nationally helped us to understand the combination of teacher education policies, learning opportunities, and levels of mathematics knowledge that future teachers reach in those countries where pupils show high mathematics achievement vis-à-vis those who do not. As we have said in previous articles, the intent of TEDS-M is to replace myths about when, what, and how teachers learn, with facts and conclusions backed by rigorous research (Tatto et al. 2011).

Methods

The most important challenges we encountered were methodological such as the sampling, the instrument development, and, given the diversity of programs we encountered, the approaches to describe the results. TEDS-M used comparative and survey research methods to produce correlational analyses. Original data were collected through the examination of policy documents; assessments of mathematics teaching knowledge; and questionnaires. TEDS-M implemented a two-stage sampling design: (a) selected samples representative of the national population of institutions offering initial teacher education to the target populations; (b) all programs in those institutions were included in the survey; (c) within institutions (and programs), samples of educators and of future teachers were surveyed. Samples had to reach the rigorous IEA sampling standards. Sampling errors were computed using balanced half-sample repeated replication (Fay 1989; Lohr 1999; McCarthy 1966; Tatto et al. 2012). The development of anchor points to interpret the knowledge scores in a meaningful way represented both a challenge and an important step forward in teacher education research. Anchor points can assist teacher preparation programs worldwide to establish benchmarks of performance for their graduates using TEDS-M assessments and analyses. These assessment tools were developed collaboratively and represent meaningful international standards (Tatto et al. 2012).

Data Sources

Policy and context data were collected using country reports, questionnaires, and interviews. TEDS-M conducted (a) surveys of the teacher education institutions using an institutional program questionnaire; (b) surveys of educators and mentors of future teachers in the institutions using a teacher educator questionnaire; and (c) surveys of future teachers in the sampled institutions. Questions on future teacher knowledge of mathematics and mathematics pedagogy were investigated via assessments developed for that purpose.

Results

The results of our study are presented in detail in the TEDS-M international report: Policy, Practice, and Readiness to Teach Primary and Secondary Mathematics in 17 Countries (Tatto et al. 2012), which is available for download from the TEDS-M webpage http://teds.educ.msu.edu/, or from the IEA webpage at http://www.iea.nl3.

For this presentation, we will only briefly highlight the key international findings from the mathematics knowledge assessments at the primary and lower secondary levels and discuss patterns in the organization of teacher preparation programs that indicate promising directions for policy.

Tables 2 and 3 show the descriptive statistics for mathematics content knowledge (MCK), by program group for the future teachers participating in the study at the primary and lower secondary levels. The tables show a key analysis strategy employed in TEDS-M: that is the way results were presented by "program groups" in order to cater for the different structures of teacher education systems. Table 2 reveals the variation in MCK scores across and within program groups. Given the international mean set at 500 and the standard deviation at 100 it can be seen that the difference in mean MCK scores between some countries, even in the same program group, was between one and two standard deviations. Here it will be helpful to illustrate the use of the anchor points—see above—to interpret TEDS-M results. In the high-scoring countries within each program group, the majority of future teachers had scores at or above the higher MCK anchor point. Differences between countries within program groups tended to be larger among the secondary groups (Table 3) than among the primary groups (Table 2). The results in the United States of America illustrate these differences.

Table 2 shows that in the USA more than 90 % of future primary teachers reach Anchor Point 1, but only 50 % reach Anchor Point 2, whether generalists or specialists; this places the USA below Taiwan, Singapore, and Switzerland in Group 2: primary generalists, and well below Poland, Singapore, Germany, and Thailand in Group 4: primary specialists. Table 3 shows the results of the secondary groups. Close to 70 % of the USA teachers do not even reach Anchor Point 1 in Group 5: lower secondary teachers preparing to teach to Grade 10, placing them below Singapore, Switzerland, Poland, Germany, and Norway. USA future teachers, however, do better in the program Group 6: lower and upper secondary teachers prepared to teach Grade 11 and above in reaching Anchor Point 1, yet they still score well below the future teachers from Taiwan, Singapore, Germany, and the Russian Federation. While in all of these other countries more than 60 % of future teachers reach Anchor Point 2, more than 55 % of USA future teachers fail to reach the same benchmark.

Table 2 Descriptive statistics for mathematics content knowledge, by program group (future teacher, primary level) *Source* Tatto et al. (2012)

Program Group	Country	Sample Size	Valid Data (N)	Percent Missing (Weighted)	Percent at or above Anchor Point 1 (SE)	Percent at or above Anchor Point 2 (SE)	Scaled Score: Mean (SE)
Group 1. Lower Primary (Grade 4 Maximum)	Georgia	506	506	0.0	11.9 (1.4)	0.9 (0.5)	345 (4)
	Germany	935	907	2.4	86.4 (1.3)	43.9 (2.1)	501 (3)
	Poland[a]	1812	1799	0.9	67.9 (1.3)	16.8 (1.2)	456 (2)
	Russian Federation[b]	2266	2260	0.2	89.7 (2.3)	57.3 (4.6)	536 (10)
	Switzerland[c]	121	121	0.0	90.5 (2.7)	44.2 (5.4)	512 (6)
Group 2. Primary (Grade 6 Maximum)	Chinese Taipei	923	923	0.0	99.4 (0.3)	93.2 (1.4)	623 (4)
	Philippines	592	592	0.0	60.7 (5.1)	6.3 (0.9)	440 (8)
	Singapore	263	262	0.4	100.0	82.5 (2.3)	586 (4)
	Spain	1093	1093	0.0	83.4 (1.6)	26.2 (1.6)	481 (3)
	Switzerland	815	815	0.0	97.2 (0.6)	70.6 (1.4)	548 (2)
	† USA[d]	1310	951	28.6	92.9 (1.2)	50.0 (3.2)	518 (5)
Group 3. Primary / Secondary (Grade 10 Maximum)	Botswana[e]	86	86	0.0	60.6 (5.3)	7.1 (2.8)	441 (6)
	Chile[f]	657	654	0.4	39.5 (1.8)	4.0 (0.7)	413 (2)
	Norway (ALU)[g]	392	392	0.0	88.5 (1.5)	46.9 (2.3)	509 (3)
	Norway (ALU+)[g]	159	159	0.0	96.5 (1.4)	68.7 (3.1)	553 (4)
Group 4. Primary Specialists	Germany	97	97	0.0	96.0 (2.1)	71.7 (7.0)	555 (8)
	Malaysia	576	574	0.4	88.7 (1.1)	28.1 (1.3)	488 (2)
	Poland[a]	300	300	0.0	97.9 (1.0)	91.0 (1.6)	614 (5)
	Singapore	117	117	0.0	98.3 (1.2)	87.3 (2.8)	600 (8)
	Thailand	660	660	0.0	91.7 (0.9)	56.2 (1.4)	528 (2)
	† USA[d]	191	132	33.2	94.9 (1.7)	48.1 (6.5)	520 (7)

Future Teachers (Primary) Mathematics Content Knowledge

1. This table and chart must be read with awareness of the annotations listed earlier.
2. The dagger symbol (†) is used to alert readers to situations where data were available from less than 85% of respondents.
3. The shaded areas identify data that, for reasons explained in the annotations, can be compared with data from other countries with caution.
4. The solid vertical lines on the chart show the two Anchor Points (431 and 516).

Percentiles

5th   25th   75th   95th

Mean and Confidence Interval (±2SE)

Table 3 Descriptive statistics for mathematics content knowledge, by program group (future teacher, lower secondary level) *Source* Tatto et al. (2012)

Program Group	Country	Sample Size	Valid Data (N)	Percent Missing (Weighted)	Percent at or above Anchor Point 1 (SE)	Percent at or above Anchor Point 2 (SE)	Scaled Score: Mean (SE)	Future Teachers (Secondary) Mathematics Content Knowledge
Group 5. Lower Secondary (to Grade 10 Maximum)	Botswana [a]	34	34	0.0	6.0 (4.2)	0.0	436 (7)	
	Chile [b]	746	741	0.6	1.2 (0.4)	0.0	354 (3)	
	Germany	408	406	0.3	53.5 (3.4)	12.6 (2.2)	483 (5)	
	Philippines	733	733	0.0	14.0 (3.0)	0.2 (0.1)	442 (5)	
	Poland [c]	158	158	0.0	75.6 (3.5)	34.7 (3.2)	529 (4)	
	Singapore	142	142	0.0	86.9 (3.1)	36.6 (4.3)	544 (4)	
	Switzerland [d]	141	141	0.0	79.7 (3.4)	26.7 (3.2)	531 (4)	
	Norway (ALU) [e]	356	344	3.9	36.1 (3.7)	2.3 (1.4)	435 (3)	
	Norway (ALU +) [e]	151	148	1.9	19.3 (1.6)	0.8 (0.4)	461 (5)	
	† USA [f]	169	121	32.7	33.5 (2.2)	2.1 (1.3)	468 (4)	
Group 6. Lower & Upper Secondary (to Grade 11 and above)	Botswana a	19	19	0.0	21.1 (7.4)	0.0	449 (8)	
	Chinese Taipei	365	365	0.0	98.6 (0.8)	95.6 (1.0)	667 (4)	
	Georgia [g]	78	78	0.0	18.2 (4.4)	5.0 (2.6)	424 (9)	
	Germany	363	362	0.1	93.4 (1.5)	62.1 (2.9)	585 (4)	
	Malaysia	389	388	0.2	57.1 (2.3)	6.9 (0.9)	493 (2)	
	Oman	268	268	0.0	37.1 (2.7)	1.8 (0.6)	472 (2)	
	Poland	140	139	0.8	85.7 (2.6)	35.7 (2.7)	549 (4)	
	Russian Federation [h]	2141	2139	0.1	88.8 (1.7)	61.1 (4.3)	594 (13)	
	Singapore	251	251	0.0	97.6 (1.0)	62.9 (2.6)	587 (4)	
	Thailand	652	652	0.0	41.0 (1.5)	8.4 (1.1)	479 (2)	
	Norway (PPU & Masters) [e]	65	65	0.0	57.8 (7.9)	16.0 (4.6)	503 (8)	
	† USA [f]	438	354.0	21.3	87.1 (2.0)	44.5 (3.9)	553 (5)	

Mean and Confidence Interval (±2SE)

Notes:
1. This table and chart must be read with awareness of the annotations listed earlier.
2. The dagger symbol (†) is used to alert readers to situations where data were available from less than 85% of respondents.
3. The shaded areas identify data that, for reasons explained in the annotations, can be compared with data from other countries with caution.
4. The solid vertical lines on the chart show the two Anchor Points (490 and 559).

What may help explain these results? Our study shows that the design of teacher education programs and curricula content and orientation may have substantial effects on the level of knowledge that future teachers are able to acquire. In general, programs where future teachers are more successful in our assessments have rigorous standards in selecting those who enter the program, they have a demanding and sequential (versus repetitive) university and school mathematics curriculum, frequent formative evaluations (written and oral), and stringent graduation requirements. A conceptual, problem solving, and active learning orientation seems to characterize the views of mathematics among those future teachers who score higher in our assessments, likely reflecting the way they themselves learned mathematics and the views that their programs espouse (Tatto et al. 2012; Tatto et al. in press).

What could be some of the policy implications emerging from TEDS-M? Teacher education programs can increase their effectiveness by selecting future teachers according to their characteristics (e.g., previous school performance) and strengthening formative and summative evaluation as they progress through their program. In fact previous performance in school, gender and socioeconomic status are characteristics that seemed to explain in some degree the knowledge that future teachers demonstrate at the end of their formal initial teacher education (Tatto et al. in press).

A general conclusion of our analysis is that future teachers, who did well in their previous schooling, and specifically in high school, perform better in our mathematics knowledge for teaching assessments (Tatto et al. in press). In all countries, opportunities to learn university level mathematics and mathematics of the school curriculum, and reading research on teaching and mathematics were related to future teachers' knowledge as measured in our assessments. The more traditional view of mathematics as a finished product has given way to a more contemporary view of mathematics as a process of inquiry (Ernest 1989, p. 250), and to the idea that mathematics is better learned through a conceptual and inquiry-based form of learning. In general, successful programs seemed to be more coherently organized around the idea of what effective teachers need to know (Tatto et al. in press).

For primary programs, the most important positive influence of teacher education on mathematics knowledge for teaching is the opportunity to learn school level mathematics, specifically in the areas of function, probability, and calculus (Tatto et al. in press). Another important yet negative association with knowledge as measured by our assessment was found among future teachers who as a group hold the exclusive view that can be summarized as "mathematics is a collection of rules and procedures that prescribe how to solve a problem". This is a view that stands in contrast with the more accepted view, supported by cognitive science research on learning that, "in addition to getting a right answer in mathematics, it is important to understand why the answer is correct" and that in addition to learning basic facts, "teachers should allow pupils to figure out their own ways to solve a mathematical problem." While the first is a view that may be espoused by teacher education programs, it could also be a "naïve view" held by future teachers based on commonly held "cultural norms" and which remains unchallenged and unchanged by their program. In other words, the program may end up reinforcing traditional ways of teaching and learning, already acquired by future teachers in their own schooling (Tatto 1999).

For secondary programs the most important influence on knowledge for teaching is the opportunity to learn university level mathematics, specifically geometry, and the opportunity to read research in teaching and learning (Tatto et al. in press). As in the primary programs the exclusive view that "mathematics is a collection of rules and procedures that prescribe how to solve a problem" had a negative association with performance in our assessment.

One conclusion of this study is that teacher education programs' quality of opportunities to learn—as measured by their association with high levels of mathematics teaching knowledge, coherence on program philosophy and approaches, and internal and external quality assurance and accountability mechanisms, are all features that seem to contribute to increased levels of mathematics knowledge for teaching among future teachers. While the TEDS-M study is limited in how much it can tell us about the effects of high quality teacher education on initial teaching practice, it provides the basis for the development of further inquiry into this unexplored yet essential question: what elements contribute to the development of high quality teachers?

A further study, FIRSTMATH, will attempt to answer this question. This is a study of novice teachers' development of mathematical knowledge for teaching and the influence of previous preparation, school context and opportunities to learn-on-the-job, on that knowledge. FIRSTMATH will explore the connections between initial teacher education and what is learned on the job as it concerns knowledge, skills, and curricular content; and the degree to which standards, accountability, and other similar mechanisms operate to regulate the support that beginning teachers of mathematics receive during their first years of teaching. For more information on TEDS-M and FIRSTMATH consult the following websites: http://teds.educ.msu.edu/ and http://firstmath.educ.msu.edu/.

Views from Audience Representatives

Finally, the chair asked two Audience representatives their view on "how mathematics (teacher) educators in their country value TEDS-M as a contribution to research."

Mellony Graven (South Africa) highlighted that her country did not participate in TEDS-M (but did in the preceding MT21 study), partially for cost reasons. In South Africa, many teacher educators are unaware of the study, and the local literature on mathematics (teacher) education shows little take up or mention of the study.

Deborah Loewenberg Ball (USA) stressed the importance of TEDS-M: it has advanced the international conversation about what it means to be mathematically well-prepared for teaching, it has raised questions about the degree to which common measures of mathematical knowledge for teaching can be developed, and it has made possible more common research about selection, education, and effects on initial teaching across countries.

The Panel closed with concluding words by the Panel members, expressing thanks to the IPC including the Panel-liaison *Gabriele Kaiser* (Germany) and the local organizers including *Mi-Kyung Ju* (South Korea) as the presider of the Panel.

Open Access This chapter is distributed under the terms of the Creative Commons Attribution Noncommercial License, which permits any noncommercial use, distribution, and reproduction in any medium, provided the original author(s) and source are credited.

References

Adler, J., Ball, D. L., Krainer, K., Lin, F.-L., & Novotna, J. (2005). Mirror images of an emerging field: Researching mathematics teacher education. *Educational Studies in Mathematics, 60*(3), 359-381.

Anthony, G., & Walshaw, M. (2009). *Effective pedagogy in mathematics: Educational practices series, 19.* International Academy of Education, International Bureau of Education & UNESCO. Available online from the UNESCO website.

Baumert, J., Kunter, M., Blum, W., Brunner, M., Voss, T., Jordan, A., Klusmann, U., Krauss, S., Neubrand, M., & Tsai, Y.-M. (2010). Teachers' mathematical knowledge, cognitive activation in the classroom, and student progress. *American Educational Research Journal, 47*(1), 133-180.

Blömeke, S. (2011). WYSIWYG: Von nicht erfüllten Erwartungen und übererfüllten Hoffnungen – Organisationsstrukturen der Lehrerbildung aus internationaler Perspektive. *Erziehungswissenschaft, 43*, 13-31.

Blömeke, S., Kaiser, G., Lehmann, R., & Schmidt, W. H. (2008). Introduction to the issue on empirical research on mathematics teachers and their education. *ZDM – The International Journal on Mathematics Education, 40*(5), 715-717.

Conference Board of Mathematical Sciences (2001). *The mathematical education of teachers (Vol. 2).* Washington, DC: American Mathematical Society.

Connelly, F. M., & Clandinin, D. J. (1988). *Teachers as curriculum planners: Narratives of experience.* New York, USA: Teachers College Press.

Education Committee of the EMS (2012). It is Necessary that Teachers are Mathematically Proficient, but is it Sufficient? Solid Findings in Mathematics Education on Teacher Knowledge. *Newsletter of the European Mathematical Society, March 2012*, Issue 83, 46-50.

Ernest, P. (1989). The impact of beliefs on the teaching of mathematics. In: Ernest, E. (Ed.), *Mathematics Teaching: The State of the Art* (pp. 249-254). New York, USA: Falmer Press.

Fay, R. E. (1989). Theoretical Application of Weighting for Variance Calculation. *Proceedings of the Section on Survey Research Methods of the American Statistical Association*, 212-217.

Fullan, M. (1993). *Change forces. Probing the depths of educational reform.* London: Falmer Press.

Garden, R., Lie, S., Robitaille, D. F., Angell, C., Martin, M. O., Mullis, I. V. S., et al. (2006). *TIMSS Advanced 2008 assessment frameworks.* Chestnut Hill, MA: Boston College.

Hattie, J. A. (2003). *Teachers make a difference: What is the research evidence?* Australian Council for Educational Research Annual Conference on: Building Teacher Quality.

Hill, H. C., Rowan, B., & Ball, D. L. (2005). Effects of Teachers' Mathematical Knowledge for Teaching on Student Achievement. *American Educational Research Journal, 42*(2), 371-406.

Hill, H. C., Sleep, L., Lewis, J. M., & Ball, D. L. (2007). Assessing teachers' mathematical knowledge: What matters and what evidence counts? In F. K. Lester (Ed.), *Second Handbook of Research on Mathematics Teaching and Learning* (pp. 111-156). Charlotte, NC: Information Age.

Hsieh, F.-J., Lin, P.-J., & Shy, H.-Y. (2012a). Mathematics teacher education in Taiwan. In Tso, T. Y. (Ed.), *Proc. 36th Conf. of the Int. Group for the Psychology of Mathematics Education, Vol. 1* (pp. 187-206). Taipei, Taiwan: PME.

Hsieh, F.-J., Lin, P.-J., & Wang, T.-Y. (2012b). Mathematics related teaching competence of Taiwanese Primary Future Teachers: Evidence from the TEDS-M. *ZDM – The International Journal on Mathematics Education*. DOI: 10.1007/s11858-011-0377-7 http://www.springerlink.com/content/074647770n574812/fulltext.pdf.

Huckstep, P., Rowland, T., & Thwaites, A. (2003). Primary teachers' mathematics content knowledge: what does it look like in the classroom? *Education-line database*. Accessed 1 April 2012 < http://www.leeds.ac.uk/educol/documents/00002534.htm>

Krainer, K. (2011). Teachers as stakeholders in mathematics education research. In B. Ubuz (Ed.), *Proc. 35th Conf. of the Int. Group for the Psychology of Mathematics Education, Volume 1* (pp. 47-62). Ankara, Turkey: Middle East Technical University.

Loewenberg-Ball, D., Blömeke, S., Delaney, S., & Kaiser, G. (Eds.) (2012). Measuring teacher knowledge - approaches and results from a cross-national perspective. *ZDM - The International Journal on Mathematics Education, 44*(2). (See http://www.springer.com/education+%26 +language/mathematics+education/journal/11858).

Lohr, L. S. (1999). *Sampling: Design and Analysis.* Pacific Grove, CA: Duxbury Press.

McCarthy, P. J. (1966). Replication: An Approach to the Analysis of Data from Complex Surveys. *Vital and Health Statistics, Series 2, No. 14.* Hyattsville, MD: National Center for Health Statistics.

Mullis, I. V. S., Martin, M. O., Ruddock, G. J., O'Sullivan, C. Y., Arora, A., & Erberber, E. (2005). *TIMSS 2007 Assessment Frameworks.* Chestnut Hill, MA: Boston College.

National Research Council (2010). *Preparing teachers: Building evidence for sound policy (Committee on the Study of Teacher Preparation Programs in the United States, Division of Behavioral and Social Sciences and Education).* Washington, DC: National Academy Press.

Schmidt, W., Blömeke, S., & Tatto, M. T. (2011). *Teacher education matters. A study of middle school mathematics teacher preparation in six countries.* New York, NY: Teachers College Press.

Senk, S. L., Tatto, M. T., Reckase, M., Rowley, G., Peck, R., & Bankov, K. (2012). Knowledge of future primary teachers for teaching mathematics: An international comparative study, *ZDM – The International Journal on Mathematics Education.* DOI: 10.1007/s11858-012-0400-7 http://www.springerlink.com/content/f881483261201gu4/

Shulman, L. S. (1986). Those who understand: Knowledge growth in teaching. *Educational Researcher, 15*(3), 4-14.

Tatto, M. T. (2008). Teacher policy: A framework for comparative analysis. *Prospects: Quarterly Review of Comparative Education, 38*, 487-508.

Tatto, M. T. (1999). The socializing influence of normative cohesive teacher education on teachers' beliefs about instructional choice. *Teachers and Teaching, 5*, 111-134.

Tatto, M. T. (Ed.) (2013, published in 2014). *The Teacher Education and Development Study in Mathematics (TEDS-M). Policy, Practice, and Readiness to Teach Primary and Secondary Mathematics in 17 Countries: Technical Report.* Amsterdam, The Netherlands: International Association for the Evaluation of Student Achievement.

Tatto, M.T., Lerman, S., & Novotná, J. (2010). The organization of the mathematics preparation and development of teachers: A report from the ICMI Study 15. *Journal of Mathematics Teacher Education, 13*, 313-324.

Tatto, M. T., Rodriguez, M. C., & Lu, Y. (in press). *The Influence of Teacher Education on Mathematics Teaching Knowledge: Local Implementation of Global Ideals.* In G. K. LeTendre & A. W. Wiseman (Eds.), Promoting and Sustaining a Quality Teacher Workforce Worldwide. Bingley, United Kingdom: Emerald.

Tatto, M. T., Schwille, J., Senk, S. L., Ingvarson, L., Rowley, G., Peck, R., Bankov, K., Rodriguez, M., & Reckase, M. (2012). *Policy, Practice, and Readiness to Teach Primary and Secondary Mathematics in 17 Countries. Findings from the IEA Teacher Education and Development Study in Mathematics (TEDS-M)*. Amsterdam, The Netherlands: International Association for the Evaluation of Student Achievement.

Tatto, M. T., & Senk, S., Rowley, G., & Peck, R. (2011). The Mathematics Education of Future Primary and Secondary Teachers: Methods and findings from the Teacher Education and Development Study in Mathematics. *Journal of Teacher Education, 62*(2), 121-137.

Wu, M. L., Adams, R. J., Wilson, M. R., & Haldane, S. A. (2007). *ACER ConQuest Version 2: Generalised item response modelling software [computer program]*. Camberwell, Victoria: Australian Council for Educational Research.

Mathematics Education in East Asia

Frederick K.S. Leung, Kyungmee Park, Yoshinori Shimizu and Binyan Xu

Abstract Students in East Asia have been performing extremely well in international studies of mathematics achievements such as TIMSS and PISA. On the other hand, education practices in East Asian countries look different from Western practices, and some practices look very backward and contradictory to what are considered as good practices. Given these intriguing phenomena, this plenary panel aims to discuss different aspects of mathematics education in these East Asian countries, and illustrate its salient features with examples. These aspects include classroom teaching in regular schools and tutorial schools, and pre-service and in-service teacher education and development. The reasons behind the distinctive features of mathematics education in East Asia are then explored, and it is argued that the common Confucian Heritage Culture (CHC) that these countries share best explain these features. This panel presentation is not meant to promote the superior student achievement or good educational practices in East Asia. Rather, it highlights the cultural differences between CHC and Western cultures, rather than the superiority of one over the other. A cultural explanation also means that simple transplant of educational policies and practices from one culture to another will not work. The panel points to the important role culture plays in accounting for educational practices and student achievement.

F.K.S. Leung (✉)
The University of Hong Kong, Pokfulam, Hong Kong
e-mail: frederickleung@hku.hk

K. Park
Hongik University, Seoul, South Korea
e-mail: kpark@hongik.ac.kr

Y. Shimizu
University of Tsukuba, Tsukuba, Japan
e-mail: yshimizu@human.tsukuba.ac.jp

B. Xu
East China Normal University, Shanghai, China
e-mail: xubinyan650@hotmail.com

© The Author(s) 2015
S.J. Cho (ed.), *The Proceedings of the 12th International Congress on Mathematical Education*, DOI 10.1007/978-3-319-12688-3_11

123

Introduction

Students in East Asia have been performing extremely well in international studies of mathematics achievements such as TIMSS and PISA (Beaton et al. 1996; Mullis et al. 1997, 2000, 2004, 2008, 2012; OECD 2001, 2003, 2004, 2010). On the other hand, classroom studies show that mathematics teaching in these countries is rather backward and traditional. International studies on teacher education and development also show that practices in East Asian countries are markedly different from those in "western" countries. Furthermore, comparative studies in teacher knowledge seem to suggest that mathematics teachers in East Asia have more solid understanding of the subject matter as well.

Given these intriguing phenomena, this plenary panel aims to present the current picture of different aspects of mathematics education in these East Asian countries more vividly, and to explore into the reasons behind these distinctive features of mathematics education. In this panel presentation, East Asia is a cultural rather than geographic demarcation. East Asian "countries" refer to systems or economies that are under the influence of the Confucian Heritage Culture, or CHC in short. They include China, Hong Kong, Japan, Korea, Singapore, and Taiwan. The classroom practices, teacher education and development, as well as the educational and socio-cultural contexts in these East Asian countries will be discussed and illustrated with examples.

Classroom Teaching in East Asia

Classroom Teaching in Regular Schools

There have been many studies about the features of mathematics classroom teaching in East Asia. For example, Zhang et al. (2004) stated that the most coherent and visible principle for mathematics instruction in China is emphasizing the importance of foundations, and the principle of "basic knowledge and basic skills" was explicitly put forward for the teaching of mathematics. Gu et al. (2004) claimed that teaching with variation is a Chinese way of promoting effective mathematics learning. According to Gu et al. (2004) which was based on a series of longitudinal mathematics teaching experiments in China, meaningful learning enables learners to establish a substantial and non-arbitrary connection between their new knowledge and previous knowledge. Classroom activities can be developed to help students establish this kind of connection by experiencing certain dimensions of variation. The theory suggests that two types of variation are helpful for meaningful learning, "conceptual variation" and "procedural variation" (Gu et al. 2004).

A number of comparative studies of classroom teaching in East Asian countries and western countries have been conducted, and among them Leung's study provided the most comprehensive interpretation of mathematics teaching in East Asia. In an attempt to search for an East Asian identify in mathematics education, Leung (2001) characterized the salient features of classroom teaching in East Asia and those in the West. He presented six dichotomies of teaching and learning: product (content) versus process; rote learning versus meaningful learning; studying hard versus pleasurable learning; extrinsic versus intrinsic motivation; whole class teaching versus individualized learning; and competence of teachers: subject matter versus pedagogy. Among the six dichotomies, product (content) versus process and whole class teaching versus individualized learning capture best the essence of the differences in mathematics teaching between East Asia and the West.

Two lesson videos were analyzed and discussed in the plenary panel session. As a representative East Asian lesson, an 11th grade Chinese lesson in Shanghai dealing with trigonometric ratio was chosen. In this review lesson, the teacher arranged the mathematics content on trigonometric ratio according to the structure of the knowledge which had already been dealt with in the class, and students accepted and internalized the knowledge structure and reflected on their own understanding. The Chinese lesson shows heavy dependence on teacher's explanation, and the teacher emphasized acquiring mathematics knowledge. Mathematics teaching was analogous to getting the body of knowledge across from the teacher to the students.

For the Western lesson, an 8th grade US lesson in San Diego dealing with linear function was chosen as a representative one. This lesson was characterized as a 'guided development lesson' by the local researchers. The lesson started with some individual activities on exploring the characteristics of functions, and then the teacher invited a student to share his opinion with his classmates. Students were given ample activities and investigations. This lesson seems to support the contemporary Western view that the critical attribute of mathematics is its distinctive way or process of dealing with reality. This process gives rise to a body of knowledge, which is also worthwhile subject matter for study. Since the critical attribute is the process, it is more important to get hold of the process rather than the content arising out of the process.

The Chinese lesson is affirming the importance of the teacher and the subject matter, while student-centered education is the basic tenor in the US lesson. We are not implying that all East Asian countries are on one side of the dichotomies and all western countries are on the other side. In fact, it is a matter of the relative positions of the two cultures on a continuum rather than two incompatible standpoints.

Teaching in Tutorial Schools

It is well known that there are various types of tutorial schools outside the formal educational system in East Asia. These tutorial schools provide supplementary help in academic subjects both for following-up what is taught in regular schools and for preparing for entrance examinations to the next school levels. The content of the

Table 1 Expenditure for private tutoring (2012 data taken from Asian Developmental Bank)

Country	Total expenditure for private tutoring (billion)
Hong Kong	US$0.255
Singapore	US$0.682
Japan	US$12.1
Korea	US$17.3

courses in those schools can be remedial or accelerated. Tutorial schools range from two or three students meeting in the home of a teacher to hundreds of students in dozens of classes in campuses all over the country.

A huge amount of money is involved in private tutoring. The expenditure for some countries in East Asia is shown in Table 1.

There are both advantages and drawbacks in having such institutions. First, tutorial schools help students to learn, and thus extend their human capital which can in turn contribute to economic development. On the other hand, tutorial schools usually maintain or exacerbate social and economic inequalities. Also, tutorial schools may dominate students' lives and restrict their leisure time in ways that are psychologically and educationally undesirable.

Tutorial Schools or Private Tutoring in Japan

Table 2 shows the percentages of Japanese students in grades 6 and 9 who attended tutorial (*Juku*) schools, including lessons with private tutors (Ministry of Education, Science, Sports, and Culture, 2010). Roughly half of grade 6 students attended some form of outside school education and more than 62 % of grade 9 students attended tutorials. In reality, there are some differences between the urban areas and small cities or rural areas in students' attendance. In urban areas, there are large *Juku* schools with a competitive atmosphere mostly attended by students preparing for the university entrance examination. On the other hand, many rural *Juku* schools for elementary and junior high schools are more informal, and basically aim to provide immediate improvement of school performance. Besides *Juku* schools

Table 2 Attendance of grade 6 and grade 9 Japanese students in *Juku* schools, including lessons with private tutors (National Institute for Educational Policy Research 2010)

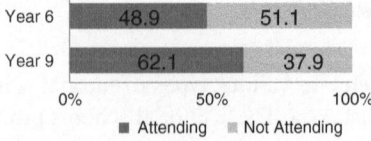

Table 3 Learning in *Juku* schools (National Institute for Educational Policy Research 2010)

Do you study in *Juku* schools (including private tutors)?	6th graders	9th graders
(1) Not attending	52.1	37.9
(2) Learning advanced content or difficult topics	23.5	18.1
(3) Learning the topic taught but not well-understood in schools	7.5	10.0
(4) Both (2) and (3)	8.5	25.9
(5) Others	8.2	7.9

which provide supplementary help in academic subjects, there are enrichment classes on other activities such as swimming, piano, or abacus.

Table 3 shows the various purposes for attending *Juku* schools. As Table 3 shows, in general learning advanced content or difficult topics is the major purpose of the Japanese students' attendance.

Two Japanese tutorial schools were described in the plenary panel session, one mainly for elementary and junior high school students, and the other mainly for senior high school students. They have different courses and systems. The first tutorial school is a *Juku* School in Tsukuba City, and the number of students is roughly 400. The school offers "afterschool classes" in weekday evenings for teaching advanced topics, and they provide a bus service to pick up students. The school runs a "Study Camp" every year during the summer vacation, where students stay in a hotel for a few days and learn together.

The other school belongs to an affiliated group of tutorial schools of more than 120 schools all over the country. The school is for university intended senior high school students who prepare for the entrance examination to universities. It provides students with an ICT-enhanced self-learning system that emphasizes a PDCA (Plan-Do-Check-Action) cycle for learning with immediate feedback. All the lectures are delivered through a Local Area Network. Each student comes to the tutorial school after their regular school class and learns with a computer. The progress of their learning is monitored by the teachers at the school and the students have the opportunities for consulting with the teachers periodically to discuss about their choice of intended university and so on.

Tutorial Schools in Korea

Korea conducts a national survey annually on tutorial schools. Based on the survey done in 2011 with 46,000 students and parents, 50.2 % of elementary and secondary students were participating in mathematics tutorial schools. This rate was the highest among all the subjects.

There are a variety of tutorial schools in Korea according to the achievement levels of the students, their purposes of attending tutorial schools, etc.:

- Repetition of school mathematics content
- Accelerated learning

- Preparing for mathematics contests or the Mathematics Olympiad
- Preparing for entrance examinations of gifted schools

To reduce the country's addiction to private, after-hours tutoring academies (called *hagwons*), the authorities have begun enforcing a curfew to stop children from studying in *hagwons* after 10 p.m. (TIME magazine, 25 Sep 2011).

Teacher Education and Development

"The success of any plan for improving educational outcomes depends on the teachers who carry it out and thus on the abilities of those attracted to the field and their preparation" (National Research Council 2010, p. 1). In East Asia, respecting teachers and attaching importance to education are an unchanging theme and a traditional virtue (Wang 2012). Teachers play the role of a guide, and instruction is teacher dominated and student involvement is minimal (Leung 2001). On the other hand teachers try to understand their students' learning and want their students to be happy in the future, which means that they need to work hard in school (Ferreras et al. 2010). They bear the responsibility if students do not study hard or work well. One of the Chinese idioms illustrates this typical characteristic of teachers in East Asia: Unpolished jade never shines; To teach without severity is the teacher's laziness (玉不琢, 不成器; 教不严, 师之惰).

In the following section, how teacher preparation and development in East Asia are carried out will be presented.

Pre-service Education: How to Become a Mathematics Teacher in East Asia

There are diversities in terms of the mechanism for preparing teachers. Some East Asian systems (such as in Korea or in Mainland China) provide an integrated approach where prospective teachers acquire a teacher certificate through a four-year bachelor degree program at a comprehensive university or teacher education university. Some systems (such as Hong Kong or Japan) adopt an end-on approach where prospective teachers complete a bachelor degree and then take a one- or two-year Post Graduate Certificate in Education program. Notwithstanding these differences, some similar characteristics of pre-service teacher training in teachers colleges and normal universities can be summarized as follows (Li et al. 2008, p. 70):

- Providing prospective teachers with a solid foundation of mathematical knowledge and advanced mathematical literacy;

- Emphasizing the review and study of elementary mathematics. It is believed that a profound understanding of elementary mathematics and strong problem-solving abilities in this field are crucial to becoming a qualified mathematics teacher.

The model in each system has its own strengths and weaknesses with regard to acquiring subject matter knowledge, pedagogical knowledge and teaching skills, but they share similar characteristics. The contents of the mathematics teacher preparation programs in some selected institutions are shown in Table 4.

As can be seen in Table 4, the Korean (minimum 30 %), Chinese (41 %) and Japanese (33 %) programs emphasize the foundations of mathematics knowledge in terms of its systematic structure, and the demand for logical reasoning. These features could reflect the belief that high quality teaching requires that teachers have a deep knowledge of the subject matter. But, the ways such a belief is reflected in practice depend on the specific contexts found in different countries.

Most of the systems require prospective teachers to obtain a government-issued certificate or license signifying that the candidates have completed the required professional preparation. In many systems, candidates also need to take a teacher employment test, and there is an emphasis on subject matter knowledge in this test in different countries.

Table 4 Outline of Teacher preparation courses for secondary mathematics majors by selected institutions

	Mathematics (%) (required and elective) (e.g. Linear algebra, number theory, real analysis, complex analysis, differential geometry, topology, probability and statistics)	Mathematics education (%) (e.g. Methodology of mathematics education, curriculum in mathematics education, problem solving and mathematics competition)	General pedagogy (%) (e.g. Philosophy of educational and history of education, curriculum and evaluation, educational method and technology, educational psychology)	Teaching practicum (%)	General or other courses (%) (e.g. Foreign language, health and sports subjects)
China[1]	41	8	10	12	29
Japan[2]	33	15	16	10	26
Korea[3]	30	6	13	3	48

[1] East China Normal University
[2] Hiroshima University
[3] Specified by the MOE of Korea (minimum units. Most students take more mathematics, mathematics education and general education courses)

Teacher Employment Test (TET) in Korea

In Korea, to be employed by national and public schools, a certified teacher must pass the teacher employment test administered by the 16 Metropolitan and Provincial Offices of Education (Ingersoll 2009, p. 58). The competition rates for mathematics in TET differ from one school district to another, but the average competition rate is higher than 10:1, i.e., more than 10 candidates compete for one place.

In the TET administered by the MOE of Korea, the core subjects are 'mathematics', 'mathematics education', and 'general pedagogy'. To examine whether a prospective teacher has successfully developed the practical competency to teach in the classroom, the TET consists of three stages. Table 5 shows the core subjects in the three stages of the examination.

In the first stage, the TET includes 26 questions about mathematics (52 %), 14 questions about mathematics education (28 %), and 40 questions about general pedagogy (20 %) in the form of multiple choice items. In the second stage, the test

Table 5 The core subjects and three stages of the Korean TET

Area	Contents	Relevant knowledge	Percent		
			Stage 1 (%)	Stage 2 (%)	Stage 3 (%)
Mathematics	Linear algebra	Content knowledge	52	55–60	0
	Abstract algebra				
	Number theory				
	Real analysis				
	Complex analysis				
	Differential geometry				
	Topology				
	Probability and statistics				
	Discrete mathematics				
Mathematics education	Mathematics curriculum and Evaluation, History of mathematics education, Theory of instruction in mathematics, psychology of teaching mathematics	Pedagogical content knowledge	28	35–40	60
General pedagogy	Philosophy of education and history of education, curriculum and evaluation. educational method and technology educational psychology, educational sociology, educational administration and management	General pedagogical knowledge	20	0	40

sets four questions from mathematics (60–65 %) and mathematics education (35–40 %) in the form of essay items; there are no questions about general pedagogy.

In the final stage, the TET assesses candidates by in-depth interview and micro-teaching. In the interview, candidates are given a set of questions related to practical issues involving school teaching such as class management and administration issues. In micro-teaching, candidates are asked to develop a teaching plan for a given mathematical topic. They are required to integrate certain instruction features such as using ICT and collaborative learning into the plan. After they set up their plans, they conduct micro-teaching based on the plans for 20 min. The final decision of teacher selection is based on the cumulative scores through the three stages.

Employment Test in Japan

Due to a decline in the school age population in Japan in recent years, the job opportunities for prospective teachers are limited and only about 30–40 % of graduates of teacher training colleges are able to secure employment in public schools. In principle, mathematics teachers at secondary schools teach only mathematics, whereas teachers at elementary schools teach most subjects. Because of this difference, more courses in pedagogy are required for those intending to teach at the lower grade levels, whereas those intending to teach at the upper grade levels are required to take more mathematics. In addition to the academic course work, teacher-training programs include a practicum (teaching practice). Prospective elementary school teachers are required to spend at least four weeks in a school for teaching practice and those for lower and upper secondary school are required to spend at least two weeks. The practicum is usually preceded and followed by a total of 15–30 h of related guidance and reflections. The national universities for teacher training have affiliated schools for the purpose of teaching practice.

The board of education of each prefecture gives a teacher certificate to a person who has completed the prescribed basic qualifications and credits at the authorized colleges and universities. The competition rates of Teacher Employment differ among school levels and from one school district (prefecture) to another, but the average competition rate was about 6:1 in 2011 (see Table 6).

Table 6 Applicants, employees, and competition rate in 2011 by school levels (*data source* Ministry of Education of Japan, as of 1 June 2011)

School level	Applicants	Those who took the test	Employees	Competition rate
Primary school	63,800	57,817	12,882	4.5
Lower secondary school	71,212	63,125	8,068	7.8
Upper secondary school	42,506	37,629	4,904	7.7

For some prefectures, the average competition rate is more than 10 (Iwate 13.6; Nagasaki 13.3), for others, less than 10 (Tokyo 5.7; Toyama 3.7). Each prefecture prepares and conducts an employment test that is conducted at two phases. The first phase is a paper and pencil test (one day in July), and the test subjects consist of general education, mathematics, and mathematics education. The second phase is an interview and micro-teaching (around October).

Teaching Skills Competition for Prospective Mathematics Teachers in China

In China, the mathematicians in teacher education institutions still value the structure and nature of mathematics, and hope to provide students with a refined and profound mathematics foundation, a broad and concise mathematics background, and further try to help students to master mathematics more easily and properly. And they leave the responsibility of connecting higher mathematics to elementary mathematics and the responsibility of providing high quality mathematics pedagogical knowledge to mathematics educators. Furthermore, enhancing the teaching skills of prospective teachers becomes an important part of the teacher preparation program.

At the end of 1996, the Ministry of Education issued "Suggestions on Teacher Education Reform and Development", emphasizing curriculum reform in order to face the challenges of the 21st Century. Much importance was attached to the cultivation of scientific thinking and methods, as well as the practical and creative abilities of students, to establish stable bases for teaching practices (Yang et al. 2012, p. 212). Since then different kinds of practice-oriented pre-service programs have been launched and carried out.

Since 2008, the Department of International Cooperation and Exchanges, and the Department of Teacher Education of the Ministry of Education in China, together with Toshiba Company, have been organizing annual competitions on "practice in innovative teaching skills". Students from normal universities/colleges can participate in this competition, but they should first win the local competitions organized by their universities. Only a few students have the honor to take part in the national competition. This competition includes three parts: a lesson plan (*jiao an*) is designed, the candidates teach a lesson (*mo ni ke*), and after that they should explain the didactical concepts of their lesson (*shuo ke*). Through such competitions, most prospective teachers engage in being trained in teaching skills. Many universities/colleges invite excellent school mathematics teachers to tutor the prospective teachers for these competitions (Fu and Han 2010).

In-service Teacher Education and Development

The success of an education system depends on the appropriate preparation and continuous development of highly qualified teachers. It is widely recognized in East

Asian education communities that learning to teach in the classroom is a life-long process for teachers. As pointed out above, for becoming a mathematics teacher in this region, it is necessary to acquire a teacher certificate for a particular type of schools by completing credits in teacher training courses offered by universities and colleges. Besides these formal systems of teacher preparation, there are other important aspects in the process of mathematics teacher education in East Asia (Leung and Li 2010; Li and Shimizu 2009). In this section, some characteristics of in-service mathematics teacher education and development in East Asia are described.

Stigler and Hiebert (1999) suggested that it is important to examine and learn the ways employed to improve the quality of mathematics classroom instruction in high-achieving education systems in East Asia. A good example is lesson study, which is now familiar to educators around the world. Lesson study is an important practice utilized in Japan to improve the quality of mathematics instruction and to develop teaching competence by promoting collaboration among teachers (Fernandez and Yoshida 2004). There are many other approaches developed and used in the pursuit of excellence in teacher development in different education systems in East Asia. For example, the model of exemplary lesson development is developed and used in mainland China (Huang and Bao 2006). Instructional contests are organized to identify and promote excellent mathematics instruction in several educational systems (e.g., Li and Li 2009; Lin and Li 2009). Master teachers are also an important part of the teaching culture in some education systems in East Asia, and play an important role in nurturing that culture (Li et al. 2008). Some examples of these approaches are provided below.

Lesson Study in Japan

Lesson study, originated in Japan, is a common element in approaches to professional developments whereby a group of teachers collaborate to study the subject matter, instruction, and how students think and understand in the classroom. The original term for lesson study, *jugyo kenkyu* in Japanese, literally means the study of lesson. The origin of lesson study can be traced back to late 1890s, when teachers at elementary schools affiliated to the normal schools started to study lessons by observing and examining them critically (Inagaki 1995). Groups of teachers started to have study meetings on newly proposed teaching methods. The original way of observing and examining lessons has spread nationwide with some major refinements and improvements. The activities of lesson study include planning and implementing the "research lesson" as the core of the whole activity, followed by post-lesson discussion and reflection by participants. A lesson plan plays the key role as a medium for the teachers to share and discuss the ideas to be examined through the process of lesson study.

Lesson study takes place in various contexts (Shimizu 2002). Pre-service teacher-training programs at universities and colleges, for example, include lesson study as a crucial and challenging part in the final week of student teaching practice.

In-service teachers also have opportunities to participate in lesson study. It may be held within their schools, outside their schools but in the same school district, city or prefecture, and even at the national level. Teachers at university-affiliated schools that have a mission to develop a new approach to teaching often open their lesson study for demonstrating an approach or new teaching materials they developed.

Lesson study is a problem solving process whereby a group of teachers work on a problem related to a certain theme. The theme can be related to examining the ways for teaching a new content or for using new teaching materials in relation to the revision of the national curriculum guidelines or to assessing students' learning of a certain difficult topic in mathematics such as common fractions or ratio. The first step of lesson study is defining the problem. In some cases, teachers themselves pose a problem to be solved, such as how to introduce the concept of common fractions, or what is an effective way to motivate students to learn mathematics. Second, planning lesson follows after the problem is defined. A group of teachers collaboratively develop a lesson plan. A lesson plan typically includes analysis of the task to be presented and of the mathematical connections both between the current topic and previous topics (and forthcoming ones in some cases) and within the topic, anticipation of students' approaches to the task, and planning of instructional activities based on them. The third step is a research lesson in which a teacher teaches the planned lesson with observation by colleagues. In most cases, a detailed record of teacher and student utterances is taken by the observers for discussion in a post-lesson meeting. Evaluation of the lesson in the post-lesson meeting focuses on issues such as the role of the implemented tasks, students' responses to the tasks, appropriateness of the teacher's questioning, and so on. Based on the evaluation of the lesson, a revised lesson plan is developed, and the lesson is taught again in another class. These entire process forms a cycle of lesson study.

In lesson study, an outside expert is often invited as an advisor who facilitates and makes comments on the improvement of the lesson in the post-lesson discussion (Fernandez and Yoshida 2004). The expert may be an experienced teacher, a supervisor, a principal of a different school, or a professor from a nearby university. In some cases, the expert is not only invited as a commentator in the discussion on site, the group of teachers may meet with him/her several times prior to conducting the research lesson to discuss issues such as reshaping the objective of the lesson, clarifying the role of the task to be posed in the classroom, anticipating students' responses to the task, and so on. In this context, the outside expert can be a collaborator who shares the responsibility for the quality of the lesson with the teachers, and not just an authority who directs the team of teachers.

After researchers in the U.S. introduced lesson study to the mathematics education community during the late 1990s, the term "lesson study" spread among researchers and educators in the U.S. and later around the world (Lewis 2002). One of the most influential books that discusses about lesson study is *The Teaching Gap* (Stigler and Hiebert 1999). Since then, school teachers in different countries have been trying to implement lesson study in their own education systems. A central question in the "adoption" of the lesson study approach in other places has been raised from the perspective of teaching as a cultural activity.

In the Japanese education system, improvement of teaching and learning through lesson study over a long period of time can take place within a context in which clear learning goals for students are shared among teachers in relation to the national curriculum standards as well as the voluntary hard efforts of the teachers with the support of the administrators. There are challenges to be resolved in practice and research possibilities to be explored in each context.

Teaching Research Groups and Mentorship in China

In the Chinese mainland, almost all mathematics teachers are involved in teaching research activities from the first day of their service, in order to obtain practical knowledge and achieve in-service professional development. This is guaranteed by the policy of "the four-level teaching research network comprising about 100,000 officers" (Yang et al. 2012, p. 216). These officers play an important role in China's education system in managing and guiding school-based teaching research activities on the one hand, and bridging the gap between teaching theories and practice on the other.

The basic units of teaching research network activities are teaching research groups and a mentoring system. They cater to the practical needs and professional development of in-service teachers.

Teaching Research Groups in China

Chinese teachers have a tradition of discussing and reviewing each other's lessons, and gradually it has become a unique culture of opening up one's classroom and discussing one's teaching with others. All the schools in China have teaching research groups, and teachers observing and discussing each other's lessons is commonly guaranteed by the teaching research system. There is more than 50 years' history since a school-based teaching research system was set up in China. In *Secondary School Teaching Research Group Rulebook (draft)* issued by MOE in 1957, the study function of the teaching research group was emphasized: "A Teaching Research Group is an organization to research teaching. It is not an administrative department. Its task is to organize teachers to do teaching research in order to improve the quality of education, and not to deal with administrative affairs" (Ministry of Education 1957).

Facing challenges of curriculum reform since the 21st Century, the school-based teaching research system is experiencing changes. The changes result not only from changes in the way of teaching and the way of research, but also from changes in the way of learning and the way of experiencing for teachers. The current essential activities of teaching research groups include:

- Action research on classroom teaching to improve effectiveness, whereby several practical research methods are developed, such as analyzing crucial teaching events (Yang 2009), classroom observation (Huang and Zhang 2011), and so on.

- Development of a distinctive teaching research culture to build up a teacher community through promoting helping each other and inquiring cooperatively (Yao 2010), or to construct a learning environment to promote teachers' professional development in teaching practice through the learning of teaching theories and the analysis of classroom teaching case studies (Gu and Wang 2003).
- Discussion of mathematics contents and corresponding teaching methods to deepen understand and to modify teaching plans [even though this is one of the typical activities, it is facing new challenge because of students' development (Wang 2011)].

Mentoring for Mathematics Teaching in China

Chinese schools have a tradition of arranging for an experienced teacher to be the mentor for a young teacher when the later just begins the teaching career. In this mentoring system, sometimes a new teacher has two mentors: one provides instructions on teaching and another provides guidance on tutoring students. The experienced teacher (mentor) should undertake the responsibility to discuss teaching methods, teaching contents and students' learning styles, etc., with the novice teacher supervised by him/her. The new teacher is expected to observe the mentor's lessons frequently and learn from him/her enthusiastically and humbly. The school encourages new teachers to conduct open lessons regularly and to participate in teaching contests (Yang et al. 2012). The mentor should try to do co-teaching and hold lesson discussion meetings with the mentee, and to suggest alternative teaching practices and ideas (Mao and Yue 2011). In some schools, a ceremony is even held to honor mentors of new teachers and to award them with mentoring certificates.

Mathematics Festival in Korea

The Ministry of Education in Korea provides compulsory in-service teacher training programs, which Korean teachers should take when they are in the 4th or 5th year of teaching. However many teachers are not satisfied with this teacher training program because it is not very relevant to their classroom teaching. Thus mathematics teacher organizations set up their own teacher training program called 'mathematics festival', and this program has been very successful. Mathematics teachers pay the participation fees from their own pocket.

Mathematics festival is a four-day program, and it consists of a variety of lectures and workshops. The lectures mostly combine theory with its application to classroom teaching. Workshops deal with practical teaching ideas including teaching/learning material, manipulatives, teaching tips, etc. Here are examples of lectures and workshops in the 2012 mathematics festival held in January.

- .How to teach circumcenter in grade 8
- Harmonics of saxophone from the perspective of mathematics
- Interdisciplinary approach: STEAM (Science, Technology, Engineering, Art, and Mathematics)
- Mathematics magic
- Geogebra, GSP 5.0, Cabri 3D
- Lecture about millennium problems
- Lecture about pentomino with participants' hands-on experience
- Lecture about real world situation (height of shoes)
- Lecture about mathematics and music with musical performance
- Computer session with Geogebra
- Computer session with Cabri 3D
- Hands-on experience to make a traditional 3-dim figure
- Zonodom.

Discussion

As mentioned earlier, this presentation is not meant to show that all East Asian countries are on one side of the extreme and all Western countries are on the other side. But the presentations above do show that there are distinctive features in the classroom teaching and teacher education and development in East Asian countries which are markedly different from the corresponding practices in Western countries. What are the causes of these differences?

Confucian Heritage Culture

There are obviously factors at the personal and institutional levels that have caused the differences. But explanation won't be complete without resorting to factors at the socio-cultural levels. China, Korea, Japan share a common culture, the Confucian Heritage Culture (CHC) (Biggs 1996a). A major characteristic of CHC is the social orientation of its people, in contrast to individual orientation typically found in Western societies. Social Orientation is a "tendency to act in accordance with external expectations or social norms, rather than with internal wishes or personal integrity" (Yang 1981, p. 161). It emphasizes integration and harmony, in contrast to independence and individualism in Western cultures (Taylor 1987, p. 235). People in CHC treasure the community, much more so than the individual. Related characteristics of CHC include compliance, obedience, respect for superiors, and filial piety (孝).

Another more relevant characteristic in the Confucian culture is its emphasis on education, and CHC parents are known to attach great importance to the education and achievement of their children. This rests upon the Confucian presumption that everyone is educable [differences in intelligence... do not inhibit one's educability (Lee 1996, pp. 28–29)] and perfectible ["sagehood is a state that any man can achieve by cumulative effort" (Chai 1965)]. This in turn motivates CHC learners to exercise their effort and will power in their study.

On CHC's emphasis on the community, of course it is the individual who learns, so effective teaching must address the needs of the individual. But too much stress on the individual may exaggerate and aggravate the individual differences that already exist. Also, human beings are social beings, and learning almost always takes place in a social context. Western societies may have gone too far in their attempt to care for the individual, and an optimal emphasis on the individual's role in the community may provide important incentives to learn.

Characteristics of CHC Related to Mathematics Learning

Examination Culture

China is the first country in the world where a national examination system was instituted (Sui Dynasty, A.D. 600). Examinations have always been the route for upward social mobility, and there is a great trust in examination as a fair method of differentiating between the able and the less able. Examination has acquired the status of something of value in itself and becomes an important incentive for studying.

Belief in Effort

In CHC, studying is considered a hardship: one should persevere in order to succeed, and is not supposed to "enjoy" the studying. "Asian parents teach their children early that the route to success lies in hard work" (Stevenson 1987), and this is consistent with the old Chinese saying that "Diligence compensates for stupidity" (以勤补拙). There is a much stronger attribution of success and failure to internal and controllable factors (effort) rather than incontrollable ones (innate ability). This is consistent with the strong belief in effort (or *Gambaru*, which means pushing on, persisting, not giving up) in Japan. Japanese teachers invariably tell parents that "it would be good if the child would just *gambaru* a little more" (White 1987, p. 30).

The Japanese also emphasize on self-discipline (*Kuro*). The idea of self-discipline in Japan is slightly different from that in the West. One should do one's best and keep on struggling, even when being unsuccessful in the end. But this is not a pointless sacrifice. In Japan pushing on, persisting and not giving up are in themselves considered important. The way something is done is more important than the accomplishment in the end.

Stress on Memorization and Practice

Liu (1986) observed the following beliefs in CHC:

> If the purpose is to acquire the knowledge contained in an article, then the best strategy is to memorize the article. ... If the purpose is to acquire any new cognitive skill, then the best strategy is to practice repeatedly (Liu 1986, pp. 80–82).

This however does not imply rote learning or rule out creativity. As Biggs observed, "the Chinese believe in skill development first, which typically involves repetitive, as opposed to rote learning, after which there is something to be creative with" (Biggs 1996a, p. 55).

Stress on Reflection

In the Confucian tradition of learning, there is a also strong emphasis on reflection, as the saying "Seeking knowledge without thinking is labour lost; thinking without seeking knowledge is perilous" (学而不思则罔, 思而不学则殆) shows. A true Confucian scholar is one who dedicates himself to studying or seeking knowledge through a lot of practice and memorization. But he also constantly reflects upon what he is practicing and memorizing until he fully grasps the knowledge.

Discussion

Students should enjoy their studies, but they should be taught to rediscover the satisfaction which comes only after hard work. Practice, examination and memorization, when done properly, may each have a place in education. Practice and memorization should not be equated with rote learning, and examination is not a necessary evil. If conducted properly, it provides a good incentive for studying.

The Chinese Language

The Japanese and Korean languages are strongly influenced by the Chinese language. For example, the Japanese language still uses a lot of *Kanji* (Chinese characters). There are features of the Chinese language which are favourable to the learning of mathematics. For example, the Chinese language uses classifiers between every cardinal number and the objects being quantified. This "unscramble the confusion that otherwise surrounds conservation of numbers ... explicitness and pragmatic retention of the essential semantic elements in the vocabulary it uses for mathematics" (Brimer and Griffin 1985, p. 23). The regular number system in Chinese also enhances the learning of arithmetic.

As for spoken Chinese, it is a monosyllabic language, where one syllable constitutes one morpheme. In particular, the short pronunciation of the numbers zero to ten makes it easy to process. As Hoosain observed, "the shortest average pronunciation duration of a Chinese number is 265 ms, significantly shorter than the corresponding average of 321 ms in pronouncing a number in English" (Hoosain 1984).

For written Chinese, it is logographic in nature. Chinese words are represented by a large number of different visual symbols known as characters, which are made up of components (radicals), and have an imaginary square as a basic writing unit. Chinese characters put emphasis on the spatial layout of strokes, and the orthography of Chinese is based on the spatial organization of the components of the characters. Lai (2008) pointed out that Chinese characters possess visual properties such as connectivity, closure, linearity and symmetry which are faster and easier to be captured by vision. Studies show that there is a close relationship between the visual-spatial properties of Chinese characters and Chinese people's childhood experience with learning the Chinese orthography. Lai (2008) found that 5 year old Chinese children in Hong Kong, compared to English speaking 5 year olds in Australia, have higher visual perceptual and geometric skills, and higher visual-motor integration skills than motor-reduced visual perceptual skills. Lai used both the motor control theory and the psychogeometric theory of Chinese-character writing to account for the surprising results. It seems that the experience of writing Chinese characters influences one's visual perceptual skills.

Implications

The superior performance of East Asian students in international studies in mathematics naturally prompts one to ask what can be learned from it, especially when one is facing grave problems in mathematics education in the home country. Some education practices in East Asian countries look different from Western practices, and some practices look very backward and contradictory to what are considered as good practices. Biggs (1996b) introduced the term Chinese Learner's Paradox to describe this contradictory phenomenon. But the phenomenon is a paradox only for someone who does not understand the culture. For someone in the culture, education is so important an endeavour that of course students are expected to do well. Compared to students in some other cultures, CHC students work relatively hard, and it is just natural that they do better in these international studies.

This panel presentation is not meant to promote the high achievement of East Asian students, or good educational practices in East Asia, or the superiority of the CHC. It is meant to highlight the cultural differences between CHC and Western cultures, rather than the superiority of one over the other. Theoretically, it hints at the important role culture could play in accounting for educational practices and student achievement. Practically, it provides references for educators in other cultures on education policies and practices. But if culture does impact upon educational practices and student achievement, a cultural explanation also means that simple transplant of educational policies and practices from one culture to another won't work. One can imitate the practices, but cannot transplant the culture, and most practices are effective only in the culture concerned.

Conclusion

In learning from another country, it is important to take any cultural differences that may exist into consideration, and then determine how much can or cannot be learned from another culture. There is a Chinese saying, "Knowing yourself and knowing others, then you will win every battle" (知己知彼, 百战百胜). In learning from another country or region, we should "know others"—not just the student achievement, not just the educational practices, but also the cultural values behind the practices. One should also know oneself—knowing or reflecting upon one's own cultural values. Then one will win any battle in this war of improving mathematics education in one's own country.

Open Access This chapter is distributed under the terms of the Creative Commons Attribution Noncommercial License, which permits any noncommercial use, distribution, and reproduction in any medium, provided the original author(s) and source are credited.

References

Beaton, A.E., Mullis, I.V.S., Martin, M.O., Gonzalez, E.J., Kelly, D.L. & Smith, T.A. (1996). *Mathematics Achievement in the Middle School Years*. Chestnut Hill, MA: International Study Center, Boston College.

Biggs, J.B. (1996a). Western Misconceptions of the Confucian-Heritage Learning Culture. In D.A. Watkins & J.B. Biggs (eds.). *The Chinese Learner*. Hong Kong: Comparative Education Research Centre: 45-67.

Biggs, J.B. (1996b) Approaches to Learning of Asian Students: A multiple paradox. In J. Pandy, D. Sinha & P.S. Bhawuk (eds.), *Asian Contributions to Cross-Cultural Psychology*. New Delhi : Sage: 180-199.

Brimer, A. & Griffin, P. (1985). *Mathematics Achievement in Hong Kong Secondary Schools*. Hong Kong: Centre of Asian Studies, University of Hong Kong.

Chai, C. (ed. & tr.) (1965). *The Humanist Way in Ancient China: Essential works of Confucianism*. New York: Bantam Books.

Fernandez, C. & Yoshida, M. (2004). *Lesson Study: A Japanese approach to improving mathematics teaching and learning*. Mahwah, NJ: Lawrence Erlbaum Associates.

Ferreras, A., Olson, S. & Sztein, E. (2010). *The Teacher Development Continuum in the United States and China: Summary of a Workshop*. U.S. National Commission on Mathematics Instruction. Washington, DC: National Academy Press. http://www.nap.edu/catalog/12874. html

Fu, G.S. & Han, G.F. (2010). A Research on Current Situation and Problems in Normalien Teaching Ability – Teaching ability competition of Normalien as a sample. *Contemporary Teacher Education*, 3,3: 26-30.

Gu, L., Huang, R. & Marton, F. (2004). Teaching with Variation: A Chinese way of promoting effective mathematics learning. In L. Fan, N. Wong, J. Cai, & S. Li (eds.), *How Chinese Learn Mathematics: Perspectives from insiders*. River Edge, NJ: World Scientific Publishing Co.: 309-347.

Gu, L.Y. & Wang, J. (2003). Teachers' Professional Development in Action Education. *Curriculum-Textbook-Pedagogy*, 1-2: 2-10.

Hoosain, R. (1984). Experiments on Digit Spans in the Chinese and English Languages. In Kao, H.S.R. & Hoosain, R. (eds.), *Psychological Studies of the Chinese Language*. Hong Kong: Chinese Language Society of Hong Kong.

Huang, R., & Bao, J. (2006). Towards a Model for Teacher Professional Development in China: Introducing *keli. Journal of Mathematics Teacher Education*, 9: 279-298.

Huang, Y.L. & Zhang, M. (2011). Improve the Effectiveness of Mathematics Teaching – Based on the Action Study of Classroom Observation by Research Group. *Journal of Mathematics Education*, 20,3: 67-70.

Inagaki, T. (1995). *A Historical Research on the Theory of Teaching in Meiji-Era* (in Japanese). Tokyo: Hyuuron-Sya.

Ingersoll, R. M. (2009). *A Comparative Study of Teacher Preparation and Qualifications in Six Nations*. The Consortium for Policy Research in Education (CPRE), 58.

Lai, M.Y. (2008). *An Exploratory Study into Chinese and English Speaking Children's Visual Perception*. Ph.D. Thesis, The University of Hong Kong.

Lee, W.O. (1996). The Cultural Context for Chinese Learners: Conceptions of Learning in the Confucian Tradition. In D.A. Watkins & J.B. Biggs, *The Chinese Learner*. Hong Kong: CERC and ACER: 25-41.

Leung, F.K.S. (2001). In Search of an East Asian Identity in Mathematics Education. *Educational Studies in Mathematics*, 47,1: 35-51.

Leung, F.K.S. & Li, Y. (eds.) (2010). *Reforms and Issues in School Mathematics in East Asia: Sharing and understanding mathematics education policies and practices*. Rotterdam: Sense Publishers.

Lewis, C. (2002). *Lesson Study: A handbook of teacher-led instructional change*. Philadelphia: Research for Better Schools.

Li, S.Q., Huang, R.J. & Shin, H. (2008). Discipline Knowledge Preparation for Prospective Secondary Mathematics Teachers: An East Asian Perspective. In: P. Sullivan & T. Wood (eds). *International Handbook of Mathematics Teacher Education: Knowledge and Beliefs in Mathematics Teaching and Teaching Development*. Rotterdam: Sense Publishers: 63-86.

Li, Y., & Li, J. (2009). Mathematics Classroom Instruction Excellence through the Platform of Teaching Contests. *ZDM - The International Journal on Mathematics Education,* 41,3: 263-277.

Li, Y. & Shimizu, Y. (2009). Exemplary Mathematics Instruction and Its Development in Selected Education Systems in East Asia. *ZDM - The International Journal of Mathematics Education*, 41,3: 257–262.

Lin, P. J., & Li, Y. (2009). Searching for Good Mathematics Instruction at Primary School Level Valued in Taiwan. *ZDM - The International Journal on Mathematics Education*, 41,3: 363-378.

Liu, I.M. (1986). Chinese Cognition. In Bond, M.N. (ed.), *The Psychology of the Chinese People*. Hong Kong: Oxford University Press.

Mao, Q.M. & Yue, K. (2011). Dilemma and Solution of Teacher Learning Based on "Apprenticeship". *Research in Educational Development*, 22: 58-62.

Ministry of Education (1957). *Secondary School Teaching Research Group Rulebook (draft)*. http://www.ncct.gov.cn/plus/view.php?aid=354 (2013.01.01)

Mullis, I.V.S., Martin, M.O., Beaton, A.E., Gonzalez, E.J., Kelly, D.L. & Smith, T.A. (1997). *Mathematics Achievement in the Primary School Years*. Chestnut Hill, MA: International Study Center, Boston College.

Mullis, I.V.S., Martin, M.O., Gonzalez, E.J., Gregory, K.D., Garden, R.A., O'Connor, K.M., Chrostowski, S.J. & Smith T.A. (2000). *TIMSS 1999 International Mathematics Report*. Chestnut Hill, MA: International Study Center, Boston College.

Mullis, I.V.S., Martin, M.O., Gonzalez, E.J., & Chrostowski, S.J. (2004). *TIMSS 2003 International Mathematics Report*. Chestnut Hill, MA: TIMSS & PIRLS International Study Center, Boston College.

Mullis, I.V.S., Martin, M.O. & Foy, P. (2008). *TIMSS 2007 International Mathematics Report*. Chestnut Hill, MA: TIMSS & PIRLS International Study Center, Boston College.

Mullis, I.V.S., Martin, M.O., Foy, P. & Arora, A. (2012). *TIMSS 2011 International Results in Mathematics*. Chestnut Hill, MA: TIMSS & PIRLS International Study Center, Boston College.

National Institute for Educational Policy Research (2010). *Results from 2010 National Assessment of Students' Academic Achievements and Learning Environments*. Tokyo.

National Research Council (2010). *Preparing Teachers: Building Evidence for Sound Policy*. Committee on the Study of Teacher Preparation Programs in the United States. Washington, DC: National Academy Press. http://www.nap.edu/catalog/12882.html

Organisation for Economic Co-operation and Development (2001). *Knowledge and Skills for Life: First Results from PISA 2000*. Paris: OECD Publications.

Organisation for Economic Co-operation and Development (2003). *Literacy Skills for the World of Tomorrow — Further Results from PISA 2000*. Paris: OECD Publications.

Organisation for Economic Co-operation and Development (2004). *Learning for Tomorrow's World – First Results from PISA 2003*, OECD.

Organisation for Economic Co-operation and Development (2010). *PISA 2009 Results: What Students Know and Can Do*, OECD.

Shimizu, Y. (2002). Lesson Study: what, why, and how? In H. Bass, Z.P. Usiskin & G. Burrill (eds.), *Studying Classroom Teaching as a Medium for Professional Development: Proceedings of a U.S.-Japan workshop*. Washington DC: National Academy Press: 53-57, 154-156.

Stevenson, H.W. (1987). America's Math Problems. *Educational Leadership*, 45: 4-10.

Stigler, J.W. & Hiebert, J. (1999). *The Teaching Gap*. New York: Free Press.

Taylor, M. J. (1987). *Chinese Pupils in Britain*. Windsor: NFER-Nelson.

Wang, J.P. (2012). *Mathematics Education in China: Tradition and Reality*. Singapore: Cengage Learning Asia Pte Ltd.

Wang, X.R. (2011). Inquiring and Reflecting Teaching Research Group Activities in Primary and Secondary Schools – A thematic discussion within mathematics teaching research group in one school. *Forum on Education Science*, 4: 32-34.

White, M. (1987). *The Japanese Education Challenge: A commitment to children*. New York: The Free Press.

Yang, K.S. (1981). Social Orientation and Individual Modernity among Chinese Students in Taiwan, *Journal of Social Psychology*, 113: 159-70.

Yang, Y.D. (2009). Capturing the Crucial Incidents in Teaching Research. *People's Education*, 1: 48-49.

Yang, Y.D., Li, J., Gao, H. & Xu, Q.F. (2012). Teacher Education and the Professional Development of Mathematics Teachers. In J.P. Wang, *Mathematics Education in China: Tradition and Reality*. Singapore: Cengage Learning Asia Pte Ltd.: 205-238.

Yao, H. (2010). Teaching Research Activities – Based on constructing teaching research culture. *Primary and Secondary School Management*, 9: 35-36.

Zhang, D., Li, S. & Tang R. (2004). The "Two Basics": Mathematics teaching and learning in mainland China. In L. Fan, N. Wong, J. Cai, & S. Li (eds.), *How Chinese Learn Mathematics: Perspectives from insiders*. River Edge, NJ: World Scientific Publishing Co.: 189-207.

Gender and Mathematics Education Revisited

Gilah C. Leder

Introduction

Beginning in the early 1970s, systematic documentation in many countries of subtle, yet consistent gender differences in mathematics performance and participation in post compulsory mathematics courses in favor of males served as a catalyst for action. In these settings, new legislation and special interventions were introduced to redress demonstrated achievement disparities in mathematics. An important aim of the panel session was to describe the current situation in countries where gender equity is enshrined in legislation at the political level, and, by drawing on recent research and contemporary data gathering tools, to document whether or not inequities have been removed in practice or continue to exist in countries where concern and action about gender differences in mathematics learning have a long standing history.

There are also a significant number of countries where gender and mathematics learning issues have typically been ignored, are still not well recognized by their governments or valued in the wider community. To document the situation in those countries and highlight what progress has been made in those settings were also central aims of the panel's presentation.

The notions of gender parity and gender equality are a unifying thread weaved throughout the presentation. The former is described by UNESCO (2012) as "aim (ing) at achieving equal participation for girls and boys in education", while

> gender equality is understood more broadly as the right to gain access and participate in education, as well as to benefit from gender-sensitive and gender-responsive educational environments and to obtain meaningful education outcomes that ensure that education

G.C. Leder (✉)
Monash University, Melbourne, Australia
e-mail: Gilah.Leder@monash.edu

© The Author(s) 2015
S.J. Cho (ed.), *The Proceedings of the 12th International Congress on Mathematical Education*, DOI 10.1007/978-3-319-12688-3_12

benefits translate into greater participation in social, economic and political development of their societies. Achieving gender parity is therefore understood as only a first step towards gender equality. (UNESCO 2012, p. 21)

In brief, the areas covered in the session reflected the different perspectives and geographic diversity of the panelists. Attention was given to regions where issues about gender and mathematics education remain barely on the agenda and relatively little is known outside those countries about work and research that have been undertaken. The more widely disseminated research findings and common assumptions about gender and mathematics learning, based on research particularly in Western countries, were also revisited and updated.

The order of presentations was part of our overall message. We therefore started off with presentations from regions where gender and mathematics is not widely seen as a primary issue of concern and/or about which relatively little is known in Western countries—whose research is disseminated widely—and moved to surveys of areas where gender equity is enshrined in legislation at the political level, but in practice inequities continue to exist.

To begin, data referring to India were presented by Jayasree Subramanian. This was followed by Nouzha El Yacoubi whose presentation also covered a large region where concern and progress re-gender and mathematics are still not well known or recognized in the wider research community, and then by Maria Trigueros Gaisman who focused on Mexico. The final three presentations also covered wide geographic areas, in alphabetical order: Australia, Europe, and the United States. Pertinent research and issues were presented respectively by Helen Forgasz, Lovisa Sumpter and Sarah Lubienski.

Each panelist sketched realities, achievements, and outcomes in mathematics education and gender in the area in which she lives and works and of which she has first hand knowledge. Reference was also made to examples of dissonance between theory and practice with respect to mathematics education and gender. Highlighted, too, were pressing next step(s) to improve the situation in the context represented by each speaker. If translated into a realistic and focused research agenda, and if taken up, these steps can move the field forwards.

Reference

UNESCO (2012). *World atlas of gender equality in education*. Paris: Author.

Gender and Mathematics Education in Africa

Nouzha El Yacoubi
University Mohammed V Rabat-Agdal. Morocco
e-mail: n.elyacoubi@yahoo.fr; nelyacoubi@fsr.ac.ma

Introduction

Even in the developed countries, where equity in Education was reached a long time ago, the rates of enrollment of girls in mathematics courses are relatively low. The gender problem and mathematics education has been studied since 1970 and some factors of that representativeness have been identified, in particular in the developed countries. But this area of research is still unexplored in the developing countries. In Africa, specifically, little research has been done until now on Gender and mathematics education despite the millennium goals recommending equity in education and the encouragement of African females to choose mathematics studies and to embrace scientific and technological careers.

Nevertheless, the role of women in the scientific development of Africa has been definitively recognized as a crucial and determining factor in building and reinforcing the continent's scientific and technological capacities, because no African country can afford to leave 50 % of its population, out of its development process.

It is evident that Education in general in Africa was, and is till now, seriously affected by poverty, but with respect to the education of girls, history, religion and culture were, and they remain, important influencing factors.

These socio-cultural barriers are more pronounced when they come to scientific, technical and vocational education and, are unfortunately, tragic when they concern mathematics education.

The Current Situation in Africa

According to the UNESCO Institute for Statistics report published in September 2010, the lowest literacy rates were observed in sub-Saharan Africa, where the adult literacy rate for males is 71.6 and 53.6 % for females and in Northern Africa it is respectively 76.7 and 58.1 %. It should be highlighted that more than half of the adult population is still illiterate in the ten following countries: Gambia (55 %), Senegal (58 %), Benin (59 %), Sierra Leone (60 %), Guinea (62 %), Ethiopia (64 %), Chad (67 %), Burkina Faso (71 %), Niger (71 %), and Mali (74 %).

The net enrolment ratio in the primary school age population in sub-Saharan Africa countries is around 52.3 % girls (and 60.7 % boys), except in a very few countries where almost all girls of primary school age are enrolled at schools.

But there is a substantial drop out among girls at the secondary school level; it is due to socio-cultural (early marriage), financial reasons, institutional barriers and poor performance of girls. The Trends in Mathematics and Science Study (TIMSS) reported that between 68 and 90 % of African students in grade eight failed to reach the low benchmark in mathematics (IEA 2003). And unfortunately no significant progress was registered in TIMSS 2007. It is a pity that Africa was so poorly represented in such an important international assessment of the mathematics and science knowledge of fourth and eighth grade students. For example in TIMSS

2007, only six African countries have participated among 59 Countries namely: Algeria-Botswana-Egypt-Ghana-Morocco and Tunisia, and there was no African country among the 8 Benchmarking participants. The African countries participating in TIMSS 1995 through 2007 are as follows:

Country	Grade 4			Grade 8			
	1995	2003	2007	1995	1999	2003	2007
Algeria			x				x
Botswana						x	x
Egypt						x	x
Ghana						x	x
Morocco		x	x		x	x	x
Tunisia		x	x		x	x	x

As for upper secondary school, the enrollment ratio of girls is just about 17 % in Sub-Saharan Africa, so only a few girls have the opportunity to be enrolled in scientific classes, and among that population very few choose Mathematics courses. The best registered percentage for enrollment of girls in Mathematics at that level is about 30 % (Huggins and Randell 2007) and this percentage decreases with grade level and is about 10 % for the tertiary level.

The Causes

The factors identified in contributing to the gender problem in mathematics education in the developed countries remain valid for Africa, but other factors should be added like negative socio-cultural attitudes, household tasks at home, gender biased curriculum, poor didactic materials, lack of school facilities (dormitories), lack of sponsorship, unmotivated and unqualified mathematics teachers, lack of moral and financial parental support, lack of self confidence among the girls, poor performance in exams, and so on.

Interventions Introduced

First, the African Union (UN) has set up mechanisms and special committees at the ministerial level for monitoring progress towards attainment for Education For All (EFA). Gender mainstreaming has been identified and adopted as a strategy for achieving gender equity. In particular, special projects were launched with the aim of increasing the enrollment of African girls in Science, mathematics and technology, and to encourage African women to embrace scientific and technological careers. The programs included: "Special Project on Scientific, Technical and

Vocational Education of Girls in Africa in the framework of the UNESCO's Medium-Term Strategy" (1996–2001); "Africa's Science and Technology" project launched in 2007 by the African Union Summit of the Heads of State and Government; "Africa and Gender Equity" including "Science, technology and engineering education" in the UNESCO Medium-Term 2008–2013, as well as other initiatives sponsored by the World Bank, USAID, NEPAD (New Partnership for Africa's Development), UNICEF, and some non-governmental organizations (NGO's).

A special program for reducing gender disparities in science, technology and innovation has also been undertaken by the United Nations Economic Commission for the East African Community member countries. This, Huggins and Randell (2007) advocated, should serve as a case study for the other African regions.

There have been various other activities, for example, international conferences on Gender, Science and Technology were held in: Beijing (1995), Arusha (1997), Harare (1997) where national surveys of 21 African countries, assessing the participation of girls and women in scientific education and vocational training, were given, (Hoffmann-Barthes and Malpede 1997), Dakar (2000), Cairo (2006), Bamako and Ségou (2009), Paris (2010): UNESCO Expert Group Meeting.

Some camps and competitions for African girls have been organized through Africa, including: Camp of Excellence in Sciences and Mathematics for Young African Girls held, since 2000, in Mali and other African countries; Girls STEM (Science, Technology, Engineering and Mathematics) Camp initiative (Abuja 2011), Miss Mathématique (created in Ivory Coast and recently in Benin) and so on.

Conclusion and Suggestions

Despite these initiatives, females' participation in Africa, in Science, and Technology, and in particular in Mathematics, from primary through tertiary education to the career level is still very low. This could be explained by, among other factors, the persistent socio-cultural barriers, lack of clear policy guidelines for increasing the rates of enrollment of African girls in mathematics, lack of assessment and follow up of the various undertaken initiatives, lack of gender analysis expertise and so on.

A valorized image of African women in mathematics education and mathematics careers should be promoted and gender stereotypes with regard to mathematics careers should be countered by parents, teachers and all other actors in the school and societal environments.

Interventions for females should aim to achieve equity of outcomes rather than just equal access to educational opportunities in mathematics. So permanent assessment and relevant follow up are key elements in any undertaken initiative.

References

Aiken, L. (1970). *Attitudes Toward Mathematics*, Review of Educational Research, 40(1): 551-596.

Fennema, E. and Sherman, J. (1977). *Sex-related differences in mathematics achievement, spatial visualisation and affective factors*, American Educational Research Journal, 14(1): 51-71.

Walden, R. and Walkerdine, V (1985). *Girls and mathematics*. London: University of London.

Burton, L. (1990). *Gender and mathematics: An international perspective*. London: Cassell.

Kaiser-Messmer, G. (1993). *Results of an empirical study into gender differences in attitudes toward Mathematics*, Educational Studies in Mathematics, 25(3): 209-233.

Fennema, E.L and Leder, G (1995). *Mathematics and Gender*. New York: Teacher College Press.

Kiania, A.M. (1995). *Gender and mathematics achievement parity: Evidence from Post-Secondary Education*. Vol. 116 (4) p.586 – 591

B.Grevholm and G.Hanna (1995). *Gender and Mathematics Education*. Sweden, Lund University Press

Lamb, S (1997). *Gender difference in mathematics participation: An Australian participation*. Vol. 23(1); pp.105 -115.

Hoffmann-Barthes, & Malpede (1997). *Scientific, technical and vocational education of girls in Africa*. Retrieved from http://www.unesco.org/education/.../girls/reports.pdf

Working document1999 of UNESCO : Scientific, technical and vocational education for girls in Africa

TIMSS 2003 International Report on achievement in the Mathematics Cognitive Domain IEA (2004). Retrieved from http://timss.bc.edu/pdf/t03_download/t03mcogdrpt.pdf

Xin Ma (2004). *Current Trend in Gender Differences in Mathematics Performance: An International Update*. ICME 2004, TSG 26.

Asimeng-Boahene, L (2005). Gender Inequity in Science and Mathematics Education in Africa: The Causes, Consequences and Solutions. Education, Vol. 126, No. 4, PP. 711-72

Huggins, A., & Randell, S.K. (2007). Gender Equality in Education in Rwanda: What is happening to our girls? Retrieved from http://www.nuffic.nl/international-organizations/international-education-monitor/country-monitor/africa/rwanda/documents

D. Fisher, R. Koul&S. Wanpen (2008): Science, *mathematics and technology Education:Beyond cultural boundaries*. 5th SMTE Proceeding

V.S Mullis, M.O Martin and P.Foy (2008). *TIMSS 2007 International Report*. Retrieved from http://www.timss.bc.edu/timss2007/intl_reports.htm

Roland G. Fryer, Jr and Steven D. Levitt (2009). *An Empirical Analysis of the Gender Gap in Mathematics*. NBER Working Paper No. 15430. JEL No. I20

N.M. Else-Quest, J.S. Hyde and M.C.Linn (2010). *Cross-National Patterns of Gender Differences in mathematics*. A Meta-Analysis. Psychological Bulletin, Vol 136. No 1, 103-127.

UNESCO Institute for Statistics (2010). *Adult and youth literacy: global trends in gender parity*. UIS Fact Sheet. (September, 2010, No. 3).

United Nations. Economic and Social Council. Commission for Africa (2011). Mainstreaming gender in Science, Technology and Innovation Systems in the East African Community. Retrieved from http://www.uneca.org/.../codist-iireportexecutivesumma

Hanna David (2011). *Overcoming the gender gap in math,Science and technology : A 21ˢᵗ Century View*. Journal of Education and Social Research. Vol (1).

Nouzha El Yacoubi (2011). *Problème du Genre et Mathématiques en Afrique*. EDIMath, IMU-CANP, ICMI workshop, Bamako September 2011. (Could be provided by the author : n.elyacoubi@yahoo.fr)

Gender and Mathematics Education in Mexico

María Trigueros Gaisman
Instituto Tecnológico Autónomo de México, ITAM
e-mail: trigue@itam.mx

Introduction

In the area of Mathematics Education in Mexico, research on gender has produced interesting findings. Some studies have analyzed gender differences in relation to results attained on performance tests, while others have focused on more specific topics, such as spatial visualization, the differential relations that mathematics teachers may establish with female and male students at various educational levels, the distinct attitudes of girls and boys towards mathematics and towards the use of technology as an aid in teaching and learning mathematics.

At the same time there has been an emerging trend on the development of educational policies to reduce the gender gap in education at all levels, and to foster equity in academic work.

Results of Gender Studies at Elementary Education

Since the first study (Bosch and Trigueros 1996) no substantial gender differences have been observed in different tests in primary school (González 2003; Rivera 2003; Ursini, et al. 2010). However, PISA results indicate that gender differences favoring boys appear in the transition to secondary school. Studies on students' attitudes towards the subject (Ursini et al. 2004, 2007; Campos 2006; Ursini and Sánchez 2008; Ursini 2010) show that self-confidence favoring boys, and perception of mathematic as a male domain, start to develop at around 13 years of age, with boys attributing good performance to intelligence or skills and girls to effort and obedience. Interestingly, teachers were found to characterize differences in children's performance in the same terms (Ramirez 2006; Ramirez and Ursini 2008).

Regarding the use of technology in the learning of Mathematics, Ursini and Sánchez (2008) found that boys held a pragmatic view of technology while girls considered it as a resource to construct knowledge. They found that the use of technology helped to develop positive attitudes towards mathematics, particularly among girls, and suggested that using technology with guiding activities to foster group-work and discussion, helped to modify certain cultural patterns of conduct which can foster equity.

The use of technology also modified teachers conception of Mathematics learning (Trigueros and Lozano 2008; Rodriguez and Ursini 2008) with females focusing more on exploration and investigation to develop students' self-confidence, independence and creativity and males on developing skills needed by students to move forward in their education.

Results of Gender Studies at Higher Education

As at the elementary school level, in higher education no specific gender differences have been found in different studies in mathematics grades and the gender inequality in access to higher education detected in earlier studies (Bosch and Trigueros 1996) has been constantly decreasing. The largest university in Mexico reported in 2009 (Saavedra 2010) that the percentage of female students was larger than that of male students and that graduation percentages also favored women (56 % of women graduated against 50 % of men). However, there is still a severe under-representation of women in mathematics. Only 38 % of women enroll in mathematics programs, and 43 % of all students who graduate from these programs are women. The gender gap is greater when considering access to post-graduate education. In 2008 only 30 % of students in postgraduate programs were women, although in programs related to mathematics education female students comprised 45 %.

In a study involving university professors (Espinosa 2007), it was found that they considered male students to be more proficient in mathematics than females. They expressed the same beliefs as those found among teachers in elementary school about women being successful in mathematics because of their effort and discipline. Observation of classes detected a more passive attitude of female students and a tendency of male students to be more participative.

Although results show that, in general, female students are more perseverant in their studies, it seems that they still consider mathematics as a male domain, too competitive for women and that professors' beliefs tend to reinforce this conception.

Results of Gender Studies on Faculty

In the last few years there has been a large increase in the academic profession in Mexico, but problems related to gender in the access to work at universities are still present. Only 40 % of professors are women. This gap widens in the case of mathematics departments where women represent less than 25 % of all professors and many of them work in mathematics education (Saavedra 2010).

In terms of research, according to 2009 data from the National System of Researchers, women researchers in the area corresponding to physics and mathematics, which is the largest area of the system, represent only 19 % of all researchers with 23 % of them investigating in mathematics. Percentages of female researchers diminish as levels related to productivity rate increase, with only 3 % of women at the top level.

Some of these differences can be related to perception of mathematics as an occupation which is difficult to combine with family life, but results show gender as a determinant of the choice of mathematics as a field of study independently of school achievement.

Policies to Reduce the Gender Gap and Stereotypes

The ministry of Education has developed several initiatives since 2008 to incorporate the gender perspective in all the educational programs to help to change stereotypes that contribute to gender inequity. Among the more important are a revision of content of all the mandatory primary school textbooks from a gender perspective to foster a change in socio-cultural patterns, and the distribution of books on gender equity and prevention of violence for teachers and students. Together with international organizations, the ministry has developed projects for school communities where people participate in activities designed to reflect on gender stereotypes and their change. Technology is used to show different behavior patterns in particular situations together with questions asking users to reflect, comment and discuss if they find those behaviors appropriate or not and why.

A revision of the published policies from different universities in Mexico reveals that in the last 10 years there has been an increase in policies intended to foster women's access to higher education and to reduce the barriers for female faculty. Most of the universities nowadays have developed innovative programs to reduce inequalities for women researchers, teachers and students. These include mandatory seminars to discuss gender issues, awards designed for women faculty and students and specific programs to recruit women as faculty. However, only a few of them have been designed specifically to increase the number of women researchers in STEM related careers or to strengthen the academic position of women researchers and their participation in academic activities.

Some of these policies have shown some positive impact, however, their implementation is unequal in different regions of the country, and some of them have had implementation problems in practice. The effective advancement of women as faculty, in particular, seems to be prevented by everyday practices that tend to ignore policies, or at least to apply them in a limited way.

Conclusions

This review of studies on gender and mathematics in Mexico shows that although some advance in reducing the gender gap in mathematics has been achieved, there is still much work to be done in terms of policies and programs to change socio-cultural perceptions which inhibit the development of women in mathematics and mathematics related areas. More efforts are also needed to increase participation of women as faculty and as decision makers in areas related to mathematics, science and technology.

References

Bosch, C. & Trigueros, M. (1996). *Gender and mathematics in Mexico. In G. Hanna (Ed.). Towards gender equity in mathematics education. An ICMI Study.* (pp. 277-284). Kluwer Academic Publishers, Dordrecht, Boston, London.

Campos, C. (2006). *Actitud hacia las matemáticas: Diferencias de género entre estudiantes de sexto de primaria y tercer grado de secundaria.* Unpublished Masters Thesis, Department of Mathematics Education, Cinvestav-IPN, México.

Espinosa, C. (2007). *Estudio de las interacciones en el aula desde una perspectiva de género.* Unpublished Masters Thesis, Department of Mathematics Education, Cinvestav-IPN, México.

González, R.M. (2003). *Diferencias de género en el desempeño matemático. Educación Matemática. 15* (2), pp.129-161.

Rivera, M. (2003). *Diferencia de género en la visualización espacial: un estudio exploratorio con estudiantes de 2° de secundaria.* Unpublished Master Thesis, Department of Mathematics Education, Cinvestav-IPN, México.

Ramirez, M. P. and Ursini, S. (2008). *Influence of the female teachers' gender vision on the type of interactions they establish with boys and girls in the mathematics classroom.* Paper presented at ICME 11, Topic Study Group 32: Gender and mathematics education, Monterrey, México.

Rodriguez, C. & Ursini, S. (2008). *Social representation and gender in the teaching of mathematics with multimedia devices.* ICME 11, Topic Study Group 32: Gender and mathematics education, Monterrey, México.

Saavedra, P. (2010). *Trayectoria académica de las mujeres matemáticas en México.* http://docencia.izt.uam.mx/psb/ciencia.pdf (March 3rd, 2012).

Trigueros, M. & Lozano, M.D. (2008). *Teachers' assessment practices in mathematics courses. Does gender make a difference?* ICME 11, Topic Study Group 32: Gender and mathematics education, Monterrey, México.

Ursini, S. (2010). Diferencias de género en la representación social de las matemáticas: Un estudio con estudiantes de secundaria. In N. Blazquez, y F. Flores (Eds.). *Investigación feminista. Epistemología, metodología y representaciones sociales. Colección Debate y Reflexión. CEIICH.* Facultad de Psicología. UNAM. México.

Ursini, S., Ramírez, M. P., Rodríguez, C., Trigueros, M., and Lozano, M. D. (2010). Studies in Mexico on Gender and Mathematics. In H.J. Forgasz, J. Rossi Becker, K.H. Lee and O. Steinthorsdottir (Eds.) *International Perspectives on Gender and Mathematics Education, 7,* 147-172.

Ursini, S. & Sánchez, G. (2008). Gender, technology and attitudes towards mathematics: A comparative longitudinal study with Mexican students. ZDM *The International Journal on Mathematic Education. 40,* 559-577.

Ursini, S., Ramirez, M. P. & Sánchez, G. (2007). Using technology in the mathematics class: How this affects students' achievement and attitudes. *Proceedings of the 8th ICTMT,* (Integration of ICT into Learning Processes). Czech Republic, University of Hradec Králové, [CD-ROM]

Ursini S., Sánchez G., & Orendain, M. (2004). Validación y confiabilidad de una escala de actitudes hacia las matemáticas y hacia las matemáticas enseñada con computadora. *Educación Matemática, 16*(3), 59-78.

Gender and Mathematics in Australia: A Downward Trajectory

Helen J. Forgasz
Monash University, Australia
e-mail: Helen.forgasz@monash.edu

Introduction

In this paper I draw attention to four areas in which gender equity in mathematics education has yet to be fully achieved in Australia, and where indications are that we are going backwards: (i) achievement in TIMSS and PISA; (ii) participation and achievement in higher level mathematics; (iii) use of technologies for mathematics learning; and (iv) public perceptions of gender issues in mathematics.

Australian Context

Despite laws and government policy decrying inequity, the realities of gender equity have not yet been fully realized in Australia. This is evident with respect to educational levels, occupations and salaries. Despite higher proportions of women than men having Year 12 or equivalent qualifications, bachelor-level degrees, and higher literacy and numeracy skill levels (Australian Bureau of Statistics 2012), graduate median starting salaries still show a $2,000 difference in favor of men, a consistent pattern over the past decade. When it comes to educational pathways leading to career options, males remain dominant in the physical sciences, and females in the humanities and social sciences.

TIMSS and PISA Results

Australian results in all years of TIMSS and PISA are shown in Table 1. The data reveal a disturbing pattern. Mean scores on TIMSS for grade 4 and grade 8 show an increasing gender gap favoring males, with the 2007 grade 8 score differences reaching statistical significance. For the PISA results, the gender gap in mean scores favors males in all years, but in 2006 and 2009, the score differences were also statistically significant.

Thomson et al. (2011, p. 299) claimed that "the re-emergence of gender difference as shown in PISA since 2006 are a salutary reminder to (Australian) schools and systems that this is still a significant issue and that if Australia is to improve its performance in mathematics, girls' scores must improve".

Participation and Achievement in Grade 12 Mathematics

The Victorian (Australia) grade 12 mathematics subject enrolment figures reveal a consistent pattern over time. Three mathematics options are offered at grade 12: Specialist Mathematics (most challenging, calculus-based), Mathematical Methods

Table 1 TIMSS (1995–2007) and PISA (2000–2009) results for Australia

	TIMSS 1995[a]	TIMSS 1999	TIMSS 2003	TIMSS 2007
Grade 4	F = 545, M = 547	No Grade 4	F = 497, M = 500	F = 513, M = 519
	2 points (M > F)		3 points (M > F)	6 points (M > F)
Grade 8	F = 532, M = 527	F = 524, M = 526	F = 499, M = 511	F = 488, M = 504
	5 points (F > M)	2 points (M > F)	12 points (M > F)	16 points (M > F)*
Final year of schooling	F = 510, M = 540			
	30 points (M > F)*			
15 year olds	F = 527, M = 539	F = 522, M = 527	F = 513, M = 527	F = 509, M = 519
	12 points (M > F)	5 points (M > F)	14 points (M > F)*	10 points (M > F)*
	PISA 2000	PISA 2003	PISA 2006	PISA 2009
15 year olds	F = 527, M = 539	F = 522, M = 527	F = 513, M = 527	F = 509, M = 519
	12 points (M > F)	5 points (M > F)	14 points (M > F)*	10 points (M > F)*

Legend: *F* female; *M* male; *statistical significant difference

Data sourced from various IEA, OECD, and Australian Council for Educational Research reports of TIMSS and PISA results

[a] Gill et al. (2002). Student achievement in England. Results in reading, mathematical and scientific literacy among 15-year-olds from OECD PISA 2000 study (p. 47). London: The Stationery Office (HMSO)

(includes calculus, pre-requisite for many university-level science-related courses), and Further Mathematics (least challenging, with an emphasis on statistics). The data in Fig. 1 reveal that enrolments have declined over time in Specialist mathematics while increasing in Further Mathematics. Yet, consistently, there have been higher proportions of males than females enrolled in all three options.

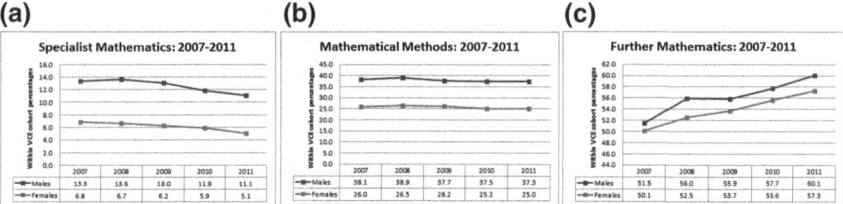

Fig. 1 Enrolment trends 2007–2009 in VCE mathematics subjects

Table 2 Highest achievers (top 2 %) in VCE mathematics (2007–2009)

Subject	Gender	2007 (N = 65)		2008 (N = 60)		2009 (N = 59)	
		n	%	n	%	n	%
Specialist mathematics	Female	15	23.1	14	23.3	14	23.7
	Male	49	75.4	44	73.3	45	76.3
	Unknown	1	1.5	2	3.3	–	
Mathematical methods	Female	50	25.1	53	25.7	67	33.7
	Male	133	66.8	150	72.8	131	65.8
	Unknown	16	8.0	3	1.5	1	0.5
Further mathematics	Female	114	36.5	114	35.5	139	42.1
	Male	187	59.9	205	63.9	191	57.9
	Unknown	11	3.5	2	0.6		

An even more disturbing trend is found when the very highest achievers in these three mathematics options are considered, that is, the top 2 %. It is found that males outperform females at a rate that is disproportionate to their enrolments in these subjects (see Table 2 for data from 2007 to 2009). The data in Table 2 reveal that more than 50 % of the highest achievers in each of the three VCE subjects were male and that this pattern persisted over the three year period, 2007–2009.

Technologies for Mathematics Learning

The adoption of computers and calculators in mathematics classrooms has received much research attention in Australasia; less common is research incorporating gender as a variable—see Geiger et al. (2012) for an overview of recent Australasian research. Technology (and ICT), like mathematics, is considered a male domain. Hence, when technology is brought into the mathematics classroom, the effect of this combination with respect to gender issues clearly demands greater research interest than is evident. Researchers examining computer and/or sophisticated calculator use for mathematics learning and gender have found that those who appear to benefit more from the use of the technologies are those who are comfortable with the technology, that is, it is more likely to be boys than girls, but not necessarily boys with the highest mathematical capabilities. Much of the work on mathematics, technology, and gender has focused on the affective domain. Here it is clear that boys' confidence and competence levels with the technologies are more positive than girls', that boys more strongly than girls say they enjoy learning mathematics with technology, and that this is also the expectation of teachers and parents.

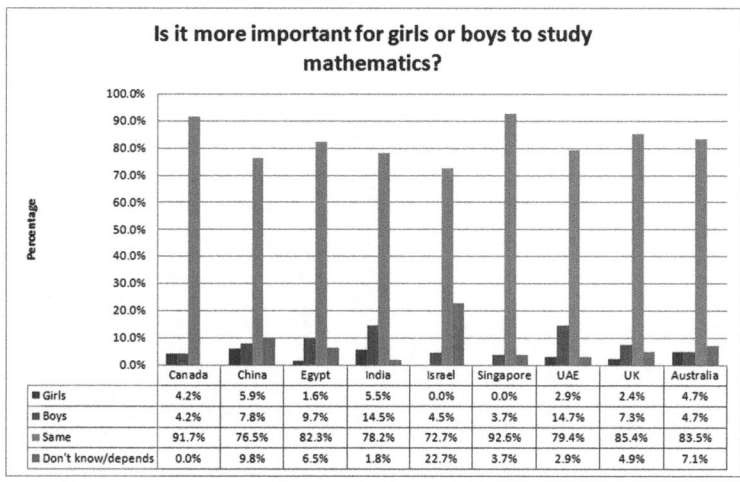

Fig. 2 Response frequency by country: is it more important for girls or boys to study mathematics?

Public Perceptions of Gender Issues in Mathematics

Early explanatory models for gender differences in mathematics learning incorporated the views of society at large as critical contributing influences. Until recently, however, the views of the general public have rarely been sought. Very recent survey data reveal that the male stereotype is alive and well in the views of the Australian public and elsewhere in the world (e.g., Forgasz et al. 2012).

The extent of the view that mathematics is a male domain varies across the globe. In many countries, a large proportion of respondents to an online survey indicated that it is equally important for boys and girls to study mathematics (see Fig. 2). However, compared to girls, many believed that: boys are better at mathematics (see Fig. 3) and that parents and teachers also believe this, that boys are better with calculators and computers (see Fig. 4), and that boys are more suited to careers in science-related and computer occupational fields. As can be seen in Figs. 2, 3 and 4, Australian respondents' views on these issues fell somewhere between the extremes, with respect to response frequencies.

Final Words

The picture portrayed in the four brief snapshots above reveal a gendered world of mathematics learning that has changed little over the thirty year period in which research into this area began. The apparent gains made to reduce the gender gap

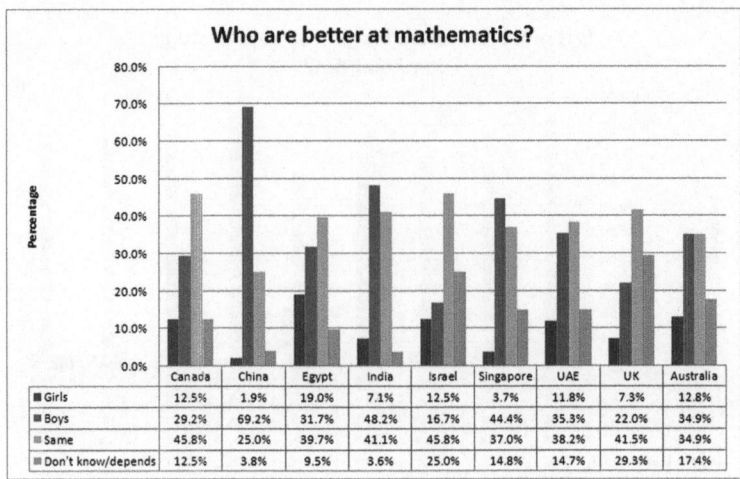

	Canada	China	Egypt	India	Israel	Singapore	UAE	UK	Australia
■ Girls	12.5%	1.9%	19.0%	7.1%	12.5%	3.7%	11.8%	7.3%	12.8%
■ Boys	29.2%	69.2%	31.7%	48.2%	16.7%	44.4%	35.3%	22.0%	34.9%
■ Same	45.8%	25.0%	39.7%	41.1%	45.8%	37.0%	38.2%	41.5%	34.9%
■ Don't know/depends	12.5%	3.8%	9.5%	3.6%	25.0%	14.8%	14.7%	29.3%	17.4%

Fig. 3 Response frequency by country: who are better at mathematics?

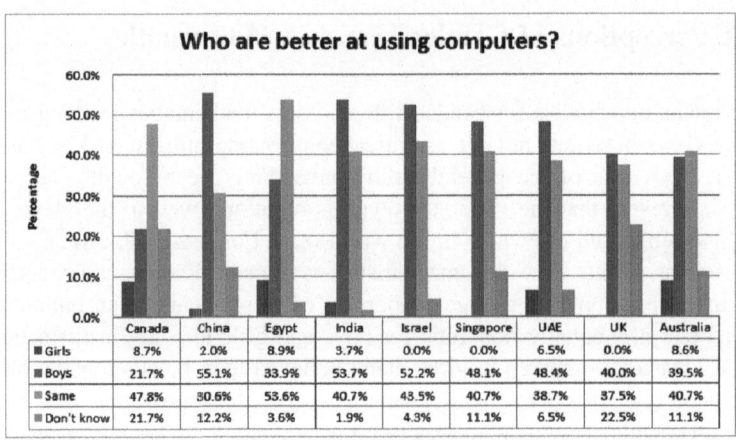

	Canada	China	Egypt	India	Israel	Singapore	UAE	UK	Australia
■ Girls	8.7%	2.0%	8.9%	3.7%	0.0%	0.0%	6.5%	0.0%	8.6%
■ Boys	21.7%	55.1%	33.9%	53.7%	52.2%	48.1%	48.4%	40.0%	39.5%
■ Same	47.8%	30.6%	53.6%	40.7%	43.5%	40.7%	38.7%	37.5%	40.7%
■ Don't know	21.7%	12.2%	3.6%	1.9%	4.3%	11.1%	6.5%	22.5%	11.1%

Fig. 4 Response frequency by country: who are better at using computers, girls or boys?

favoring males in participation, achievement, and attitudes during the 1980s and 1990s, appears to have been eroded to the point of a clear backward trajectory emerging in Australia. Believing that there was no longer a "girl problem" with respect to mathematics, with the consequential reduction in vigilance as curricula and practices have changed, may be largely to blame.

References

Australian Bureau of Statistics. (2012). *Gender indicators, Australia*. Retrieved from http://www.abs.gov.au/ausstats/abs@.nsf/mf/4125.0

Forgasz, H., Leder, G., & Gómez-Chacón, I. (2012, accepted). *Young pedestrians' gendering of mathematics: Australia and Spain*. Paper accepted for presentation at the annual conference of the Mathematics Education Research Group of Australia, Singapore.

Geiger, V., Forgasz, H., Tan, H., Calder, N., & Hill, J. (2012). Technology in mathematics education. In B. Perry, T. Lowrie, T. Logan, A. MacDonald, & J. Greenlees (Eds.), *Research in mathematics education in Australasia, 2008-2011* (pp. 111-142). Rotterdam, The Netherlands: Sense Publishers.

Thomson, S., De Bortoli, L., Nicholas, M., Hillman, K., & Buckley, S. (2011). *PISA in brief. Highlights from the full Australian report: Challenges for Australian education: Results from PISA 2009*. Retrieved from http://www.acer.edu.au/documents/PISA-2009-In-Brief.pdf

Taking a European Perspective

Lovisa Sumpter
School of Education and Humanities, Dalarna University, Sweden
e-mail: lsm@du.se

Taking a European Perspective

In this paper I look at how gender and mathematics education has been studied in Europe with the aim of highlighting trends but also discussing emerging themes. The main question posed in this paper is: What research focus in gender and mathematics can we find in papers that have been published during the years of 2007–2011? Gender is here defined as an "analytic category which humans think about and organize their social activity rather than as a natural consequence of sex difference" (Harding 1986, p. 17), emphasizing gender as something individuals do and create rather than something you have as a person. In order to talk about different foci of research on gender and mathematics, I follow Bjerrum Nielsen (2003) and use the following four aspects of gender: (1) structural gender, e.g. research of different groups within structures such as professions, level of education or social background; (2) symbolic gender e.g. studies looking at symbols and discourses that are attributed to a specific gender creating norms telling us what is normal and what is deviant; (3) personal gender e.g. studies on how girls and boys feel or think about various items or studies looking at individual's development of gender; and, (4) interactional gender e.g. research looking at how people interact with each other or how the social context is created. By using these four aspects different parts of the concept 'gender' can be emphasized.

Method

The data that constitute the base for the analysis were generated from the ERIC database, February 2012. The search terms were 'mathematics' and 'gender', peer-reviewed journal articles published within the last 5 years. By choosing only mathematics and not 'math' or 'maths' some papers were inevitably not included. The number of papers resulting from this search was 585. Thereafter I classified what could be considered European research; defined here as data collected in at least one European country, although the author/s could be positioned in any country. The list was narrowed down to 181 papers. Using Harding's (1986) definition of gender means that I have excluded all papers *only* using gender to denote division of sex, e.g. studies looking at sex-differences in performance (total 51 papers). I also excluded papers not on mathematics (e.g. using mathematics as a notion of intelligence or focusing on another subject e.g. chemistry, 23 papers) and papers that have a general international scope (11 papers). Most papers within this category were large-scale comparisons, e.g. results from international tests. Finally, four papers (all from Turkey) were not available online and therefore could not be analyzed. This left a total of 92 papers. The papers were divided into the four categories. If a paper dealt with multiple aspects, the main focus was selected. This is a simple division and it should be stressed that most papers are more complex and touch several aspects either in the background to the study, factors in the analysis and/or in the discussion of results. However, this division provided information for discussing main trends and themes.

Results and Discussion

The results were summarized in tables. Table 3 shows the number of papers produced by the different European countries in alphabetical order and the aspect of gender.

One paper has been marked as 'Europe' since the focus of the paper was evenly distributed among the participating countries; Garcia-Aracil (2008) compared college major and earning gaps in seven European countries. The countries that produced most papers during this period are UK and Germany followed by Finland. There are differences between the countries in which aspects of gender have been studied. Papers from UK, Finland, Sweden and Israel covered all aspects of gender whereas there was no paper focusing on interactional gender from Germany or Turkey. Looking closer at the papers from Germany, all of them were quantitative

Table 3 Aspect of gender and number of papers by country

Country	Number	Gender aspect[c]
Europe[a]	1	1
Belgium	2	1, 1
Croatia	1	3
Cyprus	1	1
Estonia	0.5[b]	4
Finland	12.5[b]	1, 1, 1, 1, 1, 2, 2, 2, 3, 3, 3, 4, 4
France	3	2, 3, 3
Germany	16	1, 1, 2, 2, 2, 2, 2, 2, 3, 3, 3, 3, 3, 3, 3, 3
Greece	1	3
Iceland	2	1, 1
Ireland	2	1, 3
Israel	5	1, 1, 2, 3, 4
Italy	2	2, 2
The Netherlands	7	1, 1, 1, 3, 3, 3, 3
Norway	2	1, 1
Spain	3	2, 3, 3
Sweden	6	1, 1, 2, 3, 3, 4
Turkey	7	1, 1, 1, 2, 2, 3, 3
UK	17	1, 1, 1, 2, 3, 3, 3, 3, 3, 3, 3, 3, 3, 3, 3, 4, 4

Note The number of papers is 92

[a] Seven European countries

[b] Comparative study Finland and Estonia

[c] Gender aspect: *1* structural; *2* symbolic; *3* personal; *4* interactional

studies, often large-scale, and most of them (10 of 12) were published in a journal not specific for mathematics education.

Let us look at the main focus of the selected papers. This is the number of papers covering different aspects of gender: structural, 30 papers (33 %); symbolic, 18 papers (20 %); personal, 38 papers (41 %); interactional, 6 papers (7 %). Most papers focused on structural gender or personal gender, whereas only six papers were on interactional gender. What these six papers have in common is that all of them looked at people's conceptions in relation to each other or to a development, e.g. Francis (2008) who studied interactions in different classes, where one of the classes presented is a math class. The majority of papers in structural and symbolic gender were quantitative studies, e.g., Ammann et al. (2010) who studied the number of students enrolled in undergraduate mathematics courses and Räty and Kärkkainen (2011) who looked at parents' stereotyping. We find a bigger variation of methods for data collection moving to the category 'personal gender', e.g. Mendick (2008) who used interviews when studying two students' conceptions about transitions between levels. Four papers focused on mathematics at preschool level. Klein et al. (2010) studied pre-school teachers' attributions of children's achievements in mathematics, and Ojala and Talts (2007) looked at pre-school teachers' evaluations of achievements. Palmer (2009, 2010) studied pre-school teacher education when writing about alternative mathematical practices.

As mentioned earlier, German papers were mainly found in non-mathematics education journals. This seemed to be a general trend. The top five journals in terms of publications relevant for this review were: *British Educational Research Journal*, 7 papers (8 %); *European Journal Psychology Education*, 5 papers (5 %); *Gender and Education*, 5 papers (5 %); *International Journal of Mathematical Education in Science and Technology*, 5 papers (5 %); *Scandinavian Journal of Educational Research*, 4 papers (4 %). The discussion about mathematics and gender mainly took place in journals that do not aim specifically towards mathematics education.

With respect to selecting areas for future research, the first topic I see as an emerging theme is research focusing on interactional gender. Four of the six papers on this aspect were published in 2010, possibly indicating an upcoming topic. Overall, there were few studies looking at "doing gender" in educational settings compared to the number of papers studying people "having gender". The most common type of paper was one reporting a large-scale quantitative study focusing on conceptions of different kinds, often related to mathematical achievement. Very few projects drew on qualitative measures in order to find out more about what 'doing gender' implies at various levels. Also, not many papers had a strong mathematical focus. A second theme for future is research looking at more content specific issues. The third area I see as an area that as yet has not been addressed in detail is research focusing on children under the age of five. There were only four papers aiming at pre-school mathematics, but not a single paper focused on pre-school students themselves. If we are to understand how personal gender is constructed, we need to know more about the process from the very beginning.

References

Ammann, C., Frauendiener, J. & Holton, D. (2010). German undergraduate mathematics enrolment numbers: background and change. *International Journal of Mathematical Education in Science and Technology, 41(4)*, 435-449

Bjerrum Nielsen, H. (2003). *One of the boys? Doing gender in Scouting.* Génève: World Organization of the Scout Movement.

Francis, B. 2010. Re/theorising gender: Female masculinity and male femininity in the classroom? *Gender and Education, 22 (5)*, 477–490.

Garcia-Aracil, A., (2008). College major and the gender earnings gap: a multi-country examination of postgraduate labour market outcomes. *Research in Higher Education 49*, 733–757.

Harding, S. (1986). *The Science Question in Feminism.* Ithica, NY: Cornell University Press.

Klein, P.S., Adi-Japha, E., & Hakak-Benizri, S. (2010). Mathematical thinking of kindergarten boys and girls: similar achievement, different contributing process. *Educational Studies in Mathematics, 73*, 233-246.

Mendick, H. (2008). Subtracting difference: troubling transitions from GCSE to AS-level mathematics. *British Educational Research Journal 34 (6).* 711–732.

Ojala, M., & Talts, L. (2007). Preschool achievement in Finland and Estonia: Cross-cultural comparison between the cities of Helsinki and Tallinn. *Scandinavian Journal of Educational Research 51(2)*, 205–221.

Palmer, A. (2009). "I'm not a "maths-person"! Reconstituting mathematical subjectivities in aesthetic teaching practices. *Gender and Education 21(4)*, 387–404.

Palmer, A. (2010). 'Let's dance!' Theorising alternative mathematical practices in earlychildhood teacher education. *Contemporary Issues in Early Childhood 11 (2)*, 130–143.

Räty, H. & Kärkkainen, R. (2011). Are parents' academic gender stereotypes and changes in them related to their perceptions of their child's mathematical competence? *Educational Studies, 37(3)*, 371-374.

Gender and Mathematics Education in the United States

Sarah Lubienski
University of Illinois at Urbana-Champaign, USA
e-mail: stl@illinois.edu

Introduction

Over the past several decades, the United States has made considerable progress toward gender equity in education. Substantial achievements have been made, such as the closure of gender gaps in high school mathematics course taking and college attendance (Lacampagne et al. 2007). In fact, some U.S. writers now argue that girls are more advantaged than boys, given that girls tend to score higher in reading, get better grades in school, and complete more bachelor degrees (e.g., Sommers 2000). However, gaps remain in mathematics achievement, affect, and ultimately the pursuit of high-status STEM careers.

Achievement

U.S. gender disparities in secondary mathematics achievement generally favor boys and are similar in size to those of many other industrialized nations (Else-Quest et al. 2010; OECD). However, TIMSS data suggest that significant mathematics score gaps favoring boys occur earlier in the U.S. than in most participating countries

(Mullis et al. 2008). Most recently, studies using data from the U.S. Early Childhood Longitudinal Study (ECLS), indicate that U.S. boys' and girls' mathematics proficiency is similar at the start of school (roughly age 5), but a significant male advantage emerges by age 8 (Robinson and Lubienski 2011). Regardless of grade level or dataset, U.S. mathematics gender gaps tend to be largest at the upper end of the achievement distribution (McGraw et al. 2006; Robinson and Lubienski 2011).

Affect

As in most countries participating in TIMSS and PISA, girls in the U.S. report having substantially less mathematical confidence than boys (Else-Quest et al. 2010). Recent analyses of ECLS data reveal that this trend exists already in U.S. primary schools, with gaps in confidence being substantially larger than gaps in both actual performance and interest in mathematics. Moreover young students' confidence predicts later gains in both mathematics achievement and interest (Lubienski et al. 2012).

Careers

Although women in the U.S. are at least as likely as men to pursue many science-related careers (e.g., biology), women remain under-represented in higher-paying, mathematics-intensive fields, such as engineering and computer science, in which women earn less than 20 % of bachelor's degrees (Snyder and Dillow 2011). These career patterns are a primary factor underlying earnings disparities among male and female college graduates, with U.S. women earning only 69 % of comparable men's salaries (Dey and Hill 2007).

Teachers and Students

U.S. girls are more compliant than boys in school (Rathbun et al. 2004), and boys are more likely than girls to exhibit a performance goal orientation, striving to "show off" their knowledge (Kenney-Benson et al. 2006). These patterns could cause boys to use more bold, invented methods during problem solving and could shape teachers' and students' views of who is "smart" (Fennema et al. 1998). Past research has revealed ways in which U.S. teachers attend more to boys than to girls (Sadker and Sadker 1986), and to attribute boys' mathematics success to ability and girls' success to effort (Fennema et al. 1990). More recent research reveals that U.S. elementary teachers rate boys' proficiency in mathematics—but not in reading—

higher than that of girls with equal test scores and similar classroom behavior (Robinson et al. 2012).

The Field of Mathematics

Recent research highlights subtle barriers to women's participation in mathematical fields. Lacampagne et al. (2007) emphasize the importance of women having a sense of belonging in mathematics, good relationships with faculty, flexibility in negotiating family responsibilities, and mathematical confidence. However, U.S. males remain more confident of their mathematical abilities relative to females with equal test scores (Correll 2001). Given that the opposite is true for reading, societal views about mathematics and gender likely influence students' perceptions of their own abilities.

Lingering Questions

The findings summarized thus far raise several questions. For example, why do girls report less mathematical confidence than their achievement merits? Why do U.S. teachers under-rate girls' competence in mathematics but not in literacy, relative to boys with similar behavior and achievement? (Robinson et al. 2012).

And finally, why do gaps in mathematics-related STEM fields remain so substantial despite the closure of key gaps in U.S. mathematics course-taking and college mathematics majors? One U.S. study provides an interesting insight. Males were nearly four times as likely to choose a quantitative college major than females with equal mathematics achievement, but this pattern was largely due to women's relatively strong verbal abilities (Correll 2001). In other words, women had other options, consistent with Eccles' (1986) argument that women make reasoned choices and do not simply avoid math. Interventions could fruitfully target girls' knowledge about ways in which a combination of mathematics and verbal skills could be a powerful asset in meaningful, STEM-related careers.

A Final Word About Research Methods for Studying Gender and Mathematics

The findings synthesized above are from a wide variety of qualitative and quantitative studies. Given the continued development of more sophisticated statistical methods, as well as the availability of large-scale, longitudinal datasets containing hundreds of variables, quantitative research on gender can go far beyond simply confirming the persistence of gaps in mathematics performance (Lubienski 2008). However, qualitative studies are continually needed to explore the factors underlying relationships found in large-scale data, as well as to develop the most important variables to be added to future, large-scale efforts.

References

Correll, S. J. (2001). Gender and the career choice process: The role of biased self-assessments. *American Journal of Sociology, 106*(6), 1691-1730.

Dey, J.G. & Hill, C. (2007). *Beyond the pay gap*. Washington, DC: American Association of University Women Educational Foundation.

Eccles, J.S. (1986). Gender-roles and women's achievement. *Educational Researcher, 15*(6), 15- 19.

Else-Quest, N.M., Hyde, J.S., & Linn, M.C. (2010). Cross-national patterns of gender differences in mathematics: A meta-analysis. *Psychological Bulletin, 136* (1), 101-127.

Fennema, E., Carpenter, T. P., Jacobs, V. R., Franke, M. L., & Levi, L. W. (1998). A longitudinal study of gender differences in young children's mathematical thinking. *Educational Researcher, 27*(5), 6-11.

Fennema, E., Peterson, P. L., Carpenter, T. P., & Lubinski, C. A. (1990). Teachers' attribution and beliefs about girls, boys, and mathematics. *Educational Studies in Mathematics, 21*, 55-69.

Kenney-Benson, G.A., Pomerantz, E.M., Ryan, A.M., & Patrick, H. (2006). Sex differences in math performance: The role of children's approaches to schoolwork. *Developmental Psychology, 42*(1), 11-26.

Lacampagne, C. B., Campbell, P. B., Herzig, A. H., Damarin, S., & Vogt, C. M. (2007). Gender equity in mathematics. In S. S. Klein (Ed.), *Handbook for achieving gender equity through education* (2nd ed., pp. 235–252). Florence, KY: Taylor and Francis.

Lubienski, S. T. (2008). On "gap gazing" in mathematics education: The need for gaps analyses. *Journal for Research in Mathematics Education, 39(4), 350-356.*

Lubienski, S. T., & Ganley, C., & Crane, C. (2012). Unwarranted uncertainty: Gender patterns in early mathematical confidence, interest and achievement. Paper to be presented at the American Educational Research Association, Vancouver.

McGraw, R., Lubienski, S. T., & Strutchens, M. E. (2006). A closer look at gender in NAEP mathematics achievement and affect data: Intersections with achievement, race and socio-economic status. *Journal for Research in Mathematics Education, 37*(2), 129-150.

Mullis, I. V. S., Martin, M. O., & Foy, P. (2008). TIMSS 2007 international mathematics report. Chestnut Hill, MA: Boston College.

Organisation for Economic Co-operation and Development. (2009). *Equally prepared for life? How 15-year-old boys and girls perform in school*. Paris, France: Author.

Rathbun, A. H., West, J., & Germino-Hausken, E. (2004). *From kindergarten through third grade: Children's beginning school experiences* (NCES 2004-007). Washington, DC: National Center for Education Statistics.

Robinson, J.P., & Lubienski, S.T. (2011). The development of gender achievement gaps in mathematics and reading during elementary and middle school: Examining direct cognitive assessments and teacher ratings. *American Educational Research Journal, 48*(2), 268-302.

Robinson, J. P., Lubienski, S. T., & Copur-Gencturk, Y. (2012). Gender-biased perceptions fuel early mathematics gender gap. Paper to be presented at the American Educational Research Association, Vancouver.

Sadker M., & Sadker, D. (1986). Sexism in the classroom: From grade school to graduate school. *Phi Delta Kappan, 68*, 512-515.

Sommers, C. H. (2000). *The war against boys: How misguided feminism is harming our young men*. New York, NY: Touchstone Books.

Snyder, T. D. & Dillow, S. A. (2011). *Digest of Education Statistics 2010 (NCES 2011-015)*. National Center for Education Statistics, Institute of Education Sciences, U.S. Department of Education. Washington DC.

Panel on "Gender and Mathematics Education Revisited"—Final Comments

Gilah C. Leder
Monash University, Australia
e-mail: Gilah.Leder@monash.edu

> In our culture … being "good in math" is 'being bright', and being bright in mathematics is associated with control, mastery, quick understanding, leadership. Unsuccessful mathematics implies the opposite … (Reisman and Kaufman 1980, p. 36)

The journey into the field of gender and mathematics education provided by the panelists served as a return visit to the field for some of the audience but signified a new, previously untraveled journey for others. Given the importance in many countries attached to mathematics, it is an intellectual journey well worth the effort. So what have we learnt?

Irrespective of the theoretical stance taken, it seems that there is considerable commonality in the external factors likely to facilitate or impede the pathway towards achieving gender parity and gender equality: the cultural, social, political and economic environments, systemic factors, historical precedents and community expectations.

Similarities permeate the different presentations. Despite decades of research it seems that evidence is still found of subtle but consistent gender differences in favor of males, particularly in mathematics performance and participation in post compulsory and advanced mathematics courses, on selected mathematical tasks on standardized or large scale tests, and among high performing students.

Some of the special interventions introduced in Western countries to redress demonstrated achievement disparities in mathematics learning have been taken up more widely, directly or with realistic adaptations.

Unanticipated between country differences were also reported. For example, research from Mexico suggested that girls are advantaged by technology—a finding not replicated in Australia. Perceptions (by the public in Australia) that teachers believe boys and girls are equally good at mathematics are seemingly at variance with reports from the USA that teachers rate boys and girls differently with respect to mathematics achievement.

Clearly, challenges remain before the goals of gender parity and gender equality are achieved, or even principally achieved, in an enlarged number of countries. The more modest goal of improved access for all, including females, to mathematics learning also remains elusive.

Constructive and contextually relevant recommendations have been made in the various panel presentations. The claim that "feminism has made its greatest contributions by asking new questions, often at odds with fundamental assumptions in a discipline" (Schiebinger 2001, p. 187) provokes a set of further questions which sharpen areas worthy of renewed and careful scrutiny. For example: Who, in our different countries, decides who should benefit from education; what mathematics should be taught, and to whom? Who determines educational and scientific priorities promoted for short and longer term funding? These are among practical starting points. For any changes in the current answers to be achieved, followed by constructive practical interventions, close cooperation between individuals and organizations is required. How well this challenge is met warrants careful and persistent monitoring.

References

Reisman, F. K., & Kauffman, S. (1980). *Teaching mathematics to children with special needs*. Columbus, OH: Charles E. Merrill Publishing Company.

Schiebinger, L. (2001). *Has feminism changed science?* Cambridge, Massachusetts: Harvard University Press.

Open Access This chapter is distributed under the terms of the Creative Commons Attribution Noncommercial License, which permits any noncommercial use, distribution, and reproduction in any medium, provided the original author(s) and source are credited.

Part IV
Awardees

Teaching Mathematics in Tomorrow's Society: A Case for an Oncoming Counter Paradigm

Yves Chevallard

Abstract The historical analysis of mathematics teaching at secondary level shows the succession in time of different school paradigms. The present paper describes and tries to analyse a new didactic paradigm, still at an early age, the paradigm "of questioning the world", which relies heavily on four interrelated concepts, that of inquiry and of being "Herbartian", "procognitive", and "exoteric". It is the author's ambition to show, however succinctly, how the present crisis in mathematics education could hopefully be solved along these lines, which preclude recurring to strategies seeking only to patch up the old, still dominant paradigm "of visiting works".

Keywords Anthropological theory of the didactic · Inquiry · Mathematics · Paradigm of questioning the world · Research and study path

The Anthropological Theory of the Didactic

I formally began working on mathematics education when I joined the Institute for research on mathematics teaching (IREM) in Marseilles (France) more than forty years ago—in February of 1972 to be precise. I write these lines qua 2009 recipient of the Hans Freudenthal Medal, an honour of which I am immensely proud. It is thus my wish to respond to it by indulging in a quick outline of the main conclusions at which I have arrived, letting interested readers judge for themselves the cogency of such views.

First of all, I must say that this presentation will draw upon the theoretical framework which my name has come to be associated with, I mean ATD, i.e. the *anthropological theory of the didactic*. Just as there are economic or political facts, there are *didactic* facts, which I will refer to as a whole as *the didactic*. The didactic is

Y. Chevallard (✉)
Aix-Marseille University, Marseille, France
e-mail: y.chevallard@free.fr

© The Author(s) 2015

S.J. Cho (ed.), *The Proceedings of the 12th International Congress on Mathematical Education*, DOI 10.1007/978-3-319-12688-3_13

a vital dimension of human societies. In a slightly simplified way, one can say that it is made up of the motley host of social situations in which some person does something—or even manifests an intention to do so—so that some person may "study"—and "learn"—something. The something to be studied (and learnt) is known as the *didactic stake* in the situation. As you can see, this formulation formally refers to *two* persons. I will use the letter y to denote the first person, and the letter x to denote the second, so that we can say that y does, or intends to do, something to help x study (and learn) something. Of course, at times, y and x can be one and the same person. In such a (fundamental) case of self-directed learning, x helps him/herself study the didactic stake. The "something" that y does or intends to do is metaphorically called a *didactic gesture* and is part of the didactic as a whole.

Basically, didactics is the science studying the conditions that govern such "didactic situations", i.e. social situations which hinge on some "didactic triplet" comprising some x, some y, and some didactic stake O. The didactics of mathematics is concerned with those cases in which the didactic stake O is regarded as pertaining to mathematics. More generally speaking, O is what is called, in ATD, a "work", i.e. anything, material or immaterial, created by deliberate human action, with a view to achieving definite functions. To obtain more generality, let me substitute a set X of persons for the person x, arriving thus at the "didactic triplet" (X, y, O), which can model a typical high-school class—X being the group of students, and y the teacher to whom it befalls to teach the work O. Naturally, we can also consider triplets of the form (X, Y, O), where Y is a team of didactic "helpers" that may include a full-fledged teacher alongside "assistants" of different kinds. Let me add here that, in ATD, a condition is said to be a *constraint* for a person or an institution if it cannot be modified by this person or institution, at least in the short run. Now the basic question in didactics is somewhat the following: given a set of constraints K imposed upon a didactic triplet (x, y, O), what conditions can x and y create or modify—i.e. what *didactic gestures* can they make—in order for x to achieve some determined relation to O? This will be the starting point for what follows.

The Paradigm of Visiting Works and Its Shortcomings

The prospective view on the didactic dimension in our societies that I wish to make explicit—and, I hope, clear—can be encapsulated in a crucial historical fact: the old didactic paradigm still flourishing in so many scholastic institutions is bound to give way to a new paradigm still taking its first steps. To cut a longer story short, I define a didactic paradigm as a set of rules prescribing, however implicitly, what is to be studied—what the didactic stakes O can be—and what the forms of studying them are.

The "old" paradigm I've just mentioned has been preceded by a number of distinct, sometimes long-forgotten paradigms. The most archaic of these didactic paradigms disappeared, in many countries, during the nineteenth century. In the field of mathematics as well as in many other fields of knowledge, it was organised around the study of *doctrines* or *systems*—of mathematics, of philosophy, etc.—approached from outside and considered as outstanding achievements in the history of human creation. Within this paradigm, one used to study Euclid's *Elements* in the way most of us may still study (or aspire to study) Plato's or Hegel's systems of philosophy. This initial paradigm—which I call the paradigm of "hailing and studying authorities and masterpieces"—has gradually given way to the school paradigm that nowadays all of us, willingly or not, are supposed to revel in, which evolved in the course of centuries from the older paradigm of studying "grand systems". The "great men" supposed to have authored those systems were waved aside and the systems crushed into smaller pieces of knowledge of which the authorised labels—Pythagoras, Thales, Euclid, Gauss, etc., as far as mathematics is concerned—still record their origins.

In the framework of the anthropological theory of the didactic, this paradigm is known as the paradigm of "visiting works" or—according to a metaphor used in ATD—"of visiting monuments", for each of those pieces of knowledge—e.g., Heron's formula for the area of a triangle—is approached as a monument that stands on its own, that students are expected to admire and enjoy, even when they know next to nothing about its *raisons d'être*, now or in the past.

In spite of the long-standing devotion of so many teachers and educators to this unending intellectual pilgrimage, notwithstanding the often admirable docility of so many students in accepting the teacher as a guide, this once pervasive paradigm is currently on the wane. This has come to be so, it can be argued, because the paradigm of visiting monuments tends both to make little sense of the works thus visited—"Why does this one happen to be here?", "What is its utility?" remain generally unanswered questions. The interested reader may want to check how this applies to a number of mathematical entities. For example, what purpose does the notion of *reflex* angle serve? The same question can be raised about angles in general, and also about parallel lines, intersecting lines, rays, line segments, and so on. Of course, the same goes for the reduction of fractions or polynomial expansion, with the notion of decimal number, and what have you. In what situations can this mathematical entity prove useful, if not utterly unavoidable, and how? Because these questions are usually hushed up—visiting a monument is no place to raise "What for?" or "So what?" questions—, students are reduced to almost mere spectators, even when educators passionately urge them to "enjoy" the pure spectacle of mathematical works.

A number of factors explain at least partially the long dominance of the paradigm of visiting works as monuments as well as its present decline—and, I suggest, its impending demise. Historically, the first cause seems to be the congruity of this paradigm with the social structure of formerly undemocratic countries or, since more recent times, weakly or incompletely democratic. Such societies are founded on an all-pervasive pattern inseparably linking those in command positions, on the one hand, and those in obedience positions, on the other hand. Almost all institutions (be they families, schools, or nations) hinge on some replica of this fundamental, dualistic pattern. I shall not go into debate, here, about this age-old social structuring. I only want to emphasise the specific risks that the functioning of this ubiquitous power structure easily generates, in the form of abuses of authority, power, or rank—call them as you like. The existence of a dualistic configuration with one in authority and one in obedience may for sure be vindicated, on a "technical" basis, as needed to keep institutions going. But such a technically justified twofold structure is normally limited in time and, above all, *in scope*. Authority is, or should be, restricted to a specified number of specific situations, and should therefore refrain from encroaching on every aspect of life—unless it changes into tyranny. But respecting this rule is not everyone's forte. The classical paradigm of visiting "monuments of knowledge," however small, suffers today, at many levels, from the constant abuses of pedagogic power that its historical kinship with the dualistic pattern of power mechanically generates.

The consequences of this historical situation are many. First and foremost, I shall mention a consequence already alluded to: the resistless evolution of the school mathematics curriculum towards a form of epistemological "monumentalism" in which knowledge comes in chunks and bits sanctified by tradition and whose supposed "beauty" has been enhanced by the patina of age; that students have to visit, bow to, enjoy, have fun with and even "love". All this of course is but a daydream, as far as the mass of students—not the happy few, who need very little attention—is concerned.

The main effect of this long-term situation is the growing tendency among students to develop a relation to "official", scholastic knowledge in agreement with what I shall term the "Recycle bin/Empty recycle bin" principle: all the knowledge taught may legitimately be forgotten or, more exactly, *ignored*, as soon as exams have been passed. Of course this is presumably as old as the school-and-exam system. But it has shaped a relation to knowledge as driven by institutional, short-term, and labile motives, which stands away from the functional approach to knowledge based on its real-world utility—to understand a situation, be it mathematical or not, make a decision, or postpone it to allow for further study of the problem addressed.

A correlate, if not properly a consequence, is to be found in a yet more challenging fact: what little knowledge remains after the school years is rarely regarded as something that could bear on situations one might face outside school—and this seems particularly true in the case of mathematical knowledge. School-generated knowledge tends therefore to be unusable, in that its "remnants" are unable to perform their specific function. But there is more to it than that. Visiting a monument basically boils down to listening to a report or account made by the teacher-guide about the monument visited—what we call in the French of ATD an *exposé*, a word from whose meaning the negative connotation it has acquired in English must be expelled in this context. By its very nature, any account, a report, or an exposé skips "details", i.e. aspects that, more or less arbitrarily, choice-makers have ignored or altogether discarded. To give just one example, in the French curriculum—as is the case, I presume, in many other mathematics curricula across the world—, tradition has it that the algebraic solving of cubic equations is overlooked, while quadratic equations are emphatically considered. In his/her scholastic visit of the mathematical universe, the student thus reaches an endpoint beyond which lie mathematical _territories that, more often than not, will remain indefinitely terra incognita to him/her. What will be of this student if, in later life, they need to know what a cubic equation is and how it can possibly be solved? School education along the lines of the current paradigm has no clear answer to that question, it seems.

The relation to knowledge and ignorance thus associated with the visiting of mathematical works has become increasingly unsuited to people's needs and wants, up to the point that there currently exists a widespread belief that mathematical knowledge is something one can almost altogether dispense with—whereas, in a not so remote past, mathematics could be regarded as the key to a vast number of individual as well as collective problems. In this respect, the chief flaw in the paradigm of visiting monuments, which relates to the undemocratic ethos in which this paradigm originated, has to do with the choice of "monuments" to visit at school. As we know, this choice is usually the combined result of a long-lasting tradition, on the one hand, and of irregularly spaced, hectic reforms, on the other. In no way, it seems, the decisions made go beyond what the people in charge of this choice-making think opportune, fit, or even "good" for the edification of the mounting generations. In no way, it seems, is the choice of the monuments to be visited made on an experimental basis or at least on a large and supposedly relevant experiential basis. In what follow, I will try to adduce evidence that such a "feat" can be achieved provided we opt for the emerging didactic paradigm I call the "paradigm of questioning the world".

Questioning the World: Towards a New Didactic Paradigm

Up to a point, we might soon discard the current didactic world in favour of a new paradigm which, when contrasted with the old one, looks like a *counterparadigm*— although, as we shall see, it isn't doomed to break off all contact with its predecessor. The main changes that I shall stress are few but radical. Let us consider again a triplet (X, Y, O). An almost inconspicuous but crucial tenet of traditional education is that the members x of X are children or adolescents: traditionally, the educational endeavour is about young people, before they attain maturity. When maturity has been reached, everyone is supposed to be educated—well or badly, that is another question. In contrast with this view of education, in the didactic paradigm of questioning the world, education is a lifelong process. The x in the triplet (x, y, O) can be a toddler as well as a mature adult or an older person. A society's didactic endeavour is regarded (and assessed) as applying to all—to citizens no less than to future citizens. Consequently, the assessment of this crucial endeavour can no longer focus on young people only: not only should we explore what 15-year olds happen to know, but we should extend this quest to people aged 30 to (at least) 70. More than anything, society's didactic effort is not simply known by what people know: it should be appraised on the basis of what they can *learn*—and *how* they can do so.

A second, central tenet of the paradigm of questioning the world is that, in order to learn something about some work O, x has to *study* O, often with the help of some y. You don't learn to solve a cubic equation by chance; you have to stop and consider the question that arises before you. In today's common culture, many people, it seems, have a propensity to shun every question to which the answer is not obvious to them. What the new didactic paradigm aims to create is a new cognitive ethos in which, when any question Q arises, x will consider it, and, as often as possible, will *study* it in order to arrive at a valuable answer A, in many cases with a little help from some y. In other words, x is supposed not to systematically balk at situations involving problems that he/she never came across or never solved. For reasons I shall not comment on, I call *Herbartian*—after the German philosopher and founder of pedagogy Johann Heinrich Herbart (1776–1841)—this receptive attitude towards yet unanswered questions and unsolved problems, which is normally the scientist's attitude in his field of research and should become the citizen's in every domain of activity.

The new didactic paradigm wants the future as well as the full-blown citizen to become Herbartian. Let me give three easy, miscellaneous examples of possibly impending "open" questions. First example: many people engaged in social science research but who have had little contact with statistics during their school or college years may come across Pearson chi-squared test, bump into the elusive notion of degrees of freedom, and become obsessed with the question "What does the expression 'degrees of freedom' mean exactly?" Second example: physics students may be upset about having to use the curious symbol "proportional to" ($\propto$), "an eight lying on its side with a piece removed" (Miller 2011), without having the slightest idea about how the manipulation of this symbol can be justified in mathematical terms, particularly as concerns the intriguing conclusion that, if a variable z is proportional to variables x and y, then z will also be proportional to their product xy. Third example: anyone interested in the question of biodiversity may stumble upon a mathematical equation such as this:

$$H_e = 1 - \frac{1}{1 + 4N_e\mu} \tag{1}$$

For the unrepentant non-mathematician, the first question will be: "What does that mean? What does that entail?" For all of us, I suppose, a second question will soon emerge: "Where does it come from? How can it be arrived at?" Of course, the pre-Herbartian citizen generally ignores all these questions because he/she usually recoils from anything seemingly mathematical. But the citizen in tune with the new didactic paradigm will face the questions, and, whenever possible, will come to grips with each of them. How is that possible?

In the didactic world shaped by the paradigm of visiting monuments, most people behave "retrocognitively". I use the word "retrocognition" not in its old parapsychological sense but simply to express the cognitive attitude that leads one to refer preferentially and almost exclusively to knowledge *already known* to one. Retrocognition in this sense is governed by the quasi-postulate according to which, once your school and college years are over, if you don't know in advance the answer to the question that faces you, then you'd better renounce all pretension to arrive at a sensible answer. This, of course, correlates with the propensity I mentioned earlier for staying away from unheard-of questions. By contrast, the paradigm of questioning the world calls for a very different attitude, that I dub *procognitive* (in a sense unrelated to the use of the word in denoting a drug that "reduces delirium or disorientation"), and which inclines one to behave as if knowledge was essentially still to discover and still to conquer—or to rediscover and conquer anew. In the retrocognitive bent, therefore, knowing is "knowing backwards"; whereas in the procognitive dedication, knowing is "knowing forwards".

In the scenario I present, how does one construct and validate an answer A to a question Q? Basically, inquiring into a question Q requires a twofold move. In the first place, the "inquirer" x will search the relevant literature for existing answers to question Q—a move traditionally banned at school, while to the contrary it is unavoidable in scientific research. In ATD it is common to denote an existing answer by the letter A with a small lozenge or diamond—a "thin" rhombus—in superscript, $A^{\diamond}$, in order to express that such an answer has been created and diffused by some institution which, in some sense, hallmarked it. Of course an answer $A^{\diamond}$ needs not be "true" or "valid"; but it is up to x to evaluate answers $A^{\diamond}$ to see if they are relevant—which also departs from school usage, in which answers provided by the teacher are guaranteed by the same token. In order to arrive at a proper answer—usually denoted by the letter A with a small heart in superscript regarded as the "maker's mark": $A^{\heartsuit}$—, the inquirer x has to use "tools", mathematical or not, i.e. works of different nature. It is from the combined study of the "hallmarked" answers $A^{\diamond}$ and of the works O (used as tools both to study answers $A^{\diamond}$ and to construct an answer $A^{\heartsuit}$) that the process of research for an answer $A^{\heartsuit}$ will get under way.

The inquiry led by x into Q opens up a path called a *research and study path* (or trail, or track, or course, etc.). To proceed along this path, the inquiry team X has to use knowledge—relating to answers $A^{\diamond}$ as well as to the other works O—hitherto unknown to its members, that the team will have to get familiar with to be able to continue on the trail towards answer $A^{\heartsuit}$. A necessary condition in this respect is for X and for every member x of X to behave *procognitively*, looking forward to meeting new knowledge—new works—without further ado.

Some more didactic aspects should be stressed here. Firstly, in the paradigm of questioning the world, encountering new knowledge or e-encountering old, half-forgotten knowledge along the research and study path is the way that inquirers x *learn*—they learn or relearn the answers $A^{\diamond}$, the working tools O and, finally, the answer $A^{\heartsuit}$. It should then be clear that the *contents* learnt, in this context, *have not been planned in advance*—contrary to what is usual in the paradigm of visiting monuments—and are determined essentially by two factors: by the question Q being studied, in the first place, and then by the research and study path covered, which in turn is determined by the $A^{\diamond}$ and the O encountered and studied in order to build up the answer $A^{\heartsuit}$. Secondly, it must be emphasised that studying a (mathematical or non-mathematical) work O—the same holds for the answers $A^{\diamond}$—is determined by the project of arriving at an answer $A^{\heartsuit}$. Contrary to the fiction forced upon x and y in the paradigm of visiting works, there is no such thing as a "normal" or "natural" study of a given work O. All exposés are special, none is exhaustive, and most fail to conceal their arbitrariness. The study of a work O in the context of an inquiry into some question Q will heavily depend, both quantitatively and qualitatively, on the use of O in the making of the answer $A^{\heartsuit}$. What should be clear in such a context-bound study of O is that the knowledge of O thus acquired by the investigators is *functionally coherent* because it is cohered by the inquiry into question Q, so that the *raisons d'être* of O that do explain its use in the case in point are readily apparent.

Society, School, and the New Paradigm

The paradigm of questioning the world and the inquiries that make it a reality do not exist in a vacuum. They must have a basis in society and in school. Once again let me stress here that the field of relevance of the didactic schema—called the *Herbartian schema*—outlined so far extends to the whole of society—it is not conceived as being restricted to school. Any person can represent x in a didactic triplet (x, y, O). [A didactic "helper" y may fail to exist, in which case it is common to write the triplet in the form $(x, \emptyset, O)$: the didactic triplet is then reduced in actual fact to a 2-tuple.] Of course it is easy to spot an outstanding difference. In many modern societies, going to school during the first part of one's life—while you're a youngster—is compulsory. Admittedly, there is no such thing as compulsory education for adults in general. In this respect, the scenario advocated here supposes a fundamental change, with the extension of the right to education into the right to *lifelong* education for all, provided by an adequate infrastructure that we could continue to call "school", but in a sense that goes back to ancient Greece and, more precisely, to the Greek word *skhole*, which originally designated spare time devoted to leisure (this was still its meaning in the time of Plato, for example), but which evolved to mean "studious leisure", "place for intellectual argument", and "time for liberal studies". The new role of the didactic in our societies thus implies the development of a ubiquitous institution that, in what follows, I shall term, more genuinely, *skhole*. Of course, school as we know it is a key component of *skhole*, even though, in its present form, it remains largely foreign to the new didactic paradigm. But *school* is not all of *skhole*. For example, for adults as well as for younger people, a good part of *skhole* takes place at home: home *skhole*ing will be, and already is, a master component of *skhole*. In what follows, *skhole* will be approached for its capacity to favour the development and flourishing of the paradigm of questioning the world—even though parts of it are still under the control of the old school paradigm.

I begin by considering the case of adults' *skhole*ing—of which today's "adults schooling", as we may call it, is but a meagre component. In truth, many citizens are already, though partially, equipped to inquire on their own into the many questions that may beset them, for example in their daily life. This being noted, what are the main constraints that hinder, and what are the conditions that might favour the development of adults' *skhole*ing? The first condition lies in the fact that, instead of fleeing when faced with questions, x duly confronts them. To do so, x has to formulate them explicitly, at least for him/herself. Simple as it may sound, such a move conflicts with a fundamental determinant of our cultures, the disjunction between "masters" and "underlings", if I may say so, that forbids the latter to raise questions about the world—natural or social—, or, as the saying goes, to put it "into question", while "masters" have alone the legitimacy to question the world and to change it. Sheer observation—but this conclusion can easily be submitted to experimentation—shows that most people get excited at daring to pose on their own the merest question. Historically, posing questions was the privilege of the mighty, although it has become a defining right of citizens; but it is a right not yet exercised as it should in a fully developed democracy.

Let us suppose that some citizen has decided to inquire into some question Q, becoming thus an inquirer x in a triplet $(x, ?, Q)$. At this stage of his/her study, two problems face him/her. On the one hand, x may think of getting help from some people Y; on the other hand, he/she will have to "search the world" for answers $A^\diamond$ to question Q and relevant works O. The first of these two problems has no systematic solution today. The second problem has a good approximate solution. It consists in the sum total of the information provided by the Internet and especially the Web. In fact, I shall refer to the Internet *sensu latissimo*—in the broadest sense—, a sense that, against current usage, includes... all the libraries in the world, because any document is either available on the Internet or can be regarded as *not yet* available on the Internet. To take here just one example, in the case of an inquiry into the mathematics of the "proportional to" symbol ($\propto$), when starting from Jeff Miller's well-known website on the *Earliest uses of symbols of relation* (2011), one is led to Florian Cajori's classic book on the history of mathematical notations (1993, vol. 1, p. 297), which in turn refers the inquirer to three older books, authored respectively by Emerson (1768), who was the introducer of the symbol $\propto$, Chrystal (1866), and Castle (1905). Today, all of these books are available online for free. Let us also observe that the Internet allows most inquirers x to find help from occasional helpers y, for example on Internet forums and discussion threads, so that the main solution to the second problem also supplies a (partial) solution to the first problem.

Making inquiries on the Internet *sensu latissimo* meets with well-recognised difficulties. First, if x is almost certain to come across at least some relevant resources, documents allowing him/her to go further and deeper into the question studied may be scarce. Second, the inquirer x can prove unable both to find out relevant documents that do exist and to make the most of what little information he/she culled. The inquirer's intellectual equipment—or more exactly the inquirer's *praxeological* equipment, in a sense of the word *praxeology* proper to ATD—thus rests on two pillars: the capacity to locate resources, online and offline, and the knowledge necessary to take advantage of them. This leads to the question of making good use of the works O gathered. Most general questions Q entail the use of works O pertaining to different branches of knowledge, so that the study of Q is bound to be a co-disciplinary pursuit, bringing together for a common endeavour tools from different "disciplines". It should be stressed at this point that what I've called a citizen is not a person reduced to being a member of a political community. But, much to the contrary, he/she is considered according to his/her accomplishments and potential, particularly as an inquirer into questions of any breed. It results from this that a citizen does not only have to be educated in many fields but, in the procognitive perspective of the new didactic paradigm, a citizen must be ready to study and learn, even from scratch, fields of knowledge new to him/her. A citizen is not only a law-abiding person; he/she also has to become a knowledgeable person, indefinitely ready to study works hitherto unknown to him/her, just because some inquiry calls for their study.

The citizen I portray here may feel unable to live up to what is thus required of him/her. This feeling essentially results from the old didactic organisation of school and society that has imposed upon us the illusion according to which, for any knowledge need we may experience, there somewhere exists a providential person who can teach us whatever we want to know. Such a puerile belief leads to passivity and submission to events outside our reach. In the paradigm of questioning the world, attending a course or a conference on some subject of interest is certainly not disregarded. But we should take them as means to a common end—learning something on some determined work O supposed to be useful in order to bring forth an answer $A^\heartsuit$ to question Q. In such a situation, because of a relation to ignorance and knowledge resulting from exposure to the old school paradigm, we are prone to feel frustrated at not having all the knowledge needed—all of history, biology, mathematics, physics, chemistry, philosophy, linguistics, sociology, and so on indefinitely. The character implicitly fantasised here is what I've come to call *an esoteric* (using thus the adjective also as a noun), who is supposed to already know all the knowledge needed (the idea most people have of "a historian", "a biologist", "a mathematician", "a physicist", etc., is commonly akin to this fantasy). By contrast, *an exoteric* has to study and learn indefinitely, and will never reach the elusive status of esoteric. Indeed, all true scholars are exoteric and should remain so in order to remain scholars: esotericism, as I define it here, is a fable.

The citizen in the new paradigm is therefore called upon to become Herbartian, procognitive, and *exoteric*. How can we promote this new citizenship? Beyond being possessed by the epistemological passion necessary to go all the way from pure ignorance to adequate knowledge, a crucial condition is, for sure, the *time* allotted to study and research in an adult's life. More often than not, it seems, this time tends to zero as years pass by. In this respect, I suggest that we repeat again and again the founding trick of the ancient Greeks—that of transmuting leisure time, which some of our contemporaries seem to enjoy so abundantly, into study and research time, in the authentic tradition of *skhole*. Such a pursuit pertains to what Freud once called *Kulturarbeit*, "civilisational work"—a radical change still to come, which is a sine qua non of the emergence of the new didactic paradigm.

The problem of the time allotted to study and research has an easy solution when it comes to ordinary schooling: youngsters go to school to study, in accordance with *skhole*'s defining principle. But in what measure does school welcome the new didactic paradigm? I shall not dwell too long on this subject. I will, however, suggest that in too many cases, the so-called "inquiry-based" teaching resorts to some form or another of "fake inquiries", most often because the generating question Q of such an inquiry is but a naive trick to get students to find and study works O that the teacher will have determined in advance. Of course, this is the plain consequence of the domination of the paradigm of visiting works, which implies that curriculum contents are defined in terms of works O. In contradistinction, in the paradigm of questioning the world, the curriculum is defined in terms of *questions Q*. However, the works O studied in consequence of inquiring into these questions Q play a central role in the process of defining and refining the curriculum: starting from a set Q of "primary" questions, the curriculum contents C eventually studied will include the questions Q and answers $A^{\heartsuit}$, together with the answers $A^{\diamondsuit}$ and the works O.

At this point two questions arise, though. The first question relates to the set Q of "primary" questions: where do these questions come from, and according to what mechanisms? In the case of a national curriculum, the set of primary questions to be studied at school constitutes the "core curriculum", and therefore the foundation of the national pact between society and school. Consequently, it is up to the nation to watchfully and democratically decide what the set Q will consist of and to periodically revise and update its contents on the basis of a careful monitoring of the curriculum's life-cycle. Because it is essential to the relationship between a society and its schooling system, the core curriculum—i.e. the "primary" questions—will play a decisive part in the society's *skhole*. But it should be obvious that the curriculum is not precisely defined by the primary questions alone. The inquiries entailed by these questions are in no way uniquely defined: as we know, an inquiry may follow different paths of study and research, and the questions inquired into as well as the other works encountered and, up to a point, studied, are indeed path-dependent. As a result, even if the *core* curriculum (in the sense defined above) has been made precise, the ensuing curriculum might well look fuzzily defined because of its built-in variability. How can this situation be managed for the better?

Let us consider didactic triplets (X, Y, O) with O a (finite!) family of questions. We can envisage two types of didactic triplets associated with a class of students. First, there is a *seminar*, in which O is a dynamic family of questions comprising the primary questions and the questions their study will generate. (Remember that the scenario delineated is supposed to apply to advanced students as well as to... toddlers, so that the words I use here must be taken in a very broad sense, which allows for their adaptation to a wide variety of concrete conditions.) This seminar will essentially be co-disciplinary, for primary questions rarely fall into a unique disciplinary domain. Second, there will be disciplinary *workshops* to study the questions and works put forward in the seminar but which pertain essentially to a given discipline—there will be for example a chemistry workshop, a mathematics workshop, a history workshop, a biology workshop, and so on. The activated workshops may vary depending on the primary questions studied in the seminar. The key fact is that, in this two-step process (seminar *plus* workshops), some works O and disciplines will be insistently recurrent, because they will be more often called upon in the inquiries, while others will be encountered erratically or will almost never turn up. This "degree of mobilization" of a work O, if averaged nationally across all the seminars held at a given school level, gives the "degree of membership" of the work O to the curriculum regarded, metaphorically, as a continually redefined fuzzy set—a view more adequate to the true nature of a real curriculum. As indicated above, and contrary to the age-old habit of imposing a curriculum founded essentially on opinion, the paradigm of questioning the world makes it possible to bring to light in an organic way which resources are really used in trying to question and know the world, both natural and social.

What Will Be the Place of Mathematics?

At a given point in time, an inquiry may come to a stop because some useful tool proves unavailable to the inquirers. One major reason for which an inquiry may thus grind to a halt is that the mastery of essential parts of some work O, ideally required to continue progress, lie well beyond the inquirers' reach. This, it should be stressed, is the common law of inquiry, be it at school or in a research team, and is definitely *not* the preserve of "low-level exoterics": it is part and parcel of the art of inquiry—such an "incident" is but one of the twists and turns in an inquirer's venture. But the path followed in a given inquiry, whatever its determinants, has crucial consequences in the didactic scenario displayed above: if a work O is very rarely drawn upon in seminars and workshops across the nation, then this work O will eventually vanish from the national curriculum. To be quite frank, this can result in the disappearing of parts of traditional school disciplines; for the place occupied by a discipline in the new curriculum will depend on its effectiveness in providing tools for inquiring into the curriculum-generated questions; it will depend no longer on any formerly or recently established hierarchy of disciplines, held to be the unquestionable legacy of the past. Traditionally flourishing disciplines should then worry about their future at school: will they continue to thrive or will they soon languish? The question is put to every discipline, and especially to mathematics.

If knowledge is valued according to what it enables us to rationally understand and achieve, the problem we are confronted with is not so much the fate of the disciplines as the value and quality of the inquiries going on in the seminars and workshops. From this point of view, the foregoing scenario can be improved substantially by allowing for the possibility to append "control questions" to any question pertaining to the curriculum. In some sense, this adds, to the bottom-up information flow emanating nationwide from the seminars and workshops, a top-down regulatory control on schools, operated by supervisory authorities. Any question Q can indeed be supplemented meaningfully by one or a series of "side questions" $Q*$ that will be touchstones for controlling the quality, thoroughness and profundity of an inquiry into question Q. It is in this way that it becomes possible to point out meaningfully—and not out of sheer pretentiousness—the utility of such and such work O to get deeper into the question studied. For example, to a question about biodiversity, one might relevantly add a question about *genetic* diversity and, in turn, a question about the meaning and interest of Eq. (1) above, a question likely to draw the inquirers' attention to the importance of... mathematics in inquiring into genetic diversity.

For mathematics as well as for a myriad of works pertaining to the most varied fields of knowledge, such a system of control questions seems indispensable to remind the x and the y that inquiring into some question may require the use of tools that will first appear, from within the cultural limits that they are precisely expected to transcend, as far removed from the matter under study. This is particularly true in the case of mathematical works. For deep-rooted historical reasons, mathematics is today both formally revered and, at the same time, energetically shunned. Numerous people flee away from mathematics as soon as they are no longer obliged to "do" mathematics. This has determined many mathematics educators to engage in a strategy of seduction, with a view to regaining the favour of "mathematical non-believers" by convincing them that, as the saying goes, "maths is fun"! Let me say tersely that this strategy has two main demerits and that, in my view, it should be as such utterly discarded. The first defect seems to be liberally ignored in today's educational world: for deep political and moral reasons, the instruction imparted at school must refrain from manipulating feelings and beliefs—we must be unimpeachable as far as the liberty of conscience of x (and y) is concerned. Consequently, mathematics educators must resist the temptation to try to induce students to "love" mathematics: their unique mission is to let them *know* mathematics, which is a bit more demanding! Love and hate are personal, intimate feelings that belong to the private sphere proper. Of course, it is highly probable that knowing mathematics better will result in some form of keenness towards mathematics. But all this entirely pertains to every single person's conscience.

The second defect of the much acclaimed seduction strategy is its very low yield, if I may say so. The problem with mathematics—as with other disciplines—is a mass problem. The root of it lies, in my view, in the process of cultural rejection that mathematics has suffered for a long time now, with the crucial consequence that, outside mathematical institutions proper, mathematics vanishes from the "lay" scene, so much so that many documents about topics not substantially foreign to mathematics can show no trace at all of mathematics, a fact which jeopardises the quality of many inquiries. Let me give here a simple example. Consider the question "Why does ice float in water?" Part of the answer is: because ice is less dense than liquid water. Now why is ice less dense than liquid water? The usual answer is that the arrangement of H_2O molecules occupies *more* space in ice than in liquid water. A closer look at this answer leads to some easy calculations (Ravera 2012). Indeed, it can be shown that, under certain conditions, the unit cell of ice has a height of 737 pm (i.e. 737×10^{-12} m), with its base a rhombus with sides of length 452 pm and an angle of 60°. The volume of the unit cell is therefore

$$V = \frac{\sqrt{3}}{2} \times 452^2 \times 737 \times 10^{-33} \text{L} \qquad (2)$$

The molar mass of water is approximately 18 g/mol. The mass of a unit cell of ice is known to be that of four molecules of water. Avogadro's number is taken here to be 6.02×10^{23} mol^{-1}. Hence the mass M of a unit cell:

$$M = \frac{4 \times 18}{6.02 \times 10^{23}} \text{g} \tag{3}$$

The density of ice is therefore:

$$d = \frac{M}{V} \approx 917 \text{g/L} \tag{4}$$

This (approximate) result confirms that ice is lighter than liquid water. The calculation uses elementary tools that are all (supposedly) mastered at age 15. In spite of this, this calculation is generally withheld from most relevant presentations available on the Internet. This is no exception to the rule. In a majority of cases, the mathematics of the topic being presented is decidedly absent, as if it had never existed. This is typically what mathematics educators must combat. In this respect, as far as mathematics is concerned, the "touchstone questions" that should be appended tentatively to any question proposed for study come down to this: "What are the mathematics of the matter, and how can awareness of them enhance the quality of your answer?"

Is this really a way out of the historic trap in which mathematics has been lured? I believe so. The seduction strategy, which is successful with an insignificant number of people, is but another pitfall. In my view, the only realistic solution will consist in trying to rationally persuade the citizens and, to begin with, the students that dispensing with mathematics may crucially impoverish our understanding and drastically reduce the quality of our involvement in both the natural and the social world. This, of course, will not be achieved through fine words only. It needs daily action, in schools as well as outside schools, especially in the leisure time given to learning by the citizenry to enrich their lives. In this pursuit, mathematics educators will play a crucial, though different, part.

For centuries, mathematics as a cultural institution thrived on a twofold self-presentation: it was understood as being composed, on the one hand, of "pure" mathematics, and, on the other hand, of "mixed" mathematics, with its pervasive ethos and slightly imperialistic touch. The "mixed" part, later called "applied" mathematics, has steadily declined at school during the last decades, while what remained of the former part—pure, though elementary, mathematics—tried to symbolise and maintain the old "empire". It is my belief that this time has now come to an end. Today, we have to revive the epistemological spirit of mixed mathematics, although without any cultural arrogance, but with the political and social will necessary to revitalise the idea that mathematics is for us, human beings, a solution, not a problem.

Open Access This chapter is distributed under the terms of the Creative Commons Attribution Noncommercial License, which permits any noncommercial use, distribution, and reproduction in any medium, provided the original author(s) and source are credited.

References

Cajori, F. (1993). A history of mathematical notations (Vols. 1-2). Mineola, N.Y.: Dover.

Castle, F. (1905). *Practical mathematics for beginners*. New York: Macmillan.

Chrystal, G. (1866). Algebra: An elementary text-book for the higher classes of secondary schools and for colleges (vol. 1). London: Adam and Charles Black.

Emerson, W. (1768). *The Doctrine of fluxions*. London: Richardson.

Miller, J. (2011, November 11). *Earliest uses of symbols of relation*. Retrieved from http://jeff560.tripod.com/relation.html

Ravera, K. (2012). Pourquoi la glace flotte sur l'eau. In Tangente (Eds.), *Mathématiques et chimie. Des liaisons insoupçonnées* (pp. 80-82). Paris: Pole.

Mathematics for All? The Case for and Against National Testing

Gilah C. Leder

Abstract National numeracy tests were introduced in Australia in 2008. Their format and scope are described and appraised in this paper. Of the various group performance trends presented in the annual national NAPLAN reports two (gender and Indigeneity) are discussed in some detail. For these, the NAPLAN findings are compared with broader international data. Recent Australian research spawned by, or benefitting from, the NAPLAN tests is also summarised. In some of this work, ways of using national test results productively and constructively are depicted.

Keywords National tests · Gender · Indigeneity

Introduction

> It should come as no surprise... that the introduction of a national regime of standardised external testing would become a lightning rod of claim and counter-claim and a battleground for competing educational philosophies. The National Assessment Program—Literacy and Numeracy (NAPLAN) is a substantial educational reform. Its introduction has been a source of debate and argument (Sidoti and Keating 2012, p. 3).

Formal assessment of achievement has a long history. Kenney and Schloemer (2001) point to the use, more than three thousand years ago, of official written examinations for selecting civil servants in China. The birth of educational assessment is, however, generally traced to the 19th century and its subsequent growth has undoubtedly been intertwined with advancements in the measurement of human talents and abilities (Lundgren 2011). Over time the development of large scale, high stake testing and explorations of its results have proliferated. "Many nations", wrote Postlethwaite and Kellaghan (2009), "have now established national assessment mechanisms with the aim of monitoring and evaluating the quality of their education systems across several time points" (p. 9). More recently, Eurydice (2011) also drew attention to the widespread practice of national testing

G.C. Leder (✉)
Monash University and La Trobe University, Melbourne, Australia
e-mail: Gilah.Leder@monash.edu; G.Leder@latrobe.edu.au

© The Author(s) 2015
S.J. Cho (ed.), *The Proceedings of the 12th International Congress on Mathematical Education*, DOI 10.1007/978-3-319-12688-3_14

189

throughout Europe, confined in some countries to a limited number of core curriculum subjects but in others comprising a broad testing regime. Large scale national assessment programs, with particular emphasis on numeracy and literacy[1], were introduced in Australia in 2008—after extensive consultation and much heated debate within and beyond educational and political circles.

The NAPLAN Numeracy Tests

Until 2007, Australian states and territories ran their own numeracy and literacy testing programs. Although much overlap could be found in the assessment instruments used in the different states, there were also variations—some subtle, others substantial—in these tests.

The first National Assessment Program—Literacy and Numeracy (NAPLAN) tests were administered in May 2008 and have been conducted annually since then. For the first time, students in Years 3, 5, 7, and 9, irrespective of their geographic location in Australia, sat for a common set of tests, administered nation-wide. The Numeracy tests contain both multiple choice and open-ended items. Their scope and content are informed by the *Statements of Learning for Mathematics* (Curriculum Corporation 2006). The 'what' students are taught is described by four broad numeracy strands. These are Algebra, function and pattern; Measurement, chance and data; Number; and Space, though some questions may overlap into more than one strand. Instructional strategy, the 'how' of mathematics is described by proficiency strands. "The proficiency strands—Understanding, Fluency, Problem solving and Reasoning—describe the way content is explored or developed through the 'thinking' and 'doing' of mathematics" (Australian Curriculum, Reporting and Assessment Authority ACARA 2010). In Years 3 and 5, the papers are expected to be completed without calculator use. Two distinct papers are set for Year 7 and 9 students—one is expected to be completed without the use of a calculator; for the other calculator usage is allowed.

The NAPLAN numeracy scores for Years 3, 5, 7, and 9 are reported on a common scale which is divided into achievement bands. For each of these year levels, the proportion of students with scores in the six proficiency bands considered appropriate for that level is shown. For Year 3, 5, 7, and 9 these are bands one to six; three to eight; bands four to nine; and bands five to ten respectively. Each year, results of the NAPLAN tests are published in considerable detail, distributed to each school, and made readily available to the public.

The advantages anticipated by the introduction of national tests to replace the variety of tests previously administered by the different Australian states and

---

[1] Sample assessment tests have been administered to selected groups of students in Years 6 and 10 in Scientific Literacy (Year 6 students only), Civics and Citizenship, and Information Communication Technology Literacy. These sample assessments were introduced respectively in 2003, 2004, 2005 and are held on rolling a three-yearly basis.

territories were similar to those commonly put forward in the wider literature (e.g., Postlethwaite and Kellaghan, 2009) as a rationale or justification for introducing national tests: assessment consistency across different constituencies, increased accountability, and a general driver for improvement.

ACARA is responsible for the development of the national assessment program and the collection, analysis, and reporting of data. The procedures followed are described clearly on the ACARA website and are consistent with those generally advocated for large scale assessment testings (Joint committee on testing practices 2004). Guidance on interpreting the vast amount of data in the National Report is provided in the document itself (ACARA, 2011a) and in multiple ancillary documents (see e.g., ACARA, 2011b; Northern Territory Government n.d). NAPLAN achievement outcomes are reported not only at the national level, but also by state and territory data; by gender; by Indigenous status; by language background status[2]; by geolocation (metropolitan, provincial, remote and very remote); and by parental educational background and parental occupation. Each of these categories which are clearly not mutually exclusive, has been shown, separately, to have an impact on students' NAPLAN score. Broad performance trends for the different groupings have been summarised as follows:

> In Australia, girls have typically performed better on tests of verbal skills…, while boys have typically performed better on tests of numerical skills… Children from remote areas, children from lower socioeconomic backgrounds and children of Indigenous background have tended to perform less well on measures of educational achievement (NAPLAN 2011b, p. 255).

It is beyond the scope of this paper to look at each of the categories mentioned above. Instead, the focus is on two groups of special interest: *girls/boys* and *Indigenous* students. What trends can be discerned in the years of NAPLAN data available at the time of writing this paper?

Trends in NAPLAN Data: Gender and Indigeneity

Data for Years 3 and 9 by gender and Indigeneity are shown in Tables 1 and 2 respectively.

From these tables it can be seen that:

Gender

- The mean NAPLAN score for males is invariably higher than that for females.
- The standard deviation for males is also consistently higher than for females, that is the range of the NAPLAN scores for males is higher than that for females.

---

[2] LBOTE, language background other than English, defined as "A student is classified as LBOTE if either the student or parents/guardians speak a language other than English at home."

Table 1 Numeracy Year 3 students, NAPLAN achievement data 2008–2011

Group \year		All	M	F	Indigenous	Non-Indigenous	Indigenous year 5[a]
2008	Mean	396.9	400.6	393.1	327.6	400.5	408.0
	S.D	70.4	72.8	67.6	70.6	68.4	65.8
	≥National min[b] (%)	95.0 %	94.6 %	95.5 %	78.6 %	96.0 %	69.2 %
2009	Mean	393.9	397.5	390.2	320.5	397.7	420.5
	S.D	72.9	75.3	70.0	76.0	70.6	66.4
	≥National min (%)	94.0 %	93.5 %	94.5 %	74.0 %	95.2 %	74.2 %
2010	Mean	395.4	397.8	392.9	325.3	399.0	416.9
	S.D.	71.8	74.0	69.3	71.2	69.8	70.5
	≥National min (%)	94.3 %	93.7 %	94.9 %	76.6 %	95.3 %	71.4 %
2011	Mean	398.1	402.6	393.5	334.4	401.7	421.1
	S.D.	70.6	73.0	67.6	65.0	69.1	64.0
	≥National min (%)	95.6 %	95.2 %	96.0	83.6 %	96.4 %	75.2 %

[a] I refer to the data in the last column later in the paper. To save space the information is included in this table

[b] National minimum standards: The second lowest band on the achievement scale represents the national minimum standard expected of students at each year level

Table 2 Numeracy Year 9 students, NAPLAN achievement data 2008–2011

Group \year		All	M	F	Indigenous	Non-Indigenous	Year 7 Non-Indigenous
2008	Mean	582.2	586.5	577.6	515.1	585.7	548.6
	S.D	70.2	72.0	68.1	65.6	68.7	71.6
	≥National min (%)	93.6 %	93.7 %	93.6 %	72.5 %	94.8 %	96.4 %
2009	Mean	589.1	592.4	585.6	520.2	592.4	547.0
	S.D	67.0	69.2	64.4	63.2	65.3	69.4
	≥National min (%)	95.0 %	94.7 %	95.2 %	75 %	96 %	95.8 %
2010	Mean	585.1	591.1	578.8	515.2	588.5	551.4
	S.D	70.4	72.7	67.4	64.7	68.8	70.8
	≥National min (%)	93.1 %	93.3 %	92.9 %	70.4 %	94.3 %	96.1 %
2011	Mean	583.4	589.3	577.3	515.8	586.7	548.5
	S.D	72.1	74.7	68.7	62.2	70.8	72.1
	≥National min (%)	93.0 %	93.0 %	93.0 %	72 %	94.1 %	95.5 %

(Data in both tables adapted from ACARA 2011a)

- At the Year 3 level a higher proportion of females than males score above the national minimum standard NAPLAN score. There is no such consistency at the Year 9 level, with a marginally higher proportion of males performing at or above the minimum level in some years (e.g., 2008, 2010) and a marginally higher proportion of females performing at or above the minimum level in other years (e.g., 2009).

Indigeneity

- Each year, non-Indigenous students do (a lot) better than Indigenous students. From Table 1 it can be seen that Year 5 Indigenous students performed just above the level of Year 3 non-Indigenous students; from Table 2 that Year 9 Indigenous students performed below the level of Year 7 non-Indigenous students.
- In 2011, there was a noticeable increase, compared with the previous years, in the percentage of Indigenous students at Year 3 who performed at or above the national minimum standard. No such increase is apparent at the other Year levels.

Also relevant are the following:

- In 2011, between 240,000 and 250,000 non-Indigenous students sat for the Years 3, 5, 7, and 9 NAPLAN papers. For the Years 3, 5, and 7 papers close to 13,000 Indigenous students participated. A smaller number, about 10,000 sat for the Year 9 paper. Thus at the different Year levels, Indigenous students comprised between 4 and 5 % of the national groups involved in the NAPLAN tests.[3]
- The exemption rates for the two groups are similar: around 2 % for Indigenous students and about 1 % for non-Indigenous students.

These summaries for gender and Indigenous performance outcomes are set against a broader context in the next sections.

Gender

In many countries, including Australia, active concern about gender differences in achievement and participation in mathematics can be traced back to the 1970s. Two reliable findings were given particular prominence: that consistent between-gender differences were invariably dwarfed by much larger within-group differences; and that students who opted out of post compulsory mathematics courses often restricted their longer term educational and career opportunities. These generalizations remain relevant.

---

[3] The proportion of school students in Australia identified as Aboriginal and/or Torres Strait Islanders has risen from 3.5 % in 2001 to almost 5 % in 2011(http://www.abs.gov.au/ausstats/abs@.nsf/Lookup/4221.0main+features402011).

Evidence of progress towards gender equity more broadly than with respect to mathematics learning specifically has been mapped in many different ways:

> Whereas the challenge of gender equality was once seen as a simple matter of increasing female enrolments, the situation is now more nuanced, and every country, developed and developing alike, faces policy issues relating to gender equality. Girls continue to face discrimination in access to primary education in some countries, and the female edge in tertiary enrolment up through the master's level disappears when it comes to PhDs and careers in research. On the other hand, once girls gain access to education their levels of persistence and attainment often surpass those of males. High repetition and dropout rates among males are significant problems (UNESCO 2012, p. 107).

As can be seen from large scale data bases such as NAPLAN, some gender differences in mathematics performance remain. What explanations for this have been proffered?

Explanatory Models

Over the years a host of, often subtly different, explanatory models for gender differences in mathematics learning outcomes have been proposed. They invariably contain a range of interacting factors—both person-related and environmental. Common to many models is an

> …emphasis on the social environment, the influence of other significant people in that environment, students' reactions to the cultural and more immediate context in which learning takes place, the cultural and personal values placed on that learning and the inclusion of learner-related affective, as well as cognitive, variables (Leder 1992, p. 609).

A comprehensive overview of research concerned with gender differences in mathematics learning is beyond the scope of this paper. Instead, some recent publications, the majority with at least a partial cross-national perspective and published in a variety of outlets, are listed to sketch the range of factors invoked as explanatory or contributing factors for the differences still captured. Included is work in which the need for a repositioning of perspective to examine gender differences, via a different theoretical (often feminist and/or socio-cultural) framework, is prosecuted, as well as several articles in which there are strong attempts to rebut the notion that gender differences persist.

Gender Differences: Possible Explanations

- Kaiser et al. (2012) found, in a large study involving over 1,200 students, that "the perception of mathematics as a male domain is still prevalent among German students, and that this perception is stronger among older students. This is either reinforced by the peer group, parents or teachers" (p. 137).
- Kane and Mertz (2012) concluded "that gender equity and other sociocultural factors, not national income, school type, or religion per se, are the primary determinants of mathematics performance at all levels of boys and girls" (p. 19).

- Stoet and Geary (2012) challenged but ultimately supported the notion of ste- reotype threat (provided it is carefully operationalized) as an explanation for the higher performance of males in mathematics, particularly at the upper end.
- Wai et al. (2010) examined 30 years of research "on sex differences in cognitive abilities" and focussed particularly on differences in favour of males found in the top 5 %. As well as highlighting the role of sociocultural factors they concluded: "Our findings are likely best explained via frameworks that examine multiple perspectives simultaneously" (p. 8).
- "Traditionally, all societies have given preference to males over females when it comes to educational opportunity, and disparities in educational attainment and literacy rates today reflect patterns which have been shaped by the social and education policies and practices of the past. As a result, virtually all countries face gender disparities of some sort" (UNESCO 2012, p. 21).

Gender Differences: Have They Disappeared?

- Else-Quest (2010) used a meta-analysis of PISA and TIMSS data to examine the efficacy of the gender stratification hypothesis (that is, societal stratification and inequality of opportunity based on gender) as an explanation for the continuing gender gap in mathematics achievement reported in some, but not in other, countries. They concluded that "considerable cross-national variability in the gender gap can be explained by important national characteristics reflecting the status and welfare of women" (p. 125) and that "the magnitude of gender differences in math also depends, in part, upon the quality of the assessment of mathematics achievement" (p. 125).
- Hyde and Mertz (2009) drew on contemporary data from within and beyond the U.S. to explore three major questions: (1) "Do gender differences in mathe- matics performance exist in the general population? (2) Do gender differences exist among the mathematically talented? (3) Do females exist who possess profound mathematical talent?" (p. 8801). They summarised respectively: (1) Yes, in the U.S. and also in some other countries; (2) Yes, there are more males than females are amongst the highest scoring students, but not consistently in all ethnic groups. Where this occurs, the higher proportion of males is "largely an artefact of changeable sociocultural factors, not (due to) immutable, innate biological differences between the sexes" (p. 8801); and (3) Yes, there are females with profound mathematical talent.

Gender Differences: Looking for New Directions

- Erchick (2012) argued that consideration of conceptual *clusters*, rather than topics in relative isolation, should lead to new questions in as yet fallow ground to be found in the field of gender differences in mathematics. Three clusters are proposed: "Feminism/Gender/Connected Social Constructs; Mathematics/ Equity/Social Justice Pedagogies; and Instruction/Perspectives on Mathematics/ Testing" (p. 10).

- Jacobsen (2012) is among many of those who argue for a reframing of the deficit model approach to gender differences in which male performance and experience are considered the norm to one recognizing the social construction of gender and accepting that females may learn in different, but not inferior, ways from males. One approach to translating this theoretical perspective into practice is also described.

In some of the publications listed (as well as in others not listed here) gender differences are minimized while in others they are given centre-stage. Collectively, a complex rather than simplistic network of interweaving and sometimes contrasting pressures emerges from this body of work. After four decades of research on gender and mathematics, there is only limited consensus on the size and direction of gender differences in performance in mathematics and stark variation in the explanations put forward to account when differences are found.

The NAPLAN scores summarised in Tables 1 and 2 also require a nuanced rather than uni-dimensional reading. When performance on the NAPLAN test is described in terms of mean scores, the small but consistent gender differences in favour of males mirror those obtained in other large scale tests such as the Trends in International Mathematics and Science Study (TIMSS) and the OECD Programme for International Student Assessment (PISA)[4]. But in terms of another set of NAPLAN achievement criteria, the percentage of students achieving above the minimum national average, the small differences reported generally favour girls in the earlier years of schooling, in each of 2008–2011 at Year 3; for three of the four years (2009–2011) for Years 5 and 7; but in only one year (2009) at the Year 9 level. Clearly, gender differences in performance on the NAPLAN tests are small, consistent or variable, depending on the measuring scale and the method of reporting used.

Assessment: Gender Neutral or not?

That gender differences in mathematics learning may be concealed or revealed by the assessment method used is not a new discovery. Else-Quest et al. (2010) judged that "the magnitude of gender differences in math also depends, in part, upon the quality of the assessment of mathematics achievement" (p. 125). Dowling and

---

[4] Differences in the samples involved in the three tests are worth noting. NAPLAN is administered to all students in Years 3, 5, 7, and 9. It is best described as a census test. The TIMSS tests, aimed at students in Years 4 and 8, and the PISA tests administered to 15-year-old students, are restricted to "a light sample (of) about 5 % of all Australian students at each year or age level" (Thomson, p. 76).

Burke (2012) pointed to the 2009 General Certificate of Secondary Education examinations in the U.K. as the first occasion in a decade for boys to perform better than girls in an external examination. "This reversal coincided with a change in the form of the examination" (p. 94), they noted.

A now somewhat dated, yet still striking, example of the impact of the format of examinations on apparent gender differences in mathematics achievement is provided by Cox et al. (2004). They tracked gender differences in performance in the high stake, end of Year 12 examinations in Victoria, Australia for the years 1994–1999, a sustained period of stability in the state's external assessment regime. Student performance in three different mathematics subjects—Further Mathematics (the easiest and most popular of the three mathematics subjects offered at Year 12), Mathematical Methods (a pre-requisite for many tertiary courses), and Specialist Mathematics (the most demanding of the three mathematics subjects)—were among the results inspected. For each of these three subjects there were three different examination components. These were common assessment task (CAT) 1 consisting of a school assessed investigative project or problem, to be completed over several weeks; CAT 2, a strictly timed examination comprising multiple choice and short answer questions; and CAT 3, also a strictly timed examination paper with problems requiring extended answers. Thus CATs 2 and 3 followed the format of traditional timed examinations.

During the period monitored, a student enrolled in a mathematics subject in Year 12 was required to complete three assessment tasks in that subject. A test of general ability was also administered to the Year 12 cohort. These combined requirements provided a unique opportunity to compare the performance of the same group of students on timed and untimed examinations and on papers with items requiring substantially and substantively different responses. In brief:

- Males invariably performed better (had a higher mean score) than females on the mathematics/science/technology component of the general ability test.
- In Further Mathematics, females outperformed males in CAT 1 and in CAT 2 in all of the six years of data considered, and on CAT 3 for five of the six years.
- In Mathematical Methods, females performed better than males in all of the six years on CAT 1; males outperformed females on CAT 2 and CAT 3 for the six years examined.
- In Specialist Mathematics, females performed better than males in all of the six years on CAT 1 and in five of the six years on CAT 3. However males outperformed females on CAT 2 for each of the six years examined.

Thus whether as a group males or females could be considered to be "better" at mathematics depends on which subject or which test component is highlighted. If the least challenging and most popular mathematics subject, Further Mathematics, is referenced then the answer is females. If for all three mathematics subjects the focus is confined to the CAT 1 component, the investigative project or problem assessment task, done partly at school and partly at home, then again the answer is females. But if the focus is on the high stake Mathematical Methods subject, the subject which often serves as a prerequisite for tertiary courses, and on the traditional examination formats of CAT 2 and CAT 3 in that subject, then the answer is males. Collectively these data illustrate that the form of assessment employed can influence which group, males or females, will have the higher mean performance score in mathematics. Would the small but consistent differences found in favour of males' mean performance on the NAPLAN papers disappear if the tests were changed from their traditional strictly timed, multiple choice and short answer format to one resembling the CAT 1 requirements?

Changes to the Year 12 assessment procedures in Victoria were introduced in 2000, seemingly in response to concerns about student and teacher workload and to issues related to the authentication of student work for the teacher-assessed CATs. The changes were described by Forgasz and Leder (2001) as follows:

> For the three VCE mathematics subjects the assessment changes involve the CAT 1 investigative project task being replaced with (generously) timed, classroom based tasks, to be assessed by teachers but with the scores to be moderated by externally set, timed examination results. It is worth recalling that it was on the now replaced format of CAT 1, the investigative project, that females, on average, consistently outperformed males in all three mathematics studies from 1994 to 1999. Is it too cynical to speculate that this consistent pattern of superior female achievement was a tacit factor contributing to the decision to vary the assessment of the CAT 1 task? It is difficult to predict the longer term effects of the new... assessment procedures on students' overall mathematics performance and study scores. Is there likely to be a return to earlier patterns of superior male performance in mathematics? If so, will this satisfy those who are arguing that males are currently the educationally disadvantaged group? (p. 63)

Indigeneity

That there is no ambiguity about the differences in the performance on the NA-PLAN tests between Indigenous and non-Indigenous students is clearly apparent from Tables 1 and 2, and widely emphasized elsewhere. Thomson et al. (2011), for example, examined the 2009 PISA data for Australian students and reported a substantial difference between the average performance of Indigenous and non-Indigenous students on the mathematical literacy assessment component. What message is conveyed by the reporting of these differences?

Gutiérrez (2012) has compellingly used the term "gap gazing" to describe preoccupation with performance differences between selected groups of students and has argued convincingly that highlighting such differences can be counter-productive and reinforce stereotyping. "In its most simplistic form, this approach points out there is a problem but fails to offer a solution... (T)hat it is the analytic lens itself that is the problem, not just the absence of a proposed solution" (Gutiérrez 2012, p. 31) should not be ignored.

As mentioned earlier, the results of NAPLAN testings are widely disseminated and described in media outlets. Forgasz and Leder (2011) compared the more nuanced reporting of students' results on these tests in scholarly outlets with the more superficial tone of print media reports. According to these authors "media reports on students' performance in mathematics testing regimes appear to rely heavily on the executive summaries that accompany the full reports of these data... (T)he more detailed and complex analyses undertaken of entire data sets are often omitted" (p. 218). These comments apply equally to the simplified reporting of gender differences, and differences in performance between Indigenous and non-Indigenous students. It is the arguments advanced in the "more superficial tone of the print media reports" that capture the attention of the general public and shape the sociocultural norms and expectations of the broader society. These norms and expectations are, as mentioned above, among the factors identified by Hyde and Mertz (2009) (among others) as contributing to or averting the emergence of gender difference in performance in mathematics.

Unease has been expressed, both nationally and internationally, about the negative impact of high stake, national testing. Common concerns:

> range from the reliability of the tests themselves to their impact on the well-being of children. This impact includes the effect on the nature and quality of the broader learning experiences of children which may result from changes in approaches to learning and teaching, as well as to the structure and nature of the curriculum (Polesel 2012, p. 4).

Disadvantages stemming from blanket reporting of results in large scale examinations have also been widely discussed and selectively elaborated by Berliner (2011). Although his remarks were aimed at indiscriminate and shallow reporting of the PISA results of selected groups of students in the USA, many of his comments are equally applicable to the coverage of performance of Indigenous students on the Australian NAPLAN tests. Three of his concerns seem highly relevant with respect to the portrayal of the numeracy results of Indigenous students: "what was not reported", "social class", and "the rest of the curriculum".

What Was not Reported

Each year the NAPLAN data are published, the rather high proportion of Indigenous students who fail to meet the nationally prescribed minimum numeracy standard attracts the attention of educators and the wider community. As noted by Forgasz and Leder (2011), p. 213:

> The lower performance of Indigenous students, compared with the wider Australian school population, attracted sustained media attention. The discovery that Aboriginal students living in metropolitan areas as a group performed almost as well as their non-Indigenous peers received less media attention than the more startling finding that Aboriginal students living in remote communities had an extremely high failure rate of 70–80 %. 'A combination of low employment and poor social conditions were explanations offered for the distressingly poor performance... their different pass rates are the result of different schooling' (and a high level of absenteeism).

Aggregating data for all Indigenous students overlooks the large diversity within this group, the range of different needs that inevitably accompany such diversity and the fact that there are also Indigenous students who perform at the highest level on the NAPLAN test. Pang et al. (2011) identified how valuable data are lost when the performance of a multi ethnic group is described and treated as a single entity, rather than reportedly separately for each constituent group. "Educational policies and statistical practices in which achievement is measured using the (group) aggregate result in over-generalized findings" (p. 384) and hide, rather than identify, the strengths and needs of the different subgroups. These remarks are highly relevant given the many subgroups within the Indigenous community. Gross reporting of achievement outcomes fails to recognize the substantially different backgrounds, locations, needs, and capabilities of individuals within the broader group.

Social Class

There is much diversity in the home background of Indigenous students. Some live in remote areas; others in urbanized centres with access, inside and outside the home, to the same resources as non-Indigenous students. Social class related differences in performance apply to both Indigenous and non-Indigenous students. Although Indigeneity and family background are among the categories reported separately for group results on the NAPLAN test, there is no explicit information about the interactive effects of these variables on performance. To paraphrase Berliner (2011): the scores of Indigenous students, as a group, are likely to remain low, "not because of the quality of its teachers and administrators, necessarily, but because of the distribution of wealth and poverty and the associated social capital that exist in schools" (p. 83) in different metropolitan and remote communities. In the reporting of NAPLAN data for Indigenous students, the emphasis is disproportionately on those performing below expectations without sufficient recognition of confounding, contributing factors, while high performing Indigenous students remain largely invisible.

The Rest of the Curriculum

Under this heading Berliner (2011) focuses particularly on the narrowing of the curriculum, within and beyond mathematics, when the perceived scope and requirements of a national testing program overshadow other considerations and influence the delivery of educational programs. Although this criticism cannot be ignored with respect to the NAPLAN tests, I want to focus here on another, equally pervasive issue.

In recent years, many special programs for Indigenous students have been devised, and implemented with varying degrees of success. Difficulties associated with achieving a satisfactory synchrony between the intended and experienced curriculum for Indigenous students in remote communities have been discussed by Jorgensen and Perso (2012).

In the central desert context, the Indigenous people speak their home languages which are shaped by, and also shape, their worldviews. In Pitjantjatjara, for example, the language is quite restricted in terms of number concepts. The lands of the desert are quite stark with few resources so the need for a complex language for number is limited. As such, the counting system is one of 'one, two, three, big mob'. It is rare that a collection of three or more occurs so the need for a more developed number system is not apparent. Even when living in community, the need for number is limited. Few people are aware of their birthdates, and numbers in community are very limited in terms of home numbers or prices in the local store. As such, the immersion in number that is common in urban and regional centres is very limited in remote communities. Therefore, many of the taken for granted assumptions about number that are part of a standard curriculum are limited in this context. This makes teaching many mathematical/number concepts quite challenging as it is not only the teaching of mathematical concepts and processes but a process of induction into a new culture and new worldview (Jorgensen and Perso, pp. 127–128).

Many Indigenous students live and learn in conditions more closely aligned to mainstream educational life in Australia than that depicted for Pitjantjatjara. Nevertheless, this snapshot of the prevailing norms and customs of one community highlights factors that will confound a simplistic interpretation of Indigenous group performance data.

NAPLAN and Mathematics Education Research

Not surprisingly, the introduction of NAPLAN has already fuelled a variety of research projects. An overview of work referring substantively to NAPLAN data and presented at the joint conference in 2011 of the Australian Association for Mathematics Teachers (AAMT) and Mathematics Education Research Group of Australasia (MERGA) is summarized in Table 3. It provides a useful indication of the scope and diversity of these investigations.[5] It is worth noting that the 2011 conference represented the first time the two associations held a fully joint conference. According to Clark et al. (2011) it was a unique opportunity for "practitioners and researchers to discuss key issues and themes in mathematics education, so that all can benefit from the knowledge gained through rigorous research and the wisdom of practice" (p. iii). In addition to "participants from almost every university in Australia and New Zealand, teachers from government and nongovernment schools systems throughout Australia and officers from government Ministries of Education" (Clark. et al. 2011), p. iii, there were authors and presenters from a range of other countries.[6]

---

[5] Details are extracted from the published proceedings of this joint conference, comprising 130 papers. The proceedings consisted of two sets of papers: *Research papers* and *Professional papers*, reviewed respectively according to established MERGA and AAMT reviewing processes.

[6] These included Singapore, the United States of America, Papua New Guinea and the United Kingdom.

Table 3 NAPLAN related papers presented at the AAMT-MERGA conference in 2011[a]

Author and paper title	Summary of paper and findings/recommendations
Callingham mathematics assessment: everything old is new again?	Descriptive, rather than incisive, reference was made to the NAPLAN testing program in this presentation. Noted were: the contradiction between teachers generally being urged to use formative assessment and the prominence given to the external measure of numeracy provided by NAPLAN; that no significant change has been captured "across time for any grade group" from 2008 to 2010; and that the NAPLAN "results are used for accountability at the local level". A brief reference is also made to one setting where school based NAPLAN results are used to address elements on which students under-performed
Connolly refining the NAPLAN numeracy construct	An overview is provided of the development of the 2009 and 2010 NAPLAN numeracy test papers. The core content of the test is formally based on the set of nationally agreed curriculum outcomes. Avoided are topics for which there are between state variations in the time of the year they are taught. Items are reviewed multiple times with strong input from key stakeholders. Other factors taken into account in the construction of the test include: item difficulty; cognitive dimension (knowing, applying, and reasoning); item context (abstract or non-abstract); the influence of calculators on content (calculators are not allowed in the Years 3 and 5 papers but at the Year 7 and 9 levels both calculator and non-calculator papers are set); guidelines for item writing; and using accessible language. The Rasch model (Wright, 1980) is used to analyse the test results. This requires not only that certain pre-conditions are met (items are uni-dimensional, locally independent, and uniformly discriminating) but also "allows for sensible comparisons of test scores between different years"
Edmonds-Wathen locating the learner: indigenous language and mathematics education	The author describes the difficulties encountered by Indigenous language speaking students when faced with the typical development of number concepts in the curriculum in the early years of schooling and argues that a different, and group-tailored sequencing of material should be considered. The obstacles created by a "cognitive mismatch between the teacher and student" may fail to gauge accurately the students' understanding of, for example, spatial items and be reflected in low scores on such items on NAPLAN tests—invalidating simplistic comparisons between Indigenous and non-Indigenous students
Helme and Teese how inclusive is year 12 mathematics?	NAPLAN test data are part of a larger pool of material tapped to explore the mathematics learning experiences and expectations of students at schools in the northern suburbs of Melbourne, but are not discussed per se. Nevertheless the authors' conclusions are worth noting: "Perceptions of mathematics classrooms and mathematics teachers, and expectations of success, vary according to subject, (student's) gender and social background"

(continued)

Table 3 (continued)

Author and paper title	Summary of paper and findings/recommendations
Hill Gender differences in NAPLAN mathematics performance	The performance of females and males was compared on items on the Grade 3 and Grade 9 NAPLAN papers for 2008–2010. On each paper, there were some questions on which both groups performed (percentage correct) equally well. When group differences were found they more frequently favoured males than females (e.g., Year 3 paper NAPLAN 2009, no difference on 4 items, females outperformed males on 10 items, males outperformed females on 21 items; Year 9 papers NAPLAN 2010, no difference on 8 items, females outperformed males on 11 items, males outperformed females on 45 items). These trends suggest a "decline in achievement of females as they progress through their schooling"
Hurst connecting with the Australian curriculum: mathematics to integrate learning through the proficiency strands	The scope and demands of NAPLAN tests should not be allowed to dictate the content of the curriculum, nor restrict the instructional strategies used. According to the author, "NAPLAN test scores can greatly assist teachers if they are used appropriately". Rather than expanding on this theme, the author argues that teachers should "use a constructivist approach to teaching mathematics… (with) an emphasis on rich conceptual understanding as opposed to the mere acquisition of procedural knowledge" and provides some examples that support this theme
Morley Victorian Indigenous Children's responses to mathematics NAPLAN Items	Using data from the 2008 Years 5, 7, and 9 NAPLAN papers, "whether children of Indigenous background in Victoria, Australia, have different patterns of mathematical responses from the general population" is explored in this paper. Not surprisingly, both groups perform better on high facility than low facility items. Some advantage in favour of Indigenous students is found on the Space strand of the Year 7 paper but less so on the Year 9 paper. At that level, the Algebra strand appeared to be relatively more difficult for Indigenous students
Nisbet national testing of probability in years 3, 5, 7, and 9 in Australia: a critical analysis	The limits of large scale tests are discussed at some length. Often, Nisbet argues, these tests have "a bias towards mechanical processes, and away from problem solving and creativity". A focus on the probability questions in the 2009 and 2010 NAPLAN numeracy tests for Years 3, 5, 7, and 9 revealed that there were few probability items overall and that only one such item was included in each year level in the 2010 test. Furthermore, Nisbet argued, the scope and aspects of probability probed by the items seemed unacceptably constrained, "with most being multiple-choice items… (and) fundamentally recognition tasks… (to) identify the correct response"

(continued)

Table 3 (continued)

Author and paper title	Summary of paper and findings/recommendations
Pierce and Chick reacting to quantitative data: teachers' perceptions of student achievement reports	The authors use teachers' reactions, to national and school specific information, provided in table and graphical form in a NAPLAN report (usefulness and difficulty of the table and it accompanying annotations and explanations) to gauge the level of teachers' statistical literacy. "Reactions range(d) from those verging on the statistics-phobic… through to deep engagement with the issues". Many teachers preferred the graphical representation, although some welcomed the details provided in the table. Some "reacted strongly about the overwhelming complexity of the data… (and many) expressed uncertainty or confusion over some or all aspects of the data"
Sullivan and Gunningham a strategy for supporting students who have fallen behind in the learning of mathematics	Two items, of different levels of difficulty, from the 2009 NAPLAN Year 9 (no calculator) paper are used to illustrate how poorly some students are performing in mathematics. After this introduction, an out-of-class intervention (the *Getting Ready* intervention) used to prepare students from Years 3 and 8 for work being taught in their next mathematics lesson, is described. No further reference is made to NAPLAN tests
Tomazos improving mathematical flexibility in primary students: what have we learned?	It is often assumed that schools in higher socio-economic areas with students who perform well on NAPLAN tests do not need to provide extra support for their students. Data from a pilot program at such schools revealed not only that procedural approaches were often used when teaching calculation strategies but also that with "relatively little system input, experienced teachers' classroom practices can be changed" to incorporate greater use of flexible calculation strategies. NAPLAN data were again used as a measure in sample selection, but no further reference is made in the paper to NAPLAN tests
Vale, Davidson, Davies, Hooley, Loton, and Weaven using assessment data: does gender make a difference?	Students' performance on NAPLAN tests was among the measures used to determine a student's learning needs and to select students for specifically designed intervention programs. Gender related differences in performance are reported but no further reference is made to NAPLAN tests
White and Anderson teachers' use of national test data to focus numeracy instruction	The authors argue that, without wishing to advocate 'teaching to the test', much can be gained by teachers who use NAPLAN data from their own school to identify students' numeracy needs and develop instructional strategies to combat faulty practices or inadequate understanding. The approach adopted in one school is described. Whether "professional learning support (had) an impact on student learning and on teaching practice" was also examined

[a] To conform with space constraints, the entries in this table are not included separately in the reference list at the end of this paper. All can be found in Clark et al. (2011)

Reference to NAPLAN tests was made in some 10 % of the published papers. As can be seen from Table 3, aspects covered in these papers included issues pertaining to the development of the tests, interpreting the published results of the tests, using test results for curriculum development, and examining the performance of groups of interest, specifically boys and girls and Indigenous students. In some papers reference to NAPLAN data was very much secondary to the core issue explored, for example its (seemingly increasing) use as part of a series of measures to identify a specific group worthy, or in need of, further attention. What could be learnt from the NAPLAN tests about the performance and numeracy needs of high achieving students has, however, not yet attracted research attention. The finding by Pierce and Chick is particularly disturbing. When asked about the statistical and graphical summaries of NAPLAN data relevant to their students the reactions of teachers in their sample ranged "from those verging on the statistics-phobic … through to deep engagement with the issues". The NAPLAN national reports contain much valuable and potentially usable data. But how much of these are actually understood and used constructively?

Final Words

After collating information from some 70 public opinion polls in which questions about the efficacy of national tests were included, Phelps (1998) reported:

> The majorities in favor of more testing, more high-stakes testing, or higher stakes in testing have been large, often very large, and fairly consistent over the years and across polls and surveys and even across respondent groups (with the exception of some producer groups: principals, local administrators, and, occasionally, teachers) (p. 14) .

The data on which Phelps based his conclusions are now somewhat dated. How the Australian public today values national tests, and in particular the NAPLAN testing regime, is a question still waiting to be investigated. When planning future research activities, whether linked to NAPLAN, to gender and mathematics performance, to issues pertaining to Indigenous students, or to the needs of highly able students, the recommendation of Purdie and Buckley (2010) is well worth heeding:

> Although it is important to continue small, contextualised investigations of participation and engagement issues, more large-scale research is called for. Unless this occurs, advancement will be limited because sound policy and generalised practice cannot be extrapolated from findings that are based on small samples drawn from diverse communities (p. 21).

Open Access This chapter is distributed under the terms of the Creative Commons Attribution Noncommercial License, which permits any noncommercial use, distribution, and reproduction in any medium, provided the original author(s) and source are credited.

References

Australian Curriculum, Assessment and Reporting Authority [ACARA]. (2010, March). *Australian curriculum information sheet. Mathematics.* Retrieved from http://www.acara.edu.au/verve/_resources/Mathematics.pdf

ACARA. (2011a). *National assessment program. Literacy and numeracy. Achievement in reading, persuasive writing, language conventions and numeracy.* National report for 2011. Retrieved from http://www.nap.edu.au/_Documents/National%7FReport/NAPLAN_2011%7FNational_Report.pdf

ACARA. (2011b). *NAP sample assessment.* Retrieved from http://www.nap.edu.au/NAP_Sample_Assessments/index.html

Berliner, D. C. (2011). The context for interpreting the PISA results in the U.S.A.: Negativism, chauvinism, misunderstanding, and the potential to distort the educational systems of nations. In M. A. Pereyra, H-G. Korthoff, & R. Cowen (Eds.), *PISA under examination. Changing knowledge, changing tests, and changing schools* (pp. 77-96). Rotterdam, The Netherlands: Sense Publishers.

Clark J., Kissane, B., Mousley, J., Spencer, T., & Thornton, S. (Eds.) (2011). Traditions and [new] practices. *Proceedings of the AAMT–MERGA conference held in Alice Springs, 3–7 July 2011, incorporating the 23rd biennial conference of The Australian Association of Mathematics Teachers Inc. and the 34th annual conference of the Mathematics Education Research Group of Australasia Inc.* Adelaide, South Australia: AAMT & MERGA

Cox, P. J., Leder, G. C., & Forgasz, H. J. (2004). The Victorian Certificate of Education – Mathematics, science and gender. *Australian Journal of Education, 48*(1), 27-46.

Curriculum Corporation. (2006). *Statements of learning for mathematics.* Retrieved from http://www.mceetya.edu.au/verve/_resources/SOL_Maths_Copyright_update2008.pdf

Dowling, P., & Burke, J. (2012). Shall we do politics or learn some maths today? Representing and interrogating social inequality. In H. Forgasz & F. Rivera (Eds.), *Towards equity in mathematics education. Gender, culture, and diversity* (pp. 87-103). Berlin: Springer-Verlag.

Else-Quest, N. M., Hyde, J. S., & Linn, M. C. (2010). Cross-national patterns of gender differences in mathematics: A meta-analysis. *Psychological Bulletin, 136*(1), 103-127.

Erchick, D. B. (2012). Preface to "Moving towards a feminist epistemology of mathematics". In H. Forgasz & F. Rivera (Eds.), *Towards equity in mathematics education. Gender, culture, and diversity* (pp. 9 – 14). Berlin: Springer-Verlag.

Eurydice (2011). *Mathematics in Europe: Common challenges and national policies.* Brussels: Education, Audiovisual and Culture Executive Agency. Retrieved from http://eacea.ec.europa.eu/education/eurydice/documents/thematic_reports/132EN.pdf

Forgasz, H. J., & Leder, G. C. (2001). The Victorian Certificate of Education – a Gendered Affair? *Australian Educational Researcher, 28*(2), 53-66.

Forgasz, H. J., & Leder, G.C. (2011). Equity and quality of mathematics education: Research and media portrayals. In B. Atweh, M. Graven, W. Secada, & P. Valero (Eds.), *Mapping equity and quality in mathematics education* (pp. 205-222). Dordrecht, the Netherlands: Springer.

Fyde, J. S., & Meertz, J. E. (2009). Gender, culture, and matheamtics. *Proceedings of the National Academy of Sciences of the United States, 106*(22), 8801-8807.

Gutiérrez, R. (2012). Context matters: How should we conceptualize equity in mathematics education? In B. Herbel-Eisenmann, J. Choppin, D. Wagner, & D. Pimm (Eds.), *Equity in discourse for mathematics education. Theories, practices, and policies* (pp. 17-33). Dordrecht, The Netherlands: Springer.

Jacobsen, L. (2012). Preface to "Equity in mathematics education: Unions and intersections of feminist and social justice literature". In H. Forgasz & F. Rivera (Eds.) *Towards equity in mathematics education. Gender, culture, and diversity* (pp. 31-37). Berlin: Springer-Verlag.

Joint committee on testing practices. (2004). *Code of fair testing practices in education. Washington.* DC: Joint Committee on Testing Practices. Retrieved from http://www.apa.org/science/jctpweb.html

Jorgensen, R., & Perso, T. (2012). Equity and the Australian curriculum: mathematics. In B. Atweh, M. Goos, R. Jorgensen, & D. Siemon, (Eds.), *Engaging the Australian national curriculum: Mathematics – perspectives from the field* (pp. 115-133). Online Publication: Mathematics Education Research Group of Australasia. Retrieved from http://www.merga.net. au/sites/default/files/editor/books/1/Chapter%206%7F%20Jorgensen.pdf

Kaiser. G., Hoffstall, M., & Orschulik, A.B (2012) Gender role stereotypes in the perception of mathematics: An empirical study with secondary students in Germany. In H. Forgasz & F. Rivera (Eds.), *Towards equity in mathematics education. Gender, culture, and diversity* (pp. 115-140). Berlin: Springer-Verlag.

Kane, J.,M., & Mertz, J.,E. (2012). Debunking myths about gender and mathematics performance. *Notices of the AMS, 59*(1), 10-20.

Kenney, P.,A., & Schloemer, C.,G. (2001). Assessment of student achievement, overview. In S. Grinstein & S.I. Lipsey (Eds), *Encyclopedia of mathematics education* (pp. 50-56). New York: RoutledgeFalmer.

Leder, G. C. (1992). Mathematics and gender: Changing perspectives. In D. A. Grouws (Ed.), *Handbook of research on mathematics teaching and learning* (pp. 597-622). New York: Macmillan.

Lundgren, U. P. (2011). PISA as a political instrument. In M. A. Pereyra, H-G. Kotthoff, & R. Cowen (Eds.), *PISA as a political instrument (pp. 17-30).* Rotterdam, The Netherlands: Sense Publishers.

Northern Territory Government. (n.d.). 2010 NAPLAN analysis using curriculum focus for literacy. Retrieved from http://www.det.nt.gov.au/__data/assets/pdf_file/0005/15827/ NAPLAN_Analysis_Curriculum_FocusForNumeracy.pdf

Pang, V. O., Han, P. P., & Pang, J. M. (2011). Asian American and Pacific Islander students: Equity and the achievement gap. *Educational Researcher, 40*(8), 378-389.

Phelps, R. P. (1998). The demand for standardized student testing. *Educational Measurement: Issues and Practice,* Fall, 5-23.

Polesel, J., Dulfer, N., & Turnbull, M. (2012). *The experience of education: The impacts of high stake testing on students and their families.* University of Western Sydney: The Whitlam Institute. Retrieved from http://www.whitlam.org/_data/assets/pdf_file/0008/276191/High_ Stakes_Testing_Literature_Review.pdf

Postlethwaite, T. N., & Kellaghan, T. (2009). *National assessment of educational achievement.* Paris: UNESCO, International Institute for Educational Planning. Retrieved from http://www. iiep.unesco.org/fileadmin/user_upload/Info_Services_Publications/pdf/2009/EdPol9.pdf

Purdie, N., & Buckley, S. (2010). *School attendance and retention of Australian Indigenous Australian students.* Issues paper no. 1, Closing the gap clearinghouse. Retrieved from http:// www.aihw.gov.au/uploadedFiles/ClosingTheGap/Content/Publications/2010/ctg-ip01.pdf

Sidoti, E., & Keating, J. (2012). Foreword. In J. Polesel, N. Dulfer, & M. Turnbull. (*The experience of education: The impacts of high stake testing on students and their families* (p. 3). University of Western Sydney: The Whitlam Institute. Retrieved from http://www.whitlam.org/ _data/assets/pdf_file/0008/276191/High_Stakes_Testing_Literature_Review.pdf

Stoet, G., & Geary, D. C. (2012). Can stereotype threat explain the gender gap in mathematics performance and achievement? *Review of General Psychology,* doi: 10.1037/a0026617

Thomson, S. (2010). *Mathematics learning: What TIMSS and PISA can tell us about what counts for all Australian students.* Paper presented at the ACER Research conference: Teaching mathematics? Make it count. Retrieved from http://research.acer.edu.au/research_conference/RC2010/17august/6/

Thomson, S., De Bortoli, L., Nicholas, M., Hillman, K., & Buckley, S. (2011). *Challenges for Australian education: Results from PISA 2009.* Retrieved from http://www.acer.edu.au/ documents/PISA-2009-Report.pdf

UNESCO (2012). *World atlas of gender equality in education.* Paris: Author.

Wai, J., Cacchio, M., Putallaz, M., & Makel, M. C. (2010). Sex differences in the right tail of cognitive abilities: A 30 year examination. *Intelligence,* doi:10.1016/j.intell.2010.04.006.

Wright, B. (1980). Foreword. In G. Rasch, *Probabilistic models for some intelligence and attainment tests.* Chicago, IL: The University of Chicago Press.

Early Algebraic Thinking: Epistemological, Semiotic, and Developmental Issues

Luis Radford

Abstract In this article I present some findings of an ongoing 5-year longitudinal research program with young students. The chief goal of the research program is a careful and systematic investigation of the genesis of embodied, non-symbolic algebraic thinking and its progressive transition to culturally evolved forms of symbolic thinking. The investigation draws on a cultural-historical theory of teaching and learning—the theory of objectification—that emphasizes the sensible, embodied, social, and material dimension of human thinking and that articulates a cultural view of development as an unfolding dialectic process between culturally and historically constituted forms of mathematical knowing and semiotically mediated classroom activity.

Keywords Sensuous cognition · Vygotsky · Arithmetic versus algebraic thinking

Introduction

In light of the legendary difficulties that the learning of algebra presents to students, it has been suggested that a progressive introduction to algebra in the early grades may facilitate students' access to more advanced algebraic concepts later on (Carraher and Schliemann 2007). An early development of algebraic thinking may, in particular, ease students' contact with algebraic symbolism (Cai and Knuth 2011).

The theoretical grounding of this idea and its practical implementation remain, however, a matter of controversy. Traditionally, algebra has been taught only after students have had the opportunity to acquire a substantial knowledge of arithmetic.

L. Radford (✉)
Université Laurentienne, Ontario, Canada
e-mail: Lradford@laurentian.ca

© The Author(s) 2015
S.J. Cho (ed.), *The Proceedings of the 12th International Congress on Mathematical Education*, DOI 10.1007/978-3-319-12688-3_15

That is, arithmetic thinking has been assumed to be a prerequisite for the emergence and development of algebraic thinking. Clearly, an introduction to algebra in the early grades does not conform to such an assumption. Now, if this is so, if algebra needs not to come after arithmetic, the question is: What is the difference and relationship between these two disciplines? Evading these questions does not do us any favours.

In the next section, I briefly discuss the question of the relationship between algebra and arithmetic. Drawing on historical and educational research, I suggest an epistemological distinction between the forms of thinking that are required in both disciplines. Then, I present some findings of a 5-year longitudinal classroom research program where 8-year old students were followed as they moved from Grade 2 to Grade 6. I shall focus in particular on the genesis and development of embodied, non-symbolic algebraic thinking and its progressive transition to cultural forms of symbolic thinking.

Arithmetic and Algebra: Filiations and Ruptures

The question of the filiations and ruptures between arithmetic and algebra was one of the major educational research themes in the 1980s and 1990s. This question was at the heart of several research programs. It was often discussed in various PME's Working Groups and research reports (Bednarz et al. 1996; Sutherland et al. 2001).

Filloy and Rojano's (1989) work points to one of the fundamental breaks between arithmetic and algebra—what they call a *cut*. This *cut* was observed in clinical studies where students faced equations of the form $Ax + B = Cx + D$. To solve equations of this form, the arithmetic methods of "reversal operations"— which are effective to solve equations of the type $Ax + B = D$ (the students usually subtract B from D and divide by A)—are no longer applicable. The students have to resort to a truly algebraic idea: to *operate* on the unknown. In order to operate on the unknown, or on indeterminate quantities in general (e.g., variables, parameters), one has to think analytically. That is, one has to consider the indeterminate quantities as if they were something known, as if they were specific numbers (see, e.g., Kieran 1989, 1990; Filloy et al. 2007). From a genetic viewpoint, this way of thinking *analytically*—where unknown numbers are treated on a par with known numbers—distinguishes arithmetic from algebra. And it is so characteristic of algebra that French mathematician François Viète (one of the founders of modern symbolic algebra) identified algebra as an *analytic art* (Viète 1983).

A consequence of this difference between arithmetic and algebra is the following. Because of algebra's analytic nature, formulas in algebra are *deduced*. Failing to notice this central analytic characteristic of algebra may lead us to think that the production of formulas in patterns (regardless of how they were produced) is a symptom of algebraic thinking. But as Howe (2005) notes, producing a formula might merely be a question of guessing the formula and trying it. I completely agree with him that there is nothing algebraic in trying and guessing. Try-and-guess strategies are indeed based on arithmetic concepts only.

Epistemological research has also made a contribution to the conversation about the distinction between arithmetic and algebra. This research suggests that the difference between these disciplines cannot be cast in terms of *notations*, as it has often been thought. The alphanumeric algebraic symbolism that we know today is indeed a recent invention. In the west it appeared during the Renaissance, along with other forms of representation, like perspective in painting and space representation, underpinned by changes in modes of production and new forms of labour division. The birth of algebra is not the birth of its modern symbolism. In his *Elements*, Euclid resorted to letters without mobilizing algebraic ideas. Ancient Chinese mathematicians mobilized algebraic ideas to solve systems of equations without using notations. Babylonian scribes used geometric diagrams to think algebraically. As a result, the use of letters in algebra is neither a necessary nor a sufficient condition for thinking algebraically. Naturally, our modern algebraic symbolism allows us to carry out transformations of expressions that may be difficult or impossible with other forms of symbolism. However, as we shall see in a moment, the rejection of the idea that notations are a manifestation of algebraic thinking, opens up new avenues to the investigation of elementary forms of algebraic thinking in young students.

Some Background of the Research

The investigation of young students' algebraic thinking that I report here started in 2007. The decade before, I was interested in investigating adolescent and young adults' algebraic thinking. From 1998 to 2006 I had the opportunity to follow several cohorts of students from Grade 7 until the end of high school. Like many of my colleagues, I started focusing on symbolic algebra, that is, an algebraic activity mediated by alphanumeric signs. One of my goals was to understand the processes students undergo in order to build symbolic algebraic formulas. My working hypothesis was that in order to understand the manner in which students bestow meaning to alphanumeric expressions, we should pay attention to language (Radford 2000). However, during the analysis of hundreds of hours of videotaped lessons, it became apparent that our students were not resorting only to language, but also to gestures, and other sensuous modalities in ways that were far from mere byproducts of interaction. It was clear that gestures and other embodied forms of action were an integral part of the students' signifying process and cognitive functioning. The problem was to come up with suitable and theoretically articulated explanatory principles, in order to provide an interpretation of the students' algebraic thinking that would integrate those embodied elements that the video analyses put into evidence. Although by the early 2000s, some linguists and cognitive psychologists had developed interesting work around the question of embodiment

(Lakoff and Núñez 2000), their accounts were not easy to apply to such complex settings as classrooms; nor were they necessarily taking into account the historical and cultural dimension of knowledge. In the following years, with the help of some students and collaborators, I was able to refine our theoretical approach and reveal non-conventional, embodied forms of algebraic thinking (Radford 2003). In Radford et al. (2007), we reported a passage in which Grade 9 students displayed an amazing array of sensuous modalities to come up with an algebraic formula in a pattern activity. What is amazing in the reported passage is the subtle coordination of words, written signs, drawn figures, gestures, perception, and rhythm. Figure 1 presents an interesting series of gestures that a student makes while trying to perceive a mathematical structure behind the sequence. Focusing on the first term of the sequence (which is shown in the three first pictures of Fig. 1), Mimi, the student, points with her index to the first circle on the top row and says "one;" she moves the finger to the first circle on the bottom row and repeats "one." Then she moves the index to her right and makes a kind of circular indexical gesture to point to the three remaining circles, while saying "plus three." She starts again the same series of gestures, this time pointing to the second term of the sequence (see second term in Pic 4 of Fig. 1), saying now "two, two plus three." She restarts the same series of gestures in dealing with the third term (see third term of the sequence in Fig. 1, Pic 4; we have added dashed lines to the terms of the sequence to indicate the circles that Mimi points to as she makes her gestures). In doing so, Mimi reveals an embodied formula that, instead of being made up of letters, is made up of words and gestures: the formula is displayed *in concreto*: "one, one, plus three; two, two plus three; three, three, plus three." She then applied the formula to Term 10 (which was not drawn and had to be imagined): "you will have 10 dots [i.e., circles] (she makes a gesture on the desk to indicate the position of the circles), 10 dots (she makes a similar gesture), plus 3." The embodied formula rests on a use of variables and functional relations that conform to the requirement of analyticity that, as I suggested previously, is characteristic of algebra. Although the variable 'number of the term' is not represented through a letter, it appears embodied in its surrogates—the particular numbers the variable takes. The formula is then shown as the series of calculations on the instantiated variable. And, as such, the formula is algebraic. Now, our Grade 9 students did use alphanumeric symbolism and built the formula "n + n + 3," which was then transformed into "n × 2 + 3" (Radford et al. 2007). Hence, these Grade 9 students went unproblematically from an embodied form of thinking to a symbolic one.

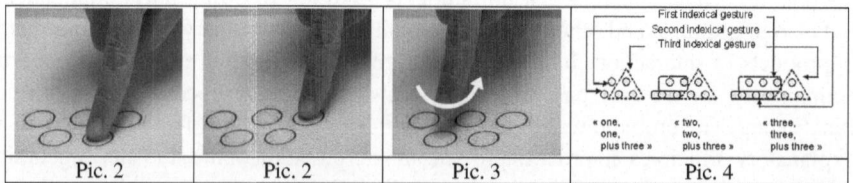

Fig. 1 A Grade 9 student displaying an impressive multimodal coordination of semiotic resources. Recostructed from the video

We came back to other published and unpublished analyses and noticed that the subtle multimodal coordination of senses and signs was a widespread phenomenon in adolescents. Then arose a research question that has kept me busy for the past 6 years: would similar embodied forms of algebraic thinking be accessible to young students? And if yes, how would these embodied forms of thinking develop as the students moved from one grade to the next? As Grade 2 students are still learning to read and write in Ontario, Grade 2 looked like a good place to start. This is how I moved to a primary school and embarked on a new longitudinal research.

Grade 2: Young Students' Non-symbolic Algebraic Thinking

The first generalizing activity in our Grade 2 class was based on the sequence shown in Fig. 2.

We asked the students to extend the sequence up to Term 6. In subsequent questions, we asked them to find out a procedure to determine the number of rectangles in Terms 12 and 25. Figure 3 shows the answers provided by two students: Carlos and James.

Contrary to what we observed in our research with adolescent students, in extending the sequence, most of our Grade 2 students focused on the numerical aspect of the terms only. Counting was the leading activity. Generally speaking, to extend a figural sequence, one needs to grasp a regularity that involves the linkage of two different structures: one *spatial* and the other *numerical*. From the spatial structure emerges a sense of the rectangles' *spatial position*, whereas their numerosity emerges

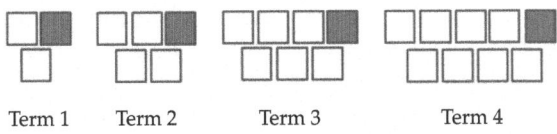

Term 1   Term 2   Term 3   Term 4

Fig. 2 The first terms of a sequence that Grade 2 students investigated in an algebra lesson

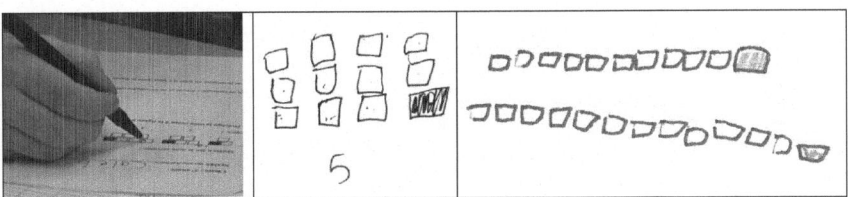

Fig. 3 To the *left*, Carlos, counting aloud, points sequentially to the squares in the top row of Term 3. In the middle, Carlos' drawing of Term 5. To the *right*, James' drawing of Terms 5 (*top*) and 6 (*bottom*)

from a numerical structure. While Carlos attends to the numerical structure in the generalizing activity, the spatial structure is not coherently emphasized. This does not mean that Carlos, James and the other students do not see the figures as composed of two horizontal rows. What this means is that the emphasis on the numerical structure somehow leaves in the background the geometric structure. We could say that the *shape* of the terms of the sequence is used to facilitate the counting process. Thus, as picture 1 in Fig. 3 shows, Carlos always counted the rectangles in a spatial orderly way. The geometric structure, however, does not come to be related to the numerical one in a meaningful and efficient way. It is not surprising within this context, then, that the students encountered difficulties in answering our questions about Terms 12 and 25. Without resorting to an efficient way of counting, the counting process of rectangles one-by-one in remote terms beyond the perceptual field became extremely difficult.

Because of their spatial connotation, it might not be surprising that, in extending the sequences, our young students did not use deictic terms, like "bottom" or "top." In the cases in which the students did succeed in linking the spatial and numerical structures, the spatial structure appeared only ostensibly, i.e., "top" and "bottom" rows were not part of the students' discourse but were made apparent through pointing and actual row counting: they remained secluded in the embodied realm of action and perception. The next day, the teacher discussed the sequence with the students and referred to the rows in an explicit manner to bring to the students' attention the linkage of the numerical and spatial structures. To do so, the teacher drew the first five terms of the sequence on the blackboard and referred to an imaginary student who counted by rows. "This student," she said to the class, "noticed that in Term 1 (she pointed to the name of the term) there is one rectangle on the bottom (and she pointed to the rectangle on the bottom), one on the top (pointing to the rectangle), plus one dark rectangle (pointing to the dark rectangle)." Next, she moved to Term 2 and repeated in a rhythmic manner the same counting process, coordinating the spatial deictics "bottom" and "top," the corresponding spatial rows of the figure, and the number of rectangles therein. To make sure that everyone was following, she started again from Term 1 and, at Term 3, she invited the students to join her in the counting process, going together up to Term 5 (see Fig. 4).

Then, the teacher asked the class about the number of squares in Term 25. Mary raised her hand and answered: "25 on the bottom, 25 on top, plus 1." The class

Fig. 4 The teacher and the students counting rhythmically say (see Pic 1) "Term 5", (Pic 2) "5 on the *bottom*", (Pic 3) "5 on *top*", (Pic 4) "plus 1."

Fig. 5 Karl explaining Term 50

spent some time dealing with "remote" terms, such as Terms 50 and 100. Figure 5 shows Karl explaining to the teacher and his group-mates what Term 50 looks like.

In picture 1, Karl moves his arm and his body from left to right in a vigorous manner to indicate the bottom row of Term 50, while saying that there would be 50 white rectangles there. He moves his arm a bit further and repeats the moving arm-gesture to signify the top row of Term 50. Then he makes a semi-circle gesture in the air to signify the dark square.

The students played for a while with remote terms. In Karl's group, one of the questions revolved around Term 500 and Term 50:

Karl   How about doing 500 plus 500?
Erica   No. Do something simpler
Karl   (*Talking almost at the same time*) 500 plus 500 equals 1000
Erica   plus 1, 1001
Karl   plus 1, equals 1001
Cindy   (*Talking about Term 50*) 50 plus 50, plus 1 equals 101

Schematically speaking, the students' answer to the question of the number of rectangles in remote particular terms was "$x + x + 1$" (where x was always a *specific* number). The formula, I argue, is algebraic in nature, even if it is not expressed in standard notations. In this case, indeterminacy and analyticity appear in an intuited form, rather than explicitly. A natural question is: Is this all that Grade 2 students are capable of? In fact, the answer is no. As we shall see in the next section, we were able to create conditions for the emergence of more sophisticated forms of algebraic thinking.

Beyond Intuited Indeterminacy: The Message Problem

On the fifth day of our pattern generalization teaching-learning sequence, the teacher came back to the sequence from the first day (Fig. 2). To recapitulate, she invited some groups to share in front of the class what they had learned about that sequence in light of previous days' classroom discussions and small group work. Then, she asked a completely new question to the class. She took a box and, in front of the students, put in it several cards, each one having a number: 5, 15, 100, 104, etc. Each one of these numbers represented the number of a term of the sequence shown in

Fig. 2. The teacher invited a student to choose randomly one of the cards and put it into an envelope, making sure that neither the student herself nor the teacher nor anybody else saw the number beforehand. The envelope, the teacher said, was going to be sent to Tristan, a student from another school. The Grade 2 students were invited to send a message that would be put in the envelope along with the card. In the message the students would tell Tristan how to quickly calculate the number of rectangles in the term indicated on the card. The number of the term was hence unknown. Would the students be able to generalize the embodied formula and engage with calculations on this unknown number? In other terms, would our Grade 2 students be able to go beyond intuited indeterminacy and its corresponding elementary form of algebraic thinking? As in the previous days, the students worked in small groups of three. The usual response was to give an example. For instance, Karl suggested: "If the number [on the card] is 50, you do 50, plus 50, plus 1." The teacher commended the students for the idea, but insisted that the number could be something else and asked if there would be another way to say it without resorting to examples. After an intense discussion, the students came up with a suggestion:

Erica   It's the number he has, the same number at the bottom, the same number at the top, plus 1...

Teacher   That is excellent, but don't forget: he doesn't have to draw [the term]. He just has to add... So, how can we say it, using this good idea?

Erica   We can use our calculator to calculate!

Teacher   Ok. And what is he going to do with the calculator?

Erica   He will put the number... (she pretends to be inserting a number into the calculator)... plus the same number, plus 1 (as she speaks, she pretends to be inserting the number again, and the number 1).

Another group suggested "twice the number plus 1." Naturally, the use of the calculator is merely virtual. In the students' real calculator, all inputs are specific numbers. Nevertheless, the calculator helped the students to bring forward the analytic dimension that was apparently missing in the students' explicit formula. Through the virtual use of the calculator, calculations are now performed on this unspecified instance of the variable—the unknown number of the figure.

Let me summarize our Grade 2 students' accomplishments during the first week that they were exposed to algebra. In the beginning, most of our students were dealing with figural sequences like the one in Fig. 1 through a focus on numerosity. Finding out the number of elements (rectangles, in the example here discussed) in remote terms was not easy. The joint counting process in which the teacher and students engaged during the second day helped the students to move to other ways of seeing sequences. The joint counting process made it possible for the students to notice and articulate new forms of mathematical generalization. In particular, they became aware of the fact that the counting process can be based on a *relational idea*: to link the number of the figure to relevant parts of it (e.g. the squares on the bottom row). This requires an altogether new perception of the number of the term and the terms themselves. The terms appear now not as a mere bunch of ordered rectangles but as something susceptible to being decomposed, the decomposed parts bearing potential clues for algebraic relationships to occur. Interestingly enough, historically speaking, the "decomposition" of geometric figures in simpler forms (e.g., straight lines) was systematically developed in the 17th century by Descartes in his *Geometry*, a central book in the development of algebraic ideas. The decomposition of figures permitted the creation of relationships between known and unknown numbers and the carrying out of calculations on them "without making a distinction between known and unknown [parts]" (Descartes 1954, p. 8). Our examples—as well as those reported by other researchers with other Grade 2 students—suggest that the linkage of spatial and numerical structures constitutes an important aspect of the development of algebraic thinking. Such a linkage rests on the cultural transformation in the manner in which sequences can be seen—a transformation that may be termed the *domestication of the eye* (Radford 2010). For the modern mathematician's eye, the complexity behind the perception of simple sequences like the one our Grade 2 students tackled remains in the background, to the extent that to see things as the mathematician's eye does, ends up seeming natural. However, as our results intimate, there is nothing natural there. To successfully attend to what is algebraically meaningful is part of learning to think algebraically. This cultural transformation of the eye is not specific to Grade 2 students. It reappears in other parts of the students' developmental trajectory. It reappears, later on, when students deal with factorization, where discerning structural *syntactic forms* become a pivotal element in recognizing common factors or prototypical expressions.

All in all, the linkage of spatial and numerical structures resulted, as we have seen, in the emergence of an elementary way of algebraic thinking that manifested itself in the embodied constitution of a formula where the variable is expressed through particular instances, which we can schematize as "x + x +1" (where x was always a *specific* number). This formula, I argued on semiotic and epistemological grounds, is genuinely algebraic. That does not mean that all formulas provided by young students are algebraic. To give an example, one of the students suggested that to find out the number of elements in Term 100, you keep adding 2, and 2 and 2 to Term 1 until you get to Term 100. This is an example of arithmetic generalization —not of an algebraic one, as there is no analyticity involved. The "Message Problem" offered the students a possibility to go beyond intuitive indeterminacy and to think, talk, and calculate explicitly on an unknown number. Although several students were able to produce an explicit formula (e.g., "the number plus the number, plus 1" or "twice the number plus 1"), other students produced a formula where the general unknown number was represented through an example. This is what Mason (1996) calls *seeing the general in or through the particular*. Both the explicit formula and the general-through-the-particular formula bear witness to a more sophisticated form of elementary algebraic thinking than the embodied one where the variable and the formula are displayed in action.

Revealing our Grade 2 students' aforementioned elementary, pre-symbolic forms of algebraic thinking responded to our first research question—i.e., whether the embodied forms of thinking that we observed in adolescents are accessible to younger students. Yet, there are differences. Adolescents in general tend to gesture, talk and symbolize in harmonious coordinated manners (often after a period of mismatch between words and gestures (Arzarello and Edwards 2005; Radford 2009a). Our young students, in contrast, tend to gesture with energetic intensity (see e.g. Fig. 5). The energetic intensity may decrease as the students become more and more aware of the variables and the relationship between known and unknown numbers. However, the energetic intensity remains relatively pronounced as compared to what we have seen in adolescents (Radford 2009a, b). This phenomenon may be a token of a problem related to our second research question, namely: How does young students' algebraic thinking *develop*?

Developmental questions are very tricky, as psychologists know very well. It is not enough to collect data year after year and merely compare what students did in Year 1, to what they did in Year 2, etc. Exposing differences *shows* something but does not *explain* anything. I struggled with the question of the development of students' mathematical thinking for about a decade when I was doing research with adolescents, and I have to confess that I was unable to come up with something satisfactory. Yet, my research with adolescents helped me to envision a sensuous and material conception of mathematical cognition (Radford 2009b) that was instrumental in tackling the developmental question. Before going further in my account of what the students did in the following years, I need to dwell on the question of development first.

Thinking and Its Development

In contrast to mental cognitive approaches, thinking, I have suggested (Radford 2009b), is not something that solely happens 'in the head.' Thinking may be considered to be made up of material and ideational components: it *is* made up of (inner and outer) speech, objectified forms of sensuous imagination, gestures, tactility, and our actual actions with cultural artifacts. Thus, in Fig. 5, for instance, Karl *is* thinking *with* and *through* the body in the same way that he *is* thinking *through* and *in* language and the arsenal of conceptual categories it provides for us to notice, highlight, and attend to things, and intend them in certain cultural topical ways. The same can be said of the teacher in Fig. 4. Although it might be argued that the teacher and the student are merely communicating ideas, I would retort that this division between thinking and communicating makes sense only within the context of a conception of the mind as a private space within us, where ideas are created, computed and only then communicated. This computational view of the mind has a long history in our Western idealist and rationalist philosophical traditions. The view that I am sketching here goes against the dualistic assumption of mind versus body or ideal versus material. Thinking appears here as a an ideal-material form of reflection and action, which does not occur solely in the head but also *in* and *through* a sophisticated semiotic coordination of speech, body, gestures, symbols and tools. This is why, during difficult conversations, rather than digging in the head first to find the ideas that we want to express, we hear ourselves thinking as we talk, and realize, at the same time as our interlocutors, what we are thinking about.

Now to say that thinking is made up of (inner and outer) speech, objectified forms of sensuous imagination, gestures, tactility, and our actual actions with cultural artifacts does not mean that thinking is a *collection* of items. If we come back to our examples, Carlos (see Fig. 3, left), while moving the upper part of his body, was resorting to pointing gestures and words to count the rectangles in the first terms of the sequence. Words and gestures were guiding his perceptual activity to deal with the numerosity of the terms. Like Carlos, Karl moved his upper body, made arm- and hand-gestures and resorted to language (Fig. 5). In stating the formula "the number plus the number, plus 1," Erica gestured as if she was pressing keys in the calculator keyboard (Radford 2011). Yet, the relationship between perception, gestures and words is not the same. What it means is that thinking is not a mere collection of items. Thinking is rather a dynamic *unity* of material and ideal components. This is why the same gesture (e.g. an indexical gesture pointing to the rectangles on top of Term 3) may mean something conceptually sophisticated or something very simple. That is, the real significance of a component of thinking can only be recognized by the role such a component plays in the context of the *unity* of which it is a part.

Now I can formulate my developmental question. If thinking is a *systemic unity* of ideational and material components, it would be wrong to study its development by focusing on one of its components only. Thus, the development of algebraic thinking cannot be reduced to the development of its symbolic component (notation use, for instance). The development of algebraic thinking must be studied as a *whole*, by taking into account the interrelated *dialectic* development of its various components (Radford 2012). If in a previous section I talked about the 'domestication of the eye,' this domestication has to be related to the 'domestication of the hand' as well. And, indeed, this is what happened in our Grade 2 class from the second day on. As we recall, the teacher (Fig. 4) made extensive use of gestures and an explicit use of rhythm, and linguistic deictics, followed later by the students, who started using their hands and their eyes in novel ways, opening up new possibilities to use efficient and evolved cultural forms of mathematical generalization that they successfully applied to other sequences with different shapes.

To sum up, it is not only the tactile, the perceptual, or the symbol-use activity that is developmentally modified. In the same way as perception develops, so do speech (e.g., through spatial deictics) and gesture (through rhythm and precision). Perception, speech, gesture, and imagination develop in an interrelated manner. They come to form a new unity of the material-ideational components of thinking, where words, gestures, and signs more generally, are used as means of objectification, or as Vygotsky (Vygotsky 1987), p. 164 put it, "as means of voluntary directing attention, as means of abstracting and isolating features, and as a means of [...] synthesizing and symbolising". Within this context, to ask the question of the development of algebraic thinking is to ask about the appearance of new systemic structuring *relationships* between the material-ideational components of thinking (e.g., gesture, inner and outer speech) and the manner in which these relationships are organized and reorganized. It is through these developmental lenses that I studied the data collected in the following years and that I summarize in the rest of this article, focusing on Grades 3 and 4.

Grade 3: Semiotic Contraction

As usual, in Grade 3 the students were presented with generalizing tasks to be tackled in small groups. The first task featured a figural sequence, S_n, having n circles horizontally and $n-1$ vertically, of which the first four terms were given. Contrary to what he did first in Grade 2, from the outset, Carlos perceived the sequence taking advantage of the spatial configuration of its terms. Talking to his teammates about Term 4 he said: "here (pointing to the vertical part) there are four. Like you take all this [i.e., the vertical part] together (he draws a line around), and you take all this [i.e., the horizontal part] together (he draws a line around; see Fig. 6, pic 1). So, we should draw 5 like that (through a vertical gesture he indicates the place where the vertical part should be drawn) and (making a horizontal gesture) 5 like that" (see Fig. 6, pics 2–3).

When the teacher came to see the group, she asked Carlos to sketch for her Term 10, then Term 50. The first answer was given using unspecified deictics and gestures. He quickly said: "10 like this (vertical gesture) and 10 like that" (horizontal gesture). The specific deictic term "vertical" was used in answering the question about Figure 50. He said: "50 on the vertical… and 49…" When the teacher left, the students kept discussing how to write the answer to the question about Term 6. Carlos wrote: "6 vertical and 5 horizontal."

In developmental terms, we see the evolution of the unity of ideational-material components of algebraic thinking. Now, Carlos by himself and with great ease coordinates gestures, perception, and speech. The coordination of these outer components of thinking is much more refined compared to what we observed in Grade 2. This refinement is what we have called a *semiotic contraction* (Radford 2008a), that is, a genetic process in the course of which choices are made between what counts as relevant and irrelevant; it leads to a contraction of previous semiotic

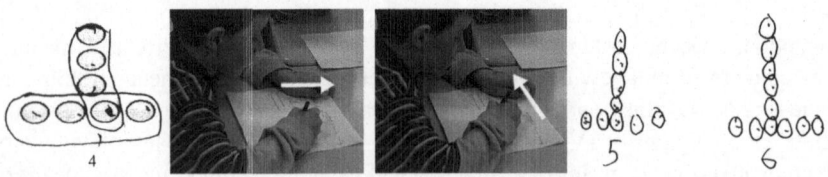

Fig. 6 To the *left*, Term 4 of the given sequence. Middle, Carlos's vertical and horizontal gestures while imagining and talking about the still to be drawn Term 5. To the *right*, Carlos's drawings of Terms 5 and 6

activity, resulting in a more refined linkage of semiotic resources. It entails a deeper level of consciousness and intelligibility of the problem at hand and is a symptom of learning and conceptual development.

Grade 4: The Domestication of the Hand

To check developmental questions, in Grade 4 we gave the students the sequence with which they started in Grade 2 (see Fig. 2). This time, from the outset, Carlos perceived the terms as being divided into two rows. Talking to his teammates and referring to the top row of Term 5, he said as if talking about something banal: "5 white squares, 'cause in Term 1, there is 1 white square (making a quick pointing gesture)… Term 2, 2 [squares] (making another quick pointing gesture); 3, (another quick pointing gesture) 3." He drew the five white squares on the top row of Term 5 and added: "after that you add a dark square." Then, referring to the bottom row of Term 4: "there are 4; there [Term 5] there are 5." When the teacher came to see their work, Carlos and his teammates explained "We looked at Term 2, it's the same thing [i.e., 2 white squares on top]… Term 6 will have 6 white squares."

There was a question in the activity in which the students were required to explain to an imaginary student (Pierre) how to build a big term of the sequence (the "Big Term Problem"). In Grade 2, the students chose systematically a particular term. This time, Carlos wrote: "He needs [to put as many white squares as] the number of the term on top and on the bottom, plus a dark square on top."

The "Message Problem" Again

At the end of the lesson, the students tackled the "Message Problem" again. As opposed to the lengthy process that, in Grade 2, preceded the building of a message without particular examples (Radford 2011), this time the answer was produced quicker:

David   The number of the term you calculate twice and add one. That's it!

Carlos   (*Rephrasing David's idea*) twice the number plus one

The activity finished with a new challenge. The teacher asked the students to add to the written message a "mathematical formula." After a discussion in Carlos's group concerning the difference between a phrase and a mathematical formula, the students agreed that a formula should include operations only. Carlos's formula is shown in pic 3 of Fig. 7.

From a developmental perspective, we see how Carlos's use of language has been refined. In Grade 2 he was resorting to particular terms (Term 1,000) to answer the same question about the "big term." Here he deals with indeterminacy in an

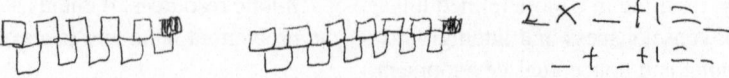

Fig. 7 *Left*, Carlos' drawings of Terms 5 and 6. *Right*, Carlos's formulas

easy way, through the expression "the number of the term." He even goes further
and produces two symbolic expressions to calculate the total of squares in the
unspecified term (Fig. 7, right). The semiotic activities of perceiving, gesturing,
languaging, and symbolizing have developed to a greater extent. They have reached
an interrelational refinement and consistency that was not present in Grade 2 and
was not fully developed in Grade 3. This cognitive developmental refinement
became even more apparent when the teacher led the students to the world of
notations, as we shall now see.

The Introduction to Notations

The introduction to notations occurred when the students discussed their answers to
homework based on the sequence shown in Fig. 8. The discussion took place right
after the general discussion about the "Message Problem" alluded to in the previous
sub-section.

The teacher gave the students the opportunity to compare and discuss their
answers to the homework by working in small groups. In Carlos' group, the terms
of the sequence were perceived as made up of two rows, each one having the same
number as the number of the term plus an addition of two squares at the end (see pic
2 in Fig. 8). As Carlos suggests, referring to Term 15, "15 on top, 15 at the bottom,
plus 2, that is 32." Or alternatively, as Celia, one of Carlos' teammates, explains,
"15 + 1 equals 16, then 16 + 16... which makes 32." After about 10 min of small-
group discussion, the teacher encouraged the students to produce a formula like the
one that they just provided for the "Message Problem." Then, the class moved to a
general discussion where various groups presented their findings. Erica went to the
Interactive White Smart Board (ISB) and suggested the following formula:
"1 + 1 + 2x__ = __" The teacher asked whether it would be possible to write,
instead of the underscores, something else. One student suggested putting an

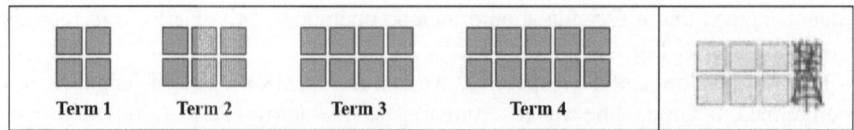

Fig. 8 Pic 1 (*left*), the sequence of the homework. Pic 2 (*right*), Carlos' decomposition of Term 3

interrogation mark. The teacher acknowledged that interrogation mark could also be used, and asked for other ideas. Samantha answered with a question:

Samantha	A letter?
Teacher	Ah! Could I write one plus one plus two times n? What does n mean?
A student	A number...
Teacher	Could we write that (i.e., one plus one plus two times n) equals n? (Some students answered yes, others no; talking to Erica who is at the whiteboard) Ok. Write it, write your formula (Erica writes $1 + 1 + 2 \times n = n$)
Carlos	No, because n (*meaning the first one*) is not equal to n (*meaning the second one*)
Teacher	Ah! Why do you say that n is not equal to n?
Carlos	Because if you do 2 times n, that will not equal [the second] n
Teacher	Wow!

In order not to rush the students into the world of notations, the teacher decided to delay the question of using a second letter to designate the total. As we shall see, this question will arise in the next activity. In the meantime, the formula was left as $1 + 1 + 2 \times n = \_\_$.

The next activity started right away. The students were provided with the new activity sheet that featured the sequence shown in Fig. 9. The students were encouraged to come up with as many formulas as possible to determine the number of squares in any term of the sequence.

During the small-group discussion, William offers a way to perceive the terms. Talking to Carlos, and referring to Term 6, which they drew on the activity sheet, William says (talking about the top row): "There are 8 [squares], because $6 + 2 = 8$. You see, on the bottom it's always the number of the term, you see?" His utterance is accompanied by a precise two-finger gesture through which he indicates the bottom row (see Fig. 10, left). He continues: "then, on the top, it's always plus 2" (making the gesture shown in Fig. 10, right).

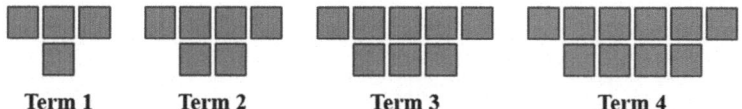

Term 1   **Term 2**   **Term 3**   **Term 4**

Fig. 9 The featured sequence of the new activity

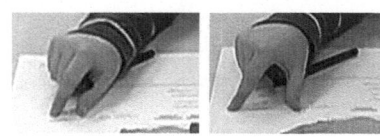

Fig. 10 William making precise gestures to refer to Term 6

The answer to the "Message Problem" was provided without difficulties. Without hesitation, Carlos said: "Ok. Double the number and add 2." The class moved to a general discussion, which was a space to discuss different forms of perceiving the sequence and of writing a formula. Marianne went to the ISB and suggested that the terms could be imagined as divided into two equal rows and that one square is added to the left and one to the right of the top row. In Fig. 11, referring to Term 3, she points first to the top row (imagined as made up of three squares; see Fig. 11, Pic 1). Then she points to the bottom row (Pic. 2), then to the extra square at the top right (Pic. 3) and to the extra square at the top left (Pic. 4). Celia proposed that a term was the same as the previous one to which two squares are added at the right end. In Fig. 11, Pic 5 and 6, she hides the two rightmost squares in Terms 2 and 3 to show that what remains in each case is the previous term. The developmental sophistication that the perception-gesture-language systemic unity has achieved is very clear.

Then, the students presented their formulas. Carlos presented the following formula: $\underline{N} + \underline{N} + 2 = \_$. The place for the variable in the formula is symbolized with a letter *and* the underscore sign. Letters in Carlos's formula appear timidly drawn, still bearing the vestiges of previous symbolizations (see Fig. 7, right).

The teacher asked if it would be possible to use another letter to designate the result:

Teacher   Well, we started with letters [in your formula]. Maybe we could continue
            with letters?
Carlos   No!
Teacher   Why not?
Carlos   An r?
Teacher   Why r?
Caleb   The answer (in French, *la réponse*)

Carlos completed the formula as follows: : $\underline{N} + \underline{N} + 2 = \underline{R}$. Other formulas were provided, as shown in Fig. 12:

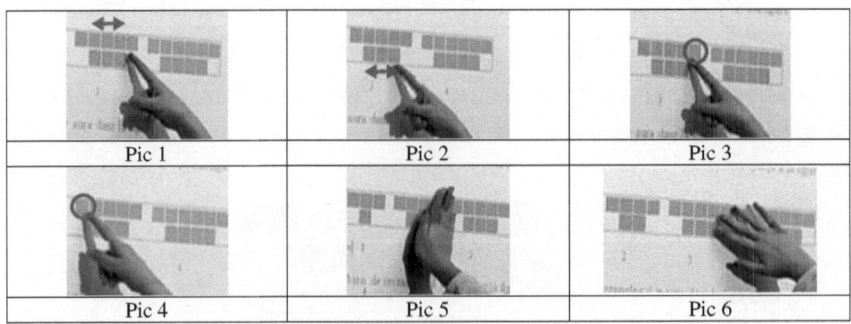

Fig. 11 Marianne's (Pic. 1–4) and Celia's (Pic. 5–6) gestures

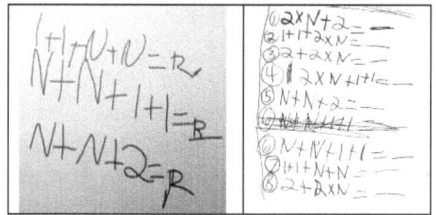

Fig. 12 *Left*, some formulas from the classroom discussion. *Right*, formulas from Erica's group

Synthesis and Concluding Remarks

In the first part of the article I suggested that algebraic thinking cannot be reduced to an activity mediated by notations. As I argued in previous work, a formula to calculate the number of rectangles in sequences like the one presented in Fig. 2, such as "2n + 1," can be attained by arithmetic trial-and-error methods. Algebraic thinking, I suggested, is rather characterized by the *analytic* manner in which it deals with indeterminate numbers. A rigorous video analysis convinced us that students signify indeterminate numbers through recourse to a plethora of semiotic embodied resources that, rather than being merely a by-product of thinking, constitute the very sensible texture of it. From this sensuous perspective on human cognition, it is not difficult to appreciate that 7–8-year-old students can effectively start thinking algebraically. In the second part of the article I dealt with the question of the development of algebraic thinking. Algebraic thinking—like all cultural forms of thinking (e.g., aesthetic, legal, political, artistic)—is a theoretical form that has emerged, evolved and refined in the course of cultural history. It pre-existed in a developed ideal form before the students engaged in our classroom activities. The greatest characteristic of child development consists in how this ideal form exerts a real influence on the child's thinking. But how can this ideal form exert such an influence on the child? Vygotsky's answer is: *under particular conditions of interaction between the ideal form and the child* (1994). In our case, the particular conditions of interaction between algebraic thinking as a historical ideal form and our Grade 2 students were constituted by a sequence of activities that were intentional bearers of this ideal form. Naturally, the students cannot discern the theoretical intention behind our questions, as this cultural ideal form that we call algebraic thinking has still to be encountered and cognized. The lengthy, creative, and gradual processes through which the students encounter, and become acquainted with historically constituted cultural meanings and forms of (in our case algebraic) reasoning and action is what I have termed, following Hegel, *objectification* (Radford 2008b).

The objectification of ideal forms requires a *temporal continuity* and stability of the knowledge that is being objectified. The objectification of ideal forms requires also the mutual emotional and ethical engagement of teacher and students in the joint activity of teaching-learning (Radford and Roth 2011; Roth and Radford 2011).

Drawing on the aforementioned idea of sensuous cognition and development, I suggested that the development of algebraic thinking can be studied in terms of the appearance of new systemic structuring *relationships* between the material-ideational components of thinking (e.g., gesture, inner and outer speech) and the manner in which these relationships are organized and reorganized in the course of the students' engagement in activity. The analysis of our experimental data focused on revealing those relationships and their progressive refinement. We saw how, for instance, the development of perception is consubstantial with the development of gestural and symbolic activity.

The whole story, however, is much more complex. As Vygotsky (1994) argued forcefully development can only be understood if we take into consideration the manner in which the student is actually emotionally experiencing the world. The emotional experience [*perezhivanie*] is, the Russian psychologist contended in a lecture given at the end of his life, the link between the subject and his/her surrounding, between the always changing subject (the perpetual being in the process of becoming) and his/her always conceptually, politically, ideologically moving societal environment. The explicit and meaningful insertion of *perezhivanie* into developmental accounts is, I suppose, still a trickier problem to conceptualize and investigate—an open research problem for sure.

Acknowledgments This article is a result of various research programs funded by the Social Sciences and Humanities Research Council of Canada (SSHRC/CRSH). I wish to thank Giorgio Santi and Chiara Andrà for their comments and suggestions on a previous version of this paper.

Open Access This chapter is distributed under the terms of the Creative Commons Attribution Noncommercial License, which permits any noncommercial use, distribution, and reproduction in any medium, provided the original author(s) and source are credited.

References

Arzarello, F., & Edwards, L. (2005). Gesture and the construction of mathematical meaning. In *Proceedings of the 29th PME conference* (pp. 123-54). Melbourne: PME.

Bednarz, N, Kieran, C. & Lee, L. (1996). *Approaches to algebra. Perspectives for research and teaching*. Dordrecht Boston London: Kluwer.

Cai, J., & Knuth, E. (2011). *Early algebraization*. New York: Springer.

Carraher, D. W., & Schliemann, A. (2007). Early algebra and algebraic reasoning. In F. K. Lester (Ed.), *Second handbook of research on mathematics teaching and learning* (pp. 669-705). Greenwich, CT: Information Age Publishing.

Descartes, R. (1954). *The geometry*. New York: Dover. (Original work published 1637)

Filloy, E., & Rojano, T. (1989). Solving equations: The transition from arithmetic to algebra. *For the Learning of Mathematics, 9*(2), 19-25.

Filloy, E., Rojano, T., & Puig, L. (2007). *Educational algebra: A theoretical and empirical approach*. New York: Springer Verlag.

Howe, R. (2005). *Comments on NAEP algebra problems*. Retrieved on 24.03.12 http://www.brookings.edu/~/media/Files/events/2005/0914_algebra/Howe_Presentation.pdf

Kieran, C. (1989). A perspective on algebraic thinking. *Proceedings of the 13th PME conference* (v. 2, pp. 163-171). Paris: PME

Kieran, C. (1990). A procedural-structural perspective on algebra research. In F. Furinghetti (Ed.), *Proceedings of the 15th PME conference* (pp. 245-53). Assisi: PME.

Lakoff, G., & Núñez, R. (2000). *Where mathematics comes from.* New York: Basic Books.

Mason, J. (1996). Expressing generality and roots of algebra. In N. Bednarz, C. Kieran, & L. Lee (Eds.), *Approaches to algebra* (pp. 65-86). Dordrecht: Kluwer.

Radford, L. (2000). Signs and meanings in students' emergent algebraic thinking: A semiotic analysis. *Educational Studies in Mathematics, 42*(3), 237-268.

Radford, L. (2003). Gestures, speech and the sprouting of signs. *Mathematical Thinking and Learning, 5*(1), 37-70.

Radford, L. (2008a). Iconicity and contraction. *ZDM - the International Journal on Mathematics Education, 40*(1), 83-96.

Radford, L. (2008b). The ethics of being and knowing: Towards a cultural theory of learning. In L. Radford, G. Schubring, & F. Seeger (Eds.), *Semiotics in mathematics education* (pp. 215-34). Rotterdam: Sense Publishers.

Radford, L. (2009a). "No! He starts walking backwards!" *ZDM - the International Journal on Mathematics Education, 41*, 467–480.

Radford, L. (2009b). Why do gestures matter? Sensuous cognition and the palpability of mathematical meanings. *Educational Studies in Mathematics, 70*(2), 111-126.

Radford, L. (2010). The eye as a theoretician: Seeing structures in generalizing activities. *For the Learning of Mathematics, 30*(2), 2-7.

Radford, L. (2011). Grade 2 students' non-symbolic algebraic thinking. In J. Cai & E. Knuth (Eds.), *Early algebraization* (pp. 303-22). Berlin: Springer-Verlag.

Radford, L. (2012). On the development of early algebraic thinking. *PNA, 6*(4), 117-133.

Radford, L., & Roth, W. -. (2011). Intercorporeality and ethical commitment. *Educational Studies in Mathematics, 77*(2-3), 227-245.

Radford, L., Bardini, C., & Sabena, C. (2007). Perceiving the general. *Journal for Research in Mathematics Education, 38*, 507-530.

Roth, W. -, & Radford, L. (2011). *A cultural historical perspective on teaching and learning.* Rotterdam: Sense Publishers.

Sutherland, R., Rojano, T., Bell, A., & Lins, R. (2001). *Perspectives on school algebra.* Dordrecht: Kluwer.

Viète, F. (1983). *The analytic art.* New York: Dover. (Original work published 1591)

Vygotsky, L. S. (1987). *Collected works (vol. 1).* New York: Plenum.

Vygotsky, L. S. (1994). The problem of the environment. In V. d. Veer & J. Valsiner (Eds.), *The Vygotsky reader* (pp. 338-54). Oxford: Blackwell. (Original work published 1934)

How We Think: A Theory of Human Decision-Making, with a Focus on Teaching

Alan H. Schoenfeld

Abstract Suppose a person is engaged in a complex activity, such as teaching. What determines what that person does, on a moment-by-moment basis, as he or she engages in that activity? What resources does the person draw upon, and why? What shapes the choices the person makes? I claim that if you know enough about a teacher's knowledge, goals, and beliefs, you can explain every decision he or she makes, in the midst of teaching. In this paper I give examples showing what shapes teachers' decision-making, and explain the theory.

Keywords Decision-making · Teaching · Theory

Introduction

I became a mathematician for the simple reason that I love mathematics. Doing mathematics can be a source of great pleasure: when you come to understand it, the subject fits together beautifully. Here I am not necessarily referring to advanced mathematics. The child who notices that every time she adds two odd numbers the result is even, wonders why, and the figures out the reason why:

> Each odd number is made up of a number of pairs, and one 'extra.' When you add two odd numbers together, the extras make a pair. That means that the sum is made up of pairs, so it's even!

is doing real mathematics. It was that kind of experience that led me into mathematics in the first place.

Sadly, very few people develop this kind of understanding, or this kind of pleasure in doing mathematics. It was this realization, and the thought that it might be possible to do something to change it, that led me into mathematics education.

A.H. Schoenfeld (✉)
University of California at Berkeley, Berkeley, CA, USA
e-mail: alans@berkeley.edu

© The Author(s) 2015

S.J. Cho (ed.), *The Proceedings of the 12th International Congress on Mathematical Education*, DOI 10.1007/978-3-319-12688-3_16

229

For more than 35 years I have pursued the question, "How can we develop deeper understandings of mathematical thinking, problem solving, and teaching, so that we can help more children experience the pleasures of doing mathematics?"

My early work was devoted to mathematical problem solving. I read Pólya's (1945) book *How to Solve It* early in my mathematical career, and it resonated. Pólya said that mathematicians used a wide range of problem solving strategies, which he called heuristics. When he described them, I recognized them—I used them too! I wondered, though, why I had not explicitly been taught those strategies. The answer, I learned, was that when people tried to teach the strategies described in Pólya's books, students did not learn to use them effectively. This was disappointing, but it also represented a lovely challenge. Could we understand such problem solving strategies well enough so that we could help students learn to use them effectively?

Thus began a decade's worth of work in which I tried to develop an understanding of problem solving: What do effective problem solvers do, which enables them to solve difficult problems? What do ineffective problem solvers do, that causes them to fail in their problem solving attempts? What can we do, as teachers, to help students become more effective problem solvers? My answers to those questions, which are summarized very briefly below, were published in my 1985 book *Mathematical Problem Solving*. The book resulted from a decade of simultaneous research on and teaching of problem solving, in which my theoretical ideas were tried in the classroom, and my experience in the classroom gave rise to more theoretical ideas.

Mathematical Problem Solving represented a solid first step in a research agenda. By the time it was written, I knew enough about problem solving to help students become more effective problem solvers. A next, logical goal was to help mathematics teachers to help their students develop deeper understandings of mathematics. In many ways, of course, teaching is an act of problem solving—but it is so much more. The challenge was, could I develop a theoretical understanding of teaching in ways that allowed me to understand how and why teachers make the choices they do, as they teach? Could that understanding then be used to help teachers become more effective? Moreover, to the degree that teaching is typical of knowledge-intensive decision making, could the theoretical descriptions of teaching be used to characterize decision making in other areas as well?

Those questions have been at the core of my research agenda for the past 25 years. My answers to them now exist, in a new book, *How We Think* (Schoenfeld 2010). The purpose of this paper is to illustrate and explain the main ideas in the book. Because my current research has evolved from my earlier problem solving work, I set the stage for the discussion that follows with a brief description of that work—what it showed and, more importantly, the questions that it did not answer. That will allow me to describe what a complete theory should be able to accomplish. I then turn to the main body of this paper, three studies of teaching. In those examples I show how, under certain circumstances, it is possible to model the act of teaching, to the point where one can provide a grounded explanation of every decision that a teacher makes during an extended episode of teaching. Following that, I give some other examples to suggest that the theory is general, and I make a few concluding comments.

The Challenge

Suppose that you are in the middle of some "well practiced" activity, something you have done often so that it is familiar to you. Depending on who you are, it might be

- cooking a meal
- fixing a car
- teaching a class
- doing medical diagnosis or brain surgery.

The challenge is this: If I know "enough" about you, can I explain (i.e., build a cognitive model that explains) every single action you take and every decision you make?

My goal for this paper is to describe an analytic structure that does just that—an analytic structure that explains how and why people act the way they do, on a moment-by-moment basis, in the midst of complex, often social activities such as teaching.

My major claim is this: *People's in-the-moment decision making when they teach, and when they engage in other well practiced, knowledge intensive activities, is a function of their knowledge and resources, goals, and beliefs and orientations. Their decisions and actions can be "captured" (explained and modeled) in detail using only these constructs.*

The main substance of this paper (as in the book) consists of three analyses of teaching, to convey the flavor of the work. Of course, it is no accident that I chose mathematics teaching as the focal area for my analyses. I am, after all, a mathematics educator! But more to the point, teaching is a knowledge intensive, highly interactive, dynamic activity. If it is possible to validate a theory that explains teachers' decision making in a wide range of circumstances, then that theory should serve to explain all well practiced behavior.

Background: Problem Solving

As discussed above, my current work is an outgrowth of my earlier research on mathematical problem solving. Here I want to summarize the core findings of that work, to show how it lays the groundwork for my current research.

My major argument about mathematical problem solving (see Schoenfeld 1985, for detail) was that it is possible to explain someone's success or failure in trying to solve problems on the basis of the following four things:

1. *Knowledge (or more broadly, resources)*. This is not exactly shocking—but, knowing what knowledge and resources a problem solver has *potentially* at his or her disposal is important.
2. *Problem solving strategies, also known as "heuristics."* We know from Pólya's work that mathematicians use heuristic strategies, "rules of thumb for making

progress when you do not know a direct way to a solution." Faculty pick up these strategies by themselves, through experience. Typically, students don't use them. But, my research showed that students can learn to use them.

3. *"Metacognition," or "Monitoring and self-regulation."* Effective problem solvers plan, and they keep track of how well things are going as they implement their plans. If they seem to be making progress, they continue; if there are difficulties, they re-evaluate and consider alternatives. Ineffective problem solvers (including most students) do not do this. As a result, they can fail to solve problems that they *could* solve. Students can learn to be more effective at these kinds of behaviors.

4. *Beliefs.* Students' beliefs about themselves and the nature of the mathematical enterprise, derived from their experiences with mathematics, shape the knowledge they draw upon during problem solving and the ways they do or do not use that knowledge. For example, students who believe that "all problems can be solved in 5 min or less" will stop working on problems even though, had they persevered, they might have solved them. Students who believe that "proof has nothing to do with discovery or invention" will, in the context of "discovery" problems, make conjectures that contradict results they have just proven. (see Schoenfeld 1985).

In sum: By 1985 we know what "counted" in mathematical problem solving, in the sense that we could explain, post hoc, what accounted for success or failure. As the ensuing 25 years have shown, this applied to all "goal-oriented" or problem solving domains, including mathematics, physics, electronic trouble-shooting, and writing.

BUT… There was a lot that the framework that I have just described did not do. In the research I conducted for *Mathematical Problem Solving*, people worked in isolation on problems that I gave them to solve. Thus: the goals were established (i.e., "solve this problem"); the tasks didn't change while people worked on them; and social interactions and considerations were negligible.

In addition, *Mathematical Problem Solving* offered a *framework*, not a theory. Above and beyond pointing out what is important—which is what a framework does—a theory should provide rigorous explanations of how and why things fit together. That is what my current work is about. What I have been working on for the past 25 years is a theoretical approach that explains how and why people make the choices they do, while working on issues they care about and have some experience with, amidst dynamically changing social environments.

I can think of no better domain to study than teaching. Teaching is knowledge intensive. It calls for instant decision making in a dynamically changing environment. It's highly social. And, if you can model teaching, you can model just about anything! I will argue that if you can model teaching, you can model: shopping; preparing a meal; an ordinary day at work; automobile mechanics; brain surgery (or any other medical practice), and other comparably complex, "well practiced" behaviors. All of these activities involve goal-oriented behavior—drawing on available resources (not the least of which is knowledge) and making decisions in order to achieve outcomes you value.

The goal of my work, and this paper, is to describe a theoretical architecture that explains people's decision-making during such activities.

How Things Work

My main theoretical claim is that goal-oriented "acting in the moment"—including problem solving, tutoring, teaching, cooking, and brain surgery—can be explained and modeled by a theoretical architecture in which the following are represented: Resources (especially knowledge); Goals; Orientations (an abstraction of beliefs, including values, preferences, etc.); and Decision-Making (which can be modeled as a form of subjective cost-benefit analysis). For substantiation, in excruciating detail, please see my book, *How we Think.* To briefly provide substantiation I will provide three examples in what follows. But first, a top-level view of how things work is given in Fig. 1. The basic structure is recursive: Individuals orient to situations and decide (on the basis of beliefs and available resources) how to pursue their goals. If the situation is familiar, they implement familiar routines; if things are

---

How Things Work

- An individual enters into a particular context with a specific body of resources, goals, and orientations.
- The individual takes in and orients to the situation. Certain pieces of information and knowledge become salient and are activated.
- Goals are established (or reinforced if they pre-existed).
- Decisions consistent with these goals are made, consciously or unconsciously, regarding what directions to pursue and what resources to use:
  - If the situation is familiar, then the process may be relatively automatic, where the action(s) taken are in essence the access and implementation of scripts, frames, routines, or schemata.
  - If the situation is not familiar or there is something non-routine about it, then decision-making is made by a mechanism that can be modeled by (i.e., is consistent with the results of) using the subjective expected values of available options, given the orientations of the individual.
- Implementation begins.
- Monitoring (whether it is effective or not) takes place on an ongoing basis.
- This process is iterative, down to the level of individual utterances or actions:
  - Routines aimed at particular goals have sub-routines, which have their own subgoals;
  - If a subgoal is satisfied, the individual proceeds to another goal or subgoal;
  - If a goal is achieved, new goals kick in via decision-making;
  - If the process is interrupted or things don't seem to be going well, decision-making kicks into action once again. This may or may not result in a change of goals and/or the pathways used to try to achieve them.

Fig. 1 How things work, in outline. From Schoenfeld (2010), p. 18, with permission

unfamiliar or problematic, they reconsider. It may seem surprising, but if you know enough about an individual's resources, goals, and beliefs, this approach allows you to model their behavior (after a huge amount of work!) on a line-by-line basis.

First Teaching Example, Mark Nelson

Mark Nelson is a beginning teacher. In an elementary algebra class, Nelson has worked through problems like, $x^5/x^3 = ?$ Now he has assigned

$$\text{(a)}\, m^6/m^2, \quad \text{(b)}\, x^3y^7/x^2y^6, \quad \text{and} \quad \text{(c)}\, x^5/x^5$$

for the class to work. Nelson expects the students to have little trouble with m^6/m^2 and x^3y^7/x^2y^6, but to be "confused" about x^5/x^5; he plans to "work through" their confusion. Here is what happens.

Nelson calls on students to give answers to the first two examples. He has a straightforward method for doing so:

- He asks the students what they got for the answer, and confirms that it is correct.
- He asks how they got the answer.
- Then he elaborates on their responses.

Thus, for example, when a student says the answer to problem (b) is xy, Nelson asks "why did you get xy?" When the student says that he subtracted, Nelson asks, "What did you subtract? When the student says "3 minus 2," Nelson elaborates:

OK. You looked at the x's [pointing to x-terms in numerator and denominator] and [pointing to exponents] you subtracted 3 minus 2. That gave you x to the first [writes x on the board]. And then [points to y terms] you looked at the y's and said [points to the exponents] 7 minus 6, gives you y to the first [writes y on board].

He then asks what to do with x^5/x^5. They expand and "cancel." The board shows
$\frac{\cancel{xxxxx}}{\cancel{xxxxx}}$. Pointing to that expression, he says, "what do I have?" The responses are

"zero," "zip," "nada," and "nothing" … not what he wants them to see! He tries various ways to get the students to see that "cancelling" results in a "1", for example,

Nelson: "What's 5/5?"
Students: "1."
Nelson: "But I cancelled. If there's a 1 there [in 5/5], isn't there a 1 there [pointing to the cancelled expression]?"
Students: "No."

Defeated, he slumps at the board while students argue there's "nothing there." He looks as if there is nothing he can say or do that will make sense to the students.

He tries again. He points to the expression $\frac{\cancel{xxxxx}}{\cancel{xxxxx}}$ and asks what the answer is.

A student says "x to the zero over 1." Interestingly, Nelson *mis-hears* this as "x to the zero equals 1," which is the correct answer. Relieved, he tells the class,

"That's right. Get this in your notes: $x^5/x^5 = x^0 = 1$."
Any number to the zero power equals 1."

To put things simply, this is *very* strange. Nelson certainly knew enough mathematics to be able to explain that if $x \neq 0$,

$$\left(\frac{x^5}{x^5}\right) = \left(\frac{x}{x}\right)^5 = 1^5 = 1,$$

but he didn't do so. WHY?

There is a simple answer, although it took us a long time to understand it. The issue has to do with Nelson's beliefs and orientations about teaching. One of Nelson's central beliefs about teaching—the belief that *the ideas you discuss must be generated by the students*—shaped what knowledge he did and did not use.

In the first example above (reducing the fraction x^3y^7/x^2y^6), a student said he had subtracted. The fact that the student mentioned subtraction gave Nelson "permission" to explain, which he did: "OK. You looked at the x's and you subtracted 3 minus 2. That gave you x to the first. And then you looked at the y's and said 7 minus 6, gives you y to the first."

But in the case of example (c), x^5/x^5, he was stymied—when he pointed to the expression $\frac{\cancel{xxxxx}}{\cancel{xxxxx}}$ and asked "what do I have?" the only answers from the students were "zero," "zip," "nada," and "nothing." Nobody said "1." And because of his belief that he had to "build on" what students say, Nelson felt he could not proceed with the explanation. Only later, when he *mis-heard* what a student said, was he able to finish up his explanation.

[Note: This brief explanation may or may not seem convincing. I note that full detail is given in the book, and that Nelson was part of the team that analyzed his videotape. So there is strong evidence that the claims I make here are justified.]

Second Teaching Example, Jim Minstrell

Here too I provide just a very brief description.

Jim Minstrell is an award-winning teacher who is very thoughtful about his teaching. It is the beginning of the school year, and he is teaching an introductory lesson that involves the use of mean, median, and mode. But, the main point of the lesson is that Minstrell wants the students to see that such formulas need to be used *sensibly*.

The previous day eight students measured the width of a table. They obtained these values:

$$106.8; \ 107.0; \ 107.0; \ 107.5; \ 107.0; \ 107.0; \ 106.5; \ 106.0 \text{ cm.}$$

Minstrell wants the students to discuss the "best number" to represent the width of the table. His plan is for the lesson to have three parts:

1. Which numbers (all or some?) should they use?
2. How should they combine them?
3. With what precision should they report the answer?

Minstrell gave us a tape of the lesson, which we analyzed. The analysis proceeded in stages. We decomposed the lesson into smaller and smaller "episodes," noting for each episode which goals were present, and observing how transitions corresponded to changes in goals. In this way, we decomposed the entire lesson—starting with the lesson as a whole, and ultimately characterizing what happened on a line-by-line basis. See Figs. 2 and 3 (next pages) for an example of analytic detail. Figure 2 shows the whole lesson, and then breaks it into major episodes (lesson

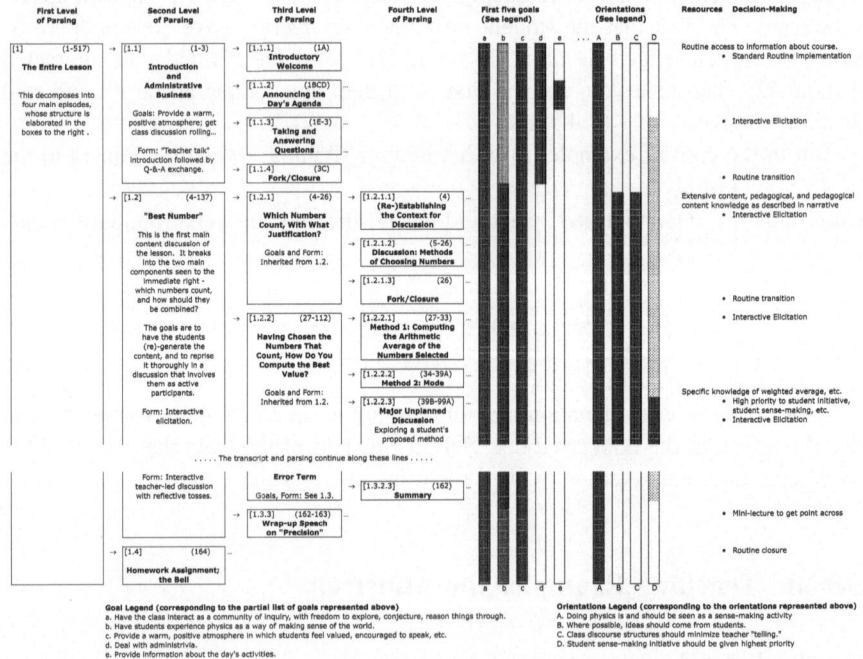

Fig. 2 A "top level" view of Minstrell's lesson, "unfolding" in levels of detail. (With permission, from Schoenfeld 2010, pp. 96–97)

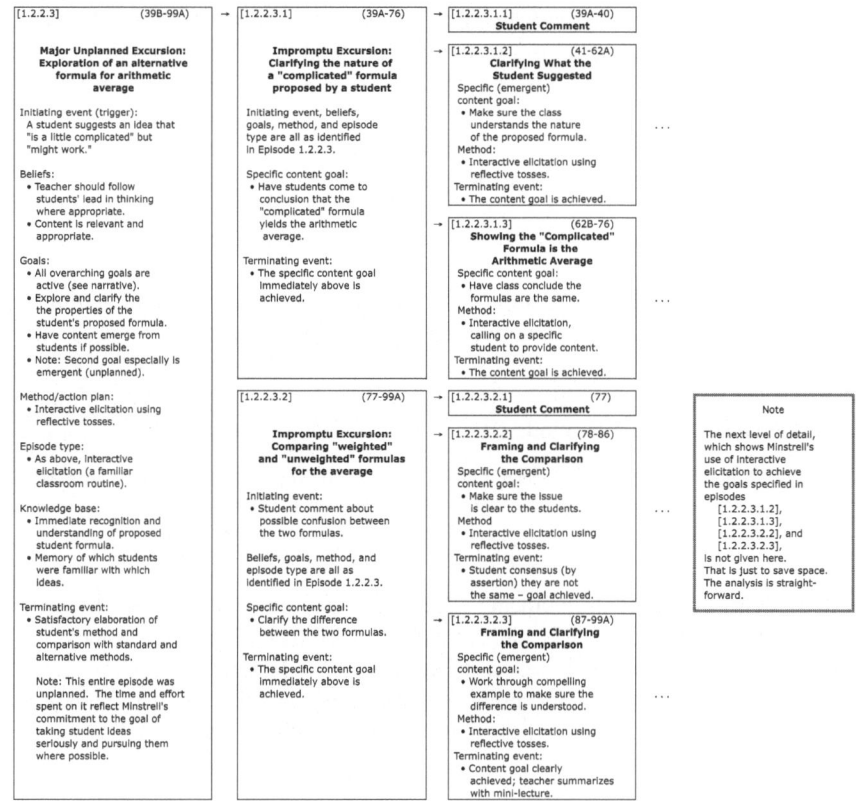

Fig. 3 A more fine-grained parsing of Episode [1.2.2.3]. (From Schoenfeld 2010, pp. 116–117, with permission)

segments), each of which has its own internal structure. Most of the lesson was very simple to analyze in this way.

Minstrell has a flexible "script" for each part of the lesson:

- He will raise the issue;
- He will ask the class for a suggestion;
- He will clarify and pursue the student suggestion by asking questions, inserting some content if necessary.

Once the suggestion has been worked through, he will ask for more suggestions. When students run out of ideas, he may inject more ideas, or move to the next part of the lesson.

In this way, the lesson unfolds naturally, and it is easy to "capture" it—see Fig. 2 for a "top level" summary of how the lesson unfolded. The episodes in the second and third columns, which correspond to an analysis of the lesson as taught, show that Minstrell did cover the big topics as planned.

A line-by-line analysis (see Schoenfeld 1998, 2010) shows that when Minstrell was dealing with expected subject matter, he followed the "script" described above very closely. So, it is easy to model Minstrell's behavior when he is on familiar ground.

But what about unusual events? Remember the data: The eight values the students had obtained for the width of the table were

106.8; 107.0; 107.0; 107.5; 107.0; 107.0; 106.5; 106.0 cm.

As the lesson unfolded, Minstrell asked the students about "a way of getting the best value." (see box 1.2.2 in the third column of Fig. 2.) As the class proceeded, one student mentioned the idea of using the "average" and, when asked by Minstrell, provided a definition. (Box 1.2.2.1 in the fourth column of Fig. 2.) Another student mentioned mode (Box 1.2.2.2). Then, a student said:

> This is a little complicated but I mean it might work. If you see that 107 shows up 4 times, you give it a coefficient of 4, and then 107.5 only shows up one time, you give it a coefficient of one, you add all those up and then you divide by the number of coefficients you have.

This is an unexpected comment, which does not fit directly with Minstrell's flexible script. The question is, can we say what Minstrell would do when something unexpected, like this, arises in the middle of his lesson?

Before proceeding, I want to point out that there is a wide range of responses, which teachers might produce. I have seen responses like all of the following:

> That's a very interesting question. I'll talk to you about it after class.

> Excellent question. I need to get through today's plans so you can do tonight's assigned homework, but I'll discuss it tomorrow.

> That's neat. What you've just described is known as the 'weighted average.' Let me briefly explain how you can work with that...

> Let me write that up as a formula and see what folks think of it.

> Let's make sure we all understand what you've suggested, and then explore it.

So, teachers might do very different things. Is it possible to know what Minstrell will do? According to our model of Minstrell, (1) His fundamental *orientation* toward teaching is that physics is a sense-making activity and that students should experience it as such; (2) One of his major *goals* is to support inquiry and to honor student attempts at figuring things out; (3) His *resource base* includes favored techniques such as "reflective tosses"—asking questions that get students to explain/elaborate on what they said.

Thus, the model predicts that he will pursue the last option—making sure that the students understand the issue that the student has raised (including the ambiguity about how you add the coefficients; do you divide by 5 or 8?) and pursuing it. He will do so by asking the students questions and working with the ideas they produce.

This is, in fact, what Minstrell did. Figure 3 shows how that segment of the lesson evolved. It is an elaboration of Box 1.2.2.3 in Fig. 2.

As noted above, it is possible to model Minstrell's decision. The model shows that, when faced with options such as those listed above, Minstrell is by far most likely to pursue the one I have indicated. The computations take about seven pages of text, so I will spare you the detail! More generally:

We have found that we were able to capture Minstrell's routine decision-making, on a line-by-line basis, by characterizing his knowledge/resources and modeling them as described in Fig. 1, "How Things Work;" and,

We were able to model Minstrell's non-routine decision-making using a form of subjective expected value computation, where we considered the various alternatives and looked at how consistent they were with Minstrell's beliefs and values (his orientations).

In summary, we were able to model every decision Minstrell made during the hour-long class.

Third Teaching Example, Deborah Ball

Some years ago, at a meeting, Deborah Ball showed a video of a third grade classroom lesson she had taught. The lesson was amazing—and it was controversial. In it,

- Third graders argued on solid mathematical grounds;
- The discussion agenda evolved as a function of classroom conversations;
- The teacher seemed at times to play a negligible role, and she made at least one decision that people said was not sensible.

In addition, I had little or no intuition about what happened. Thus, this was a perfect tape to study! There were major differences from cases 1 and 2:

- the students were third graders instead of high school students;
- psychological (developmental) issues differed because of the children's age;
- the "control structure" for the classroom was much more "organic";
- the teacher played a less obvious "directing" role.

The question was, could I model what happened in this lesson? If so, then the theory covered an extremely wide range of examples, which would comprise compelling evidence of its general validity. If not, then I would understand the limits of the theory. (Perhaps, for example, it would only apply to teacher-directed lessons at the high school level.)

Here is what happened during the lesson. Ball's third grade class had been studying combinations of integers, and they had been thinking about the fact that, for example, the sum of two even numbers always seemed to be even. The previous day Ball's students had met with some 4th graders, to discuss the properties of even numbers, odd numbers, and zero. Ball had wanted her students to see that these were complex issues and that even the "big" fourth graders were struggling with them. The day after the meeting (the day of this lesson), Ball started the class by asking what the students thought about the meeting:

- How do they think about that experience?
- How do they think about their own thinking and learning?

Ball had students come up to the board to discuss "what they learned from the meeting." The discussion (a transcript of which is given in full in Schoenfeld 2008, 2010) covered a lot of territory, with Ball seemingly playing a small role as students argued about the properties of zero (is it even? odd? "special"?). For the most part, Ball kept her students focused on the "meta-level" question: what did they learn about their own thinking from the meeting with the fourth graders the previous day?

But then, after a student made a comment, Ball interrupted him to ask a *mathematical* question about the student's understanding. This question, which took almost 3 min to resolve, completely disrupted the flow of the lesson. Many people, when watching the tape of the lesson, call that decision a "mistake." How could Ball, who is a very careful, thoughtful, and experienced teacher, do such a thing? If the decision was arbitrary or capricious in some way, that is a problem for the theory. If highly experienced teachers make arbitrary decisions, it would be impossible to model teachers' decision making in general.

In sum, this part of the lesson seems to unfold without Ball playing a directive role in its development—and she made an unusual decision to interrupt the flow of conversation. Can this be modeled? The answer is yes. A fine-grained analysis reveals that Ball has a "debriefing routine" that consists of asking questions and fleshing out answers. That routine is given in Fig. 4.

In fact, Ball uses that routine five times in the first 6 min of class. Moreover, once you understand Ball's plans for the lesson, her unexpected decision—what has been called her "mistake" by some—can be seen as entirely reasonable and consistent with her agenda. This has been modeled in great detail. For the full analysis, see Schoenfeld 2010; for an analytic diagram showing the full analysis, download Appendix E from my web page, http://www-gse.berkeley.edu/faculty/AHSchoenfeld/AHSchoenfeld. html.

To sum things up: As in the two previous cases, (1) We were able to model Ball's routine decision-making, on a line-by-line basis, by characterizing her knowledge/resources and modeling them as described in Fig. 1. (2) We were able to model Ball's non-routine decision-making as a form of subjective expected value computation.

In short, we were able to model every move Ball made during the lesson segment.

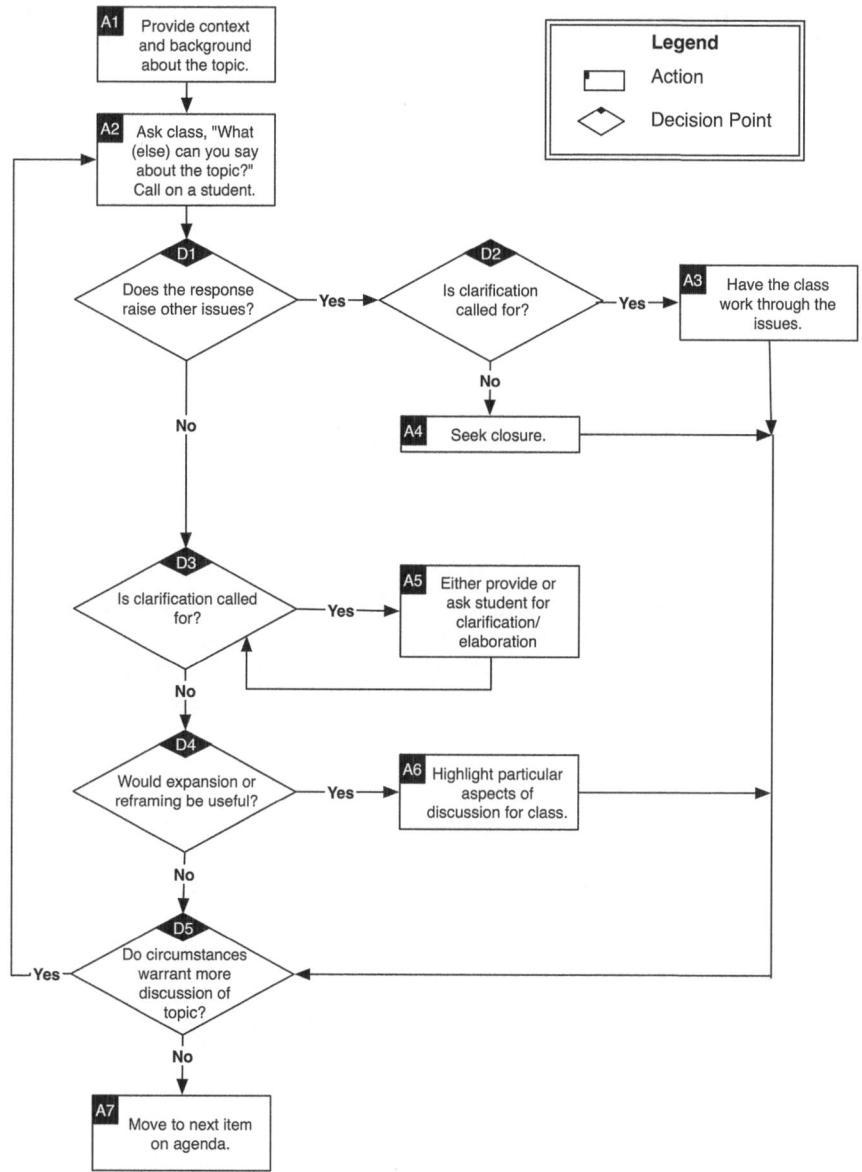

Fig. 4 A flexible, interruptible routine for discussing a topic. (From Schoenfeld 2010, p. 129, with permission)

Yet More Examples

Making Breakfast (or Any Other Meal)

If you look at Fig. 1, you can see that it would be easy to model decision-making during cooking. Usually we have fixed routines for cooking familiar meals. And if something changes (for example, when my daughter asks me to make a fancy breakfast), that calls for a "non-routine" decision, which can also be modeled. Readers might enjoy creating models of their own cooking practices and decision making.

Routine Medical Diagnosis and Practice

To see if my ideas worked outside of the classroom, I asked my doctor if I could tape and analyze one of my office visits with her. She said yes; an analysis of our conversation is given in *How We Think*. The conversation was easy to model, because the doctor follows a straightforward (and flexible) script. Modeling a two-person interaction is a lot easier than modeling a classroom; it is more like modeling a tutoring interaction. When the person being modeled (in this case, the doctor) only has to pay attention to one other person (instead of the 30 children a teacher has to pay attention to), decision-making is comparatively simple—and simple to model.

I should also note that there is a very large artificial intelligence literature on modeling doctors' decision making—there are computer programs that make diagnoses, etc. (The field is well established: see, e.g., Clancey and Shortliffe 1984). So, the idea that it is possible to capture doctors' routine decision making is not new. More recent, and also consistent with my emphasis on beliefs as shaping behavior, there are studies (e.g., Groopman 2007) of how doctors' stereotypes (beliefs and orientations) regarding patient behavior lead them to miss what should be straightforward diagnoses.

Discussion

The approach I have outlined in this paper "covers" routine and non-routine problem solving, routine and non-routine teaching, cooking, and brain surgery—and every other example of "well practiced," knowledge-based behavior that I can think of. All told, I believe it works pretty well as a theory of "how we think."

Readers have the right to ask, why would someone spend 25 years trying to build and test a theory like this? Here is my response.

First, theory building and testing should be central parts of doing research in mathematics education. That is how we make progress.

Second, the more we understand something the better we can make it work; when we understand how something skillful is done we can help others do it. This was the idea behind my problem solving work, where an understanding of problem solving helped me to help students become better problem solvers. I believe that a comparably deep understanding of teachers' decision making can be used to help mathematics teachers become more effective.

Third, this approach has the potential to provide tools for describing developmental trajectories of teachers. Beginning teachers, for example, often struggle with issues of classroom "management"—of creating an orderly classroom environment in which their students can learn productively. While teachers are struggling at this, they have little time or attention to devote to some of the more subtle aspects of expert teaching, such as teaching responsively—listening carefully to what their students say, diagnosing what the students understand and misunderstand, and shaping the lesson so that it helps move the students forward mathematically. The more we understand what teachers understand at particular points in their careers, the more we will be able to provide relevant professional development activities for them. An understanding of teachers' developmental trajectories can help us help teachers get better at helping their students learn. (see Chap. 8 of Schoenfeld 2010, for detail.)

Fourth and finally, it's fun! The challenge of understanding human behavior has proved itself to be every bit as interesting and intellectually rewarding as the challenge of understanding mathematics. It has occupied me for the past 35 years, and I look forward to many more years of explorations. Exploring questions of how teachers' understandings develop, and of how and when one can foster the development of mathematics teachers' expertise, are intellectually challenging. Equally important, addressing them can, over the long run, lead to improvements in mathematics teaching and learning.

Open Access This chapter is distributed under the terms of the Creative Commons Attribution Noncommercial License, which permits any noncommercial use, distribution, and reproduction in any medium, provided the original author(s) and source are credited.

References

Clancey, W. J., & Shortliffe, E. H. (Eds). (1984). *Readings In Medical AI: The First Decade*. Reading, Ma: Addison-Wesley.
Groopman, J. (2007). *How Doctors Think*. Boston: Houghton Mifflin.
Pólya, G. (1945; 2nd edition, 1957). *How To Solve It*. Princeton: Princeton University Press.
Schoenfeld, A. H. (1985) *Mathematical Problem Solving*. Orlando, Fl: Academic Press.
Schoenfeld, A. H. (1998). Toward A Theory Of Teaching-In-Context. *Issues In Education*, 4(1), 1 - 94.
Schoenfeld, A. H. (Ed). (2008) *A Study Of Teaching: Multiple Lenses, Multiple Views*. Journal For Research In Mathematics Education. Monograph Series # 14). Reston, Va: National Council Of Teachers Of Mathematics,
Schoenfeld, A. H. (2010). *How We Think: A Theory Of Goal-Oriented Decision Making And Its Educational Applications*. New York: Routledge.
Schoenfeld, A. H. (in press). How We Think. *Revistas Tópicos Educacionais*.

Part V
Survey Teams

Curriculum and the Role of Research

Gail Burrill, Glenda Lappan and Funda Gonulates

Abstract The survey team collected information on the development and use of curriculum from 11 diverse countries around the world. The data show that a common set of mathematics learning goals are established in almost all countries. However, only a few countries report a substantial role for research in designing and monitoring the development of their curriculum. The data also suggest great variation among countries at the implementation level.

Keywords Standards · Curriculum · Technology · Tracking · Textbooks · Research · Teacher support

Introduction

This report is based on an analysis of responses to survey questions on curriculum standards and goals from 11 countries: Australia, Brazil, Egypt, England, China, Honduras, Indonesia, Japan, Namibia, Peru, and six states in the United States.[1] The paper is organized in five sections: standards/curricular goals; relation of standards to the status quo, the role of textbooks in enacting the curriculum, the role of technology in classrooms, and teacher support related to standards/curricular goals.[2]

---

[1] See end of report for list of response teams from each country.
[2] Survey Team: Chair Glenda Lappan (USA), Jiansheng Bao (China), Karen D'Emiljo (Namibia), Keiko Hino (Japan), Vinício de Macedo Santos (Brazil), Malcolm Swan (England), IPC Liaison: Gail Burrill (USA).

G. Burrill (✉) · G. Lappan · F. Gonulates
Michigan State University, East Lansing, USA
e-mail: burrill@msu.edu

G. Lappan
e-mail: glappan@math.msu.edu

F. Gonulates
e-mail: fgonulates@gmail.com

© The Author(s) 2015
S.J. Cho (ed.), *The Proceedings of the 12th International Congress on Mathematical Education*, DOI 10.1007/978-3-319-12688-3_17

247

The intent of the report is to allow others to examine their standards/curriculum goals relative to those of other countries across the world.

Standards/Curricular Goals

Who Is Responsible for the Development of Standards/ Curricular Goals?

In most countries the ministry of education establishes curricular standards. In the United States, however, control of education is a state's right, and in many states, for example, Montana, state constitutions give control of education to local districts. The federal government influences education through funding initiatives, such as the No Child Left Behind Act in 2001. The 2010 Common Core State Standards (CCSS) initiative is not a federal program but has been adopted and is being implemented by 45 of the 50 states and the District of Columbia. China also does not have a mandated national curriculum. China Mainland, including Shanghai, has common standards; Hong Kong, Taiwan and Macau create their own standards/ curriculum goals.

In many countries, standards/curricular goals are set by historical tradition or cultural norms. For example, Namibia used the Cambridge curriculum when they became independent in 1990 and only recently has begun to develop their its own standards. Brazil 's standards are attributed to the history of the discipline, the prescribed curricula, and the comparative analysis among national documents from different historical periods and national and international documents. Some countries base their standards and guidelines on those of countries with high achievement scores on recent international exams. For example, both England and the United States cite countries such those from the Pacific Rim and Finland as resources for their new standards. Peru noted that an analysis of documents from other countries in South American and from TIMSS, Programme for International Student Assessment (PISA), and National Council of Teachers of Mathematics (NCTM) contributed to the development of their Diseño Curricular Nacional (CND) (National Curricular Design) (2009).

Why Standards?

Over time, many countries have changed from local standards to national standards. For example, Brazil found that the lack of national standards contributed to unequal opportunity for education. For much the same reason, the documented difference in the rigor and quality of individual state standards, the state governors in the United

States supported the development and adoption of the CCSS. The new US standards are intended to be substantially more focused and coherent.

Standards are viewed as political: i.e., Brazil suggests that mathematics curricular goals depend more on political timing, election campaigns and government administrations, where "the logic of an education agenda that transcends governments and politicians' mandates, set as a goal for a democratic and developed society, is not the rule" (Response to ICME 12 Curriculum Survey 2011, p. 6). In the United States the two major political parties have different views on education, its funding and its goals. This has recently given rise to the creation of publicly funded schools governed by a group or organization with a legislative contract or charter from a state or jurisdiction that exempts the school from selected state or local regulations in keeping with its charter. Hong Kong also reported that writing standards seems to be more politically based than research based. Many of the changes in England's National Curriculum (NC) are the result of criticism from the current government that the NC is over-prescriptive, includes non-essential material, and specifies teaching method rather than content. In Peru each new curricular proposal is viewed as an adjustment to the prior curriculum. In this process, radical changes do occur, such as changing the curriculum by capabilities (CND 2005) to the curriculum by competencies (CND 2008) in the secondary education level. These decisions are often the result of a policy change with each new government.

In most countries surveyed, a diverse team, including mathematics education researchers, ministry of education staff, curriculum supervisors, and representatives of boards of education are responsible for developing the standards/goals. In some countries (Japan, Australia) teachers are involved, but in others the design teams are primarily experts from universities, teaching universities or the ministry of education (Indonesia, Egypt). The design of the framework for the National Curriculum in England is carried out by a panel of four, not necessarily mathematics educators, charged to reflect the view of the broader mathematics education community including teachers.

What Is the Role of Research?

Research has different interpretations and meanings in relation to the development and implementation of standards or curricula guidelines. One common response in the surveys was to cite as research the resources used in preparing standards (for example, other countries' standards). In addition, the degree to which research is used in compiling the standards often depends on the vision, perspectives and beliefs of the team responsible for the development.

The use of research related to student learning in developing standards/curricular goals is not common among the countries surveyed. A typical description of the process was given by Hong Kong, where the development team might do a literature review and refer to documents of other countries, but the process is not necessarily well structured and often depends on the expertise of the team members.

England, however, noted that the first version of their National Curriculum (NC) was largely based on the Concepts in Secondary Mathematics and Science project, (Hart 1981) that sought to formulate hierarchies of understanding in 10 mathematical topics normally taught in British secondary schools based on the results of testing 10,000 children in 1976 and 1977. The NC was also based on the ILEA Checkpoints (1979) and the Graded Assessment in Mathematics (1988–1990) projects. The original research-based design of the NC had many unintended consequences. Although the attainment targets were intended to measure learning outcomes on particular tasks, the levels were used to define the order in which topics should be taught, rather than paying attention to the development of concepts over time. The processes of mathematics, originally called "Using and applying mathematics" were defined in a general way related to progressions and levels that made interpretation difficult. As a consequence, the NC was revised several times and as of summer 2012 was again in the process of revision.

After a 1996 survey showed that social segmentation in Brazil seemed to be an obstacle to access to a quality education, research led to the development of the National Curricular Parameters in Brazil (1997). The Board of National Standards for Education (*Badan Standar Nasional Pendidikan*) in Indonesia examined the national needs for education, the vision of the country, societal demands, challenges for the future, and used their findings in developing the curriculum (Ministry of National Education 2006).

What Is the Nature of Standards?

In Brazil, Indonesia, Namibia and Peru, the standards/curricular frameworks are general and provide overarching guidelines for the development of discipline specific content. In the United States, Australia, and Japan, the mathematical standards essentially stand alone, although supporting documents may illustrate how the maths standards fit into the larger national education philosophy and perspective. Some standards include process goals. For example, Australia includes standards for four proficiencies (understanding, fluency, problem solving and reasoning) based on those described in *Adding It Up* (Kilpatrick et al. 2001). The new Australian standards want students to see that mathematics is about creating connections, developing strategies, and effective communication, as well as following rules and procedures. The United States CCSS has mathematical practice standards specifying eight "habits of mind" students should have when doing mathematics. In Brazil ideas such as "learn to learn", "promote independence", "learn to solve problems" are being incorporated into new curricula. In Peru and Indonesia the emphasis is primarily on the processes of problem solving, reasoning and proof, and mathematical communication.

In some cases standards reinforce the role of education in responding to the needs of the country. For example, the Curriculum for Basic Education (1st–9th grade) in Honduras (Department of Education 2003) was developed under three axes: personal, national and cultural identity, and democracy and work. The four pillars of lifelong learning defined by Delors (1996) (personal fulfilment, active citizenship, social inclusion and employability/adaptability) were used to define the mathematical content and methodological guides with problem solving as the central umbrella. Namibia's National Curriculum for a Basic Education outlines the aims of a basic education for the society of the future and specifies a few very general learning outcomes for each educational level (Namibia MoE 2008).

Standards span different sets of school grades or levels and differ in generality. Some countries have grade specific standards for what students should know throughout their primary and secondary schooling (i.e., US, Japan). Australia specifies a common curriculum for grades 1–10 and course options for students in upper secondary. Egypt and Honduras have curricular goals for students in grades 1–9 (age 14). At the high school level, Honduras focuses on post high school preparation with more than 53 career- focused schools for students.

The development of fractions in Australia by the Australian Curriculum and Assessment Reporting Authority (ACARA 2011), the Japanese Ministry of Education, Culture, Sports, Science and Technology (MEXT 2008), the Ministry of Education in Namibia (MoE 2005, 2006), and the US (CCSS 2010) illustrates the difference in standards across countries In grade 1, the standards/goals in the US, Namibia and Australia introduce words such as half, quarter and whole; this happens in grade 2 in Japan. Both US and Japan treat fraction as a number on the number line beginning in grade 3, emphasize equal partitioning of a unit and consider a fraction as composed of unit fractions: $4/3 = 4$ units of $1/3$. Australia suggests relating fractions to a number line only for unit fractions in grade 3, while Namibia does not mention fractions in relation to the number line. Equivalent fractions are taught in grade 4 in US, Japan, and Australia and in grade 6 in Namibia. Addition and subtraction of fractions with like denominators occurs in grade 4 in Japan, with unlike denominators in grade 5 in the US and Japan, and grade 7 in Namibia and Australia. Australia and Namibia have fractions as parts of collections in grade 2 and again in grade 4 in Namibia, but fractions as subsets of a collection are not mentioned in the standards/goals in the US and Japan. Students are expected to multiply and divide fractions in grade 5 in the US (with the exception of division of a fraction by a fraction, which happens in grade 6), in grade 6 in Japan, and in grade 7 in Australia and Namibia.

The next section describes what is taught in classrooms and how this relates to the standards/curricular goals of the country.

Examining the Status Quo

How Are Standards/Goals Related to the Implemented Curriculum?

Standards play different roles in shaping curriculum. For example, as described above, Peru does not have National Standards, but the mathematics learning goals for students are set out in the Curriculum National Design. With this as a guide, each of the country's regions develops a regional curriculum that considers the diversity of cultures and languages. Similarly, since 2005 Indonesia has National Standards for Education, which include standards for content in each subject area and curriculum structure. Based on these and competency standards, every school develops their own curriculum considering the vision of the school, local culture and students' background. In many of the US states, for example Massachusetts, standards provide a framework with the details of the curriculum, including the materials used for teaching and learning established at the district and school level. Japanese schools base their curricula on the national Course of Study (CS), a "Teaching Guide," resources and guidelines developed by local boards of education in the prefecture, and planning guides from textbook companies. Adaptions are sometimes made based on the situation of the school and its students. When the prefectural or the municipal boards of education develop their own model plans, such as the "nine year schooling system" (ShoChu-Ikkan-Kyoiku), the school in the prefecture or the municipality follows those plans and makes revisions to the CS accordingly.

In some instances, countries turn to other countries with more resources for support in implementing the standards. For example, the Japan International Cooperation Agency supported Honduras in developing curriculum and resources for teachers. Macau uses resources from China Mainland, Hong Kong and Canada.

What Drives the Implemented Curriculum?

Standards, textbooks, or high-stakes examinations seem to drive what happens in classrooms in the countries surveyed. While Hong Kong indicated that standards play that role, teachers in Brazil, Taiwan, Egypt, Honduras, and Japan rely on textbooks, and China mainland cited both textbooks and practice books.

In several countries high stakes examinations are significant in determining what teachers actually teach. In the United States, with the exception of Montana, the states surveyed indicated they followed the curriculum based on the state standards, but in reality most teachers teach only to what they know from experience will be tested (Au 2007). The implemented curriculum in England also seems to be shaped by what is assessed, which determines the nature of the tasks students meet in classrooms. The curriculum in Indonesia is determined both by textbooks and the national examination. Entrance examinations of leading universities impact the curriculum in Brazil and Macau (95 % of the students in Macau attend private schools to prepare for university).

How Do Countries Monitor Implementation of the Curriculum?

Countries use several strategies for monitoring and evaluating the enacted curriculum: large scale research studies conducted by the government or a private agency, small focused research studies on what is being taught and learned, student achievement on high stakes assessments, and approval of textbooks teachers use to deliver the curriculum. Relatively large-scale research studies on students' achievement are carried out in Honduras under the auspices of the Inter-American Development Bank and USAID. The Ministry of Education in Brazil investigated the incorporation of the National Curricular Parameters (PCN) into textbooks and other materials supporting teachers' work, but little research has been dedicated to any of the various stages in the process of curriculum development including the curriculum enacted in classrooms.

Japan administers national assessments on a regular basis in mathematics and Japanese for students in the sixth year of elementary school and the third year of lower secondary school. The results often reveal challenges in knowledge and skill utilization, which lead to revisions in educational policies and classroom lesson plans. These assessments are viewed as invaluable in monitoring and revising the curriculum.

In the United States, perhaps the most significant change in the last decade has been the increasing role of high stakes assessments measuring student achievement in elementary/secondary education. Every year each state assesses each student in grades 3–8 and assesses students once in grades 9–12 using a common state assessment, typically consisting of multiple-choice procedural questions. The results are used to evaluate teachers, administrators, and the curriculum. Little or no evidence exists correlating success on these tests with curriculum (or any other factor). This has not deterred federal and state levels policy makers from making use of the assessment results in these ways. The emphasis on high stakes assessment and accountability are seen in England as well, although it is not clear that the results have contributed to changes in the curriculum or standards.

How Are Changes Made to the Standards/Curricular Goals?

Change occurs in different ways. In the US, the most recent change was brought about by entities outside of the government and teachers. Japan bases changes in goals/standards on research examining student learning. Standards teams summarize, examine, and investigate the results of research studies on what has been achieved though the current Course of Study (CS) and the results of pilot trials of new goals/standards in designated "research schools" (Kenkyu-Kaihatsu-Gakko). They monitor emerging trends, societal needs and international assessments. For example, the most recent revisions to the CS in Japan for elementary and lower secondary schools were in March 2008 and for upper secondary and special needs education in March 2009. In this CS, the aim of mathematics education stresses the student's abilities to express their thinking and utilize mathematics in daily social life. In the CS for lower secondary schools, a new curricular strand "Use of Data" was added to enrich the content of statistics in the compulsory education. International mathematics assessments have helped statistics became a requirement in upper secondary schools. Taiwan and Hong Kong use some research supported by the government to construct and modify the curriculum as well as to inform teacher professional development and resource materials.

The Role of Textbooks

Survey responses indicated commercial publishers, private organizations, and government related organizations were involved in textbook development and distribution but to different degrees. The use of supplementary materials or teacher created worksheets was common in many of the countries. Many countries mentioned national standards/curricular guidelines as tools used in textbook development.

What is the approval or vetting process for textbooks?

In most of the countries with the exception of England and some of the states in the United States, some formal approval is necessary before texts can be used. For example, in Japan, textbooks are edited for adherence to the national curriculum and must be examined and authorized by MEXT. However, each textbook company can design and develop a textbook series with a final draft submitted to MEXT for examination and subsequent revision. During the development process, professionals (such as university researchers and teachers) play a large role in textbook design and development.

Many countries (China, Indonesia, Australia) have multiple textbook options for each grade level. Textbook adoption procedures vary, with decisions made at the national level (Brazil), state level (North Carolina), district level (Japan for elementary and lower secondary), school level (Japan for upper secondary) or even at an individual level (Taiwan). For the most part, the content would be the same across textbook options for each grade level since standards were the main drivers of the textbook development. Textbooks differ in the extent to which the contents are ordered and compiled but often have a similar style. Teachers in England make less use of textbooks than many other countries, and there is no uniform adoption procedure (Askew et al. 2010). In addition, public examination bodies produce textbooks that contain exercises from compilations of past examination questions that are popular with British teachers who see them as preparation for high-stakes assessment.

What Is the Role of Research in the Development of Textbooks?

Most countries mentioned an indirect or no use of research in textbook development. In the United States and England textbooks that are developed through large projects typically involve some research. In the United States, some curriculum materials (such as CMP 2012) are research based and developed with government or other sources of funding. Designers study trialling in classrooms, identify issues that emerge, what is working and not working to inform the next iteration of materials. The cycle may have several iterations, depending on funding and on commercial sales. (If the materials market poorly, the development is quickly terminated.)

Textbooks authored by individual teachers or commercial publishers did not seem to be noticeably influenced by pilot studies, research or research related to learning. In organizing textbook content, Japan makes use of research on high stakes assessment (the National Assessment of Academic Ability and other assessments implemented by local governments), the content and sequence of the old textbooks, and information obtained from teachers on the usability of the textbook and on the students' responses to the textbook problems during the lesson. In Brazil, some authors of mathematics textbooks use research, or rely on research results, to develop books.

Focused research projects on aspects of the curriculum, supplements to illustrate the standards, pilot studies of initiatives, action research and/or small seed projects are common in Hong Kong and Japan. In the United States, research studies on student learning typically focus on specific content areas or the development of a single concept, such as understanding cardinality (i.e., Clements 2012) and have little direct connection to the curriculum. Graduate students carry out many such projects in the United States and in other countries such as Brazil, England and Australia.

The Role of Technology in the Curriculum

What Is the Relationship Between Standards/Curricular Goals and Technology?

From a broad perspective, interacting with technology is seen in most countries as a critical life skill. In Peru, for example, the aim is to develop students' "skills and attitudes that will enable them to use and benefit from ICT ... thus enhancing the autonomous learning throughout life" (MoE 2009, p. 17). The National Curricular

Parameters (1997) in Brazil cite the value of technology as important for preparing students for their work outside of school. Australia defines Information and Communication Technology (ICT) as one of seven basic capabilities, i.e., the "skills, behaviours and dispositions that, together with curriculum content in each learning area and the cross curriculum priorities, will assist students to live and work successfully in the twenty-first century" (ACARA 2012, p. 10) Namibia has much the same statement in their National Curriculum for Basic Education emphasizing creating and learning to use software such as Word or Excel. Hong Kong's Technology Learning Targets calls for technology to enhance learning and teaching; provide platforms for discussions; help students construct knowledge; and engage students in an active role in the learning process, understanding, visualizing and exploring math, experiencing the excitement and joy of learning maths.

Some countries such as Namibia and Peru do not outline how technology should be used in the mathematics curriculum. Others describe the use of technology in mathematics classrooms in very general terms. Indonesia, for example, calls for the use of technology to develop understanding of abstract ideas by simulation and animation. In mainland China, the Nine Year Compulsory Education Mathematics Curriculum Standards emphasized the use of technology to benefit student understanding of the nature of mathematics. In Macau the standards call for educators to consider the impact of computers and calculators on the content and approaches in mathematics teaching and learning. In Taiwan, technology should support understanding, facilitate instruction, and enhance connections to the real world. England's curriculum documents are more specific, consistently encouraging the use of appropriate ICT tools to solve numerical and graphical problems, to represent and manipulate geometrical configurations and to present and analyse data.

The standards/curricular goals of some countries provide general goals for incorporating technology into the curriculum and then describe specific instances. For example, the United States Common Core State Standards (2010) for mathematical practices call for students to visualize the results of varying assumptions, exploring consequences, and comparing predictions; engage students in activities that deepen understanding of concepts; create opportunities for and learning— comparing and contrasting solutions and strategies, creating patterns, generating simulations of problem situations. These generalizations are followed by statements throughout, such as in grade 7, "Draw (freehand, with ruler and protractor, and with technology) geometric shapes with given conditions" (p. 50) or in algebra, "find the solutions approximately, e.g., using technology to graph the functions, make tables of values, or find successive approximations" (p. 66). The new Australian Mathematics Curriculum specifically calls for the use of calculators to check solutions beginning in grade 3 and, by year 10 includes general statements about the use of technology, "Digital technologies, such as spreadsheets, dynamic geometry software and computer algebra software, can engage students and promote understanding of key concepts (p. 11)". The curriculum provides specific examples: i.e., students should "Solve linear simultaneous equations, using algebraic and graphical techniques including using digital technology (p. 61)."

Japan has explicit learning goals for the use of technology and its Course of Study provides a guide for teachers that describes how calculators and computers can be used, with specific grade level examples under three headings; (1) as tools for calculation, (2) as teaching materials, and (3) as information/communication networks.

How Is Technology Used in Classrooms?

Respondents cited general issues related to the use of ICT. In England, for example, inspection reports based on evidence from 192 schools between 2005 and 2007 criticized schools' use of ICT, finding effective usage was decreasing and the potential of ICT to enhance the learning of mathematics rarely realized. In Brazil, the number of schools equipped with technological resources is increasing; however, programs using the technology are still restricted to pilot projects.

In Japan a 2010 survey on ICT facilities found that computers (98.7 %), digital cameras (98.1 %), and CD players (95.2 %) were used almost daily or at least two to three times a week (MEXT 2011). Yet, results from international studies such as TIMSS indicate little actual computer use in Japanese mathematics classrooms. At least one computer is typically available in classrooms in Egypt, Peru, China mainland and Macau but rarely used for mathematics instruction. Honduras has a one laptop per child program, but the lack of suitable mathematics related activities limits the use of laptops in classrooms. This was also identified as a problem in England. Brazil reported that a preliminary analysis of research conducted in the country suggests that technologies are used very little. Teachers are uncomfortable with laptops and have few resources for using them.

The availability of technological tools for students varied among countries and within countries. Some have class sets of calculators available; others expect students to provide their own (China Mainland, Macau, Hong Kong). Some schools have computer labs; some have class sets of laptops, while others use a single computer with overheard display (common in China Mainland). Many schools in England have a separate computer suite, where pupils learn to use ICT as a mathematical tool, for example using spreadsheets to generate number patterns or present statistical information but their use to enhance mathematics learning is limited.

Some use computers to provide practice procedures and skills (England, Macau, North Carolina). Some (China mainland, Taiwan, North Carolina) use technology as a way to differentiate instruction. North Caroline describes using interactive sites that allow the learner to manipulate data and objects and then provide immediate feedback; video, games, and other learning activities for struggling students, and providing advanced students with online activities that challenge and invite further learning; real world math practice using tools like Google Earth for measurement, stock market simulations, digital cameras for capturing real-life examples of geometric figures, Skype or other conferencing tools to interact with scientists and mathematicians. Formative and summative assessment was also indicated as a way of bringing technology into the classroom.

Interactive whiteboards are becoming increasingly common, although their role in learning mathematics is not well documented. They are heavily used in Great Britain (in about 75 % of schools) (Schachter 2010), and usage is growing in Japan from 16,403 in 2009 to 60,474 in 2011 (MEXT 2011) and the United States with 51 % of classrooms (Gray 2010). According to England an advantages of interactive white boards include high-quality, diagrams and relevant software to support learning through, for example, construction of graphs or visualization of transformations. A negative effect of interactive whiteboards seemed to be a reduction in pupils' use of concrete manipulatives.

Teacher Support

What Support Is Provided to Teachers to Help Them Know the Curriculum?

The survey results from Brazil and Egypt indicated minimum support is provided to teachers to help them learn about the curriculum. Brazil noted the materials are distributed to teachers usually without any actions involving the teachers. The other countries surveyed provide some form of support for teachers although the amount and form as well as who was in charge of providing support differs. Some countries (i.e., England, China, Japan) have ministry driven efforts to help teachers learn about the curriculum. For example, in Japan, once a new course of study (CS) is determined, the Ministry of Education, using a "trainer of trainers" process, conducts "transmission lectures" (Dentatsu-Koshu) on the principles and content of the new CS to superintendents on the prefectural boards of education who in turn give lectures to the superintendents on the municipal boards of education. The local superintendents then give lectures to all schoolteachers within a period of three years. The Ministry makes information available to teachers by showing concrete teaching examples, especially for large changes from an old to a new course of study. A variety of research meetings and conferences as well as lectures and symposiums are offered to educate teachers on the new CS.

A similar trainer of trainers process organized by the Ministry is also used in Honduras and Peru, although in Peru, some question the effectiveness of the process, given the results of five evaluations available on the web page of the Ministry of Education. Since 2010 the Ministry of Education in Mainland China has invested considerable resources to help teachers (over 1.1 million teachers at the primary level) understand the basic ideas of the curriculum standards and main content of the curriculum. The work is organized and financed by the Ministry but carried out at the local level. In Hong Kong, the Ministry of Education organized a professional development series, "Understanding the Curriculum", to explain the breadth and width of the curriculum. Exemplars, usually a product of collaborative research with schools, are used for illustration.

Other countries have a blend of ministry designed strategies and local initiatives. In Indonesia, the local (district and province) as well as central governments facilitate in-service training for teachers helping them to understand more about the curriculum. District school supervisors, advisors and/or experts from universities do the training and aim to improve the understanding of the Standards of Content, Process and Evaluation. Workshops and sessions on the standards are often organized and provided at the local level by university educators, school districts, curriculum consortia, and non-profit partners for all educators in a region of a state. Web based resources are provided in several countries (Honduras, China Mainland, Hong Kong, Japan). North Carolina provides webinars on the structure, organization, and content of the state standards, and Ohio provides online resources and disseminates curriculum models and other support documents to districts.

What Support Is Provided to Teachers to Help Them Enact the Curriculum?

In some countries support for instruction related to curriculum comes from the ministry of education (China Mainland, Hong King, England, Peru, some states in the United States) and in others it is provided through a combination of ministry of education and local initiatives or at the local level. Support primarily takes three forms: resources, professional development and mentoring.

1. Resources: Supplemental resources, materials created by outside research-based projects, and documents based on the state/national curriculum or standards are often designed and delivered through university programs. In some areas in Brazil, teachers are given written supporting material, videos, and learning resources, and technical pedagogical teams often help teachers in the implementation of the curriculum.

2. Professional Development: A variety of forms of professional development were also cited as ways to help teachers enact the curriculum. In Taiwan the curriculum development council provides lectures at the school level, instruction counselling groups and in-service workshops. Teacher training in Indonesia helps teachers develop teaching plans and provides strategies, methods, and approaches that have been adopted from the current research and theory. Honduras uses a "learn by doing" model for in-service, and many districts in the United States support mathematics "learning communities". Some form of collaborative lesson planning is typical in several of the countries (Japan, Macau, some states in the United States). In many countries (i.e., Hong Kong, United States) universities offer a variety of programs for in-service teacher education; graduate programs are sites for teachers' professional development.

Publishers also organize and deliver professional development workshops (China Mainland, United States).

Japan has a structured system of support. Local boards of education provide training for beginning teachers and for those with five, 10 and 20 years of teaching experiences as well as a variety of professional, non-mandatory training courses to enhance teaching ability and skills; for example, the Tochigi prefectural board of education offers 50 courses a year. Recently, a new teacher training/licensing system has been employed. Ordinary and special licenses are valid for 10 years; teachers need to renew their licenses by attending training courses every 10 years, given by general universities and teacher-training universities. These training courses are required to offer information based on the most recent research.

3. Mentoring: A third form of support in some countries is individualized, such as the Strategic Program for Learning Achievements in Peru where, since 2010, classroom teachers working with children up through the first two years of Basic Education (grades 6–8) receive advice from a specialist teacher. In the United States, many local districts have mathematics coaches who work with teachers, particularly at the elementary level. Hong Kong has dedicated "research schools" that mentor other schools in the implementation of the curriculum. A slightly different strategy is used in Honduras where teachers travel to Japan to see how the curriculum is enacted in classrooms and to learn about mathematics education.

While some cite a research base for professional development, the connection to research is often very limited (Hong Kong, Massachusetts and North Carolina in the United States). England provided ministry organized teacher support designed with a research perspective and later studies investigated the success of the implementation. The National Strategies (DFE 2011) were, from 1998 until 2011, the main delivery vehicle for supporting teachers to understand and implement government teaching and learning priorities. The programme, originally called the National Numeracy Strategy (NNS), was aimed at primary education but was later expanded to include secondary schools with the National Mathematics Strategy (NMS). The National strategies conducted a massive professional development programme, running courses and providing publications, advice and professional development materials such as videos to schools. These also included guidance on course planning, teaching and learning, assessment, subject leadership, inclusion, intervention and mathematics specific content. Detailed assessment guidance, lesson plans, and intervention programs were all provided (DFE 2011). An annotated bibliography of research evidence claimed to underpin the National Strategies (Reynolds and Muijs 1999). However, the research evidence was described as ambivalent and relatively scarce (Brown et al. 2003).

Evaluations of the implementation of the NNS were carried out and indicated some success, but this was contested by many who asserted the gains on National Tests attributed to the programme may be attributed to a careful choice of statistical baseline and to teachers' increasing tendency to orient their teaching towards the tests. When alternative tests were used, smaller gains were noted. Teaching seemed to have changed mainly in superficial ways, and some evidence suggested that in almost no cases were there 'deep' changes. (Brown et al. 2003, p. 668). In 2008 an inspection service found weaknesses in basic teaching skills and had difficultly assessing which initiatives worked and which did not. The frequent introduction of new initiatives, materials and guidance led to overload and diminished the potential effectiveness of each individual initiative (Ofsted 2010). As of March 2012, the Coalition Government abolished the National Strategies programme, and future professional development is decentralized and in the hands of individual schools.

Concluding Remarks

The survey data shows us that a common set of mathematics learning goals are established in almost all countries with a very minor role for research in designing and monitoring the development of their curriculum. Standards, textbooks, or high-stakes examinations seem to drive what happens in classrooms. Countries vary greatly in the amount of support provided to teachers in learning about and implementing the curriculum specified in their standards/goals.

Survey Responders

Australia: Peter Sullivan (Monash University)
Brazil: This report is a result of the collaboration between the Group of Studies and Research on Mathematical Education and Education (USP) & Organization, Curriculum Development and Teacher Education (PUCSP)

Vinício de Macedo Santos (University of Sao Paulo),
Célia Maria Carolino Pires (Pontifícia Universidade Católica de São Paulo),
Elenilton Vieira Godoy (Pontifícia Universidade Católica de São Paulo and Centro Universitário Fundação Santo André),
João Acácio Busquini (Secretaria de Estado da Educação de São Paulo),
José Carlos Oliveira (Costa Centro Universitário Fundação Santo André).

China: China Mainland—Jiansheng Bao, Xuefen Gao, Likun Sun & Xiaoli Ju (East China Normal University, Shanghai)

Taiwan—Hsin-Mei E. Huang (Taipei Municipal University of Education)
Hong Kong—Polly Lao (Hong Kong Bureau)
Macau—Chunlian Jiang (University of Macau)

Egypt: Fayez Mina (Ain Shams University)
Honduras: Libni Berenice Castellón (Universidad Pedagógica Nacional Francisco Morazán.)

Indonesia: Edy Tri Baskoro (Board of National Standard for Education)
Japan: Keiko Hino (Utsunomiya University)
Namibia: Karen D'Emiljo (Otjiwarongo Secondary School)
Peru: Martha Rosa Villavicencio Ubillus (National University San Marcos); Olimpia Rosa Castro Mora (Ministry of Education)
United Kingdom, England: Malcolm Swan, Sheila Evans (University of Nottingham)

Open Access This chapter is distributed under the terms of the Creative Commons Attribution Noncommercial License, which permits any noncommercial use, distribution, and reproduction in any medium, provided the original author(s) and source are credited.

References

Askew, M., Hodgen, J., Hossain, S., & Bretscher, N, (2010). *Values and variables: Mathematics education in high-performing countries*. London: Nuffield Foundation.

Au, W. (2007). High-stakes testing and curricular control: A qualitative metasynthesis.

Australian Curriculum and Assessment Reporting Authority. (2012). The Australian Curriculum: Mathematics.

BRASIL. Secretaria de Educação Fundamental. (1997). Parâmetros Curriculares Nacionais: Matemática (National Curriculum Parameters: Mathematics). Secretaria de Educação Fundamental. Brasília: MEC/SEF.

Brown, M., Askew, M., Millett, A., & Rhodes, V. (2003). The key role of educational research in the development and evaluation of the National Numeracy Strategy. *British Educational Research Journal, 29*(5): 655-667.

Clements, D. (2012). Address at the Association of Mathematics Teacher Educators Annual Meeting. Fort Worth TX

Common Core State Standards. (2010). Council of Chief State School Officers & National Governor's Association.

Connected Mathematics Project website http://connectedmath.msu.edu

Currículo National Básico (2003). Department of Education Honduras. www.se.gob.hn/index.php?a=Webpage&url=curriculo

Delors, J. (1996) Learning: The treasure within. Report to UNESCO of the International Commission on Education for the Twenty-first Century, UNESCO.

DFE (2011). The National Strategies, from http://nationalstrategies.standards.dcsf.gov.uk/

Diseño Curricular Nacional (National Curricular Design). (2005). Lima, Peru: Ministry of Education.

Diseño Curricular Nacional (National Curricular Design). (2008). Lima, Peru: Ministry of Education.

Diseño Curricular Nacional. (National Curricular Design). (2009). Lima, Peru: Ministry of Education.

Graded assessment in mathematics. (1988–1990). Basingstoke Hants: Macmillan Education.

Gray, L., Thomas, N., & Lewis, L. (2010). *Teachers' use of educational technology in U.S. public schools: 2009* (NCES 2010-040). Washington DC: National Center for Education Statistics, Institute of Education Sciences, U.S. Department of Education.

Hart, K. (Ed.). (1981). *Children's understanding of mathematics 11-16*. London: John Murray.

Inner London Education Authority (ILEA). (1979). *Checkpoints assessment cards*. Inner London Education Authority Learning Materials Service.

Kilpatrick, J., Swafford, J., & Findell, B. (Eds.). (2001). *Adding it up: Helping children learn mathematics*. National Research Council. Washington, DC: National Academy Press.

Ministry of Education, Culture, Sports, Science and Technology, Japan, (2008). Elementary school teaching guide for the Japanese course of study: Mathematics (English translation Japanese mathematics curricula in the course of study, March, 2008 by Asia-Pacific Mathematics and Science Education Collaborative at DePaul University, Chicago IL, USA)

Ministry of Education, Culture, Sports, Science and Technology, Japan, (2011). Results of the survey on the states of educational use of information technology in schools, 2011. (in Japanese).

Ministry of National Education (2006). Tentang standar kompetensi lulusan untuk satuan pendidikan dasar dan menengah (Graduate competency standards for basic and secondary educations), Republic of Indonesia.

Namibia Ministry of Education. (2006). Mathematics Syllabus Upper Primary Phase Grades 5 – 7. National Institute for Educational Development

Namibia Ministry of Education. Curriculum for the Lower Primary Phase Grades 1-4 (2005). National Institute for Educational Development

Namibia Minstry of Education. (2008). National Curriculum for Basic Education. National Institute for Educational Development.

Ofsted (2010). *The national strategies: A review of impact*. From www.ofsted.gov.uk/Ofsted-home/Publications-and-research/Browse-all-by/Documents-by-type/Thematic-reports/The-National-Strategies-a-review-of-impact/(language)/eng-GB.

Reynolds, D., & D. Muijs (1999). National numeracy strategy: An annotated bibliography for teachers and schools. London.

Schachter, R, (2010). Whiteboards: Learning from Great Britain. *Scholastic Administrator*. www.scholastic.com/browse/article.jsp?id=3753768

Key Mathematical Concepts
in the Transition from Secondary
School to University

Mike O.J. Thomas, Iole de Freitas Druck, Danielle Huillet, Mi-Kyung Ju, Elena Nardi, Chris Rasmussen and Jinxing Xie

This report[1] from the ICME12 Survey Team 4 examines issues in the transition from secondary school to university mathematics with a particular focus on mathematical concepts and aspects of mathematical thinking. It comprises a survey of the recent research related to: calculus and analysis; the algebra of generalised arithmetic and abstract algebra; linear algebra; reasoning, argumentation and proof; and modelling, applications and applied mathematics. This revealed a multi-faceted web of cognitive, curricular and pedagogical issues both within and across the mathematical topics above. In addition we conducted an international survey of those engaged in teaching in university mathematics departments. Specifically, we aimed to elicit perspectives on: what topics are taught, and how, in the early parts of university-level mathematical studies; whether the transition should be smooth; student preparedness for university mathematics studies; and, what university departments do to assist those with limited preparedness. We present a summary of the survey results from 79 respondents from 21 countries.

---

[1] *A fuller version of this report is available from* http://www.math.auckland.ac.nz/ ~ thomas/ST4. pdf.

M.O.J. Thomas (✉)
Auckland University, Auckland, New Zealand
e-mail: moj.thomas@auckland.ac.nz

I. de Freitas Druck
University of Sao Paolo, São Paulo, Brazil

D. Huillet
Eduardo Mondlane University, Maputo, Mozambique

M.-K. Ju
Hanyang University, Seoul, South Korea

E. Nardi
University of East Anglia, Norwich, UK

C. Rasmussen
San Diego State University, San Diego California, USA

J. Xie
Tsinghua University, Beijing, China

© The Author(s) 2015
S.J. Cho (ed.), *The Proceedings of the 12th International Congress on Mathematical Education*, DOI 10.1007/978-3-319-12688-3_18

265

Background

Changing mathematics curricula and their emphases, lower numbers of student enrolments in undergraduate mathematics programmes (Barton and Sheryn 2009; and http://www.mathunion.org/icmi/other-activities/pipeline-project/) and changes due to an enlarged tertiary entrant profile (Hockman 2005; Hoyles et al. 2001), have provoked some international concern about the mathematical ability of students entering university (PCAST 2012; Smith 2004) and the traumatic effect of the transition on some of them (Engelbrecht 2010). Decreasing levels of mathematical competency have been reported with regard to essential technical facility, analytical powers, and perceptions of the place of precision and proof in mathematics (Brandell et al. 2008; Hourigan and O'Donoghue 2007; Kajander and Lovric 2005; Luk 2005; Selden 2005). The shifting profile of students who take service mathematics courses has produced a consequent decline in mathematical standards (Gill et al. 2010; Jennings 2009). However, not all studies agree on the extent of the problem (Engelbrecht and Harding 2008; Engelbrecht et al. 2005) and James et al. (2008) found that standards had been maintained. The recent President's Council of Advisors on Science and Technology (PCAST) (2012) states that in the USA alone there is a need to produce, over the next decade, around 1 million more college graduates in Science, Technology, Engineering, and Mathematics (STEM) fields than currently expected and recommends funding around 200 experiments at an average level of $500,000 each to address mathematics preparation issues. This helps to place the transition situation above in context and emphasises the importance of addressing the issues arising.

We found relatively few papers in the recent literature related directly with our brief to consider the role of mathematical thinking and concepts related to transition. Hence we also reviewed literature analysing the learning of mathematics on one or both sides of the transition boundary. To achieve this we formed the somewhat arbitrary division of this mathematics into: calculus and analysis; abstract algebra; linear algebra; reasoning, argumentation and proof; and modelling, applications and applied mathematics, and report findings related to each of these fields. We were aware that other fields such as geometry and statistics and probability should have been included, but were not able to do so.

The Survey

We considered it important to obtain data on transition from university mathematics departments. We wanted to know what topics are taught and how, if the faculty think the transition should be smooth, or not, their opinions on whether their students are well prepared mathematically, and what university departments do to assist those who are not. Hence, we constructed an anonymous questionnaire on transition using an Adobe Acrobat pdf form and sent it internationally by email to members of mathematics departments. The 79 responses from 21 countries were collected electronically. The sample comprised 56 males and 23 females with a mean of 21.9 years of academic teaching. Of these 45 were at the level of associate professor, reader or full professor, and 30 were assistant professors, lecturers or senior lecturers. There were 5 or more responses from each of South Africa, USA, New Zealand and Brazil.

Clearly the experience for beginning university students varies considerably depending on the country and the university that they attend. For example, while the majority teaches pre-calculus (53, 67.1 %), calculus (76, 96.2 %) and linear algebra (49, 62 %) in their first year, minorities teach complex analysis (1), topology (3), group theory (1), real analysis (5), number theory (9), graph theory (12), logic (15), set theory (17) and geometry (18), among other topics. Further, in response to 'Is the approach in **first** year mathematics at your university: Symbolic, Procedural; Axiomatic, Formal; Either, depending on the course.' 21 (26.6 %) answered that their departments introduce symbolic and procedural approaches in first year mathematics courses, while 6 replied that their departments adapt axiomatic formal approaches. Most of the respondents (50, 63.3 %) replied that their approach depended on the course.

When asked 'Do you think students have any problems in moving from school to university mathematics?' 72 (91.1 %) responded "Yes" and 6 responded "No". One third of those who answered "Yes" described these problems as coming from a lack of preparation in high school, supported by comments such as "They don't have a sufficiently good grasp of the expected school-mathematics skills that they need." Further, two thirds of those who answered "Yes" described the problems as arising from the differences, such as class size and work load, between high school classes and university, with many specifically citing the conceptual nature of university mathematics as being different from the procedural nature of high school mathematics. Comments here included "university is much more theoretical" and "Move from procedural to formal and rigourous [sic], introduction to proof, importance of definitions and conditions of theorems/rules/statements/formulas." There is also a need to "...deal with misconceptions which students developed in secondary school...We also have to review secondary school concepts and procedures from an adequate mathematical point of view." Other responses cited: students' weak algebra skills (12.5 %); that university classes are harder (5 %); personal difficulties in adjusting (10 %); poor placement (3 %); and, poor teaching at university (1 %).

Looking at specific mathematical knowledge, we enquired 'How would you rate first year students' mathematical understanding of each of the following on entry to university?' With a maximum score of 5 for high, the mean scores of the responses were algebra or generalised arithmetic (3.0), functions (2.8), real numbers (2.7), differentiation (2.5), complex numbers (1.9), definitions (1.9), vectors (1.9), sequences and series (1.9), Riemann integration (1.8), matrix algebra (1.7), limits (1.7) and proof (1.6). The mathematicians were specifically asked whether students were well prepared for calculus study. Those whose students did not study calculus at school rated their students' preparation for calculus at 2.1 out of 5. Those whose students did, rated secondary school calculus as preparation to study calculus at university at 2.4, and as preparation to study analysis at university at 1.5. These results suggest that there is some room for improvement in school preparation for university study of calculus and analysis.

Since the view has been expressed (e.g., Clark and Lovric 2009) that, rather than being 'smooth', the transition to university should require some measure of struggle by students, we asked 'Do you think the transition from secondary to university education in mathematics should be smooth?' Here, 54 (68.4 %) responded "Yes" and 22 (27.8 %) responded "No". Of those who responded "No", many of the comments were similar to the following, expressing the belief that change is a necessary part of the transition: "Not necessarily smooth, because it is for most students a huge change to become more independent as learners." and "To learn mathematics is sometimes hard." Those who answered yes were then asked 'what could be done to make the transition from secondary to university education in mathematics smoother?' The majority of responses mentioned changes that could be made at the high school level, such as: encourage students to think independently and abstractly; change the secondary courses; have better trained secondary teachers; and, have less focus in secondary school on standardised tests and procedures. A few mentioned changes that could be made at the university, such as: better placement of students in classes; increasing the communication between secondary and tertiary teachers; and, addressing student expectations at each level. This lack of communication between the two sectors was highlighted as a major area requiring attention by the two-year study led by Thomas (Hong et al. 2009).

Since one would expect that, seeing students with difficulties in transition, universities would respond in an appropriate manner (see e.g. Hockman 2005), we asked 'Does your department periodically change the typical content of your first year programme?' 33 (41.8 %) responded "Yes" and 44 (55.7 %) "No". The responses to the question 'How does your department decide on appropriate content for the first year mathematics programme for students?' by those who answered yes to the previous question showed that departments change the content of the first year programme based on the decision of committees on a university or department level. Some respondents said that they change the course content for the first year students based on a decision by an individual member of faculty who diagnoses student needs and background. 15 of the 35 responded that their universities try to integrate student, industry, and national needs into first year mathematics courses. The follow-up question 'How has the content of your first year mathematics courses changed in the last 5 years?' showed that 35 had changed their courses in the last 5 years, but 10 of these said that the change was not significant. 17 out of the 35 respondents reported that their departments changed first year mathematics courses by removing complex topics, or by introducing practical mathematical topics. In some of the courses, students were encouraged to use tools for calculation and visualisation. In contrast, six departments *increased* the complexity and the rigour of their first year mathematics courses.

The survey considered the notion of proof in several questions. In response to 'How important do you think definitions are in **first** year mathematics?' 52 (65.8 %)

replied that definitions are important in first year mathematics, while 15 presented their responses as neutral. Only 8 respondents replied that definitions are not important in first year mathematics. Responses to the question 'Do you have a course that explicitly teaches methods of proof construction?' were evenly split with 49.4 % answering each of "Yes" and "No". Of those who responded "Yes", 15 (38.4 %) replied that they teach methods of proof construction during the first year, 23 (58.9 %) during the second year and 5 (12.8 %) in either third or fourth year. While some had separate courses (e.g. proof method and logic course) for teaching methods of proofs, many departments teach methods of proofs traditionally, by introducing examples of proof and exercises in mathematics class. Some respondents replied that they teach methods of proof construction in interactive contexts, citing having the course taught as a seminar, with students constructing proofs, presenting them to the class, and discussing/critiquing them in small size class. One respondent used the modified Moore method in interactive lectures. Looking at some specific methods of introducing students to proof construction was the question 'How useful do you think that a course that includes assistance with the following would be for students?' Four possibilities were listed, with mean levels of agreement out of 5 (high) being: Learning how to read a proof, 3.7; Working on counterexamples, 3.8; Building conjectures, 3.7; Constructing definitions, 3.6. These responses appear to show a good level of agreement with employing the suggested approaches as components of a course on proof construction. It may be that these are ideas that the 49.4 % of universities that currently do not have a course explicitly teaching proof construction could consider implementing as a way to assist transition.

Mathematical modelling in universities was another topic our survey addressed. In response to the questions "Does your university have a mathematical course/activity dedicated to mathematical modeling and applications?" and "Are mathematical modelling and applications contents/activities integrated into other mathematical courses?", 44 replied that their departments offer dedicated courses for modelling, while 41 said they integrate teaching of modelling into mathematics courses such as calculus, differential equations, statistics, etc. and 7 answered that their university does not offer mathematics courses for mathematical modelling and applications. Reasons given for choosing dedicated courses include: the majority of all mathematics students will end up doing something other than mathematics so applications are far more important to them than are detailed theoretical developments; most of the mathematics teaching is service teaching for students not majoring in mathematics so it is appropriate to provide a relevant course of modelling and applications that meets the needs of the target audience; if modelling is treated as an add-on then students may not learn mathematical modelling methods. Those who chose integrated courses did so because students need to be equipped with a wide array of mathematical techniques and solid knowledge base. Hence, it is appropriate for earlier mathematics courses to contain some theory, proofs, concepts and skills, as well as applications.

Considering what happens in upper secondary schools, 26 (33 %) reported that secondary schools in their location have mathematical modelling and applications integrated into other mathematical courses, with only 4 having dedicated courses. 44 (56 %) said that there were no such modelling courses in their area. When asked for their opinion on how modelling should be taught in schools, most of the answers stated that it should be integrated into other mathematical courses. The main reasons presented for this were: the many facets of mathematics; topics too specialised to form dedicated courses; to allow cross flow of ideas, avoid compartmentalization; and students need to see the connection between theory and practice, build meaning, appropriate knowledge. The question 'What do you see as the key differences between the teaching and learning of modelling and applications in secondary schools and university, if any?' was answered by 33 (42 %) of respondents. The key differences pointed out by those answering this question were: at school, modelling is poor, too basic and mechanical, often close implementation of simple statistics tests; students have less understanding of application areas; university students are more independent; they have bigger range of mathematical tools, more techniques; they are concerned with rigour and proof. Asked 'What are the key difficulties for student transition from secondary school to university in the field of mathematical modelling and applications, if any?' the 35 (44 %) university respondents cited: lack of knowledge (mathematical theory, others subjects such as physics, chemistry, biology, ecology); difficulties in formulating precise mathematical problems/ interpreting word problems/understanding processes, representations, use of parameters; poor mathematical skills, lack of logical thinking; no experience from secondary schools; and lack of support. One message for transition is to construct more realistic modeling applications for students to study in schools.

In order to investigate how universities respond to assist students with transition problems we enquired "Do you have any academic support structures to assist students in the transition from school to university? (e.g., workshops, bridging courses, mentoring, etc.).", and 56 (71 %) replied 'Yes' and 22 'No'. Of those saying yes, 34 % have a bridging course, 25 % some form of tutoring arrangement, while 23 % mentioned mentoring, with one describing it as a "Personal academic mentoring program throughout degree for all mathematics students" and another saying "We tried a mentoring system once, but there was almost no uptake by students." Other support structures mentioned included 'study skills courses', 'maths clinics', 'support workshops', 'pre-course', 'remedial mathematics unit', and a 'Mathematics Learning Service (centrally situated), consulting & assignment help room (School of Maths). The MLS has a drop-in help room, and runs a series of seminars on Maths skills. These are also available to students on the web.' Others talked of small group peer study, assisted study sessions, individual consultations, daily help sessions, orientation programmes and remedial courses. There is some evidence that bridging courses can assist in transition (Varsavsky 2010), by addressing skill deficiencies in basic mathematical topics (Tempelaar et al. 2012) and building student confidence (Carmichael and Taylor 2005). Other successful transition courses (e.g., Leviatan 2008) introduce students to the mathematical "culture" and its typical activities (generalizations, deductions, definitions, proofs, etc.), as well as central concepts and tools.

Overall the survey confirmed that students do have some difficulties in transition and these are occasionally related to a deficit in student preparation or mathematical knowledge. However, there are also a number of areas that universities could address to assist students, such as adjusting the content of first year courses, and instituting a course on proving and proof (where this doesn't already exist) and constructing appropriate bridging courses.

Literature Review

A number of different lenses have been used to analyse the mathematical transition from school to university. Some have been summarised well elsewhere (see e.g., Winsløw 2010) but we preface our discussion with a brief list of the major theoretical perspectives we found in the transition-related literature. One theory that is in common use is the Anthropological Theory of Didactics (ATD) based on the ideas of Chevallard (1985), with its concept of a *praxeology* comprising task, technique, technology, theory. ATD focuses on analysis of the organisation of praxeologies relative to institutions and the diachronic development of didactic systems. A second common perspective is the Theory of Didactical Situations (TDS) of Brousseau (1997), where *didactical situations* are constructed in which the teacher orchestrates elements of the didactical milieu under the constraints of a dynamic didactical contract. Other research uses the action-process-object-schema (APOS) framework of Dubinsky (e.g. Dubinsky and McDonald 2001) for studying learning. This describes how a process can be constructed from actions by reflective abstraction, and subsequently an object is formed by encapsulation of the process. The Three Worlds of Mathematics (TWM) framework of Tall (2008) is also considered useful by some. This describes thinking and learning as taking place in three worlds: the embodied; the symbolic; and the formal. In the embodied world we build mental conceptions using visual and physical attributes of concepts and enactive sensual experiences. In the symbolic world symbolic representations of concepts are acted upon, or manipulated, and the formal world is where properties of objects are formalized as axioms, with learning comprising building and proving of theorems by logical deduction from these axioms. We use the acronyms above to refer to each of these frameworks in the text below.

Calculus and Analysis

A number of epistemological and mathematical obstacles have been identified in the study of the transition from calculus to analysis. These include:

Functions: Students have a limited understanding of the concept of function (Junior 2006) and need to be able to switch between local and global perspectives (Artigue 2009; Rogalski 2008; Vandebrouck 2011). Using a TWM lens Vandebrouck

(2011) suggests a need to reconceptualise the concept of function in terms of its multiple registers and process-object duality. The formal axiomatic world of university mathematics requires students to adopt a local perspective on functions, whereas only pointwise (functions considered as a correspondence between two sets of numbers) and global points of view (representations are tables of variation) are constructed at secondary school. An ATD-based study of the transition from concrete to abstract perspectives in real analysis by Winsløw (2008) suggests that in secondary schools the focus is on practical-theoretical blocks of concrete analysis, while at university level the focus is on more complex praxeologies of concrete analysis and on abstract analysis.

Limits: Students need to work with limits, especially of infinite sequences or series. Two obstacles regarding the concept of infinite sum are the intuitive and natural idea that the sum of infinity of terms should also be infinite, and the conception that an infinite process must go through each step, one after the other and without stopping, which leads to the potential infinity concept (González-Martín 2009; González-Martín et al. 2011). According to Oehrtman (2009), students' reasoning about limit concepts appears to be influenced by metaphorical application of experiential conceptual domains, including collapse, approximation, proximity, infinity as number and physical limitation metaphors. However, only physical limitation metaphors were consistently detrimental to students' understanding. One approach to building thinking about limits, suggested by Mamona-Downs (2010), is the set-oriented characterization of convergence behaviour of sequences of that supports the metaphor of 'arbitrary closeness' to a point. Another, employing a TDS framework (Ghedamsi 2008) developed situations that allowed students to connect productively the intuitive, perceptual and formal dimensions of the limit concept.

Institutional factors: An aspect of transition highlighted by the ATD is that praxeologies exist in relation to institutions. Employing the affordances of ATD, Praslon (2000) showed that by the end of high school in France a substantial institutional relationship with the concept of derivative is already established. Hence, for this concept, he claims that the secondary-tertiary transition is not about intuitive and proceptual perspectives moving towards formal perspectives, as TWM might suggest, but is more complex, involving an accumulation of micro-breaches and changes in balance according several dimensions (tool/object dimensions, particular/general objects, autonomy given in the solving process, role of proofs, etc.). Building on this work Bloch and Ghedamsi (2004) identified nine factors contributing to a discontinuity between high school and university in analysis and Bosch et al. (2004) show the existence of strong discontinuities in the praxeological organization between high school and university, and build specific tools for qualifying and quantifying these. Also employing an institutional approach, Dias et al. (2008; see also Artigue 2008) conducted a comparative ATD study of the secondary-tertiary transition in Brazil and France, using the concept of function as a filter. They conclude that although contextual influences tend to remain invisible there is a need for those inside a given educational system to become aware of them in order to envisage productive collaborative work and evolution of the system.

Other areas: One TDS-based research project examined a succession of situations for introducing the notions of interior and closure of a set and open and closed set (Bridoux 2010), using meta-mathematical discourse and graphical representations to assist students to develop an intuitive insight that allowed the teacher to characterise them in a formal language. Another examined the notion of completeness (Bergé 2008), analysing whether students have an operational or conceptual view, or if it is taken for granted. The conclusion was that many students have a weak understanding of ideas such as the suprema of bounded subsets, convergence of Cauchy sequences and the completeness of **R**.

Some possible ways to assist the calculus-analysis transition have been considered. For example, Gyöngyösi et al. (2011) report an experiment using Maple CAS-based work to ease the transition from calculus to real analysis. A similar use of graphing calculator technology in consideration of the Fundamental Theorem of Calculus by Scucuglia (2006) made it possible for the students to become gradually engaged in deductive mathematical discussions based on results obtained from experiments. In addition, Biehler et al. (2011) propose that blending traditional courses with systematic e-learning can facilitate bridging of school and university mathematics.

Abstract Algebra

Understanding the constructs, principles, and eventually axioms, of the algebra of generalised arithmetic could be a way to assist students in the transition to study of more general algebraic structures. Focusing on students' work on solving a parametric system of simultaneous equations and the difficulties they experience with working with variables, parameters and unknowns, Stadler (2011) describes their experience of the transition from school to university mathematics as an often perplexing re-visiting of content and ways of working. The study showed that constructs of number, symbolic literals, operators, the '=' symbol itself, and the formal equivalence relation, as well as the principles of arithmetic, all contribute to building a deep understanding of equation. This agrees with the observations of Godfrey and Thomas (2008), who, using the TWM framework, provided evidence that many students have a surface structure view of equation and fail to integrate the properties of the object with that surface structure.

Students' encounter with abstract algebra at university marks a significant point in the transition to advanced mathematical formalism and abstraction, with concepts introduced abstractly, defined and presented by their properties, and deduction of facts from these properties alone. The role of verbalisation in this process, as a semantic mediator between symbolic and visual mathematical expression, may require a level of verbalisation skills that Nardi (2008, 2011) notes is often lacking in first year undergraduates.

Studies that focus on the student experience in their first encounters with key concepts in abstract algebra describe a number of difficulties. While some have suggested that an over-reliance on concrete examples of groups leading to a lack of skills in proof production, others, such as Burn (1996), recommend reversing the order of presentation, using examples and applications to stimulate the discovery of definitions and theorems through permutation and symmetry. An example of reducing group theory's high levels of abstraction (Hazzan 2001) is to ask students to construct the operation table for low order groups. This was also implemented by Larsen (2009) as a series of tasks exploring symmetries of an equilateral triangle, constructing low order group multiplication tables and culminating in negotiating preliminary understandings of group structure, the order of a group and isomorphism.

In an analysis of student responses to introductory group theory problem sheets, Nardi (2000) identified student difficulties with the order of an element, group operation, and the notions of coset and isomorphism. The duality underlying the concept of group and its binary operation, were also discussed by Iannone and Nardi (2002). They offer evidence of a student tendency to ignore the binary operation, consider the group axioms as properties of the group elements and omit checking axioms perceived as obvious, such as associativity. In addition, research by Ioannou (see Ioannou and Nardi 2009, 2010; Ioannou and Iannone 2011) considers students' first encounter with abstract algebra, focusing on the Subgroup Test, symmetries of a cube, equivalence relations, and employing the notions of kernel and image in the First Isomorphism Theorem. Provisional conclusions are that students' overall problematic experience of the transition to abstract algebra is characterised by the strong interplay between strictly conceptual matters, affective issues and those germane to first year students' wider study skills and coping strategies.

Linear Algebra

A sizeable amount of research in linear algebra has documented students' transition difficulties, particularly as these relate to students' intuitive or geometric ways of reasoning and the formal mathematics of linear algebra (e.g. Dogan-Dunlap 2010). The theoretical framework of Hillel (2000) for understanding student reasoning in linear algebra that identified geometric, algebraic, and abstract modes of description is valuable. For example, the relationship between linear algebra and geometry were at the core of Gueudet's research programme (2004, 2008; Gueudet-Chartier, 2004) that identified specific views on student difficulties. She claims that the epistemological view leads to a focus on linear algebra as an axiomatic theory, which is very abstract for the students and identifies a need for various forms of flexibility, in particular between dimensions. Further work at the geometry-formalism boundary by Portnoy et al. (2006) and Britton and Henderson (2009) has demonstrated some difficulties. First, pre-service teachers who engaged with transformations as geometric processes still had difficulty writing proofs involving linear transformations, and second, students experienced problems moving between a formal understanding of subspace and algebraic problem statements due to an insufficient understanding of the symbols used in the questions and in the formal definition of subspace.

Employing a framework using APOS theory in conjunction with TWM, Stewart and Thomas (2009, 2010; Thomas and Stewart 2011) analysed student understanding of various concepts in linear algebra, including linear independence, eigenvectors, span and basis. The authors found that generally students do not think of these concepts from an embodied standpoint, but instead rely upon a symbolic, process-oriented matrix manipulation manner of reasoning. However, employing a course that introduced students to embodied, geometric representations in linear algebra, along with the formal and the symbolic, appeared to enrich student understanding of the concepts and allowed them to bridge between them more effectively than with just symbolic processes.

Another aspect that has been investigated is students' intuitive thinking in linear algebra. Working with modelling and APOS frameworks Possani et al. (2010) leveraged students' intuitive ways of thinking through a genetic composition of linear independence and systems of equations. Student use of different modes of representation in making sense of the formal notion of subspace was analysed by Wawro, Sweeney and Rabin (2011a), and their results suggest that in generating explanations for the definition, students rely on their intuitive understandings of subspace, which can be problematic but can also help develop a more comprehensive understanding of subspace.

Some research teams have spearheaded innovations in the teaching and learning of linear algebra. For example, Cooley et al. (2007) developed a linear algebra course combined with learning about APOS theory and found the focus on a theory for how mathematical knowledge is generated enriched understanding of linear algebra. Another group of researchers used a design research approach simultaneously creating instructional sequences and examining students' reasoning about key concepts such as eigenvectors and eigenvalues, linear independence, linear dependence, span, and linear transformation (Henderson et al. 2010; Larson et al. 2008; Sweeney 2011). They argue that knowledge of student thinking prior to formal instruction is essential for developing thoughtful teaching that builds on and extends student thinking. In a study on tasks for developing student reasoning they (Wawro et al. 2011b) report how an innovative instructional sequence beginning with vector equations rather than systems of equations successfully leveraged students' intuitive imagery of vectors as movement to develop formal definitions.

Proof and Proving[2]

The transition to university mathematics includes a requirement for understanding and producing proofs. This requires logical deductive reasoning (Engelbrecht 2010) and rigour (Leviatan 2008). Research highlighting examples of this includes

---

[2] *At the time of writing the book* Proof and Proving in Mathematics Education: *The 19th ICMI study–Hanna & de Villiers,* 2012, *was still in press.*

conceptualisation related to the use of quantifiers (Chellougui 2004), the relationship between syntax and semantics in the proving process (Barrier 2009; Blossier et al. 2009) and logical competencies (Durand-Guerrier and Njomgang Ngansop 2010).

One recommendation is the need for more explicit teaching of proof, both in school and university (Balacheff 2008; Hanna and de Villiers 2008; Hemmi 2008), with some (e.g., Stylianides and Stylianides 2007; Hanna and Barbeau 2008) arguing for it to be made a central topic in both institutions. A possible introduction to proof, suggested by Harel (2008) and Palla et al. (2012) is proof by mathematical induction. However, they propose that it should be introduced slowly, building on students' own pre-existing epistemological resources (Solomon 2006) valuing both ways of understanding and thinking (Harel 2008), and distinguishing between proof schemes and proofs.

A number of potential difficulties in any attempt to place proving and proof more prominently in the transition years have been identified. These include the role of definitions, and the problem of student met-befores (Tall and Mejia-Ramos 2006). Using definitions as the basis of deductive reasoning in schools is likely to meet serious problems (Harel 2008; Hemmi 2008) since this form of reasoning is generally not available to school students, and Hemmi (2008) advocates the principle of *transparency*, which makes the difference between empirical evidence and deductive argument visible to students. In addition, the influence of student met-before can be strong, with Cartiglia et al. (2004) showing that the most recent met-before for university students, a formal approach, had a strong influence on their reasoning. A further difficulty, highlighted by Iannone and Inglis (2011), is a range of weaknesses in beginning university mathematics students' ability to produce a deductive argument, even when they were aware they should do so.

Some consideration has been given to methods of bridging the gap between the fields of argumentation and proof. One pedagogical strategy that may be an effective way to introduce the learning of proof and proving is student construction and justification of conjectures. The idea of an interconnecting problem was employed by Kondratieva (2011) to get students to construct and justify conjectures. Further, conjectures may also have a role during production of indirect argumentation (Antonini and Mariotti 2008), such as that in contradiction and contraposition, by activating and bridging significant hidden cognitive processes. Another approach discussed by Pedemonte (2007, 2008) employs the construct of *structural distance*, and she argues for an abductive step in the structurant argumentation in order to assist transition by decreasing the gap between argumentation and proof. Another proposition is that pivotal, bridging or counterexamples could assist students with proof ideas (Stylianides and Stylianides 2007; Zazkis and Chernoff 2008). A potential benefit of a counterexample is to produce cognitive conflict in the student, while a pivotal example is designed to create a turning point in the learner's cognitive perception. Counterexamples may also foster deductive reasoning, since deductions are made by building models and looking for counterexamples. For Zazkis and Chernoff (2008) a counterexample is a mathematical concept, while a pivotal example is a pedagogical concept, which is within, but pushing the boundaries of the set of examples students have experienced. The role of examples also arose in research by Weber and Mejia-Ramos (2011) on proof reading by mathematicians. This suggests that students might be taught how to use examples to increase their conviction in, or understanding of, a proof. In order to know what skills to teach students, Alcock and Inglis (2008) maintain that identifying different strategies of proof construction among experts will grow knowledge of what skills to teach students, and how they can be employed.

Mathematical Modelling and Applications

Mathematical modelling and applications continues to be a central theme in mathematics education research (Blum et al. 2002), with a primary focus on practice activities. However, it appears that little or no literature exists explicitly discussing these topics with a focus on the 'transition' from the secondary to the university levels, possibly because there have been no roadmaps to sustained implementation of modelling education at all levels. Hence, recent literature relevant to the secondary-tertiary transition issue is briefly considered here.

One crucial duality, mentioned by Niss et al. (2007), is the difference between 'applications and modelling for the learning of mathematics' and 'learning mathematics for applications and modelling'. This duality is seldom made explicit in lower secondary school, and instead both orientations are simultaneously insisted on. However, at upper secondary or tertiary level the duality is often a significant one. The close relationship between modelling and problem solving is taken up by a number of authors. For example, English and Sriraman (2010) suggest that mathematical modelling is a powerful option for advancing the development of problem solving in the curriculum. In addition, according to Petocz et al. (2007), there are distinct advantages to using real world tasks in problem solving in order to model the way mathematicians work. This is supported by the research of Perrenet and Taconis (2009), who describe significant shifts in the growth of attention to metacognitive aspects in problem solving related to the change from secondary school mathematics problems to authentic mathematics problems at university. One difficulty outlined by Ärlebäck and Frejd (2010) is that upper secondary students have little experience working with real situations and modelling problems, making the incorporation of real problems from industry problematic. A second possible difficulty (Gainsburg 2008) is that teachers tend not to make many real-world connections in teaching. One possible solution is to bring together combinations of students, teachers and mathematicians to work on modelling problems (Kaiser and Schwarz 2006). This opportunity may be created through a "modelling week" (Göttlich 2010; Heilio 2010; Kaland et al. 2010), during which small groups of school or tertiary students work intensely, in a supported environment, on selected, authentic modelling problems.

There is some agreement that the secondary school curriculum could include more modelling activities, although high-stakes assessment at the secondary-tertiary interface is an unresolved problem in any implementation (Stillman 2007). Other initiatives for embedding modelling in the curriculum proposed by Stillman and Ng

(2010) include a system-wide focus emphasising an applications and modelling approach to teaching and assessing mathematical subjects in the last two years of school and interdisciplinary project work from primary through secondary school, with mathematics as the anchor subject.

Conclusion

The literature review presented here reveals a multi-faceted web of cognitive, curricular and pedagogical issues, some spanning across mathematical topics and some intrinsic to certain topics—and certainly exhibiting variation across the institutional contexts of the many countries our survey focused on. For example, most of the research we reviewed discusses the students' limited cognitive pre-paredness for the requirements of university-level formal mathematical thinking (whether this concerns the abstraction, for example, within Abstract Algebra courses or the formalism of Analysis). Within other areas, such as discrete mathematics, much of the research we reviewed highlighted that students may arrive at university with little or no awareness of certain mathematical fields.

The review presented in this report, as well as the longer version, is certainly not exhaustive. However we believe it is reasonable to claim that the bulk of research on transition is in a limited number of areas (e.g. calculus, proof) and that there is little research in other areas (e.g. discrete mathematics). While this might simply reflect curricular emphases in the various countries that our survey focused on, it also indicates directions that future research may need to pursue. Furthermore across the preceding sections a pattern seems to emerge with regard to *how*, not merely *what*, students experience in their first encounters with advanced mathematical topics, whether at school or at university. Fundamental to addressing issues of transition seems also to be the coordination and dialogue across educational levels—here mostly secondary and tertiary—and our survey revealed that at the moment this appears largely absent.

Acknowledgments We acknowledge the assistance of Michèle Artigue, and colleagues at the University Paris 7 and other universities, in the preparation of this report.

Open Access This chapter is distributed under the terms of the Creative Commons Attribution Noncommercial License, which permits any noncommercial use, distribution, and reproduction in any medium, provided the original author(s) and source are credited.

References

Alcock, L., & Inglis, M. (2008). Doctoral students' use of examples in evaluating and proving conjectures. *Educational Studies in Mathematics, 69*, 111–129.

Antonini, S., & Mariotti, M. A. (2008). Indirect proof: What is specific to this way of proving? *ZDM – The International Journal on Mathematics Education, 40*, 401–412.

Ärlebäck J. B., & Frejd P. (2010). First results from a study investigating Swedish upper secondary students' mathematical modelling competencies. In A. Araújo, A. Fernandes, A. Azevedo, & J. F. Rodrigues (Eds.), *EIMI 2010 Conference (educational interfaces between mathematics and industry) Proceedings*. Comap Inc., Bedford, MA, USA.

Artigue, M. (2008). Continu, Discontinu en mathématiques: Quelles perceptions en ont les élèves et les étudiants? In L. Viennot (Ed.), *Didactique, épistémologie et histoire des sciences. Penser l'enseignement* (pp. 151–173). Paris: Presses Universitaires de France.

Artigue, M. (2009). L'enseignement des fonctions à la transition lycée – université. In B. Grugeon (Ed.), *Actes du XVe Colloque CORFEM 2008* (pp. 25–44). Université de Cergy-Pontoise, IUFM de Versailles.

Balacheff, N. (2008). The role of the researcher's epistemology in mathematics education: An essay on the case of proof. *ZDM – The International Journal on Mathematics Education, 40*, 501–512.

Barrier, T. (2009). Quantification et Variation en Mathématiques: perspectives didactiques issues de la lecture d'un texte de Bolzano. In Kourkoulos M., Tzanakis C. (Ed.), Proceedings of the 5th International Colloquium on the Didactics of Mathematics (Vol. 2), University of Crete, Rethymnon, Greece.

Barton, B., & Sheryn, L. (2009). The mathematical needs of secondary teachers: Data from three countries. *International Journal of Mathematical Education in Science and Technology, 40*(1), 101–108.

Bergé, A. (2008). The completeness property of the set of real numbers in the transition from calculus to analysis. *Educational Studies in Mathematics, 67*(3), 217–236.

Biehler, Fischer, Hochmuth & Wassong (2011). Designing and evaluating blended learning bridging courses in mathematics. In M. Pytlak, T. Rowland, & E. Swoboda (Eds.), *Proceedings of the 7th Conference of European Researchers in Mathematics Education* (pp. 1971–1980). Rzeszow, Poland.

Bloch, I. & Ghedamsi I. (2004) The teaching of calculus at the transition between upper secondary school and the university: Factors of rupture. Communication to the Topic Study Group 12, Dans M. Niss (Eds.) *Actes de ICME10*. Copenhagen. Copenhagen: Roskilde University.

Blossier, T., Barrier, T., & Durand-Guerrier, V. (2009). Proof and quantification. In F.-L. Lin, F.-J. Hsieh, G. Hanna & M. de Villiers (Eds.), *ICMI Study 19 Conference Proceedings*, Taiwan: Taiwan University.

Blum, W., Alsina, C., Biembengut, M. S., Bouleau, N., Confrey, J., Galbraith, P., Ikeda, T., Lingefjärd, T., Muller, E., Niss, M., Verschaffel, L., Wang, S., Hodgson, B. R. & Henn, H.-W. (2002). ICMI Study 14: Applications and modelling in mathematics education – Discussion Document. *Educational Studies in Mathematics, 51*, 149–171.

Bosch, M., Fonseca, C., Gascón, J. (2004). Incompletitud de las organizaciones matemáticas locales en las instituciones escolares. *Recherches en Didactique des Mathématiques, 24/2.3*, 205–250.

Brandell, G., Hemmi, K., & Thunberg, H. (2008). The widening gap—A Swedish perspective. *Mathematics Education Research Journal, 20*(2), 38–56.

Bridoux, S. (2010). Une séquence d'introduction des notions de topologie dans l'espace R^n : de la conception à l'expérimentation. In A. Kuzniak & M. Sokhna (Eds.) *Actes du Colloque International Espace Mathématique Francophone 2009, Enseignement des mathématiques et développement, enjeux de société et de formation, Revue Internationale Francophone*.

Britton, S., & Henderson, J. (2009). Linear algebra revisited: An attempt to understand students' conceptual difficulties. *International Journal of Mathematical Education in Science and Technology, 40*, 963–974.

Brousseau, G. (1997). *Theory of didactical situations in mathematics* (N. Balacheff, M. Cooper, R. Sutherland & V. Warfield: Eds. and Trans.). Dordrecht: Kluwer Academic Publishers.

Burn, R. P. (1996). What are the fundamental concepts of Group Theory? *Educational Studies in Mathematics, 31*(4), 371–377.

Carmichael, C., & Taylor, J. A. (2005). Analysis of student beliefs in a tertiary preparatory mathematics course. *International Journal of Mathematical Education in Science and Technology, 36*(7), 713–719.

Cartiglia, M., Furinghetti, F., & Paola, D. (2004). Patterns of reasoning in classroom. In M. J. Hoines & A. B. Fuglestad (Eds.), *Proceedings of the 28th Conference of the International Group for the Psychology of Mathematics Education* (Vol. 2, pp. 287–294). Bergen, Norway: Bergen University College.

Chellougui (2004). *L'utilisation des quantificateurs universel et existentiel en première année d'université. Entre l'implicite et l'explicite*. Thèse de l'université Lyon 1.

Chevallard, Y. (1985). *La transposition didactique*. Grenoble: La Pensée Sauvage.

Clark, M. & Lovric, M. (2009). Understanding secondary–tertiary transition in mathematics. *International Journal of Mathematical Education in Science and Technology*, 40(6), 755–776.

Cooley, L., Martin, W., Vidakovic, D., & Loch, S. (2007). Coordinating learning theories with linear algebra. *International Journal for Mathematics Teaching and Learning* [online journal], University of Plymouth: U.K. http://www.cimt.plymouth.ac.uk/journal/default.htm

Dias, M., Artigue, M., Jahn A., & Campos, T. (2008). A comparative study of the secondary-tertiary transition. In M. F. Pinto & T. F. Kawasaki (Eds.). *Proceedings of the 34th Conference of the PME* (Vol. 2, pp. 129–136). Belo Horizonte, Brazil: IGPME.

Dogan-Dunlap, H. (2010). Linear algebra students' modes of reasoning: Geometric representations. *Linear Algebra and its Applications, 432*, 2141–2159.

Dubinsky, E., & McDonald, M. (2001). APOS: A constructivist theory of learning. In D. Holton (Ed.) *The teaching and learning of mathematics at university level: An ICMI study* (pp. 275–282). Dordrecht, The Netherlands: Kluwer Academic Publishers.

Durand-Guerrier, V., & Njomgang Ngansop, J. (2010). Questions de logique et de langage à la transition secondaire – supérieur. L'exemple de la négation. In A. Kuzniak & M. Sokhna (Eds.), Actes du Colloque International Espace Mathématique Francophone 2009, Enseignement des mathématiques et développement, enjeux de société et de formation, *Revue Internationale Francophone* (pp. 1043–1047). Numéro Spécial 2010.

Engelbrecht, J. (2010). Adding structure to the transition process to advanced mathematical activity. *International Journal of Mathematical Education in Science and Technology*, 41(2), 143–154.

Engelbrecht, J., & Harding, A. (2008). The impact of the transition to outcomes-based teaching on university preparedness in mathematics in South Africa, *Mathematics Education Research Journal, 20*(2), 57–70.

Engelbrecht, J., Harding, A., & Potgieter, M. (2005). Undergraduate students' performance and confidence in procedural and conceptual mathematics. *International Journal of Mathematical Education in Science and Technology, 36*(7), 701–712.

English, L. & Sriraman, B. (2010). Problem solving for the 21st Century. In B. Sriraman, & L. English (Eds.), *Theories of mathematics education, advances in mathematics education* (pp. 263–290). Berlin: Springer-Verlag. doi:10.1007/978-3-642-00742-2_27

Gainsburg, J. (2008). Real-world connections in secondary mathematics teaching. *Journal of Mathematics Teacher Education, 11*, 199–219.

Ghedamsi, I. (2008). *Enseignement du début de l'analyse réelle à l'entrée à l'université: Articuler contrôles pragmatique et formel dans des situations à dimension a-didactique*. Unpublished doctoral dissertation, University of Tunis.

Gill, O., O'Donoghue, J., Faulkner, F., & Hannigan, A. (2010). Trends in performance of science and technology students (1997–2008) in Ireland. *International Journal of Mathematical Education in Science and Technology*, 41(3), 323–339. doi:10.1080/00207390903477426

Godfrey, D., & Thomas, M. O. J. (2008). Student perspectives on equation: The transition from school to university. *Mathematics Education Research Journal, 20*(2), 71–92.

González-Martín, A. (2009). L'introduction du concept de somme infinie : une première approche à travers l'analyse des manuels. *Actes du colloque EMF 2009. Groupe de travail 7*, 1048–1061.

González-Martín, A. S., Nardi, E., & Biza, I. (2011). Conceptually-driven and visually-rich tasks in texts and teaching practice: The case of infinite series. *International Journal of Mathematical Education in Science and Technology, 42*(5), 565–589.

Göttlich, S. (2010). Modelling with students – A practical approach. In A. Araújo, A. Fernandes, A. Azevedo, J. F. Rodrigues (Eds.), *EIMI 2010 Conference (educational interfaces between mathematics and industry)* Proceedings. Comap Inc., Bedford, MA, USA.

Gueudet, G. (2004). Rôle du géométrique dans l'enseignement de l'algèbre linéaire *Recherches en Didactique des Mathématiques* 24/1, 81–114.

Gueudet, G. (2008). Investigating the secondary-tertiary transition. *Educational Studies in Mathematics, 67*(3), 237–254.

Gueudet-Chartier, G. (2004). Should we teach linear algebra through geometry? *Linear Algebra and its Applications, 379*, 491–501.

Gyöngyösi, E., Solovej, J. P., & Winsløw, C. (2011). Using CAS based work to ease the transition from calculus to real analysis. In M. Pytlak, T. Rowland, & E. Swoboda (Eds.), *Proceedings of the 7th Conference of European Researchers in Mathematics Education* (pp. 2002–2011). Rzeszow, Poland.

Hanna, G., & Barbeau, E. (2008). Proofs as bearers of mathematical knowledge. *The International Journal on Mathematics Education, 40*, 345–353.

Hanna, G., & de Villiers, M. (2008). ICMI Study 19: Proof and proving in mathematics education. *ZDM – The International Journal on Mathematics Education, 40*, 329–336.

Hanna, G., & de Villiers, M. (2012). *Proof and Proving in Mathematics Education: The 19th ICMI study*. Dordrecht: Springer.

Harel, G. (2008). DNR perspective on mathematics curriculum and instruction, Part I: focus on proving. *ZDM – The International Journal on Mathematics Education, 40*(3), 487–500.

Hazzan, O. (2001). Reducing abstraction: The case of constructing an operation table for a group. *Journal of Mathematical Behavior, 20*, 163–172.

Heilio, M. (2010). Mathematics in industry and teachers' training. In A. Araújo, A. Fernandez, A. Azevedo, & J. F. Rodrigues (Eds.), *EIMI 2010 Conference (educational interfaces between mathematics and industry) Proceedings*. Comap Inc., Bedford, MA, USA.

Hemmi, K. (2008). Students' encounter with proof: The condition of transparency. *ZDM – The International Journal on Mathematics Education, 40*, 413–426.

Henderson, F., Rasmussen, C., Zandieh, M., Wawro, M., & Sweeney, G. (2010). Symbol sense in linear algebra: A start toward eigen theory. *Proceedings of the 14th Annual Conference for Research in Undergraduate Mathematics Education*. Raleigh, N.C.

Hillel, J. (2000). Modes of description and the problem of representation in linear algebra. In J.-L. Dorier (Ed.), *On the Teaching of Linear Algebra* (Vol. 23, pp. 191–207). Springer: Netherlands.

Hockman, M. (2005). Curriculum design and tertiary education. *International Journal of Mathematical Education in Science and Technology, 36*(2–3), 175–191.

Hong, Y., Kerr, S., Klymchuk, S., McHardy, J., Murphy, P., Spencer, S., Thomas, M. O. J. & Watson, P. (2009). A comparison of teacher and lecturer perspectives on the transition from secondary to tertiary mathematics education. *International Journal of Mathematical Education in Science and Technology, 40*(7), 877–889. doi:10.1080/00207390903223754

Hourigan, M., & O'Donoghue, J. (2007). Mathematical under-preparedness: The influence of the pre-tertiary mathematics experience on students' ability to make a successful transition to tertiary level mathematics courses in Ireland. *International Journal of Mathematical Education in Science and Technology, 38*(4), 461–476.

Hoyles, C., Newman, K. & Noss, R. (2001). Changing patterns of transition from school to university mathematics. *International Journal of Mathematical Education in Science and Technology, 32*(6), 829–845. doi:10.1080/00207390110067635

Iannone, P. & Inglis, M. (2011). Undergraduate students' use of deductive arguments to solve "prove that..." tasks. In M. Pytlak, T. Rowland, & E. Swoboda (Eds.), *Proceedings of the 7th Conference of European Researchers in Mathematics Education* (pp. 2012–2021). Rzeszow, Poland.

Iannone, P., & Nardi, E. (2002). A group as a 'special set'? Implications of ignoring the role of the binary operation in the definition of a group. In A. D. Cockburn & E. Nardi (Eds.), *Proceedings of the 26th Conference of the International Group for the Psychology of Mathematics Education* (Vol. 3, pp. 121–128). Norwich, UK.

Ioannou, M. & Iannone, P. (2011). Students' affective responses to the inability to visualise cosets. *Research in Mathematics Education 13*(1), 81–82.

Ioannou, M., & Nardi, E. (2009). Engagement, abstraction and visualisation: Cognitive and emotional aspects of Year 2 mathematics undergraduates' learning experience in abstract algebra. *Proceedings of the British Society for Research into Learning Mathematics, 29*(2), 35–40.

Ioannou, M., & Nardi, E. (2010). Mathematics undergraduates' experience of visualisation in Abstract Algebra: The metacognitive need for an explicit demonstration of its significance. In *Proceedings of the 13th Special Interest Group of the Mathematical Association of America (SIGMAA) Conference on Research in Undergraduate Mathematics Education (RUME).* Available at: http://sigmaa.maa.org/rume/crume2010/Archive/Ioannou%20&%20Nardi.pdf

James, A., Montelle, C., & Williams, P. (2008). From lessons to lectures: NCEA mathematics results and first-year mathematics performance. *International Journal of Mathematical Education in Science and Technology, 39*(8), 1037–1050. doi:10.1080/00207390802136552

Jennings, M. (2009). Issues in bridging between senior secondary and first year university mathematics. In R. Hunter, B. Bicknell, & T. Burgess (Eds.), *Crossing divides* (Proceedings of the 32nd annual conference of the Mathematics Education Research Group of Australasia, Vol. 1, pp. 273–280). Palmerston North, NZ: MERGA.

Junior, O. (2006). *Compreensões de conceitos de cálculo diferencial no primeiro ano de matemática uma abordagem integrando oralidade, escrita e informática.* Unpublished PhD Thesis, Universidade Estadual Paulista, Rio Claro, Brazil.

Kaiser, G., & Schwarz, B. (2006). Mathematical modelling as bridge between school and university. *ZDM – The International Journal on Mathematics Education, 38*(2), 196–208.

Kajander, A., & Lovric, M. (2005). Transition from secondary to tertiary mathematics: McMaster University Experience. *International Journal of Mathematical Education in Science and Technology, 36*(2–3), 149–160

Kaland, K., Kaiser, K., Ortlieb, C. P., & Struckmeier, J. (2010). Authentic modelling problems in mathematics education. In A. Araújo, A. Fernandes, A. Azevedo, & J. F. Rodrigues (Eds.). *EIMI 2010 Conference (educational interfaces between mathematics and industry) Proceedings.* Comap Inc., Bedford, MA, USA.

Kondratieva, M. (2011). Designing interconnecting problems that support development of concepts and reasoning. In M. Pytlak, T. Rowland, & E. Swoboda (Eds.), *Proceedings of the 7th Conference of European Researchers in Mathematics Education* (pp. 273–282). Rzeszow, Poland.

Larsen, S. (2009). Reinventing the concepts of group and isomorphism: The case of Jessica and Sandra. *Journal of Mathematical Behavior, 28,* 119–137.

Larson, C., Zandieh, M., & Rasmussen, C. (2008). A trip through eigen-land: Where most roads lead to the direction associated with the largest eigenvalue. *Proceedings of the 11th Annual Conference for Research in Undergraduate Mathematics Education.* San Diego, CA.

Leviatan, T. (2008). Bridging a cultural gap. *Mathematics Education Research Journal, 20*(2), 105–116.

Luk, H. S. (2005). The gap between secondary school and university mathematics. *International Journal of Mathematical Education in Science and Technology, 36*(2–3), 161–174.

Mamona-Downs, J. (2010). On introducing a set perspective in the learning of limits of real sequences. *International Journal of Mathematical Education in Science and Technology, 41* (2), 277–291.

Nardi, E. (2000). Mathematics undergraduates' responses to semantic abbreviations, 'geometric' images and multi-level abstractions in Group Theory. *Educational Studies in Mathematics, 43,* 169–189.

Nardi, E. (2008). *Amongst mathematicians: Teaching and learning mathematics at university level.* New York: Springer.

Nardi, E. (2011). 'Driving noticing' yet 'risking precision': University mathematicians' pedagogical perspectives on verbalisation in mathematics. In M. Pytlak, T. Rowland, & E. Swoboda (Eds.), *Proceedings of the 7th Conference of European Researchers in Mathematics Education* (pp. 2053–2062). Rzeszow, Poland.

Niss, M., Blum, W., & Galbraith, P. L. (2007). Part 1: Introduction. In W. Blum, P. L. Galbraith, H.-W. Henn, & M. Niss (Eds.), *Modelling and applications in mathematics education. The 14th ICMI Study*. New York/etc.: Springer, New ICMI Studies series 10.

Oehrtman, M. (2009). Collapsing dimensions, physical limitation, and other student metaphors for limit concepts. *Journal for Research in Mathematics Education, 40*(4), 396–426.

Palla, M., Potari, D., & Spyrou, P. (2012). Secondary school students' understanding of mathematical induction: structural characteristics and the process of proof construction. *International Journal of Science and Mathematics Education, 10*(5), 1023–1045.

Pedemonte, B. (2007). How can the relationship between argumentation and proof be analysed? *Educational Studies in Mathematics, 66*, 23–41.

Pedemonte, B. (2008). Argumentation and algebraic proof. *ZDM – The International Journal on Mathematics Education, 40*(3), 385–400.

Perrenet, J., & Taconis, R. (2009). Mathematical enculturation from the students' perspective: Shifts in problem-solving beliefs and behaviour during the bachelor programme. *Educational Studies in Mathematics, 71*, 181–198. doi:10.1007/s10649-008-9166-9

Petocz, P., Reid, A., Wood, L. N., Smith, G. H., Mather, G., Harding, A., Engelbrecht, J., Houston, K., Hillel, J., & Perrett, G. (2007). Undergraduate students' conceptions of mathematics: An international study. *International Journal of Science and Mathematics Education, 5*, 439–459.

Portnoy, N., Grundmeier, T. A., & Graham, K. J. (2006). Students' understanding of mathematical objects in the context of transformational geometry: Implications for constructing and understanding proofs. *The Journal of Mathematical Behavior, 25*, 196–207.

Possani, E., Trigueros, M., Preciado, J., & Lozano, M. (2010). Use of models in the teaching of linear algebra. *Linear Algebra and its Applications, 432*, 2125–2140.

Praslon, F. (2000). Continuités et ruptures dans la transition Terminale S/DEUG Sciences en analyse. Le cas de la notion de dérivée et son environnement. In, T. Assude & B. Grugeon (Eds.), *Actes du Séminaire National de Didactique des Mathématiques* (pp. 185–220). IREM Paris 7.

President's Council of Advisors on Science and Technology (PCAST) (2012). *Engage to excel: Producing one million additional college graduates with Degrees in Science, Technology, Engineering, and Mathematics*. Washington, DC: The White House.

Rogalski, M. (2008). Les rapports entre local et global: mathématiques, rôle en physique élémentaire, questions didactiques. In L. Viennot (Ed.) *Didactique, épistémologie et histoire des sciences* (pp. 61–87). Paris: Presses Universitaires de France.

Scucuglia, R. (2006). *A investigação do teorema fundamental do cálculo com calculadoras gráficas*. Unpublished PhD Thesis, Universidade Estadual Paulista, Rio Claro, Brazil.

Selden, A. (2005). New developments and trends in tertiary mathematics education: Or more of the same? *International Journal of Mathematical Education in Science and Technology, 36*(2–3), 131–147.

Smith, A. (2004). *Making mathematics count*. UK: The Stationery Office Limited.

Solomon, Y. (2006). Deficit or difference? The role of students' epistemologies of mathematics in their interactions with proof. *Educational Studies in Mathematics, 61*, 373–393.

Stadler, E. (2011). The same but different – novice university students solve a textbook exercise. In M. Pytlak, T. Rowland, & E. Swoboda (Eds.), *Proceedings of the 7th Conference of European Researchers in Mathematics Education* (pp. 2083–2092). Rzeszow, Poland.

Stewart, S., & Thomas, M. O. J. (2009). A framework for mathematical thinking: The case of linear algebra. *International Journal of Mathematical Education in Science and Technology, 40*, 951–961.

Stewart, S., & Thomas, M. O. J. (2010). Student learning of basis, span and linear independence in linear algebra. *International Journal of Mathematical Education in Science and Technology, 41*, 173–188.

Stillman, G. (2007). Upper secondary perspectives on applications and modelling. In W. Blum, P. L. Galbraith, H.-W. Henn, M. & Niss (Eds.). *Modelling and applications in mathematics education. The 14th ICMI Study.* New York/etc.: Springer, New ICMI Studies series 10.

Stillman, G., & Ng, D. (2010). The other side of the coin-attempts to embed authentic real world tasks in the secondary curriculum. In A. Araújo, A. Fernandes, A. Azevedo, & J. F. Rodrigues (Eds.), *EIMI 2010 Conference (educational interfaces between mathematics and industry) Proceedings.* Comap Inc., Bedford, MA, USA.

Stylianides, A. J., & Stylianides, G. J. (2007). The mental models theory of deductive reasoning: Implications for proof instruction. *Proceedings of CERME5, the 5th Conference of European Research in Mathematics Education,* 665–674.

Sweeney, G. (2011). Classroom activity with vectors and vector equations: Integrating informal and formal ways of symbolizing R^n. *Paper presented at the 14th Conference on Research in Undergraduate Mathematics Education,* Portland, OR.

Tall, D. O. (2008). The transition to formal thinking in mathematics. *Mathematics Education Research Journal, 20*(2), 5–24.

Tall, D. O., & Mejia-Ramos, J. P. (2006). The long-term cognitive development of different types of reasoning and proof. *Conference on Explanation and Proof in Mathematics: Philosophical and Educational Perspectives,* Universität Duisburg-Essen, Campus Essen, 1–11.

Tempelaar, D. T., Rienties, B., Giesbers, B., & Schim van der Loeff, S. (2012). Effectiveness of a voluntary postsecondary remediation program in mathematics. In P. Van den Bossche et al. (Eds.), *Learning at the Crossroads of Theory and Practice,* Advances in Business Education and Training 4 (pp. 199–222), Dordrecht: Springer. doi:10.1007/978-94-007-2846-2_13

Thomas, M. O. J., & Stewart, S. (2011). Eigenvalues and eigenvectors: Embodied, symbolic and formal thinking. *Mathematics Education Research Journal, 23*(3), 275–296.

Vandebrouck, F. (2011). Students' conceptions of functions at the transition between secondary school and university. In M. Pytlak, T. Rowland, & E. Swoboda (Eds.), *Proceedings of the 7th Conference of European Researchers in Mathematics Education* (pp. 2093–2102). Rzeszow, Poland.

Varsavsky, C. (2010). Chances of success in and engagement with mathematics for students who enter university with a weak mathematics background. *International Journal of Mathematical Education in Science and Technology, 41*(8), 1037–1049. doi:10.1080/0020739X.2010.493238

Wawro, M., Sweeney, G., & Rabin, J. (2011a). Subspace in linear algebra: investigating students' concept images and interactions with the formal definition. *Educational Studies in Mathematics, 78*(1), 1–19.

Wawro, M., Zandieh, M., Sweeney, G., Larson, C., & Rasmussen, C. (2011b). *Using the emergent model heuristic to describe the evolution of student reasoning regarding span and linear independence.* Paper presented at the *14th Conference on Research in Undergraduate Mathematics Education,* Portland, OR.

Weber, K., & Mejia-Ramos, J. P. (2011). Why and how mathematicians read proofs: An exploratory study. *Educational Studies in Mathematics, 76,* 329–344.

Winsløw, C. (2008). Transformer la théorie en tâches: La transition du concret à l'abstrait en analyse réelle (Turning theory into tasks: Transition from concrete to abstract in calculus). In A. Rouchier, et al. (Eds.), *Actes de la XIIIième école d'été de didactique des mathématiques.* Grenoble: La Pensée Sauvage.

Winsløw, C. (2010). Comparing theoretical frameworks in didactics of mathematics: The GOA-model. In V. Durand-Guerrier, S. Soury-Lavergne & F. Arzarello (Eds.) *Proceedings of the Sixth Congress of the European Society for Research in Mathematics Education* (pp. 1675–1684), CERME: Lyon France.

Zazkis, R., & Chernoff, E. J. (2008). What makes a counterexample exemplary? *Educational Studies in Mathematics, 68,* 195–208.

Socioeconomic Influence on Mathematical Achievement: What Is Visible and What Is Neglected

Paola Valero, Mellony Graven, Murad Jurdak, Danny Martin,
Tamsin Meaney and Miriam Penteado

Abstract The survey team worked in two main areas: Literature review of published papers in international publications, and particular approaches to the topic considering what in the literature seems to be neglected. In this paper we offer a synoptic overview of the main points that the team finds relevant to address concerning what is known and what is neglected in research in this topic.

Keywords Poverty · Early childhood · Intersectionality of positionings · Statistical reifications · Macro-systemic perspective · History of mathematics education practices

P. Valero (✉)
Aalborg University, Aalborg, Denmark
e-mail: paola@learning.aau.dk

M. Graven
Rhodes University, Grahamstown, South Africa
e-mail: m.graven@ru.ac.za

M. Jurdak
American University of Beirut, Beirut, Lebanon
e-mail: jurdak@aub.edu.lb

D. Martin
University of Illinois at Chicago, Chicago, USA
e-mail: dbmartin@uic.edu

T. Meaney
Bergen University College, Bergen, Norway
e-mail: Tamsin.Meaney@hib.no

M. Penteado
Sao Paulo State University at Rio Claro, Rio Claro, São Paulo, Brazil
e-mail: mirgps@rc.unesp.br

© The Author(s) 2015
S.J. Cho (ed.), *The Proceedings of the 12th International Congress on Mathematical Education*, DOI 10.1007/978-3-319-12688-3_19

Introduction

It is known that socioeconomic factors have an influence on mathematical achievement. Nowadays such link has become a "fact" that researchers, teachers, administrators and politicians have at hand: "the better off you—and your family—are, the more likely you will do well in school, including mathematics". Such a statement embodies its opposite: "the worse off you—and your family—are, the more likely you will do poorly in school and mathematics". Studies defining socio-economic status (SES) and showing its relation to school performance emerged at the beginning of the 20th century. The specification of the relationship for school mathematics was enunciated as a problem for society and for research in the 1960s. However, it is only in the 1980s that such issue started to be a focus of attention of the mathematics education community. What is known so far—which may be part of a commonsense understanding of the topic—and what seems to be forgotten—which are critical readings challenging the commonsense—were the central questions that have guided the work of the survey team.

We thank Alexandre Pais, Aalborg University, Denmark; Arindam Bose, Tata Institute of Fundamental Research, India; Francisco Camelo, Bogotá's Capital District University "Francisco José de Caldas", Colombia; Hauke Straehler-Pohl, Freie University, Germany; Lindong Wang, Beijing Normal University, China; and Troels Lange, Malmö University, Sweden, for their contribution to the teamwork.

What Is Visible

A global literature review for this topic poses challenges such as the multiple languages in which research reports are made available. We gathered literature that would indicate some trends in what is known about the socioeconomic influences on mathematical achievement in different parts of the world. Most of what was reviewed was published in English.

At a general educational level, the relationship between socioeconomic factors and school achievement is inserted in the history of expansion of mass education systems and differential access to education around the world during the 20th century. Meyer et al. (1992) show that the consolidation of Modern nation states is correlated to the expansion and Modern organization of mass systems of education. Many nation states growingly focused on the socialization of citizens with a vision of progress in which the scientific rationality was an articulating element. The link between personal development and the mastery of the curriculum, and such individual mastery and the progress of the nation were established. With the expansion of mass education, the issue emerged of who has access to education and the goods of society and on the grounds of what. To know who was having effective access to education became important.

The report "Equality of Educational Opportunity" (Coleman et al. 1966) was one of the first large-scale national surveys that formulated a model to determine the extent to which educational opportunities were equally available to all citizens in the USA. It allowed individual students' socio-economic, racial and ethnic characteristics to be connected to school inputs in terms of resources available to run education, and to students' individual performance in achievement tests in different school subjects. Internationally, the International Association for the Evaluation of Educational Achievement (IEA) started providing international comparative information about how different national curricula provide different opportunities to learn, and the existence of a lack of equity between different groups of students. Since then, the measurement of educational quality was moved from an input-output model based on school resources to an individualization of the measurement of educational quality in terms of students' achievement, even in mathematics. This fundamental change in the general reports on educational access is central for connecting socio-economic influences with mathematical achievement.

The discussion on what may be the socioeconomic influences on mathematical achievement emerged from general social science research and educational research. Therefore what has become visible about the topic is found in general reports on educational systems around the world, as much as in mathematics educational research literature. Thus any talk about the topic in the realm of mathematics is bound to general discussions about social and educational disparities for different types of students.

At the level of mathematics education research the concern for this connection emerged as a research topic in the 1980s. The studies that address this issue are mainly quantitative and to some extent large scale. It is important to mention that the amount of literature testing different hypothesis about socio economic influences and achievement has increased with the growing importance given to periodic, international, standardized, comparative studies such as TIMSS and PISA since the 1990s.

In different parts of the world there are results about a society's sense of expected, normal school achievement and how different groups of students are compared to the normal expectation. While in the USA, factors that systematically generates differentiation to the expected norm are socioeconomic status (SES) and race, in other countries it is socioeconomic status as in for example in the UK and Australia, or home language and ethnicity in the case of some European countries such as Germany and Denmark, or rurality as in China or many of the African and Latin American countries. Although other factors are also present, the tendency of countries to focus on one factor has influenced the way discussions operate in these countries. In different countries the independent variables considered to be the socioeconomic influences on mathematical achievement —the dependent variable —change. What may be considered the 'socioeconomic' influences on 'mathematical achievement' depends on the systems of differentiation and stratification of the population. It is not any kind of existing, a priori characteristic of individuals and groups of students or of mathematical achievement per se.

Once the general differentiation is possible, similar statistical indicators are adopted in the studies. Prior to the existence of international comparable, standardized national data sets, the variable of socioeconomic status has been one of the most used in the studies. Since its construction in the 1920s, the measurement has been composed by a series of reliable indicators—parents' educational level, family income, possession of appliances, possession of books, etc.—which have not changed much in almost 100 years. The tendency to simplify the measurement is connected to how difficult it is to collect reliable information on this matter from children. The assignment of a socio-economic level to individual students often takes place on very thin evidence. The effect of the measurement, on the contrary, has the tendency to reify a solid state that follows individual children all through their school life. This reification has been documented in studies that have addressed how the discussion of students' differential results is dealt with in the media and public debates.

Even if many studies have a tendency to establish the relationship between a limited variables indicating differential positioning, many studies conclude that those variables intersect. This means that students whose participation in school mathematics results in low achievement experience differential positioning in schooling because they are attributed simultaneously several categories of disadvantage. For example, low achievement in mathematics in certain regions in China is explained by the intersection of rurality, parents' educational level, mother-at-work, and language (Hu and Du 2009). In other words, existing studies devise sophisticated statistical measurements to trace the factors that correlate to differential access to mathematical achievement. However, the very same statistical rationality on which those studies are based imposes a restriction for understanding how the complexity of the intersectionality of variables of disadvantage effect differential results in mathematics.

There is an over-representation of research reports addressing the socioeconomic influences on mathematical achievement in English speaking countries (USA, UK, Australia and New Zealand), while there is little research on this matter in many other places in the world. Such difference may not only be due to the extent of research in mathematics education in these countries, but also to the fact that differential achievement has not been construed as a problem. In East Asia there is little research in mathematics education investigating those who do not perform highly and why. In Taiwan research discards the focus on socio-economic variables and privileges variables such as student's learning goal orientation (Lin et al. 2009). Researchers argue that it is more meaningful to study what educators can impact positively to improve students' results. In South Korea the differentiated achievement is explained in terms of access to private tuition, which reflects a difference in resources that educational policies cannot compensate for (Kang and Hong 2008). In India, it is argued that differential achievement is due to students' mathematical aptitude, gender and urbanity/rurality, the socio-economic and cultural characteristics of communities, and the impact of child work for the lower castes and poorer communities (World Bank 2009).

Existing research both in general education and in mathematics education has constructed the positive correlation between a lower positioning of groups of students with respect to the valued norm of societies, and the results of the school mathematical experience measured in terms of achievement. Poverty, rurality, ethnicity, gender, language, culture, race, among others, have been defined as the variables that constitute socioeconomic influences on mathematical achievement. The question remains whether it is possible to interpret the meaning of "socioeconomic influences" and "mathematical achievement" in ways that allow us to go beyond the facts established in the last 50 years of research. In the following sections each one of the members of the team offers a perspective on this issue.

What Is Neglected

Paola Valero on Historicizing the Emergence of Differential Access to Mathematics Education

I discuss the historical conditions that make it possible to formulate the "socioeconomic influences on mathematical achievement" as a problem of research in mathematics education. How and when the problem has been made thinkable, up to the point that nowadays it is part of the commonsense or taken-for-granted assumptions of researchers and practitioners alike? My strategy of investigation builds on thinking the field of practice of mathematics education as a historical and discursive field. There are at least three important conditions that make the problem possible:

Education, Science and the Social Question. The social sciences and educational research are expert-based technologies for social planning. In the consolidation of Modernity and its cultural project in the 20th century, the new social sciences were seen as the secular rationality that, with its appeal to objective knowledge, should be the foundation for social engineering. Statistical tools in the social sciences allow generating constructs that identify the ills of society that science/education needs to rectify. This is an important element in how educational sciences address the differential access of children to the school system. Constructs, such as students' "socioeconomic status", later on expanded to school and communities socio-economic status, emerged in the 1920s in a moment where the newly configured social sciences started to address the "problems" of society. Educational sciences made it possible to articulate salvation narratives for facing the social problems for which education was a solution (Tröhler 2011). Measurements of intelligence, achievement and socioeconomic status were and still are technologies to provide the best match between individuals and educational and work possibilities. The double gesture of educational sciences of promoting the importance of access to education and reifying difference by constructing them as a fact inserts human beings in the calculations of power.

Mathematics and progress. During the second industrialization the justification for the need for mathematics education was formulated clearly in the first number of

L'Enseignement Mathématique. In the times of the Cold War, the justification was related to keeping the supremacy of the Capitalist West in front of the growing menace of the expansion of the Communist Soviet Union. Nowadays, professional associations and economic organizations argue that the low numbers of people in STEM fields can severely damage the competitiveness of developed nations in international, globalized markets. The narrative that connects progress, economic superiority, and development to citizen's mathematical competence is made intelligible in the 20th century. The consolidation of nation states and the full realization of the project of Modernity required forming particular types of subjects. The mathematics school curriculum in the 20th century embodied and made available cosmopolitan forms of reason, which build on the belief of science-based human reason having a universal, emancipatory capacity for changing the world and people (Popkewitz 2008). In this way, subjects are inserted in a logic of quantification that makes possible the displacement of qualitative forms of knowing into a scientific rationality based on numbers and facts for the planning of society. Thus, from the turn of the 19th century to present day, the mathematics curriculum is an important technology that inserts subjects into the forms of thinking and acting needed for people to become the ideal cosmopolitan citizen.

Mathematics for all. That high achievement in mathematics is a desired and growing demand *for all* citizens is a recent invention of mathematics education research. Between the years of reconstruction after the Second World War and the Cold War, school curricula were modernized with focus on the subject areas for the purpose of securing a qualified elite of college students. In the decade of the 1980s the new challenge of democratization and access was formulated. At the "Mathematics Education and Society" session at ICME 5 it was publicly raised the need to move towards inclusion of the growing diversity of students in school mathematics (Damerow et al. 1984). The systematic lack of success of many students was posed as a problem that mathematics education research needed to pay attention to and take care of. Mathematics education researchers, the experts in charge of understanding the teaching and learning of mathematics as well as of devising strategies to improve them, took the task of providing the technologies to bring school mathematics to the people, and not only to the elite. "Mathematics for all" can be seen as an effect of power that operates on subjects and nations alike to determine who are the individuals/nations who excel, while creating a narrative of inclusion for all those who, by the very same logic, are differentiated.

It is on the grounds of at least these three interconnected conditions that the "socioeconomic influences on mathematical achievement" has been enunciated as a problem of research in the field. I do not intend to say that underachievement is an unimportant "social construction". My intention is to offer a way of entering into the problem that makes visible the network of historical, social and political connections on which differential social and economic positioning is related to differential mathematical achievement.

Mellony Graven on Socio-Economic Status and Mathematics Performance/Learning in South African Research

South Africa's recent history of apartheid and its resultant high levels of poverty and extreme social and economic distance between rich and poor continue to manifest in the education of its learners in complex ways. The country provides a somewhat different context for exploring the relationship between SES and education than other countries. The apartheid era only ended in 1994 with our first democratic elections. Education became the vehicle for transforming South African society and a political rhetoric of equity and quality education for all emerged. Thus educational deliberations focused on redressing the inequalities of the past and major curriculum introductions and revisions were attempted. Engagement with SES and mathematics education became *foregrounded* in policy, political discourses and a range of literature since 1994 although in must be remembered that transformation of education was a priority of the eighties period of resistance and the people's education campaign (although heavily suppressed at the time). Yet for all the political will and prioritization little has been achieved in redressing the inequalities in education.

Much of the recent data available on the relationship between SES and mathematics performance can be 'mined' from large scale general education reviews. These studies provide findings indicating patterns or correlations between school performance and socio-economic context. Several indicate that correlations are exacerbated in mathematics. These reports highlight a range of factors or areas that affect learner performance, such as social disadvantage, teachers' subject knowledge, teaching time, teacher absenteeism, resources, poorly managed schools, poverty effects including malnutrition and HIV/AIDS. In general reports present a consistent picture. In South Africa, since poverty affects more than half of our learners, studies tend to focus on the poorest (but largest) SES group when looking at challenges in education. Many reports point to numeracy scores and mathematics results being consistently below other African neighbour countries with much less wealth. Furthermore, South Africa has the highest levels of between-school performance inequality in mathematics and reading among SACMEQ countries.

What might be somewhat different from other countries exploring SES and mathematics achievement is that South African poverty levels are extreme even while there is relative economic wealth. Fleisch (2008) argues that poverty must be understood in its full complexity and not only in economic terms and argues for "the need to understand the underlying structural dimensions of persistent poverty, which engages the complexities of social relations, agency and culture, and subjectivity" (p. 58). He also notes that "Poor families rather than being just a source of social and cultural deficit, are important supporters of educational success [...] poor South Africans share with the middle class an unqualified faith in the power of education. For poor families education is the way out of poverty, and as such many spend a large portion of their disposable income on school fees, uniforms and transport [...]" (p. 77)

Mathematics education research conducted in South Africa almost inevitably touches on issues of equity and redress when engaging with the context of studies. One important area is research on language and mathematics education. The overlap between language of learning with SES and mathematics achievement is referred to in almost all of the large quantitative studies above (as a correlating factor) and the data provides for a complex picture that cannot easily be explained in terms of causal relationships. Setati and collaborators (e.g., Barwell et al. 2007) urge that multilingualism needs to be reconceptualised as a resource rather than a disadvantage. In this way the deficit discourse around multilingualism and how it negatively correlates with mathematics performance should be reframed. Most language 'factors' referred to in the literature above position multilingualism as a factor that correlates with low mathematics performance but this should not be read as causal.

Recent research by Hoadley (2007) analyses how learners are given differentiated access to school knowledge in mathematics classrooms. She argues that the post-apartheid curriculum with its emphasis on everyday knowledge has had a disempowering effect in marginal groups who are not exposed to more specialised knowledge of mathematics. The result is that "the lower ability student, paradoxically, is left free to be a local individual but a failed mathematics learner" (Muller and Taylor 2000, p. 68). In its implementation teachers in low SES schools struggled to make sense of these changes resulting in even further mathematics learning gaps between 'advantaged' and 'disadvantaged' learners (Graven 2002). The result has been that "students in different social-class contexts are given access to different forms of knowledge, that context dependent meanings and everyday knowledge are privileged in the working-class context, and context-independent meanings and school knowledge predominate in the middle class schooling contexts" (Hoadley 2007, p. 682).

While studies relate poverty, class, race and access to English to differentiated learning outcomes from a variety of perspectives, most, I would argue, are not sufficiently concerned with the impact of extreme income inequality within a context of widespread and deep absolute poverty. Many poor countries achieve much better educational outcomes compared to South Africa but have lower levels of inequality. A deeper understanding of inequality as a core component of SES, and not just of the nature and impact of poverty might enrich our understanding of the relationship of SES to mathematical educational outcomes.

Murad Jurdak on a Culturally-Sensitive Equity-in-Quality Model for Mathematics Education at the Global Level

Equity, quality, and cultural relevance are independent dimensions in mathematics education. I refer to this 3-dimensional framework as *culturally-sensitive, equity-in-quality in mathematics education*. In the period 1950–2008 the agendas of equity and quality in education, and of mathematics education have moved in different directions. While the provision for universal primary education was paramount between 1950 and 2000, educational quality received low priority during that period. In the first decade of the 21st century, quality education for all has emerged as a top priority. On the other hand, mathematics education literature shows that the evolution of mathematics education was dominated by quality concerns in scholarly discourse between 1950 and 1980. The social and cultural aspects of mathematics education started to emerge as legitimate research in the 1980s. Towards the end of 1980s, equity became a major concern in mathematics education. The first decade of the 21st century witnessed the beginning of convergence towards an increased emphasis on achieving equal access to quality math education (Jurdak 2009).

In the last half of the past century, the decline of colonization was a major reason for the emergence of the two-tiered system of mathematics education. During colonization, many developing countries adopted the mathematics education of their colonial rulers. However, as colonization dismantled, the developing countries invested most of its resources in increasing coverage at the expense of the quality of education, and educational research and development. Thus developing countries did not accumulate enough 'credentials' in mathematics education to fully participate in the international mathematics education community. This situation led to the formation of a two-tiered system of math education at the global level. The upper tier, referred to as the *optimal mode of development*, includes the developed countries that are integrated in the international mathematics education community. The lower tier, referred to as the *separate mode of development*, consists of the marginalized countries which have yet to be integrated in the international activities of mathematics education.

The majority of countries having average or high quality index (measured in terms of national achievement in TIMSS 2003) and low or average inequity index (measured in terms of size of between-school variation) generally fit the optimal mode of development. These countries have high or average mathematics achievement performance, contribute significantly to international research in mathematics education, and assume leadership roles in international mathematics education organizations and conferences. On the other hand, the majority of countries having low quality index in mathematics education, irrespective of its equity index, fit in the separate mode of development. These countries have low mathematics performance, have little contribution to international research in mathematics education, and normally have humble participation in international mathematics education conferences, such as the ICME. In other words, they are marginalized by the international mathematical education community and left to follow their own path in developing their mathematics education. Some of these countries use the preservation of cultural values as an argument to rationalize the lack of their integration in the international mathematics education community. Other countries do not have the resources to participate and contribute to the international math education community.

A country classified as fitting in the separate mode of development of mathematics education is likely to be relatively poor, low in the spread and level of education among its population, and belongs to a socioeconomically developing region (Arab states, Latin America, and Sub-Saharan Africa). On the other hand, a country classified as following the optimal mode of development of mathematics education is likely to be relatively rich, high in the spread and level of education among its population, and is part of a developed region (North America, Western and Eastern Europe, East Asia and the Pacific). There seems to be a divide between developing and developed countries in mathematics education, and some of the significant factors that contribute to that divide (socioeconomic status of a country, its educational capital, and its culture) seem to be beyond the sphere of influence of local or international mathematics education communities, whereas the other factors are not. For example, policies that govern international organizations and conferences may be addressed by the international mathematics education community.

The international mathematics education community has a responsibility to find ways to encourage and enable mathematics educators to be integrated in the international mathematics education community. The participation in and contribution to international mathematics education conferences and international mathematics education journals are critical for such integration. One measure in this regard would be to favour the participation of mathematics educators from developing countries. Writing and presenting in English is a major barrier to the participation of many mathematics educators in international conferences. Some form of volunteered mentoring by their colleagues who can provide their support in reviewing and editing manuscripts could be a desirable strategy. Providing opportunities for presentations in international conferences in languages other than English would broaden access to such conferences. All these measures may hopefully help enhance the integration of more mathematics educators in the international community.

Danny Martin on Politicizing Socioeconomic Status and Mathematics Achievement

In the United States discussions about the relationships between SES and schooling processes and outcomes—persistence, achievement, success, failure, opportunity to learn, access to resources, and so on—are long and enduring. These discussions have surrounded mathematics education—more so than being generated and sustained by mathematics educators—as much of the research and policy generated to support various positions about socioeconomic status has been produced in fields like sociology, economics, critical studies, and public policy.

In many of these studies there is often a deficit-oriented narrative that is generated and reified about "poor" children and families, while normalizing certain middle- and upper-class children and families. SES is often used as a proxy for "race" but the discussions are often unwilling to explore the impact of racism in generating socioeconomic and achievement differences. The dialectic between race and social class is important. In fact, a number of dialectics are important with respect to SES as one considers its racialized, gendered, and contextual nature. The processes undergirding its formation and strata in a given historical and political context may help to explain outcomes like school achievement in ways that are more insightful than just placing human bodies into various socioeconomic strata and characterizing their achievement in relation to human bodies in other strata.

There have been recent reports that consider race, class, gender, ethnicity, and language proficiency in relation to mathematics education (e.g., Strutchens and Silver 2000; Tate 1997). They support the intuitive finding that higher socioeconomic status is associated with increased course-taking and higher achievement on various measures of mathematics achievement. However, the story is less clear when one considers that many "Asian" students from the lowest socioeconomic levels in the U.S. outscore White and other students at the highest socioeconomic levels. Moreover, many of these reports leave unexplained high achievement among African American, Latino, and Native American students, who are disproportionately represented among the lower socioeconomic levels in the U.S.

I would argue that while SES is positively correlated with achievement, mathematics education research in the U.S. context still has far to go in addressing the complexity of these issues. Tate (1997), for example, noted that in defining and operationalizing socioeconomic status, "Typically the mathematics-achievement literature is organized according to a hierarchy of classes—working class, lower-middle class, middle class, and so on. This hierarchy often objectifies high, middle, and low positions on some metric, such as socioeconomic status (SES)" (p. 663). This objectification presents SES as static and uncontested and not influenced by larger political and ideological forces.

There is complexity that goes unexplored even within the socioeconomic strata that are used. In the U.S. it is generally true that even among poor and working class "Whites" and "Blacks", within-class racism often mitigates the opportunities of Blacks. Across economic strata, the sociology and economics of schooling suggest that "Whites" often enjoy the capital associated with their "Whiteness" even in a supposed meritocracy that many claim and wish for in our society (e.g., Jensen 2006). I would argue that such considerations extend to mathematics education to affect the conditions under which students learn and in which opportunities unfold or are denied.

My particular orientation is to move "race" to a more central position in the conversation on SES within the U.S. context (Martin 2009). It might be argued that "race" is not a central concern in other national and global regional contexts. I would disagree based on the histories of nationalism, colonialism, xenophobia, anti-Muslim sentiments, and anti-multiculturalism throughout Europe, South America, and other locations. Every context, without exception, experiences a historically contingent "racial" ordering of its society that also structures its socioeconomic ordering. Research on the global contexts of racism(s), in all its forms, makes this point clear for the U.S., Europe, Brazil, Asia, and so on. So, while it may not be an issue of "White" and "Black" in a particular location, there are likely to be some other forms of "race" and "racism" that are at play (including differences that result from "lighter" and "darker" skin), whether they be manifested in the lives of Indians living in Singapore, the ideologies of the Danish People's Party (DF) in Denmark, or the rise of xenophobic nationalism throughout Europe.

We know that SES does not explain all of the variation in achievement and does not explain why some "poor" or low SES children in a given context succeed academically and why some "rich" or high SES children do less well. Analyses of SES often treat it as a static variable and often do not examine human agency or the manipulation of SES by those in power. SES is intimately linked to other variables that may impact schooling processes and achievement. These other variables include gender; geographic location; language status; immigrant status and the prevailing racial context in given society including nationalism, anti-immigrant sentiment, xenophobia; quality of health care and pre-school systems; history of colonialism; the prevailing political context and ideologies that dominate that context; larger economic system; and so on.

I argue for a more politicized view of SES that takes into account race and racism, political projects, socioeconomic projects and manipulation, among others. SES may be conceptualized differently in different contexts. The common reporting line "the more economic resources one has, the greater their achievement is likely to be" is not an interesting finding even if it gets repeated in research. It does not explain why some have more resources than others. We, in mathematics education, should continue to trouble that imbalance.

Tamsin Meaney on Back to the Future? Mathematics Education, Early Childhood Centres[1] and Children from Low Socio-Economic Backgrounds

In the last two decades, early childhood has become the focus for much discussion in regard to overcoming inequalities in educational outcomes between groups. Although there is a perception that such a connection has only been newly recognised, the history of early childhood centres shows otherwise. For example, May (2001) outlined how preschools in New Zealand have changed dramatically from being charitable organisations for the urban poor in the late nineteenth century to now being seen as essential for all children, to the extent that children who do not attend are perceived as likely to be problems for society. The right to determine the appropriate care for young children through education arose during the history of early childhood centres.

---

[1] Throughout history and across the world, different names have been given to institutions set up outside of homes for the care and education of young children. To overcome this confusion, the term early childhood centres has been adopted.

> An activity such as preschool, like most of the welfare institutions, is marked by its history. There is a clear relationship between a country's traditions in preschool and school system and its administration and integration of new challenges and demands. (Broman 2010, p. 34; own translation)

I suggest that the history of early childhood centres as carers and educators of poor children has produced different sorts of mathematical education programmes. The physical care of young children, who are seen as unable to look after themselves, always has been part of the role of early childhood centres. As well, characteristics of the child, from their character to their imagination, have been perceived as being in need of moral care. Education, including mathematics education programmes, reflected these different perceptions of moral care. Many instigators of early childhood centres have considered that education could overcome faults in children, particularly poor children. Table 1 provides a summary of the main early childhood centres for the last two hundred years and the sorts of moral care and education provided to children.

In recent years, a moral deficiency that early childhood centres are supposed to overcome is a lack of school readiness in regards to mathematics knowledge. An analysis by Greg Duncan and colleagues of six longitudinal studies suggested that early mathematics knowledge is the most powerful predictor of later learning, including the learning of reading (Duncan et al. 2007). The mathematical programmes, now being advocated in early childhood centres, reflect society's wish to

Table 1 Summary of the kind of care and education provided in early childhood centres

	Time	Care	Education	Mathematics
Robert Owen—Infants School	Early 19th century	Care of the character	Broad curriculum	Arithmetic from manipulating objects from nature
Frederick Frobel—Kindergarten	1837 to end of 19th century	Spiritual care could only occur in schools	Playful and based on children's own interests	Geometry and other math learnt through engagement with gifts and occupations
Margaret McMillan—Nursery Schools	Early 20th century	Care of the imagination	Physical and mental development through play	Math learning was incidental to using their imagination to explore the world
Maria Montessori—Children's houses	Early 20th century	Care for children's personalities	Learning though the senses, using children's interests. School preparation	Materials were math in they required comparisons
Diversity of approaches	Middle to late 20th century	Care for psychological well-being	Learning to play with other children	Experiences were valuable for later school math learning
Present day	1990s to present	Care for academic well-being	Content becomes the focus of education	Math concepts have become the focus of preschool programs

care for poor children's academic needs, which are considered to be at risk and which could result in them being non-productive workers in the future (Pence and Hix-Small 2009). If all children could receive a quality early childhood education then the risk of society having citizens with insufficient education and unable to gain jobs would be alleviated.

A consequence of the acceptance of early childhood centres' right to determine the education necessary to appropriately care for young children is leading to the imposition of a homogenised view of young children, including as young mathematics learners. Providing mathematics programs for this homogenised child can result in a lack of recognition and undervaluing of what poor children bring to early childhood centres. Although the jury is still out on the long-term effectiveness of present structured mathematics programmes, an education that does not recognise nor value children's transition back into their home communities (Meaney and Lange 2013) will result in some children becoming failures before they begin school.

Miriam Penteado on Mathematics Education and Possibilities for the Future

The Brazilian educational system is organized as shown in Table 2 below. For both basic school and the higher education there are two parallel systems: the private and the public. Concerning basic schools, in general, private schools have more status and offer better learning and teaching conditions for students. On the other hand, public schools include the majority of the Brazilian population. The teaching and learning conditions in public schools is very poor. Many schools are in bad structural condition and there are cases of no electricity, no potable water, etc. It is known that Brazilian public schools students study less content than those in private schools. Furthermore, in Brazil there is lack of teachers. It is difficult to find people who want to be educated as a teacher, and there is a set of reasons for this: low salary, low social status, and violence. The best teacher students who graduate are hired in private schools with better working conditions than in public schools.

Concerning higher education the situation is the opposite of what happens in basic schools. *Public universities* are those with the highest investment in research and teaching. In fact, in the last years part of the policy of the Brazilian government has been to increase the investment in higher education making available to the system a considerable amount of resources. It is more difficult to gain enrolment as undergraduate student in public universities than in private, especially in more

Table 2 The Brazilian educational system

Basic school	Primary and secondary level (9 years—from 6 to 14 years old)
	High school level (3 years—from 15 to 18 years old)
Higher education	Different length

prestigious courses such as medicine or engineer. For this reason, those who attend private schools are more likely to become a student at a public university. Many students from public school do not even dream of having further education at a public university. The choice (when it is the case) is to work during the day and take a course in the evening at a private faculty.

Considering the situation it is possible to state that a person with high socio-economic background follows the route: from private school to public university. One with low socioeconomic background follows the route: from public school to private faculty. There is financial governmental support for students from public schools to study in private faculties. Only a small percentage of the Brazilian population has further education at the tertiary level (private or public system). According to the OECD[2] the number of Brazilian people within 25–64 years old who has completed tertiary education has increased to 11 %. However it is still low when compared with other countries.

Public universities are trying to facilitate access for students from low SES, however it is not for any career. As an example, one can use a socio economic report of a public university in Sao Paulo State for the year 2010. The distribution of students in relation to their background (basic school in the private or the public system) in university courses such as medicine and mathematics is very different. While students who enter medicine have studied in private institutions (85.9 % of students have attended a private primary school and 94.6 % have attended a private high school), the majority of mathematics students (future mathematics teachers) have studied primary and high school in public institutions (an average of 72.5 % for public primary schools, and 74.6 % for public high schools). Thus one can see that medicine does not function as any social-ladder, while mathematics has the possibility to do so.

That socioeconomic factors influence students' educational life is common sense. Given this, one could think that there is not so much to say about the survey theme. However, this common sense could be challenged. When working with students in so-called disadvantaged context one can consider the question: What possibilities could be constructed together with the students?

It is important for a mathematics education to create new possibilities for students. Creating possibilities for students could mean thinking of the opportunities they might obtain for the future. One could think as students' possibilities for, later on in life, to participate as (critical) citizens in political issues. To consider the conditions for coming to "read and write" the world, to use an expression formulated by Paulo Freire (1972).

There might exist a tendency to consider low achievement related to the students and to their background. And from this perspective one can start discussing strategies for compensating the, say "low cultural capital". One can pay attention to the general living conditions of the students, including their conditions of getting to school. One can consider their learning with reference to their worlds and their foregrounds.

---

[2] http://www.oecd-ilibrary.org/economics/country-statistical-profile-brazil_csp-bra-table-en.

One can claim that it is an important aim for mathematics education to help to establish possibilities within the horizon of students' foregrounds (Skovsmose 2005). To make them recognise that: This could also be for me!

Open Access This chapter is distributed under the terms of the Creative Commons Attribution Noncommercial License, which permits any noncommercial use, distribution, and reproduction in any medium, provided the original author(s) and source are credited.

References

Barwell, R., Barton, B., & Setati, M. (2007). Multilingual issues in mathematics education: introduction. *Educational Studies in Mathematics, 64*(2), 113-119.

Broman, I. T. (2010). Svensk förskola - ett kvalitetsbegrepp. (Swedish preschools – a quality concept.) In B. Ridderspore & S. Persson (Eds.), *Utbildningsvetenskap för förskolan* (Teacher development for preschools) (pp. 21-38). Stockholm: Natur & Kultur.

Coleman, J. S., Campbell, E. Q., Hobson, C. J., McPartland, F., Mood, A. M., Weinfeld, F. D., et al. (1966). *Equality of educational opportunity.* Washington: U.S. Government.

Damerow, P., Dunkley, M., Nebres, B., & Werry, B. (Eds.). (1984). *Mathematics for all.* Paris: UNESCO.

Duncan, G. J., Dowsett, C. J., Claessens, A., Magnuson, K., Huston, A. C., Klebanow, P. et al. (2007). School readiness and later achievement. *Developmental Psychology, 43*(6), 1428-1446.

Fleisch, B. (2008). *Primary Education in Crisis: Why South African schoolchildren under achieve in reading and mathematics.* Cape Town: Juta & Co.

Freire, P. (1972). *Pedagogy of the Oppressed.* Harmondsworth: Penguin Books.

Graven, M. (2002) Coping with new mathematics teacher roles in a contradictory context of curriculum change. *The Mathematics Educator.* Vol 12, no. 2, pp 21-28.

Hoadley, U. (2007). The reproduction of social class inequalities through mathematics pedagogies in South African primary schools. *Journal of Curriculum Studies, 39*(6), 679-706.

Hu, Y. & Du, Y. (2009). Empirical Research on the Educational Production Function of Rural Primary Schools in Western China. *Educational Research, 2009*(7), 58-67.

Jensen, R. (2006). *The heart of Whiteness: Confronting race, racism, and White privilege.* San Francisco: City Lights.

Jurdak, M. E. (2009). *Toward equity in quality in mathematics education.* New York: Springer.

Kang, N., & Hong, M. (2008). Achieving excellence in teacher workforce and equity in learning opportunities in South Korea. *Educational Researcher, 37*(4), 200–207.

Lin, C.-J., Hung, P.-H., Lin, S.-W., Lin, B.-H., & Lin, F.-L. (2009). The power of learning goal orientation in predicting student mathematics achievement. *International Journal of Science and Mathematics Education, 7*(3), 551-573.

Martin, D. (2009). Researching race in mathematics education. *Teachers College Record, 111*(2), 295-338.

May, H. (2001). *Early Childhood Care and Education in Aotearoa – New Zealand: An Overview of history, policy and curriculum.* Keynote address to Australian Education Union, Early Childhood Roundtable, 25 October 2001.

Meaney, T. & Lange, T. (2013). Learners in transition between contexts. In K. Clements, A. J. Bishop, C. Keitel, J. Kilpatrick, & F. Leung (Eds.), *Third international handbook of mathematics education* (pp. 169-201). Springer: New York.

Meyer, J. W., Ramirez, F. O., & Soysal, Y. N. (1992). World expansion of mass education, 1870-1980. *Sociology of Education, 65*(2), 128-149.

Muller, J., & Taylor, N. (2000). Schooling and everyday life. *Reclaiming Knowledge: Social Theory, Curriculum and Education Policy* (pp. 57-74). London, New York: Routledge Falmer.

Pence, A. & Hix-Small, H. (2009). Global children in the shadow of the global child. *International Critical Childhood Studies*, 2(1). Available from: http://journals.sfu.ca/iccps/index.php/childhoods/article/viewFile/11/15.

Popkewitz, T. S. (2008). *Cosmopolitanism and the age of school reform: Science, education, and making society by making the child*. New York: Routledge.

Skovsmose, O. (2005). Foregrounds and politics of learning obstacles. *For the Learning of Mathematics, 25*(1), 4-10.

Strutchens, M. E., & Silver, E. A. (2000). NAEP findings regarding race/ethnicity: Students' performance, school experiences, and attitudes and beliefs. In E. A. Silver & P. A. Kenney (Eds.), *Results from the 7th mathematics assessment of the National Assessment of Educational Progress* (pp. 45–72). Reston, VA: NCTM.

Tate, W. F. (1997). Race–ethnicity, SES, gender, and language proficiency trends in mathematics achievement: An update. *Journal for Research in Mathematics, 28*, 652–679.

Tröhler, D. (2011). *Languages of education: Protestant legacies, national identities, and global aspirations*. New York: Routledge.

World Bank. (2009). *Secondary education in India: Universalizing opportunity*. Washington: The World Bank.

Part VI
National Presentations

National Presentation of Korea

Sun-Hwa Park

Introduction

Korea, as represented by the Korea Institute for Curriculum and Evaluation (KICE), is pleased to present a National Presentation at ICME-12. We think this is a precious opportunity where we can introduce the mathematics education in Korea. Following an overview of the mathematics education in Korea, we plan to present the policies and many efforts we devote to improve our mathematics education.

Presentations (Two Sessions)

Session I (80-min Session)

1. Overview of Mathematics education in Korea
2. The National Mathematics Curriculum of Korea
3. The Development and Characteristics of Korean Mathematics Textbooks
4. Teaching and Learning practices of mathematics classroom in Korea
5. The educational practices for the Mathematically-gifted and the underachieving students

S.-H. Park (✉)
Korea Institute for Curriculum and Evaluation (KICE), Seoul, Korea
e-mail: shpark@kice.re.kr

© The Author(s) 2015
S.J. Cho (ed.), *The Proceedings of the 12th International Congress on Mathematical Education*, DOI 10.1007/978-3-319-12688-3_20

Session II (70 min Session)

Overview of Mathematics Education in Korea

Here we provide an overview of the mathematics education in Korea that consists of five areas: mathematics curriculum, textbooks, teaching and learning, educational evaluation, and teachers' education. First, we introduce the national curriculum of mathematics and the revision that has been made throughout the history of curriculum. Second, we introduce the mathematics textbooks that are used at each school level. Third, various types of elementary and secondary classroom mathematics teaching will be reviewed. Fourth, we discuss the three kinds of mathematics assessment: National Assessment of Educational Achievement (NAEA), College Scholastic Aptitude Test (CSAT), and international student assessments. Korea has been one of the top performing countries in students' mathematics achievement in the international assessment such as PISA and TIMSS. Behind its shining accomplishments, the mathematics education in Korea also has some noticeable afflictions which remain for us to resolve. Here we attempt to obtain constructive advice from other countries on our difficulties and problems both in theoretical and in practical perspectives.

The National Mathematics Curriculum of Korea

Korea has a national curriculum system. After the curriculum is announced at the national level, the writers of textbooks begin to develop textbooks on the basis of the national curriculum. Once the development of textbooks is completed, new curriculum is implemented in the school. It takes about 2 years to apply the newly announced national curriculum to the school. The new curriculum is sequentially applied from the 1st grade.

It was 1954 that the mathematics curriculum at the national level was first announced in Korea. Since then, the national curriculum has been revised 8 times. The 7th curriculum was announced by public notification in 1997. Then in 2007 and 2009 revisions were made to respond to rapidly changing external environment in recent decades in Korea.

The latest mathematics curriculum emphasizes mathematical creativity designed to equip students with capacities on basic learning ability, divergent thinking ability, problem-solving ability, originality, and ability to create new values. It also reinforces mathematical process including problem-solving ability, reasoning ability, and communication ability. To facilitate creative class activities, it was organized to reduce more than 20 % of the existing mathematics contents, and to apply 'the grade cluster system' to enhance connection and cooperation between grades.

The Development and Characteristics of Korean Mathematics Textbooks

Mathematics textbooks in the elementary, middle and high schools are being developed based on the curriculum. Some materials suggested here are still under development, which will give us a general picture about the mathematics textbooks in Korea. In Korea, three different kinds of textbooks are used in schools: (1) Government-copyrighted textbooks, (2) government-authorized textbooks, and (3) government-approved textbooks.

Elementary school mathematics textbooks are government-copyrighted textbooks, which are developed by the institute commissioned by the government. Secondary school mathematics textbooks are government-authorized textbooks and government-approved textbooks, which are published through the authorization procedure to guarantee high quality textbooks. In addition to textbooks, student workbooks and teacher guidebooks are developed. Student workbooks are to help students' self studies and teacher guidebooks are to help teachers apply various teaching methods and guarantee quality teaching.

Teaching and Learning Practices of Mathematics Classroom in Korea

Here we investigate the teaching and learning practices that are implemented in the mathematics classrooms in Korea. Various types of teaching practices in the elementary and secondary classrooms will be reviewed. At the elementary level, we investigate the factors of similarities and differences of the teaching and learning across mathematics classrooms. Also, we explore the general characteristics of mathematics classrooms such as activity-based lessons, emphasis on cooperative learning and communication. At the secondary level, we find the characteristics of mathematics classrooms such as differentiated lessons, subject-based classroom system, preparation for university entrance exam, and the cases of teaching practices. We provide some example cases for a better understanding of the teaching and learning practices at each level.

The Educational Practices for the Mathematically-Gifted and the Underachieving Students

In this section, we investigate mathematics education for the gifted and the underachieving students in Korea. We will describe mathematics educational systems, contents, and plans of actions focused on these particular students. First, for the mathematically-gifted students, we will introduce the development of various policies and explain three types of institutions for the gifted: schools, education centers, and classes. Additionally we introduce several gifted education programs that are implemented at the elementary and secondary level schools. Second, for the mathematically underachieving students, we will explain various new policies implemented after 2009 where the government, local offices of education, and schools actively participating in supporting underachieving students. Additionally, we will introduce the institutions and programs for the underachieving and the Internet website, called Ku-Cu (www.basics.re.kr), which is operated by KICE to support education for the children with underachievement.

National Assessment of Educational Achievement

The major aims of the National Assessment of Educational Achievement (NAEA), targeted for all schools in Korea, are to (1) acquire information and implications on directions to improve curriculum, teaching, and learning methods, (2) review the educational quality, (3) diagnose and remedy each student's performance level, and (4) examine educational accountability of school education. The NAEA is implemented targeting sixth-grade elementary school students, third-grade (9th) middle school students, and second-grade (11th) high school students across the nation. Test items of the NAEA are developed in contents and behaviour areas based on the national curriculum. Here we describe in depth the NAEA such as the structure, testing time, development of assessment tool, domains of assessment, and the scoring and reporting results. We also discuss the recent trend of the assessment results.

College Scholastic Ability Test (CSAT) in Korea

In this section, we discuss the College Scholastic Ability Test (CSAT). The CSAT has been implemented since 1994, and it was adapted with the changes of the national curriculum and college recruitment systems. The current CSAT for mathematics consists of Mathematics 'GA' (Korean) type and Mathematics 'NA' (Korean) type. Students who will major in mathematics and natural science at college should take Mathematics 'GA' (Korean) type and other students should take Mathematics 'NA' (Korean) type. Test items are developed to examine students' competencies on calculation, mathematical understanding, reasoning, and problem-solving.

Starting from the 2014 school year, the CSAT will be improved to reflect the aims of the 2009 revised curriculum and reduce the importance of the CSAT in the college admission to enhance the autonomy of each college. The title, Mathematics 'GA' (Korean) type and 'NA' (Korean) type, will be changed to Mathematics A type and B type.

Achievement on International Assessment in Korea

In the international student assessment part, we will analyze the characteristics of Korean students' mathematics achievements revealed in two representative international assessments: PISA and TIMSS. Korea has continuously been ranked among the top performing countries in PISA and TIMSS, which has been the result of more students with a high level of proficiency and less students with the lowest levels of proficiency compared with other countries. More than 2/3 of Korean students have performed at the excellent level, and 98 % of them have performed above the basic level. The proportions of Korean students with the highest level of proficiency in PISA and TIMSS, however, have been decreased, which requires policy measures to deal with the situation. We also find further implications from the test results.

Mathematics Assessment at School Level

The evaluation at the school level is administered according to the curriculum. Mathematics assessment at the school level is distinguished by the student's grade. We start with introducing the principles of assessment, the types of assessments and schedule, and the assessment methods. Usually, there are diagnostic assessments at the beginning of school year, scheduled examinations such as mid-terms and final exams at the single school level and performance assessment and quizzes at the class level. Even though the national curriculum strongly recommends various assessment methods, selection type focusing on multiple choice items and constructed-response type problems focusing on short-answer types are in the majority. However, constructed-response items that require the students to create their own answer have also been treated in fair proportion and are applied to not only scheduled examinations, but also performance assessments and diagnostic assessments. We further provide information about the analysing, reporting, and application of the assessment results.

Mathematics Teacher Education in Korea

Here we will discuss about pre-service teacher education, the teacher employment test and professional development of teachers. First, we will review the curriculum of various teacher education programs for the elementary and secondary level prospective teachers in Korea, which features a strong zeal for education. We will also examine the teacher employment test including the procedure, structure, and test areas for the elementary and secondary level.

In addition, we will discuss various teacher professional development programs which are implemented by the 16 metropolitan/local education offices. Typical professional development programs include the pre-employment training program, the 'first-level teacher' training program for teachers with more than 3 years of teaching experiences, and various in-service training programs. We will also explain the master teacher system, teaching consulting programs, and the classroom assessment system, which are designed to develop teachers' professionalism.

Research Members

Sun Hwa Park	shpark@kice.re.kr
Seung Hyun Choe	jhtina@kice.re.kr
Jeom Rae Kwon	kwonjr@kice.re.kr
Yun Dong Jo	jydong05@kice.re.kr
Hee-Hyun Byun	bhhmath@kice.re.kr
Kwang Sang Lee	leeks@kice.re.kr
Seongmin Cho	csminy@kice.re.kr
Young Ju Jeon	whaljuro@kice.re.kr
Haemee Rim	rimhm@kice.re.kr
Jiseon Choi	jschoi@kice.re.kr
Jae Hong Kim	masshong@kice.re.kr

Open Access This chapter is distributed under the terms of the Creative Commons Attribution Noncommercial License, which permits any noncommercial use, distribution, and reproduction in any medium, provided the original author(s) and source are credited.

Mathematics Education in Singapore

Berinderjeet Kaur, Cheow Kian Soh, Khoon Yoong Wong, Eng Guan Tay, Tin Lam Toh, Ngan Hoe Lee, Swee Fong Ng, Jaguthsing Dindyal, Yeen Peng Yen, Mei Yoke Loh, Hwee Chiat June Tan and Lay Chin Tan

Abstract Mathematics education in Singapore is a shared responsibility of the Ministry of Education (MOE) and the National Institute of Education (NIE). The MOE overseas the intended, implemented and attained curriculum in all schools while the NIE is involved in teacher preparation and development and also research in mathematics education. Therefore this report has two sections respectively, the first describes the education system and school mathematics curricula while the second briefly provides relevant information on teacher preparation and development and mathematics education research in Singapore.

Keywords Singapore · Mathematics education · Curriculum · Teacher education · Research

Introduction

Mathematics education in Singapore is a shared responsibility of the Ministry of Education (MOE) and the National Institute of Education (NIE). MOE develops the national mathematics curriculum and oversees its implementation in all schools, while the NIE is involved in teacher preparation and development and also research in mathematics education. This report comprises two sections: the first describes the education system and school mathematics curricula while the second provides relevant information on teacher preparation and development and mathematics education research in Singapore.

B. Kaur (✉) · K.Y. Wong · E.G. Tay · T.L. Toh · N.H. Lee · S.F. Ng · J. Dindyal
National Institute of Education, Nanyang Technological University, Singapore, Singapore
e-mail: berinderjeet.kaur@nie.edu.sg

C.K. Soh (✉) · Y.P. Yen · M.Y. Loh · H.C.J. Tan · L.C. Tan
Ministry of Education, Singapore, Singapore
e-mail: SOH_Cheow_Kian@moe.edu.sg

© The Author(s) 2015
S.J. Cho (ed.), *The Proceedings of the 12th International Congress on Mathematical Education*, DOI 10.1007/978-3-319-12688-3_21

The Education System and School Mathematics Curricula

Education in Singapore has evolved through a continual process of change, improvement and refinement since the country gained independence in 1965. Today, all children receive at least 10 years of general education in over 350 primary, secondary and post-secondary schools. There are diverse pathways and opportunities for students to discover their talents, realize their potential, and develop a passion for life-long learning. Singapore's education system largely follows a 6-4-2 structure, with 6 years of primary (Grade 1–6), 4 years of secondary (Grade 7–10) and 2 years of pre-university (Grade 11–12) education (MOE 2012).

Mathematics is a compulsory subject from Primary 1 up to the end of secondary education. In the early grades, about 20 % of the school curriculum time is devoted to mathematics so that students build a strong foundation to support further learning in later years. The mathematics curriculum is centrally planned by MOE. However, flexibility is given to schools to implement the curriculum to best meet the abilities and interests of students. The mathematics curriculum is reviewed every 6 years with consultation of key stakeholders and partners to ensure that it meets the needs of the nation.

The mathematics curriculum aims to enable students to acquire and apply mathematical concepts and skills; develop cognitive and metacognitive skills through a mathematical approach to problem solving; and develop positive attitudes towards mathematics. A single mathematics curriculum framework (MOE 2007) unifies the focus of the mathematics curriculum for all levels from primary to pre-university. The focus is on developing students' mathematical problem-solving abilities through five integral components namely, concepts, skills, processes, attitudes, and metacognition.

A spiral approach is used in the design of the mathematics syllabuses from primary to pre-university. At every level, the syllabuses comprise a few content strands (e.g. number and algebra, geometry and measurement, statistics and probability), facilitating connections and inter-relationships across strands. The content in each strand is revisited and taught with increasing depth across levels. There is differentiation in the content, pace and focus among syllabuses within the same levels to cater to different student profiles.

Primary 1–4 students follow a common mathematics syllabus, covering the use of numbers in measurements, understanding of shapes and simple data analysis. At Primary 5–6, there are two syllabuses: the Standard Mathematics syllabus builds on the concepts and skills studied in Primary 1–4, whereas the Foundation Mathematics syllabus revisits some of the important concepts and skills taught earlier. At the secondary level, there are 5 different syllabuses for students in the Express, Normal (Academic) and Normal (Technical) courses. These syllabuses include concepts and skills in number and algebra, measurement and geometry, and statistics and probability. Calculus and trigonometry are covered in the additional mathematics syllabuses for Secondary 3–4 students who are more mathematically-inclined. At the pre-university level, mathematics is an optional subject. Three syllabuses

(H1, H2 and H3) are available to prepare students for different university courses and the use of graphing calculators is expected.

There are also programmes to support the slow progress students and stretch those talented in mathematics. Primary 1 students (about 5 %) who lack age-appropriate numeracy skills are given support through the Learning Support Programme for Mathematics where they are taught in small groups by specially-trained teachers. For gifted learners, there is an enriched mathematics curriculum that emphasizes problem solving, investigations, making conjectures, proofs and connections among concepts. The NUS High School of Mathematics and Science also offers mathematically talented students a broad-based 6-year programme that includes undergraduate level topics and a mathematics research component.

For the teaching of mathematics at the primary levels the Concrete-Pictorial-Abstract (C-P-A) approach, introduced in 1980, is prevalent. Since 1990s, it has been used together with activity-based learning to encourage active participation by students in the learning process. In the early 1980s, MOE also developed the model method for solving word problems at the primary level (MOE 2009). This method provides a visual tool for students to process and analyse information and develop a sequence of logical steps to solve word problems. The model method is also used with algebra to help students formulate algebraic equations to solve problems in lower secondary mathematics. This facilitates the transition from a dominantly arithmetic approach at the primary level to an algebraic one at the secondary level. At the secondary and pre-university levels, teacher-directed inquiry and direct instruction are common. These approaches are used with other activities and group work to engage students in learning mathematics.

Resources are critical to curriculum implementation and effective delivery of mathematics lessons. Textbooks are essential materials to help teachers understand the emphases and scope of the syllabuses, and for students to learn independently. In the late 1990s, MOE devolved textbook writing to commercial publishers to allow for a greater variety of textbooks. Quality is assured through a rigorous textbook authorization and approval process by MOE. Besides textbooks, MOE also produces additional materials to support teachers especially at the primary levels.

Teacher Preparation and Development, and Research in Mathematics Education

The NIE and Teacher Education

The National Institute of Education (NIE) is an autonomous institute within the Nanyang Technological University and sole teacher education institution in Singapore. It offers both pre-service and in-service education programmes ranging from diploma to doctorate levels. Its present model of Teacher Education for the 21st century (TE[21]) is unique and has six foci intended to enhance the key elements of teacher education. The foci are the Values[3], Skills and Knowledge (V^3SK) model, the Graduand Teacher Competencies (GTC) framework, strengthening the theory-practice nexus, an extended pedagogical repertoire, an assessment framework for 21st century teaching and learning, and enhanced pathways for teacher professional development (NIE 2009). In particular the V^3SK model explicates three dimensions of values for the teacher, viz. learner-centredness, teacher identity and service to the profession and community, without which the beginning teacher may easily lose her focus in an increasingly technological and knowledge-driven world. The GTC framework makes clear the competencies to which the student teacher should aspire to attain or be aware of in his studies at NIE. This is a distinct attempt to state what must be achieved in one's pre-service teacher education and also what should be reasonably accomplished only after some years of experience as a teacher.

Pre-service Education of Mathematics Teachers

Pre-service education provides the crucial initial training that can have long-term impacts on the quality of future teachers in an education system. Besides education courses and the practicum, trainee teachers take mathematics-related courses called Curriculum Studies (methodology), Subject Knowledge (deeper understanding of school mathematics), and Academic Studies (tertiary mathematics). These courses are taught by mathematicians, mathematics educators, and "mathematician educators" (those with expertise in both areas) from the same Mathematics and Mathematics Education Academic Group. These courses stress the rigour of mathematics contents and relevance to local school contexts and school mathematics, in particular, the model method used in problem solving. Locally developed resources (Lee and Lee 2009a, b) used in these courses combine local experience and research with international "best practices". Blended learning is used in teacher education courses in response to the significant roles of ICT in instruction as well as the changing characteristics of the trainee teachers. Findings from IEA's Teacher Education and Development Study in Mathematics (TEDS-M) (Tatto et al. 2012) show that NIE trainee teachers scored above international average in mathematics content and pedagogical content knowledge, and most of them expressed strong commitment to the teaching profession as their life-long career.

Professional Development of Mathematics Teachers

Since 1998 all teachers in Singapore are entitled to 100 h of training and core-upgrading courses each year to keep abreast with current knowledge and skills. The Professional Development (PD) is funded by the MOE. Teachers have different pathways to upgrade their knowledge and skills through the Professional Development Continuum Models (PCDM) of the MOE. The MOE works closely with NIE to design courses for practicing teachers. Numerous academic courses offered by NIE lead to postgraduate degrees. For example, in order to upgrade mathematics teachers' content knowledge, a unique master degree programme MSc (Mathematics for Educators) is offered by NIE. The mathematics chapter of the Academy of Singapore Teachers (AST), the Association of Mathematics Educators (AME) and the Singapore Mathematical Society (SMS) are also actively engaged in the PD of teachers. They hold relevant annual meetings, seminars and conferences for teachers. Teachers may also attend international conferences or study trips to widen their perspectives on mathematics education. Lastly, teachers are also engaging in professional learning and development by participating in research projects at the school level. Examples of two such projects are the Enhancing the pedagogy of mathematics teacher (EPMT) project (Kaur 2011) and the Think-Things-Through (T3) project (Yeap and Ho 2009).

Mathematics Education Research in Singapore

Research is undertaken by graduate students and university scholars. Since 2002, the MOE through the Office of Education Research (OER) at NIE has funded research to inform policy and practice so as to improve education in Singapore. Some of the projects in mathematics education that have been funded and completed are as follows: An exploratory study of low attainers in primary mathematics (Kaur and Ghani 2012); The Singapore mathematics assessment and pedagogy project (Wong et al. 2012); Individual differences in mathematical performance: social-cognitive and neuropsychological correlates (Lee and Ng 2011); Mathematical problem solving for everyone (Toh et al. 2011), Student perspective on effective mathematics pedagogy (Kaur 2009), and Teaching and learning mathematical word problems: A comparison of the model and symbolic methods (Lee et al. 2011). These projects were carried out by university scholars in collaboration with students, teachers and research staff at NIE. Research studies undertaken by graduate students almost always culminate in dissertations, thesis or academic reports, all of which are available at the NIE library repository.

Open Access This chapter is distributed under the terms of the Creative Commons Attribution Noncommercial License, which permits any noncommercial use, distribution, and reproduction in any medium, provided the original author(s) and source are credited.

References

Kaur, B. (2009). Characteristics of good mathematics teaching in Singapore grade eight classrooms—A juxtaposition of teachers' practice and students' perception. *ZDM—The International Journal on Mathematics Education, 41*(3), 333-347.
Kaur, B. (2011). Enhancing the pedagogy of mathematics teachers (EPMT) project: A hybrid model of professional development. *ZDM —The International Journal on Mathematics Education, 43*(7), 791-803.

Kaur, B. & Ghani, M. (Eds.). (2012). *Low attainers in primary mathematics.* Singapore: World Scientific.

Lee, K. & Ng, S.F. (2011). Neuroscience and the teaching of mathematics. *Educational Philosophy and Theory, 43*(1), 81-86.

Lee, K., Ng, S.F., Bull, R., Pe, M.L. & Ho, R.M.H. (2011). Are patterns important? An investigation of the relationships between proficiencies in patterns, computation, executive functioning, and algebraic word problems. *Journal of Educational Psychology, 103*(2), 269-281.

Lee, P. Y., & Lee, N. H. (Eds.). (2009a). *Teaching primary school mathematics: A resource book* (2nd ed.). Singapore: McGraw-Hill Education (Asia).

Lee, P. Y., & Lee, N. H. (Eds.). (2009b). *Teaching secondary school mathematics: A resource book* (2nd & updated ed.). Singapore: McGraw-Hill Education (Asia).

Ministry of Education (MOE, 2007). *Primary mathematics syllabus.* Singapore: Author

Ministry of Education (MOE, 2012). *The Singapore education landscape.* Retrieved December 3, 2012, from http://www.moe.gov.sg/education/landscape/

Ministry of Education (MOE, 2009). *The Singapore model method for learning mathematics.* Singapore: Marshall Cavendish Education

NIE (National Institute of Education) (2009). *TE21: A teacher education model for the 21st century.* Singapore: Author.

Tatto, M.T., Schwille, J., Senk, S.L., Ingvarson, L., Rowley, G., Peck, R., Bankov, K., Rodriguez, M., & Reckase, M. (2012). *Policy, practice, and readiness to teach primary and secondary mathematics in 17 countries: Findings from the IEA Teacher Education and Development Study in Mathematics (TEDS-M).* Amsterdam, The Netherlands: International Association for the Evaluation of Educational Achievement (IEA).

Toh, T.L., Quek, K.S., Leong, Y.H., Dindyal, J. & Tay, E.G. (2011). *Making mathematics practical: An approach to problem solving.* Singapore: World Scientific.

Wong, K.Y., Oh, K.S., Ng, Q.T.Y., & Cheong, J.S.K. (2012). Linking IT-based semi-automatic marking of student mathematics responses and meaningful feedback to pedagogical objectives. *Teaching Mathematics Applications, 31*(1), 57-63.

Yeap, B.H. & Ho, S.Y. (2009). Teacher change in an informal professional development programme: The 4-I model. In K.Y. Wong, P.Y. Lee, Kaur, B., P.Y. Foong, & S.F. Ng (Eds.) *Mathematics education: The Singapore journey* (pp. 130 – 149). Singapore: World Scientific.

National Presentation of the United States of America

Rick Scott

The United States of America was honored to be invited to make one of the National Presentations at the 12th International Congress on Mathematical Education in Seoul, Korea. The United States National Commission on Mathematics Instruction (USMC/MI) oversaw the U.S. participation. (The USNC/MI advises the National Academy of Sciences-National Research Council (NAS-NRC) in all matters pertaining to the International Commission on Mathematical Instruction.) Significant financial support was supplied by the National Science Foundation (NSF) and important logistical support was provided by the National Council of Teachers of Mathematics (NCTM). The main activities of the U.S.A. National Presentation were the

- National Presentation sessions,
- U.S.A. Exhibit,
- *Capsule Summary Fact Book*, and
- U.S.A. Reception.

The National Presentation Sessions

The National Presentation highlighted the uniqueness and important features of mathematics education in the U.S. in two 90-min sessions with a total of five presentations. The first three presentations provided *An Overview of Math Ed in the*

R. Scott (✉)
U.S. National Commission on Mathematics Instruction, Washington DC, USA
e-mail: pscott@nmsu.edu

© The Author(s) 2015   317
S.J. Cho (ed.), *The Proceedings of the 12th International Congress
on Mathematical Education*, DOI 10.1007/978-3-319-12688-3_22

U.S.: Curriculum Reform. The last two focused on *Teaching Mathematics in the United States.*

1. *Mathematics Education in the United States 2012*—Katherine Halvorsen, Smith College
   This presentation provided an overview on the system of education in the U.S., including the size of the educational enterprise, governance, intended curriculum, implemented curriculum, attained curriculum, the Common Core Standards in Mathematics (CCSSM), programs for special populations, and teacher education and professional development. More details can be found below in the section entitled *A Capsule Summary Fact Book*.

2. *Evolution and Revolution: From the NCTM Standards to the Common Core State Standards in the U.S.*—Michael Shaughnessy, Immediate Past President of the National Council of Teachers of Mathematics (NCTM)
   This short history on mathematics standards in the U.S. pointed to the main NCTM documents and the major themes in those documents (1989 and 2000 NCTM Standards, Curriculum Focal Points, and the high school Reasoning and Sense Making Initiative). The case was made that the CCSSM represents an evolutionary step, anchored in NCTM's Process Standards and the *Adding It Up* proficiencies, drew on Curriculum Focal Points series, and is closely tied to NCTM's Reasoning and Sense Making effort. The revolutionary step is that the CCSSM is *common*. The history of prior adoption and implementation mathematics standards in the US has been on a state by state basis, with local control and local decision making. For states to adopt a set of Common Standards is quite different for our nation.

3. *Research Perspectives on Mathematics Standards Reform in the U.S.*—Mary Kay Stein, University of Pittsburgh
   The presentation began with a short introduction that identified the roles that research can and has played in past standards-based eras. Then attention turned to some key areas in which research might shed light in this era of the CCSSM. After referring to the NSF report, *A Priority Research Agenda for Understanding the Influence of the CCSSM*, four features of the CCSSM were outlined that set them apart from past standards: Fewer, clearer, higher standards; learning progressions; the positioning of mathematical practices, and their commonness. Theories-of-action associated with how each of these features is expected to contribute to improved mathematics instruction and student learning were described. Suggestions were made regarding how research could help us monitor the extent to which the theories-of-action play out as expected and whether there are any unintended consequences.

4. *The "Mathematics Studio": Sustainable School-Based Professional Learning*—Linda Foreman, President of Teachers Development Group
   The design of the "Mathematics Studio" professional development model is guided by a robust body of research on effective mathematics learning, teaching, professional development, and school leadership. Implementation of the model over time produces a powerful school-based culture of professional learning in

which sense making about meaningful mathematics instruction is continuous for all teachers and school leaders. At the heart of this work is engaging all mathematics teachers and leaders from a school together—during live classroom teaching episodes—in publicly coached rehearsals of well-defined, research-based "mathematically productive teaching routines." Grounded by the premise that all students are capable mathematical thinkers, this work fosters leaders' and teachers' habitual use of practices that yield all students' internalization of mathematical habits of mind typified by the *Common Core State Standards for Mathematical Practice*. This presentation provided a glimpse of the Mathematics Studio model's design, implementation, and impact.

5. *Challenges of Knowing Mathematics for Teaching in the United States*— Deborah Ball, University of Michigan
   Teachers' mathematical knowledge is a concern in many countries around the world. However, several features of the U.S. educational context present special challenges for ensuring that teachers know mathematics well enough to teach it. This presentation examined the mathematical knowledge needed for teaching and analyzed the special challenges presented by the U.S. context. Questions for participants were: How unique are these U.S. challenges for mathematical knowledge for teaching? Do other countries share these challenges?

The U.S.A. Exhibit

The U.S. Exhibit showcased not only the uniqueness of math education in the U.S., but also the diversity and variety of products, key players, and stakeholders involved in the practice. The U.S. exhibit included speakers, videos, materials, and interactive experiences.

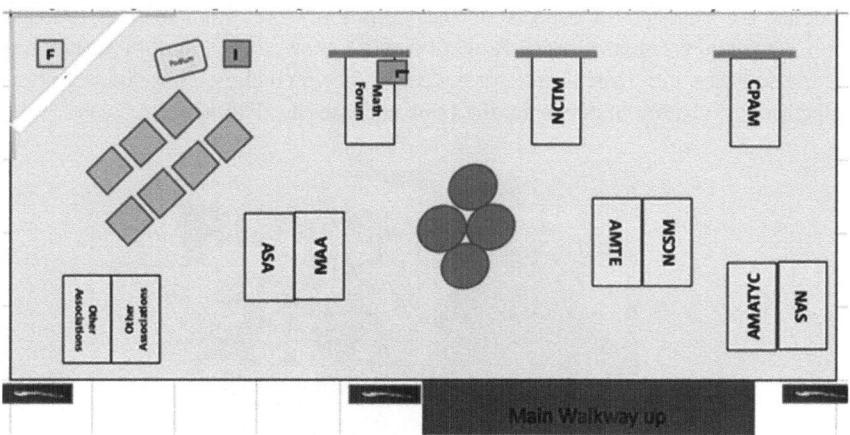

Since the number and diversity of the professional organizations devoted to various aspects of mathematics education was judged to be something unique about the United States all such organizations were asked to exhibit information about their activities. The figure on the right, which was one of the banners at the entrance to the Exhibit, indicates the organizations that participated in the U.S.A. National Exhibit.

A unique aspect of the Exhibit was that a corner was set aside with a projector and a screen so that presentations could be made. Those presentations led to many productive international conversations.

Another corner of the Exhibit area had a monitor that showed videos of mathematics classes in the United States. Coordinated by the Council of Presidential Awardees in Mathematics (http://cpam.teachersdg.org/), participants were assisted in using an iPad app called "Common Core Look-fors (CCL4s)" that can be used as a non-evaluative assessment tool to determine the extent to which teachers and students are engaged in aspects of the Common Core Mathematical Practices (http://www.corestandards.org/). Both an English transcription and a translation of the transcription into Korean were provided for the video clips. Those transcriptions helped many visitors understands the clips and use the iPad app.

Particularly popular among both teachers and the crowds of Korean students who attended ICME-12 were Zome Tools (http://zometool.com/), an innovative manipulative that challenges students to creatively explore geometry and informally introduces them to concepts of topology. In addition to working at the tables set up in the middle of the exhibit, teachers and students were given individual packets of Zome Tools so they could continue with their explorations.

Besides USNC/MI members and representatives from the professional organizations in the Exhibit, U.S. mathematics educators who had received travel grants from NCTM with funding from the National Science Foundation (NSF) also took turns at the Exhibit discussing mathematics educations with visitors to the Exhibit.

A Capsule Summary Fact Book

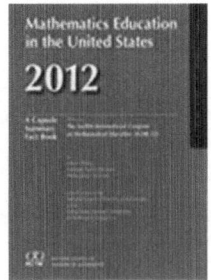

It has become a tradition for NCTM to commission for each ICME a document entitled *Mathematics Education in the United States: A Capsule Summary Fact Book.* For ICME-12 it was prepared by John Dossey, Katherine Halvorsen, and Sharon McCrone. Attendees at the National Presentation and the National Exhibit were given a copy of the *Fact Book* on a USB drive. The following two paragraphs from the "Preface" give a very good indication of what can be found in the complete report:

This document begins with some general information about education in the United States. The three kinds of curricula identified in the Second International Mathematics Study— intended, implemented, and attained—are then described (McKnight et al. 1987). A special focus is given to the emergence of a common K–grade 12 curriculum that has been adopted by forty-five states and the District of Columbia. This curriculum, the Common Core State Standards for Mathematics (CCSSM), was developed by a consortium consisting of state governors and chief state education officers (National Governors Association Center for Best Practices and Council of Chief State School Officers [NGA Center and CCSSO] 2010). The adoption of such a set of common outcomes, matching assessments, and similar instructional materials is expected to bring to U.S. mathematics education a level of uniformity that it has never before seen.

As in earlier editions, this publication has sections dealing with programs for high-achieving students, programs for mathematics teacher education, and resources for additional information about U.S. mathematics education. One message that comes through repeatedly in these descriptions is the variety of available programs and thus the inability to characterize them adequately in a brief document like this one. Another message is that all levels of the educational system exhibit great flux, and even though we have attempted to provide the latest available information, we realize that the information presented here will quickly become dated. By listing our sources, we hope to enable the interested reader to obtain updated information.

The report is available for download at www.nctm.org/about/affiliates/content.aspx?id=16955.

The U.S. Reception

A U.S. reception was held at the ICME-12 on July 10th for 150 international attendees. The reception was intended to foster international collaborations between U.S. math educators and their international peers. A Speed Networking session facilitated the networking experience between the attendees. The reception was

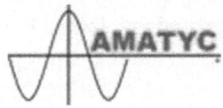

sponsored by the National Council of Teachers of Mathematics (NCTM), the American Mathematical Association of Two-Year Colleges (AMATYC), The American Statistical Association (ASA), and the Conference Board of the Mathematical Sciences (CBMS).

Open Access This chapter is distributed under the terms of the Creative Commons Attribution Noncommercial License, which permits any noncommercial use, distribution, and reproduction in any medium, provided the original author(s) and source are credited.

National Presentation of India

K. Subramaniam

The National Presentation Process

A proposal to make a national presentation on the status of mathematics education in India at ICME-12 was sent by the Indian National Science Academy to the Chairman of the International Programme Committee in February, 2011. After the proposal was accepted, a Steering Committee was formed to oversee the preparation of the National presentation. A national initiative was launched to identify initiatives and innovations in mathematics education in India and to bring diverse groups working to improve mathematics education together on a common platform. The National Initiative on Mathematics Education (NIME) aimed to develop a vision about the changes necessary in mathematics education policy, the need for research studies on mathematics education, and ways of implementing system wide improvement and transformation of the practice of mathematics education.

Under the NIME initiative five regional conferences on mathematics education were held in 2011 across India to provide a wide participatory platform. This was followed by a National Conference on mathematics education in early 2012. The aim of the conferences was to bring together the important and significant innovations and efforts to improve mathematics education in school and in higher education in the diverse regions of the country and to build awareness of such efforts in the community of mathematics educators. The proceedings of the conferences formed an input for the Indian National Presentation.

The Indian National Presentation at ICME-12 includes the following components:

1. A book on *Mathematics Education in India: Status and Outlook*, containing key articles on the themes identified for the National Presentation.

K. Subramaniam (✉)
Homi Bhabha Centre for Science Education, Mumbai, India

© The Author(s) 2015
S.J. Cho (ed.), *The Proceedings of the 12th International Congress on Mathematical Education*, DOI 10.1007/978-3-319-12688-3_23

2. The slides of the presentations made at ICME-12 in two slots: one sub-plenary slot of 90 min, and a parallel session of 90 min.
3. Four video films (1 long and 3 short) on the challenges and hope giving initiatives in mathematics education in India.
4. An exhibition covering historical aspects of mathematics and mathematics education, the challenge of diversity, basic data about mathematics education in India and information about some initiatives displayed in a stall in the exhibition area of ICME-12.

All these components, except the long video film, are available on the NIME website for free download under an open access license: http://nime.hbcse.tifr.res.in. The website also contains links to the NIME regional and national conference websites, and the research papers presented by Indian participants at ICME-12.

Mathematics Education in India

The Indian National presentation was organized under the following four major themes:

1. Historical and Cultural aspects of mathematics and mathematics education
2. Systemic and Policy aspects of education
3. Mathematics Curriculum and Pedagogy at the elementary, secondary and tertiary levels (including nurture and enrichment programmes)
4. Teacher education and development

We provide below a summary of the presentations made for the Indian National Presentation (presentation slides available on the NIME website). The book on *Mathematics Education in India: Status and Outlook*, may be consulted for more details. The first presentation session covered the first two themes, while the remaining themes were covered in the second presentation session.

Historical and Cultural Aspects of Mathematics and Mathematics Education

There were two presentations under this theme, the first on the history of Indian mathematics and the second on the history of Indian mathematics education. Ramasubramaniam presented an overview of the historical tradition of mathematics in India. He described the main contributions made by Indian mathematicians in the Ancient Period (prior to 500 CE), the Classical Age (500–1200 CE) and the Medieval Period (1250–1750 CE). The examples that he described included the knowledge of geometry used to identify the cardinal directions and methods of finding irrational quantities such as square roots in the *Śulbasūtras* (~500 BCE),

Aryabhaṭa's recurrence relation for sine values as well as the table of sine differences (499 CE), the summation of series and finding the sums of sums in the Classical Age, and the infinite series for π and the fast converging approximations developed in the Kerala school of mathematics in the Medieval Period. He pointed out that the Indian approach to mathematics laid emphasis on the development of algorithms and on practical applications. Indian mathematical texts typically illustrated a principle or rule with a large number of examples drawn from the practical world. He also pointed out the role of memory in communicating mathematics and the organization of the texts in compressed verse form (*sūtras*). He argued that inclusion of the history of mathematics in mathematics education would help eliminate euro-centrism and biases, and also introduce a cross-cultural perspective on mathematics.

Senthil Babu spoke about the indigenous traditions of mathematics education in pre-British South India. Indian merchants, traders and craftsmen were renowned for their facility in arithmetic and computational ability. They learned to carry out a variety of complex computations grounded in practical contexts in indigenous schools called "*pāṭhaśālas*". The curriculum in these schools was grounded in the needs of the economy and society. The objective of the *pāṭhaśāla* education was to produce competence and skill in dealing with numbers and letters. A primary mode of learning was recollective memory, which combined knowledge of tables and series of numbers and quantities with problem solving. Public display of competence and skill was a celebrated part of *pāṭhaśāla* learning. The encounter with the Colonial British government and the efforts to introduce modern education gradually led to the *pāṭhaśālas* being appropriated and replaced with a curriculum and education system that was disconnected from the life of the community. Babu pointed to lessons that this may hold for the problems surrounding mathematics education in contemporary India.

Systemic and Policy Aspects of Education

There were three presentations under this theme covering respectively school education, the assessment culture and nurture programs for high achievers in mathematics. Anita Rampal presented an overview of the challenges and policy initiatives in mathematics education at the school level. Although India is a country with strong mathematical traditions, it is grappling with multiple challenges emerging from endemic poverty and large numbers of children not completing school. The systemic challenges include restructuring the education system to ensure an equitable education of high quality to a huge young population with high aspirations. Rampal presented an overview of the institutional structures and organization of mathematics education at various levels in India. She described two major policy initiatives in school education—the National Curriculum Framework of 2005 and the Right to Education Act of 2009. The new curriculum framework, which emphasized learning through activity and exploration and making the child free from fear and anxiety, had major implications for mathematics education. The new Act has set in place assessment reforms that can have a major impact. The new textbooks at the primary level aim to build on how children think and integrate themes from work, crafts and cultural hertiage. Rampal argued for redesigning secondary education curricula to meet the needs of a diversity of learners and called for a culturally responsive critical pedagogy of mathematics education.

Shailesh Shirali spoke on the role of assessment in mathematics education in India. He described the high stakes, highly competitive examination environment that students in India face at the end of schooling in order to gain entry to higher education. Such intense competition has led to a pervasive culture of private coaching and has shaped assessment practices right down to the primary school level. Shirali discussed sample questions from some of the most competitive entrance examinations, which tend to emphasize procedure and manipulative skills, and heavy dependence on memorization. He described the recent initiative on "continuous and comprehensive evaluation" as promising, but as critically dependent on adequate teacher preparation. He also called for research on assessment tools and models.

Kumaresan described the training programs at different levels aimed at high achievers in mathematics in India. He grouped the training programs under three categories: (i) those at the undergraduate level (ii) those at the graduate and Phd level and (iii) those aimed at National Olympiad toppers. The most significant program at the undergraduate level is the Mathematics Training and Talent Search (MTTS) program. This program aims to move students out of a pervasive culture of rote learning towards discovering mathematics by inquiry, to awaken their thinking abilities, to expose them to the excitement of doing mathematics, and to change the teaching of mathematics in the country in the long run. The MTTS sessions are highly interactive, where students are trained to observe patterns, formulate conjectures and develop proofs. The Advanced Training in Mathematics (ATM) schools address the needs of graduate and Phd level students. Olympiad toppers are trained through a special nurture program.

Mathematics Curriculum and Pedagogy at the Elementary, Secondary and Tertiary Levels

There were four presentations on the mathematics curriculum and pedagogy at different levels of education in India. The first two presentations described the curriculum and the pedagogical challenges at the primary level. The remaining two presentations analysed the curriculum at the secondary and tertiary levels of education. Amitabha Mukherjee described the changes introduced in the primary mathematics curriculum and textbooks following the revision of the school curriculum framework by contrasting the new curriculum and the traditional curriculum along several dimensions. The new curriculum emphasises concrete experience as the basis for learning mathematics, and encourages multiple approaches to solving problems. Topics not emphasised in traditional curriculum such as shapes and space, measurement, data handling and patterns have been given space in the new curriculum. Jayasree Subramanian analysed the limitations of well-intentioned reforms in primary mathematics education. She pointed out that activity based approaches need resources in the form of teaching-learning materials, which are not available in most schools catering to children from low socio-economic backgrounds. "Drill and practice" still dominates classroom teaching of mathematics. Jayasree Subramanian cautioned that curriculum and pedagogy alone cannot ensure mathematics for all in a society fractured by several inequities.

Jonaki Ghosh presented an overview of the secondary mathematics curriculum, where the central focus is on consolidation of concepts learned earlier and exploring wider connections. There is an emphasis on the structure of mathematics as a subject and mathematical processes such as argumentation and proof, logical formulation, visualization, mathematical communication and making connections. While assessment is largely summative, a new initiative by the Central Government has shifted the emphasis towards continuous and formative assessment by removing the mandatory requirement of a final public examination at the end of Grade 10. The senior secondary stage (Grades 11–12) is dominated by the culture of high-stakes examinations, and Ghosh identified areas where a change of approach is needed, especially of making students familiar with the power of applications of mathematics in solving real world problems. She also emphasised the role of technology and the need to apply it thoughtfully to overcome the challenge of resource-poor classrooms.

Geetha Venkataraman spoke on the challenges facing mathematics education at the tertiary level where there are about 400 Universities and 18,000 colleges offering undergraduate courses in mathematics. Although syllabus reforms have taken place in the undergraduate curriculum since the late 1960s and 1970s, further reform is needed at the present time. The recommendations and model syllabus of the University Grants Commission failed to provide leadership in terms of applicability of mathematics and the use of information technology in mathematics education. The pedagogy followed is largely one of demonstrating content by stating and proving theorems with minimal student interaction. Assessment typically requires students to reproduce from memory rather than to think, analyse and solve problems. There are almost no pre-service training or inservice training programs available for faculty to learn about teaching methods and tools. Geetha Venkataraman also called for better links between the community of research mathematicians and mathematics educators.

Teacher Education and Development

There were two presentations on teacher education and development, one focused on the organization of pre and in-service teacher eduation and one on innovations and initiatives in teacher education. Ruchi Kumar presented an overview of the structures and institutions implementing teacher education in India and the place of mathematics education in the teacher education curriculum. The new school curriculum framework demands a revisiting of teachers' beliefs, a strengthening of content knowledge of mathematics and a better understanding of the psychology of young children. Ruchi Kumar gave some examples of what aspects of teacher education need change. She concluded that research on teachers' beliefs and knowledge and the relation of these to student learning was greatly needed but largely absent in the Indian context. Subramaniam spoke about the trends of change in mathematics teacher education in India and the sources of change. He cited the example of the innovative program for preparing primary mathematics teachers launched in the 1990s by the Indira Gandhi National Open University. The curriculum of this program aimed at addressing teachers' beliefs about mathematics and its learning and at giving them experiences of exploring interesting mathematics. He also mentioned some in-service teacher development initiatives as harbingers of change in mathematics teacher education. He emphasized the role of teacher associations in bringing about change in mathematics teaching at the secondary level and called for greater participation by the associations in framing curricula for pre-service teacher education.

The final presentation by Rakhi Banerjee presented a brief review of research in mathematics education in India, which is still a highly under-developed research domain in the country. Traditional studies typically follow psychometric models aiming to identify learning difficulties in mathematics or factors responsible for poor achievement in mathematics. She described some promising new trends in mathematics education research in the country which include intervention studies aimed at alternative learning trajectories for key concepts such as whole numbers, fractions or algebra. Research studies on teacher education and development are very much needed, but are nearly absent. She criticised traditional studies for failing to provide insights into the nature of the problem or possible solutions. The new research studies often lack methodological rigour or a strong theoretical framework. She also pointed to the lack of adequate support in the universities for mathematics education research and the isolation of education departments from subject disciplines as factors hindering the growth of mathematics education research in the country.

Three short video films, screened at the end of the first presentation session, had the following titles: (i) Legacy of maths at work and play (ii) Diverse learners multiple terrains (iii) Initiatives to transform maths learning. These video films and the presentation slides can be found at http://nime.hbcse.tifr.res.in.

Open Access This chapter is distributed under the terms of the Creative Commons Attribution Noncommercial License, which permits any noncommercial use, distribution, and reproduction in any medium, provided the original author(s) and source are credited.

Spanish Heritage in Mathematics and Mathematics Education

L. Rico

On the occasion of the 12th ICME, the Spanish Committee of Mathematics Education decided to prepare a national presentation entitled Spanish Cultural Heritage.

The presentation takes the form of a series of posters, each of which has a special focus, showing relevant historical events identified according to time and institutions. As a whole, the posters outline a comprehensive historical trajectory devoted to the Hispanic Heritage.

The relevance of mathematics in the relations between Spain and America has remained unbroken since its beginning 520 years ago. Julio Rey Pastor emphasizes the importance and scope of this heritage for its scientific and technological use and its benefits since the discovery of America. Since then, throughout 520 years of continuous cultural cooperation, the mathematical background shared by Spain and the American Republics, people and countries, that have remained solid and permanent.

To present the Spanish Heritage in the ICME 12 of Seoul (Korea), from the Education Commission of the CEMAT (Spanish Mathematics Committee) have been prepared 27 posters, which set out key moments, characters and events in the history of mathematics.

The list of themes chosen is as follows:

1. Spanish Heritage in Mathematics and Mathematics Education.
2. Mathematics and Science in the Discovery of America.
3. The Founding of the First American Universities.
4. First Mathematical-Scientific Publication in the New World.
5. The House of Trade: Navigation, Cartography, and Astronomy.
6. The 16th-Century Mathematics Academy: Philip II, Siliceo, Juan de Herrera.
7. Science and Technology in the 16th Century.

L. Rico (✉)
University of Granada, Granada, Spain
e-mail: lrico@ugr.es

© The Author(s) 2015
S.J. Cho (ed.), *The Proceedings of the 12th International Congress on Mathematical Education*, DOI 10.1007/978-3-319-12688-3_24

8. Mathematics in the Baroque Period in Spain.
9. Scientific Policy of the First Bourbons. The Jesuits and Mathematics.
10. Enlightenment Mathematics. The Reforms of Charles III.
11. José Celestino Mutis. An Enlightened Scientist in the New World.
12. Jorge Juan and Antonio de Ulloa. Meridian measurement in Quito.
13. Educational Reforms in Hispano-America, based on the 1812 Constitution.
14. The *Compendium of Pure and Mixed Mathematics* by D. José Mariano Vallejo.
15. 19th-Century Mathematics.
16. The Metric System in Textbooks in the Second Half of the 19th Century.
17. The Mathematicians of Scientific '98.
18. Andrés Manjón and the Ave María Schools at the end of the 19th Century.
19. The Spanish Republican Exile: the Mathematicians in America.
20. Researching Together: Return Journeys.
21. The Iberoamerican Mathematical Olympiad.
22. Research Centers.
23. Journals, research and collaboration in Mathematics Education.
24. ICME 8 Seville (Spain), July 1996.
25. Miguel de Guzmán (1936–2004) Academic, scientific, and educational legacy.
26. Mathematical Research in Ibero-America, Spain and Portugal.
27. Spanish Mathematics: the last 20 years.

The posters has been prepared by:

• M. de León and A. Timón, from the Institute of Mathematical Research (ICMAT).
• J. Peralta, from University Autonomous of Madrid.
• A. Maz; N. Adamuz; N. Jiménez-Fanjul; M. Torralbo and A. Carrillo de Albornoz, from the University of Cordoba.
• L. Rico; E. Castro-Rodríguez; J. A. Fernández-Plaza; M. Molina; M.C. Cañadas; J.F. Ruiz-Hidalgo; J.L. Lupiáñez; M. Picado; I. Segovia; I. Real and F. Ruiz, University of Granada.
• I. Gómez-Chacón; M. Castrillón and M. Gaspar, from the University Complutense of Madrid.
• M. Sierra and M.C. López, from University of Salamanca.
• B. Gómez, L. Puig and O. Monzó, University of Valencia.

The main objective of this work was to present the joint activity on mathematics and mathematics education, thought and written in Spanish, conducted by Spanish and American in more than 500 years of history and shared culture. We will stress the links established between Americans and Spaniards, as demonstrated by the information presented. We will underscore the scientific, technological, or cultural value of these events, their subsequent implications, and the social impact they produced in their time.

As there is a common language, a shared history and culture, there are ways of thinking and doing math based on that language, that culture and that history. This work aims show and claim the shared heritage in mathematics and mathematics education in this community. We have done a selection of the information presented in the posters and we will comment it here.

We have organized the posters considering five general comprehensive periods:

1. Discovery and colonization.
2. The Creole society.
3. The Century of Independence.
4. 20th Century: mutual assistance and help.
5. Current cooperation.

Summary of key moments and ideas of the above mentioned periods.

1. Mathematics and Science in the Discovery of America.

On October 12, 1492, a Spanish expedition commanded by Admiral Christopher Columbus arrived at the island of Guanahani and took possession of the land in the name of their Majesties Isabella of Castile and Ferdinand of Aragon. This act of Discovery is essential to the birth of historical modernity and of science. It marks the origin of a sociocultural community, the Ibero-American community, based on the unique relationship between Spaniards and Americans. Columbus's goal was to reach Asia, that say, the island of Cipangu (Japan), which was thought to be at the same latitude as the Canary Islands. Columbus was not trying to discover a new continent, but rather to "reach the East by sailing West." The information that Columbus used involved several significant errors and to understand them it was needed a big change of ideas. Toscanelli's map, reflects the ideas of many navigators and geographers of the period, describes the route that Columbus believed he had travelled. He thus believed that the distance from the Canary Islands to Japan was 800 leagues by west (4,500 km), when it was actually about 3,500 leagues (19,500 km). These data were sufficient grounds for undertaking his first and the subsequent voyages. When Columbus arrives in the Antilles, he was convinced that he has reached the western coast of Asia hence his naming these lands the West Indies (Fig. 1).

Institutions: the first Universities. Starting with their first years in the colonies, the Crown, the Church, and the religious orders intervene in the area of education to teach and train clergy, government officials, and the middle classes. Their knowledge was classified into study in the trivium (grammar, rhetoric, and logic) and the quadrivium (arithmetic, geometry, music, and astronomy). Founding the Universities and Colleges in America is a historical feat and cultural phenomenon of prime importance, particularly in the first half of the 16th century. The first universities were the universities of Santo Domingo, Lima and Mexico, that were respectively founded in 1538 (Santo Domingo) and in 1551 (Lima and Mexico).

MATHEMATICS AND SCIENCE IN THE DISCOVERY OF AMERICA

- Authors:
- Dr. Luis Rico. University of Granada. lrico@ugr.es Spanish ICMI Representative (CEMat)
- Dª. Elena Castro- Rodríguez, University of Granada. elecr@correo.ugr.es

On October 12, 1492, a Spanish expedition commanded by Admiral Christopher Columbus arrived at the island of Guanahani and took possession of the land in the name of their Majesties Isabella of Castile and Ferdinand of Aragon. This act of Discovery is essential to the birth of historical modernity and of science. It marks the origin of a sociocultural community, the Hispano-American community, based on the unique relationship between Spaniards and Americans.

Columbus's goal was to reach the island of Cipangu (Japan), which was thought to be at the same latitude as the Canary Islands. Columbus was not trying to discover a new continent, but rather to "reach the East by sailing west." Toscanelli's map, which reflects the ideas of many navigators and geographers of the period, describes the route that Columbus believed he had travelled.

The information that Columbus used involved several significant errors.
It assumed that the earth had a maximum circumference of 29,000 km., estimated according to Posidonius and the length of a terrestrial degree, which was based on Arabic, not the shorter Italian, mile. Posidonius calculated the earth's circumference as less than three fourths of its real size, and this figure had been accepted as scientific truth since the time of Eratosthenes.
On the other hand, interpreting the information from the voyages of Marco Polo, Columbus estimated the width of Eurasia to be 225º, subtracting only 135º of ocean.

He thus believed that the distance from the Canary Islands to Japan was 800 leagues by west (4500 km), when it was actually about 3500 leagues (19,500 kilometers). But these data were sufficient grounds for undertaking his first and the subsequent voyages. When Columbus arrives in the Antilles, he is convinced that he has reached the western coast of Asia hence his naming these lands the West Indies.

Columbus's subsequent voyages demonstrated the crucial importance of the new lands and cultures discovered for the construction and grounding of new knowledge. The exploration of the New World established a contrast between reality and speculative ancient doctrines on the size and shape of the earth's geography. This stimulated the spirit of free investigation and empirical contrast of hypotheses. Geography, cartography, cosmography, navigation, geodesy, and the physics of the globe ceased to be practical techniques and began their careers as physical-mathematical sciences.

The problem of geographical longitudes also had to be solved to respond to the need for nautical maps. The demand for greater precision led to the emergence of new techniques and instruments for measurement, which required a considerable advance in science, particularly mathematics.

Although Columbus made four voyages, explored the islands of Guadalupe and Puerto Rico, reached the mouth of the Orinoco River, navigated the islands of Chacachacare and Margarita, Tobago and Granada, and landed in region of Panama, he died without knowing that he had discovered the "New World" for Europeans. In 1507, the German cartographer Waldseemüller published a geography book that included a map. The book represented the new lands and included the tales of a Florentine navigator, Amerigo Vespucci, who stated that the discoveries were not Asian lands, but a new continent. In Vespucci's honor, the 1507 map called the new lands America.

 Spanish Heritage   ICME12 SEOUL

Fig. 1 Mathematics and science in the discovery of America

2. The Creole Society.

To talk about the Creole society we have fixed our attention in the following ideas:

- Scientific Policy of the First Bourbons. The Jesuits and Mathematics.
- Enlightenment Mathematics. The Reforms of Charles III.
- José Celestino Mutis. An Enlightened Scientist in the New World.
- Jorge Juan and Antonio de Ulloa. Meridian measurement in Quito.

At the beginning of the 18th century, the Jesuits assumed responsibility for educating the nobility, through Seminaries for Nobility, which began in Madrid. This institution's model of teaching spreads to Barcelona, Valencia, Gerona, and other cities. Based on its model, new centers are founded throughout the 18th century. In Mexico, the Royal Academy of San Carlos of the Noble Arts of New Spain is founded. These centers also trained the American elites. The Seminaries of Nobility become one of the most important centers of teaching and research in America. The Jesuits authored fertile textbooks in mathematics.

José Celestino Mutis y Bosio (Cadiz 1732; Bogotá 1808) Botanist, doctor, astronomer, and mathematician. Mutis developed important scientific work on American soil. He gave the inaugural speech for the Chair of Mathematics at the College of Our Lady of the Rosary in Santa Fe de Bogotá. There, he held the positions of rector and director. He determined the coordinates of Santa Fe de Bogotá, observed an eclipse of a satellite of Jupiter, and was one of the observers of the Transit of Venus on June, 1769.

Jorge Juan y Antonio de Ulloa. They participated in the expedition from 1735 to 1744 to measure an arc of 1° of latitude near the equator and one near the pole, to determine the lemon or orange shape of the earth.

The Royal Academy of Sciences decided to undertake the task of obtaining precise data from two meridian positions at two locations on Earth: Lapland (North Pole) and the Viceroyalty of Peru (the equator). To do this, two expeditions were organized. If the measurements obtained by both expeditions were the same, the Earth was sphere-shaped. If the measurement was greater at the pole, there was flattening at the poles.

If the polar measurement was smaller, the French were right and the lemon shape shall be the model. To carry out the expedition to the cities of Quito and Cuenca, located today in the Republic of Ecuador (Fig. 2).

3. The Century of independence. We have fixed our attention on the following subjects.

- Educational Reforms in Hispano-America, based on the 1812 Constitution.
- The *Compendium of Pure and Mixed Mathematics* by D. José Mariano Vallejo.
- 19th-Century Mathematics.
- The Metric System in Textbooks in the Second Half of the 19th Century.
- The Mathematicians of Scientific '98.

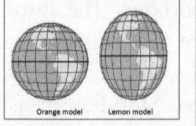
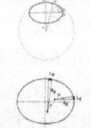
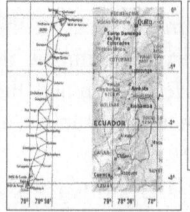
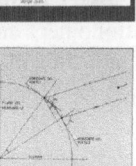

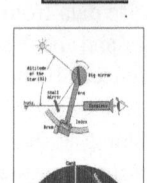
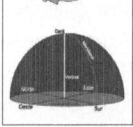

Fig. 2 Jorge Juan and Antonio de Ulloa. Meridian measurement in Quito

Conception of education. Spanish resistance to Napoleon's invasion in 1808 led to the formation of the Courts of Cadiz, which developed the Constitution of 1812, oriented ideologically to exalt and safeguard individual liberty.

The Constitution's ninth title, dedicated to education, articulated the liberal idea of education, defending the idea of general, uniform education for all citizens and the need to form a Council for Public Education. This Council prepared a report that can be considered the most representative document on liberal ideology in matters of education. The report was published in Cadiz on September 9, 1813, under the following title: *Report of the Council created by the Regency to propose the means for proceeding to regulate the various branches of Public Education.*

It was established that "education must be universal, uniform, public, and free, and it must enjoy liberty." This Report formed the basis and origin of the educational reforms put into effect throughout Hispano-America after the revolutions that led to the independence of the Spanish colonies.

The 19th century is a turbulent period in the history of Spain. It begins with the invasions of Napoleon's armies in 1808 and ends with the Spanish-American War in 1898, known as the *disaster of '98*. The beginning of the 19th century brings the independence of the former colonies in America, giving rise to the new American republics. Spain loses its status as world power. The 19th century ends with the loss of Cuba, Puerto Rico, and the Philippines and the defeat of the Spanish fleet in Santiago, Cuba. Spain concludes its political presence in America. Spain is aware of its cultural and educational decline, a feeling aggravated by the loss of its colonies. It is thought that the military defeat was caused in good part by the country's scientific and technical backwardness: *the crisis of '98* (Fig. 3).

4. 20th Century: mutual assistance and help.

Where we are? Who are the leaders? We choose five points to reflect about our common work during the 20th Century.

- Andrés Manjón and the Ave María Schools at the end of the 19th Century.
- The Spanish Republican Exile: the Mathematicians in America.
- Researching Together: Return Journeys.
- The Iberoamerican Mathematical Olympiad.
- ICME 8 Seville (Spain), July 1996. Current research and cooperation (Fig. 4).

5. Current cooperation.

To describe this point we selected the following reflections:

- Research Centers.
- Journals, research and collaboration in Mathematics Education.
- Miguel de Guzmán Ozámiz (1936–2004) Academic, scientific, and educational legacy.
- Mathematical Research in Ibero-America, Spain and Portugal.
- Spanish Mathematics: the last 20 years (Fig. 5).

19TH-CENTURY MATHEMATICS

Author: Javier Peralta. Faculty of Teacher Training and Education.
Autonomous University of Madrid

After the death of Ferdinand VII (1836), the governments of Isabel II entered a favorable period for Spanish education and science, which had lagged behind the times. Educational reforms were carried out and new institutions founded. This process gained momentum during what is known as the *six democratic years* (1868-1874), when the queen was dethroned by the Revolutionary Councils, with their democratic ideology (freedom of religion, universal suffrage...).
In 1876, the Institute of Free Teaching was created. This institution played a highly beneficial role in education, and the reform movement for renewal received new stimuli with the *Crisis of '98*.
This cultural and scientific renaissance would have a positive influence on the development of Spanish mathematics. It would also have repercussions for education in Cuba and Puerto Rico.

Isabel II   Alfonso XII

First reforms and institutions

1845: *The Pidal Plan*
. Establishes the Bachelor of Science (within the Faculty of Philosophy).
. Creates provincial institutions for Secondary Education.

1857: *Moyano Law*
First law on education in Spain: creates the Faculty of Sciences (Madrid).
1858: Degree of Bachelor of Exact Sciences created.

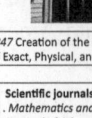

Paraninfo de la Universidad de Madrid (1852)

1847 Creation of the Royal Academy Of Exact, Physical, and Natural Sciences.

Development of mathematics

Introduction of new theories (*last third of the century*)
. S. Archilla, L. Clariana: Cauchy analysis.
. J. Echegaray: Chasles geometry, calculus of variations, determinants, t the Galois Theory, elliptic and Abelian functions.
. V. Reyes: Non-Euclidean geometries, symbolic logic.
. Z. G. de Galdeano: functions of a complex variable, substitution groups.

Scientific journals (founded in the mid-19th century)
. *Mathematics and Physics Monthly* (Cadiz, 1848).
. *Journal of Advances in the Exact, Physical, and Natural Sciences* (Madrid, 1950).
. *Journal of the Society of Teachers of Science* (Madrid, 1874).
. *Science Chronicle* (Barcelona, 1878).
. *Progress in Mathematics* (Zaragoza, 1891).
. *Archive of Pure and Applied Mathematics* (Valencia, 1896).
. *The Aspirant* (Toledo, 1897).

El Progreso Matemático   Revista de los Progresos de las Ciencias

Activity in Cuba and Puerto Rico prior to 1998

Many Cubans and Puerto Ricans finished their education in the US, although some did so on the Spanish Peninsula. Some returned to work in their birthplaces, but others remained on the Peninsula.

Born in Cuba, they receive their Bachelor's and Doctoral degrees in Exact Sciences on the Spanish Peninsula and stay to work as professors on the Peninsula: José María Villafañe and Gumersindo Vicuña.

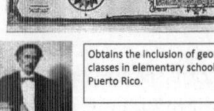

Born in Puerto Rico. Study physical-mathematic sciences in Madrid and return to work in Puerto Rico. Abolitionists. Members of Parliament:

José Julián Acosta (1825-1891)

Chair of Agriculture and Director of the Institute of San Juan, Puerto Rico.

José María Villafañe (1830-1915) Chair of Analytic Geometry at the University of Valencia and of Mathematical Analysis in Barcelona and Madrid. During the Enlightenment, his most important work was the *Treatise on Mathematical Analysis*, in three volumes.
Chair in Mathematical Physics at the University of Madrid. Academic in Science and Language and members of Parliament.

Gumersindo Vicuña (1840-1890)

Obtains the inclusion of geometry classes in elementary schools in Puerto Rico.

Ramón Baldorioty (1822-1889)

- First half of the century: Education is in the hands of religious orders and the Economic Societies of the Friends of the Nation. Around the 1940s, special laws are enacted for the islands that adapt education to the Spanish regulations and impose greater control by the State.
- 1844: publication of the *General plan for public education for the islands of Cuba and Puerto Rico*, which remains in effect from 1863 and 1865, respectively. From this time onward, the Pidal Plan is applied, but under the strict authority of the governor.
- Normal Schools are created: one in Cuba in 1857 (until 1864) and two more in 1890; and one in Puerto Rico in 1882.
- Institutions of Secondary Education are created: one in Cuba in 1863 (Cuba would ultimately have four); and one in Puerto Rico in 1882.
- University: Cuba has a university from 1728 onward. The *Royal and Papal University of San Geronimo* was founded by Dominicans. In the 19th century, this university was gradually adapted to the Spanish legislation. In 1842, it was secularized and took the name of the *Royal and Literary University of Havana*, which underwent successive reforms and whose study programs were assimilated into those on the Peninsula.
- Puerto Rico: The Athenaeum was created in 1863, with authorization to impart higher education under the supervision of the Cuban university system. Shortly after, the University District is established in Havana, for the system of public education on both islands.

 CEMat

Spanish Heritage   ICME12 SEOUL

Fig. 3 19th-century mathematics

Fig. 4 The Spanish Republican exile: the mathematician in America

JOURNALS, RESEARCH AND COLLABORATION IN MATHEMATICS EDUCATION

Alexander Maz-Machado, Noelia Jiménez-Fanjul *University of Córdoba (Spain)*

Iberoamerican countries have a wide variety of journals for both research and popularization of Mathematics Education where Spanish, Portuguese, and Iberoamerican researchers currently publish.

To collect and disseminate bibliographic Information about scientific publications produced in Iberoamerican countries was created a network of bibliographical institutions that work collaboratively: LATINDEX is the product of this cooperation,

Two Iberoamerican journals in the area are indexed in the JCR: the *Bolema-Mathematics Education Bulletin-Boletim de Educacao Matematica* (BOLEMA) and the *Journal of Latin American Research on Mathematics Education-Revista Latinoamericana de Investigación en Matemática Educativa-Relime* (RELIME).

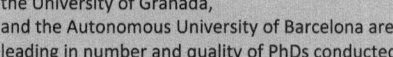

International collaboration was strengthened in 2005, with the creation of the Iberoamerican Federation of Societies for Mathematics Education. Members include Societies from Argentina, Brazil, Chile, Columbia, Spain, Peru, Portugal, and Uruguay. This Society publishes the journal *Unión*.

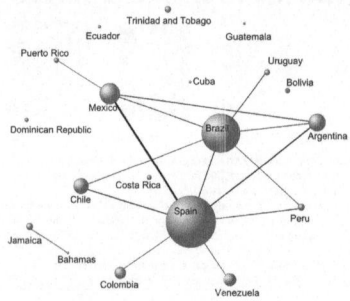

Red de colaboración en artículos de Educación Matemática indexados en SCOPUS

Doctoral Programs
In training researchers in Spanish-speaking countries, the CINVESTAV of Mexico, the University of Granada, and the Autonomous University of Barcelona are leading in number and quality of PhDs conducted.

International Congresses
A site for meeting and collaboration between Spain and Iberoamerica. Among the most significant are: the Latin American Meetings for Educational Mathematics ([RELME]), the Inter American Conference of Mathematics Education [CIAEM]), the Iberoamerican Congress of Mathematics Education [CIBEM]) and, in Spain, the Symposium for the Spanish Society for Mathematics Education (Simposio de la Sociedad Española de Educación Matemática [SEIEM])

If we examine the scientific productivity in the 770 articles indexed in SCOPUS as authored by Spanish or Iberoamerican researchers, we find systematic collaboration. This collaboration generates a network that revolves around Spain, Brazil, and Mexico.

 Spanish Heritage   ICME12 SEOUL

Fig. 5 Journals, research and collaboration in mathematic education

The goal of the presentation was to show and underline the relevance of the cultural and scientific cooperation in mathematics and mathematics education between Spain and the American Republics over the last 500 years. The presentation seeks to publicize this common mathematical heritage by emphasizing its importance and the far-reaching influence these relationships have had and continue to have for science, technology, and education in our countries.

Open Access This chapter is distributed under the terms of the Creative Commons Attribution Noncommercial License, which permits any noncommercial use, distribution, and reproduction in any medium, provided the original author(s) and source are credited.

Part VII
Regular Lectures

Part VII
Regular Lattices

Abstracts of Regular Lectures

Sung Je Cho

Understanding the Nature of the Geometric Work Through Its Development and Its Transformations*

Kuzniak Alain
Université Paris Diderot, France
alain.kuzniak@univ-paris-diderot.fr

The question of the teaching and learning of geometry has been profoundly renewed by the appearance of Dynamic Geometry Software (DGS). These new artefacts and tools have modified the nature of geometry by changing the methods of construction and validation. They also have profoundly altered the cognitive nature of student work, giving new meaning to visualisation and experimentation. In our presentation, we show how the study of some geneses (figural, instrumental and discursive) could clarify the transformation of geometric knowledge in school context. The argumentation is supported on the framework of Geometrical paradigms and Spaces for Geometric Work that articulates two basic views on a geometer's work: cognitive and epistemological.

Keywords: Geometric work, Visualization, Geometrical paradigm

---

S.J. Cho (✉)
Seoul National University, Seoul, Republic of Korea
e-mail: sungjcho@snu.ac.kr

© The Author(s) 2015
S.J. Cho (ed.), *The Proceedings of the 12th International Congress on Mathematical Education*, DOI 10.1007/978-3-319-12688-3_25

Integration of Technology into Mathematics Teaching: Past, Present and Future*

Adnan Baki
  Karadeniz Technical University, Turkey
  abaki@ktu.edu.tr

This paper deals with my endeavor as a researcher and lecturer within the world of educational computing to integrate technology into mathematics teaching. I started with the book titled "New Horizons in Educational Computing". In this book Saymor Papert enthusiastically says that computers as powerful learning tools will change tomorrow's classrooms. It is difficult to use this potential of computers for changing teacher's role and practice within an educational setting based on telling and showing. It was not easy for me to shift from traditional notions of teacher to constructivist teacher using Logo, Cabri and Geogebra as primary tools for doing and exploring mathematics in classrooms.

Investigating the Influence of Teachers' Pedagogical Beliefs and Reported Practices on Student Achievement in Basic Mathematics

Allan B.I. Bernardo
  De La Salle University, Philippines
  allan.bernardo@dlsu.edu.ph
  Auxencia A. Limjap
  De La Salle University, Philippines
  auxencia.limjap@dlsu.edu.ph

This study investigated the pedagogical beliefs of the elementary and high school mathematics teachers. It sought to find out whether their pedagogical beliefs are consistent with the School Mathematics Tradition (SMT) and Inquiry Mathematics Tradition (IMT). It determined if there are differences in the pedagogical beliefs of math teachers in high, average and low performing schools (HPS, APS, LPS) at the elementary and secondary levels. It also determined how the pedagogical beliefs of teachers are related to their reported teaching practices. Results show that there is no difference in reported teaching practices in HPS, APS and LPS. Teachers' pedagogical beliefs but not practices might be related to the performance of their students. There was a clearer link between the performance level of the schools and the teachers' pedagogical beliefs. The qualitative data suggest that many teachers hold the native view that how students learn mathematics is determined by how they teach mathematics.

Keywords: Basic mathematics, Pedagogical beliefs, School mathematics tradition, Inquiry mathematics tradition.

Developing Free Computer-Based Learning Objects for High School Mathematics: Examples, Issues and Directions*

Humberto José Bortolossi
Fluminense Federal University, Brazil
hjbortol@vm.uff.br

In late 2007, the Brazilian government launched a grant program offering 42 million dollars to support the production of digital contents to high school level in the following areas: Portuguese, biology, chemistry, physics and mathematics. Of this amount, the CDME Project (http://www.cdme.im-uff.mat.br/) of the Fluminense Federal University won 124 thousand dollars to develop educational software, manipulative materials and audio clips to the area of mathematics. In this article, we report our experience (and what we learned from it) within this project, regarding the development of educational software as learning objects. We hope that the examples, issues and directions shown here are useful for other teams concerned about cost, time and didactic quality in the development of their applications and online teaching systems.

Keywords: Learning objects in mathematics, software development technologies, visualization in the teaching and learning of mathematics.

Doing Research Within the Anthropological Theory of the Didactic: The Case of School Algebra*

Marianna Bosch
Ramon Llull University, Spain
marianna.bosch@iqs.edu

Since its emergence in the early 80s with the study of didactic transposition processes, the Anthropological Theory of the Didactic maintains a privileged relationship with school algebra and its diffusion, both in school and outside school. I have chosen this case study to introduce the main "gestures of research" promoted by this framework and the methodological tools used to help researchers detach from the dominant viewpoints of the institutions where teaching and learning processes take place or which affect these processes in the distance. The construction of alternative reference models concerning school algebra and teaching and learning processes leads to some recent teaching experiences that break down the established didactic contracts, raising new research questions that need more in-depth analysis in the way opened by the "procognitive paradigm".

Keywords: School algebra, Anthropologic theory of the didactic, Didactic transposition, Arithmetic calculation programme, Algebraization process

Curriculum Reform and Mathematics Learning: Evidence from Two Longitudinal Studies*

Jinfa Cai
  University of Delaware, USA
  jcai@math.udel.edu

Drawing on longitudinal evidence from the LieCal project, issues related to mathematics curriculum reform and student learning are discussed. The LieCal Project was designed to longitudinally investigate the impact of a reform mathematics curriculum called the Connected Mathematics Program (CMP) in the United States on teachers' teaching and students' learning. Using a three-level conceptualization of curriculum (intended, implemented and attained), a variety of evidence from the LieCal Project is presented to show the impact of mathematics curriculum reform on teachers' teaching and students' learning. The findings from the two longitudinal studies in the LieCal project serve both to show the kind of impact curriculum has on teachers' teaching and students' learning and to suggest powerful ways researchers can investigate curriculum effect on both teaching and learning.

Keywords: Curriculum, Education reform, Mathematics learning, Longitudinal studies, LieCal Project, Problem solving, Algebra, Standards

Mathematical Problem Solving Beyond School: Digital Tools and Students' Mathematical Representations*

Susana Carreira
  University of Algarve and University of Lisbon, Portugal
  scarrei@ualg.pt

By looking at the global context of two inclusive mathematical problem solving competitions, the Problem@Web Project intends to study young students' beyond-school problem solving activity. The theoretical framework is aiming to integrate a perspective on problem solving that emphasises understanding and expressing thinking with a view on the representational practices connected to students' digital mathematical performance. Two contextual problems involving motion are the basis for the analysis of students' digital answers and an opportunity to look at the ways in which their conceptualisations emerge from a blend of pictorial and schematic digital representations.

Keywords: Problem solving, Expressing thinking, Digital mathematical performance, Competitions.

Teaching Probability in Secondary School

Paulo Cezar Pinto Carvalho
  IMPA and FGV, Brazil
  pcezar@impa.br

Probability teaching in secondary school many times emphasizes computing probabilities as ratios of favourable to possible cases. Often, however, not enough attention is given to whether the possible cases are equally likely. We argue that being able to identify the adequateness of an equiprobable model for a given situation is a fundamental ability to be developed in secondary school. The role of simulation in understanding the meaning of computed probability values and realizing the power and limitations of probabilistic methods is also discussed.
  Keywords: Probability, Modelling, Simulation, Secondary school.

Mathematics is Alive!: Project Based Mathematics

Kyung Yoon Chang
  KonKuk University, Korea
  kchang@konkuk.ac.kr

Mathematics is a structured body of knowledge invented to describe properties and solve problems around us but is considered merely as a collection of facts, concepts, and procedures. Mathematics is a hard and boring subject to many. Although Korean students showed top level performances in recent international studies, their affect toward mathematics and self-confidence level were at a surprisingly low level. Projects may give vitality to school mathematics and motivate students to pursue solutions to nontrivial real problems and create products.

In this lecture I will present arguments concerning project-based learning in mathematics. Working with projects making solid dissected models will be introduced and analysed with students' endeavours. Data was collected from pre-service secondary math teachers taking a geometry course at undergraduate or graduate level for 3 years.

In the process of engaging activities, making connections among concepts, integrating process and procedural knowledge, modifying their conjectures, and experiencing reflective thinking, pre-service secondary mathematics teachers illustrated their feelings of satisfaction and confidence. Technology took a crucial role in project-based learning, because students' endeavours could not be realized as the final products without the aid of it.

Mathematics is not mere artifacts but a reality. Mathematics is alive in project!
Keywords: Project based, Geometric model, Conic section, Affect, Technology.

Weaving Exploration in the Process of Acquisition and Development of Mathematical Knowledge

Marcos Cherinda
Universidade Pedagógica Mozambique, Mozambique
mCherinda@up.ac.mz

This paper presents an experience of a process of gathering mathematical knowledge by exploration of artifacts of students' cultural environment—the mat twill weaving techniques and their resulting products as well. That process means starting from posing and reflecting on "why …", "how …", and "what if …" questions related to the existence and gestalt of such artifacts, when one is manipulating them either physically or mentally. This highlights the Gerdes' research approach in the new research field called ethnomathematics. Furthermore, the paper brings the context of all that process, the classic theoretical framework on research methods in mathematics education, illustrating what does mean "doing mathematics", and how mathematics teachers can make their students feel themselves mathematics producers and owners, just by exploring those artifacts. The experience was gain both in and out of school settings but always leaded to know about the process of acquiring mathematical knowledge by the involved subjects.

An Illustration of the Explanatory and Discovery Functions of Proof

Michael de Villiers
University of KwaZulu-Natal, South Africa
profmd1@mweb.co.za

This paper provides an illustration of the explanatory and discovery function of proof with a geometric conjecture made by a Grade 11 learner. After logically explaining (proving) the result geometrically and algebraically, the result is generalized to other polygons by further reflection on the proof(s).
Keywords: Proof, Explanation, Discovery, Generalization, Viviani's theorem.

Constructing Abstract Mathematical Knowledge in Context*

Tommy Dreyfus
Tel Aviv University, Israel
tommyd@post.tau.ac.il

Understanding how students construct abstract mathematical knowledge is a central aim of research in mathematics education. Abstraction in Context (AiC) is a theoretical-methodological framework for studying students' processes of constructing abstract mathematical knowledge as they occur in a mathematical, social, curricular and learning-environmental context. AiC builds on ideas by Freudenthal, Davydov, and others. According to AiC, processes of abstraction have three stages: need, emergence and consolidation. The emergence of new (to the student) constructs is treated by means of a model of three observable epistemic actions: Recognizing, Building-with and Constructing—the RBC-model. This paper presents a theoretical and methodological introduction to AiC including to the RBC-model, and an overview of pertinent research studies.

Keywords: Abstraction, Knowledge construction, Context, RBC-model.

Digital Technology in Mathematics Education: Why It Works (Or Doesn't)*

Paul Drijvers
  Freudenthal Institute
  Utrecht University, the Netherlands
  p.drijvers@uu.nl
The integration of digital technology confronts teachers, educators and researchers with many questions. What is the potential of ICT for learning and teaching, and which factors are decisive in making it work in the mathematics classroom? To investigate these questions, six cases from leading studies in the field are described, and decisive success factors are identified. This leads to the conclusion that crucial factors for the success of digital technology in mathematics education include the design of the digital tool and corresponding tasks exploiting the tool's pedagogical potential, the role of the teacher and the educational context.

Keywords: didactical function, digital technology, instrumentation.

Mathematical Thinking Styles in School and Across Cultures*

Rita Borromeo Ferri
  Institute of Mathematics, University of Kassel, Germany
  borromeo@mathematik.uni-kassel.de
A mathematical thinking style is the way in which an individual prefers to present, to understand and to think through, mathematical facts and connections by certain internal imaginations and/or externalized representations. In which way mathematical thinking styles (analytic, visual and integrated) are influence factors on the learning and teaching of mathematics is described on the basis of selected qualitative empirical studies from primary up to secondary school. Within the

current MaTHSCu-project the styles are measured quantitatively by comparing mathematical thinking styles in eastern and western cultures. This study is introduced and first results are shown. Finally conclusions and implications for school are drawn.

Learning to See: The Viewpoint of the Blind*

Lourdes Figueiras
    Universitat Autònoma de Barcelona, Spain
    Lourdes.figueiras@uab.cat
    Abraham Arcavi
    Weizmann Institute of Science, Israel
    abraham.arcavi@weizmann.ac.il
Visualization goes beyond "seeing". On the one hand, it includes other sensorial perceptions, relationships with previous experiences and knowledge, verbalization and more. On the other hand, visualization can develop also in the absence of vision. On the basis of these premises, we attempt to revise the processes of visualization in mathematics education by (a) analyzing learning and teaching of mathematics by blind students with an expert blind mathematics teacher, and (b) simulating blindness with mathematics teachers with normal vision.
    Keywords: Visualization, Blind

Issues and Concerns About Integration of ICT into the Teaching and Learning of Mathematics in Africa: Botswana Case*

Kgomotso Gertrude Garegae
    University of Botswana, Botswana
    garegaek@mopipi.ub.bw
This paper discusses challenges that developing countries, especially African countries, face when trying to integrate ICT into the school curriculum particularly the mathematics curriculum. The general belief that the availability of ICT gadgets in schools guarantees the ICT integration in specific subjects, is challenged. Issues such as teachers' lack of relevant skills, shortage of teaching tools and unavailability of support staff act as an impediment to ICT accessibility in classrooms. The paper describes the development of infrastructure in Botswana and experiences pertaining to the school curriculum and argues that proper preparation for a smooth implementation of ICT infusion and integration is necessary.
    Keywords: ICT integration in Africa, ICT availability and accessibility, ICT and the mathematics curriculum, ICT integration in Botswana schools

Learning Mathematics in Secondary School: The Case of Mathematical Modelling Enabled by Technology*

Jonaki B Ghosh
University of Delhi, India
jonakibghosh@gmail.com
This article describes a study which was undertaken to investigate the impact of teaching mathematics using mathematical modelling and applications at senior secondary school level in India. While traditionally the emphasis in mathematics teaching in India is on the development of procedural skills, the study shows that the use of modelling and applications enabled by technology enhanced student's understanding of concepts and led them to explore mathematical ideas beyond their level. Using this approach, a balanced use of technology and paper pencil skills led to a deeper understanding of the subject.
Keywords: Mathematical modelling, Technology, Visualization, Exploration, Paper pencil skills

Doing Mathematics in Teacher Preparation: Giving Space and Time to Think, Reflect, Share and Feel*

Frédéric Gourdeau
Université Laval, Canada
Frederic.Gourdeau@mat.ulaval.ca
Describing the minimal mathematical content knowledge needed for secondary school teachers is not the most useful way to approach the mathematical preparation of teachers. Rather, focusing on the doing of mathematics, on the quality of their engagement with mathematics, is crucial. In doing so, I argue that the role of mathematicians in the mathematical preparation of teachers is not reduced but rather enhanced: it is work of a different nature than is often argued, leaning more on the expertise of the mathematician in the doing of mathematics than on his or her knowledge of the facts of mathematics.

This talk is based on work done at Université Laval (Québec, Canada) for the past 15 years, mostly with Bernard R. Hodgson. We have developed a series of courses for teachers (in pre-service training mostly) which aim to engage them fully in the doing of mathematics. What do we mean by engaging them fully in the doing of mathematics? How do we try to achieve this? We will describe our approach through some examples, outlining important aspects which need to be taken into account: for instance, allowing genuine exploration of mathematical problems, working on the communication of ones' understanding (paying attention to words, definitions and statements) and learning to identify mathematical processes. These aspects are generally largely independent of the actual mathematical topics at hand

and, in that sense, the main objectives pursued in each of these courses have little to do with the precise mathematical content.

Even though our reflections are based on a specific curriculum in a specific setting, we believe that some of the reflections we wish to share will have resonance with many outside Canada.

Keywords: Teacher education, Doing of mathematics, Mathematics preparation, Mathematicians' role.

Resources, at the Core of Mathematics Teachers' Work*

Ghislaine Gueudet
GREAD, ESPE Bretagne UBO, France
Ghislaine.Gueudet@espe-bretagne.fr
Mathematics teachers work with resources in class and out of class. Textbooks, in particular, hold a central place in this material. Nevertheless, the available resources evolve, with an increasing amount of online resources: software, lesson plans, classroom videos etc.

This important change led us to propose a study of mathematics teachers documentation. Mathematics teachers select resources, combine them, use them, revise them, amongst others. Teachers' documentation is both this work and its outcome. Teachers' documentation work is central to their professional activity; it influences the professional activity, which evolves along what we call professional geneses.

In this conference, I introduce a specific perspective on teachers resources, which enlightens in particular the changes caused by the generalized use of Internet resources.

Keywords: Communities, Documentation, Internet, Professional development, Resources

Mathematics Education Reform Movement in Indonesia*

Sutarto Hadi
Lambung Mangkurat University, Indonesia
shadiunlam@gmail.com
The reform of mathematics education in Indonesia started in the mid-seventies. The reform movement reported in this lecture is the second attempt after the first movement to reform traditional mathematics to modern mathematics (1975–1990) was a complete failure. Several mathematicians have dedicated their expertise and experiences to rebuild mathematics education from the remnant of modern mathematics. Their concerns are focused particularly with the weakest group of students. After a long consideration they come to the decision to implement the theory of realistic mathematics education (RME) as a basic concept for developing the local

theory of mathematics teaching and learning. They have the same view that RME could be a vehicle for improving mathematics teaching and learning and at the same time as a tool for social transformation. They began with four teacher education institutes and 12 pilot schools. RME has since expanded to 23 universities that supervise over 300 schools and has trained thousands of teachers. In this process of mathematics education reform the theory of RME has been transformed into PMRI, the Indonesian version of RME, and has been widely accepted as a movement to reform mathematics education.

Keywords: Mathematics learning, RME, Teacher education, PMRI

Emotions in Problem Solving*

Markku S. Hannula
University of Helsinki, Finland
markku.hannula@helsinki.fi

Emotions are important part of non-routine problem solving. A positive disposition to mathematics has a reciprocal relationship with achievement, both enhancing the other over time. In the process of solitary problem solving, emotions have a significant role in self-regulation, focusing attention and biasing cognitive processes. In social context, additional functions of emotions become apparent, such as interpersonal relations and social coordination of collaborative action. An illustrative case study presents the role of emotions in the problem solving process of one 10-year old Finnish student when he is solving an open problem of geometrical solids. The importance of emotions should be acknowledged also in teaching. Tasks should provide optimal challenge and feeling of control. The teacher can model the appropriate enthusiasm and emotion regulation. Joking and talking with a peer are important coping strategies for students.

Keywords: Emotion, Problem solving, Coping

Freudenthal's Work Continues*

Marja van den Heuvel-Panhuizen
Freudenthal Institute, Utrecht University, the Netherlands
m.vandenheuvel-panhuizen@uu.nl

In this paper I address a number of projects on elementary mathematics education carried out at the Utrecht University. The focus is on (1) using picture books to support kindergartners' development of mathematical understanding, (2) revealing mathematical potential of special needs students, and (3) conducting textbook analyses to disclose the learning opportunities that textbooks offer.

I discuss how these projects are grounded in the foundational work of Freudenthal and his collaborators in the past and how this will be continued.

Keywords: Picture books, Special education, Subtraction, Textbook analysis, Didactics of mathematics

Hands That See, Hands That Speak: Investigating Relationships Between Sensory Activity, Forms of Communicating and Mathematical Cognition*

Lulu Healy
   Universidade Bandeirante de São Paulo, Brazil
   lulu@pq.cnpq.br

This contribution explores the role of the body's senses in the constitution of mathematical practices. It examines the mathematics activities of learners with disabilities, with the idea being that by identifying the differences and similarities in the practices of those whose knowledge of the world is mediated through different sensory channels, we might not only become better able to respond to their particular needs, but also to build more robust understandings of the relationships between experience and cognition more generally. To focus on connections between perceptual activities, material and semiotic resources and mathematical meanings, the discussion concentrates on the mathematical practices of learners who see with their hands or who speak with their hands. This discussion centres around two examples from our research with blind learners and deaf learners and, in particular, analyses the multiple roles played by their hands in mathematical activities.

   Keywords: Blind mathematics learners, Deaf mathematics learners, Embodied cognition, Gestures.

Teachers Learning Together: Pedagogical Reasoning in Mathematics Teachers' Collaborative Conversations*

Ilana Seidel Horn
   Vanderbilt University, USA
   ilana.horn@vanderbilt.edu

In the United States, teaching is an isolated profession. At the same time, ambitious forms of teaching have been shown to benefit from teacher collaboration. What is it about collegial conversations that supports teachers' ongoing professional learning? In this paper, I synthesize findings from prior studies on mathematics teachers' collaborative conversations, focusing my analysis on collective pedagogical reasoning. I examine four facets of collegial conversations that support refinements in this reasoning. These facets are: interactional organization, engagement of individual teachers in a group, epistemic stance on mathematics teaching, and locally negotiated standards of representational adequacy. Together, these aspects of teacher talk differently organize opportunities for professional learning.

Keywords: Professional learning, In-service teachers, Discourse analysis

Transforming Education Through Lesson Study: Thailand's Decade-Long Journey*

Maitree Inprasitha
Khon Kaen University, Thailand
inprasitha_crme@kku.ac.th

The development of teaching and the teaching profession are issues that countries around the world have been struggling to solve for many centuries. Lesson study, a Japanese way of professional development of teachers, dates back nearly 140 years, in 1872 the Meiji government invited foreign teachers to teach Japanese teachers about "whole class instruction" (Isoda 2007). Ironically, in 1999, Stigler and Hiebert brought back to the U.S. the same idea on how to present whole class instruction, "If you want to improve education, get teachers together to study the processes of teaching and learning in classrooms, and then devise ways to improve them" (Stigler 2004 cited in Fernandez and Yoshida 2004). Although the education reform movement around the world calls for effective reform tools or even ideas like Japanese lesson study, transferring those tools/ideas to other socio-cultural setting in other countries is not easy and always complicated. Thus, education reform movements sometimes support but sometimes hinder movement of society. Taking Japan as a case study, Japan has undergone the movement of society from agricultural to industrialized, to information, and now knowledge-based society during the two centuries since the late 18th century to the present. Not visible to outside people, an evolution in the approach to school has taken place in Japan, which supports the movement of society, which has not occurred in most developing countries, including Thailand. Thailand has looked to Japan for ideas and has been implementing Lesson Study since 2000 but with a unique approach to adapt. Thailand's experience with Lesson study has been shared with APEC member economies over the last six years and has been deemed "quite a success" in improvement of teaching and learning of mathematics.

Dialectic on the Problem Solving Approach: Illustrating Hermeneutics as the Ground Theory for Lesson Study in Mathematics Education*

Masami Isoda
    University of Tsukuba, Japan
    isoda@criced.tsukuba.ac.jp
Lesson study is the major issue in mathematics education for developing and sharing good practice and theorize a theory for teaching and curriculum development. Hermeneutic efforts are the necessary activities for sharing objectives of the lesson study and make them meaningful for further development. This paper illustrate hermeneutic efforts with two examples for understanding the mind set for lesson study. First example, the internet communication between classrooms in Japan and Australia, demonstrates four types of interpretation activities for hermeneutic effort: Understanding, Getting others' perspectives, Instruction from experience (self-understanding), and the hermeneutic circle. Using these concepts, we will illustrate the dialectic discussion amongst students in the problem solving classroom engaged in a task involving fractions.

History, Application, and Philosophy of Mathematics in Mathematics Education: Accessing and Assessing Students' Overview and Judgment*

Uffe Thomas Jankvist
    Roskilde University, Denmark
    utj@ruc.dk
The Regular Lecture addresses the three dimensions of history, application, and philosophy of mathematics in the teaching and learning of mathematics. It is discussed how students' overview and judgment—interpreted as 'sets of views' and beliefs about mathematics as a discipline—may be developed and/or changed through teaching activities embracing all three dimensions of history, application, and philosophy. More precisely, an example of such a teaching activity for upper secondary school is described along with a method for both accessing and assessing students' overview and judgment. Examples of data analysis are given based on a concrete implementation of the teaching activity.

Keywords: History, applications, and philosophy of mathematics, Overview and judgment, Students' beliefs, views, and images of mathematics as a discipline.

Teaching Mathematical Modeling in School Mathematics

Ok-Ki Kang
   SungKyunKwan University, Korea
   okkang@skku.edu
   Jihwa Noh
   University of Northern Iowa, USA
   jihwa.noh@uni.edu
   Modeling is a cyclical process of creating and modifying models of empirical situations to understand them better and improve decisions. The role of modeling and teaching mathematical modeling in school mathematics has received increasing attention as generating authentic learning and revealing the ways of thinking that produced it. In this paper and interactive lecture session, we will review a subset of the related literature, discuss benefits and challenges in teaching and learning mathematical modeling, and share our attempts to improve traditional textbook problems so that they can become more authentic modeling activities and implications for instruction and assessment as well as for research.
   Keywords: Models, Representations, Modeling activities, Teaching modeling.

Implications from Polya and Krutetskii*

Wan Kang
   Seoul National University of Education, Korea
   wkang@snue.ac.kr
   Enhancing mathematical problem solving abilities, George Polya gave tremendous contribution to mathematics educators. He identified 4 steps in the problem solving process; (1) understand the problem, (2) devise a plan, (3) carry out the plan, and (4) look back and check. For each step, Polya revealed many useful habits of thinking in forms of questions and suggestions. V. A. Krutetskii analysed mathematical abilities of school children, which suggest valuable implementation to many trying to develop effective ways of expanding mathematical problem solving abilities. Krutetskii's research was inspecting mathematical behaviour in 3 stages of information gathering, processing, and retention. He concluded that mathematically able students show strong trends to gather information in more synthetic way, to process information in more effective, economic, and flexible way and to retain indispensable information more than inessential.
   Keywords: Mathematical heuristics, Mathematical abilities, Elementary teacher education, G. Polya, V.A. Krutetskii

Derivative or Derivation?

Matthias Kawski
    Arizona State University, United States
    kawski@asu.edu

In calculus, linear algebra, statistics, and many others, final exams commonly consist largely of tasks that only demand applying algorithms that consist of only algebraic manipulations. Rather than simply complaining about such training in "mindless symbol manipulation" [MSM], we argue that it is critical to acknowledge that it is precisely this characteristic of mathematics which is responsible in part for its utility and success across so many aspects of modern societies. Modern mathematical notation and formalism allowed the figurative medieval trader to employ assistants who could effectively do the bookkeeping, do the numbers, without having to understand the meaning of the symbols and the origin of the rules. With time, arithmetic was augmented by algebra, then by calculus. Today, every year millions of students participate in this amazing endeavour of mastering the rules for doing calculus, yet many of whom never understand what they do. From a popular quote of V. I. Arnold: "Leibniz quite rapidly developed formal analysis [i.e. calculus] ... in a form especially suitable to teach analysis by people who do not understand it to people who will never understand it.

The Social Dimension of Argumentation and Proof in Mathematics Classrooms

Christine Knipping
    Universität Bremen, Germany
    knipping@math.uni-bremen.de

Argumentation and proof have received increasing attention in mathematics education in recent years. However, social dimensions of proof and argumentation have not been emphasised. These included the social, argumentational dimension of proving in academic mathematical practice, the social process that transforms mathematical proving and argumentation in the context of school mathematics classrooms, and the interplay between the socio-cultural backgrounds of students and the social expectations around proving and argumentation in schools. Without adequate attention to these dimensions, there is a danger that classroom argumentation could become a social filter, emphasising students' pre-existing advantages and disadvantages. Attention to the structures of argumentations in mathematics classrooms combined with research on social dimensions can provide a better understanding of the filtering effect of argumentation in classrooms. This could provide a basis for minimising unexpected and undesirable consequences of a greater focus on argumentation and proof.

Keywords: Argumentation, proof, mathematics classrooms, equity, socio-cultural background.

Constructionism: Theory of Learning or Theory of Design?*

Chronis Kynigos

University of Athens, Greece

kynigos@ppp.uoa.gr

Constructionism has established itself as an epistemological paradigm, a learning theory and a design framework, harnessing digital technologies as expressive media for students' generation of mathematical meanings individually and collaboratively. It was firstly elaborated in conjunction with the advent of digital media designed to be used for engagement with mathematics. Constructionist theory has since then been continually evolving dynamically and has extended its functionality from a structural set of lens to explanation and guidance for action. As a learning theory, the constructionist paradigm is unique in its attention to the ways in which meanings are generated during individual and collective bricolage with digital artefacts, influenced by negotiated changes students make to these artefacts and giving emphasis to ownership and production. The artefacts themselves constitute expressions of mathematical meanings and at the same time students continually express meanings by modulating them. As a design theory it has lent itself to a range of contexts such as the design of constructionist-minded interventions in schooling, the design of new constructionist media involving different kinds of expertise and the design of artifacts and activity plans by teachers as a means of professional development individually and in collective reflection contexts. It has also been used as a lens to study learning as a process of design. This paper will discuss some of the constructs which have or are emerging from the evolution of the theory and others which were seen as particularly useful in this process. Amongst them are the constructs of meaning generation through situated abstractions, re-structurations, half-baked microworlds, and the design and use of artifacts as boundary objects designed to facilitate crossings across community norms. It will provide examples from research in which I have been involved where the operationalization of these constructs enabled design and analysis of the data. It will further attempt to forge some connections with constructs which emerged from other theoretical frameworks in mathematics education and have not been used extensively in constructionist research, such as didactical design and guidance as seen through the lens of Anthropological Theory from the French school and the Theory of Instrumental Genesis.

Keywords: Constructionism, Design, Meaning generation, Theory networking, Digital media

Adjacent Schools with Infinite Distance—Narratives from North Korean Mathematics Classrooms

Jung Hang Lee
  Nyack College, USA
  jhehlee@yahoo.com

This research addresses mathematics education in one of the most closed countries in the world, North Korea. North Korean secondary school mathematics education is examined through review of North Korea's social and educational structures as well as its political and ideological position. In-depth interviews were conducted with dislocated secondary school mathematics teachers and former students to understand their lived experiences in secondary school mathematics in North Korea. Participants responded to questions concerning typical ways teaching and learning were carried out in mathematics classes; the Workers' Party's influence in every aspect of education, from teacher education to curriculum and textbooks issued; and the impact the March of Suffering had on the teaching and learning of mathematics as well as its lingering effects in secondary mathematics education. One of the goals of this research was to provide a more realistic picture and background of secondary school mathematics education in North Korea. The participants came from different parts of North Korea and were interviewed based on their experiences in secondary school mathematics ranging anywhere between three to 25 years ago. Therefore, this collective interview analysis presents a solid viewpoint on secondary school mathematics in North Korea.

Keywords: North Korea, Secondary mathematics education, Interview.

Mobile Linear Algebra with Sage*

Sang-Gu Lee
  Sungkyunkwan University, Korea
  sglee@skku.edu

Over the last 20 years, our learning environment for linear algebra has changed dramatically mathematical tools take an important role in our classes.

Sage is popular mathematical software which was released in 2005. This software has efficient features to adapt the internet environment and it can cover most of mathematical problems, for example, algebra, combinatorics, numerical mathematics, calculus and linear algebra.

Nowadays there are more mobile/smartphones than the number of personal computers in the world. Furthermore, the most sophisticated smartphones have almost the same processing power as personal computer and it can be connected to the internet. For example, we can connect from mobile phone to any Sage server through the internet. We have developed over the years on Mobile mathematics with Smartphone for teaching linear algebra (Ko et al. 2009; Lee and Kim 2009; Lee et al. 2001).

In this article, we introduce Sage and how we can use it in our linear algebra classes. We aim to show the mobile infrastructure of the Sage and the mobile-learning environment. We shall also introduce mobile contents for linear algebra using Sage. In fact, almost all the concepts of linear algebra can be easily covered.

Keywords: Mobile mathematics, Sage, Learning environment, Smartphone, Linear algebra

Discernment and Reasoning in Dynamic Geometry Environments*

Allen Leung
Hong Kong Baptist University, Hong Kong
aylleung@hkbu.edu.hk

Dynamic Geometry Environments (DGE) give rise to a phenomenological domain where movement and variation together with visual and sensory-motor feedback can guide discernment of geometrical properties of figures. In particular, the drag-mode in DGE has been studied in pedagogical settings and gradually understood as a pedagogical tool that is conducive to mathematical reasoning, especially in the process of conjecture formation in geometry. The epistemic potential of the drag-mode in DGE lies in its relationship with the discernment of invariants. In this lecture, I will discuss means of discernment and reasoning for DGE based on a combined perspective that puts together elements from the Theory of Variation and the Maintaining Dragging Scheme. My focus is on an idea of invariant as the fundamental object of discernment. Furthermore, an idea of instrumented abduction is proposed to frame how such reasoning can be developed. Exploring by dragging is a powerful tool supporting geometrical reasoning. At the end, I will introduce a Dragging Exploration Principle that might help to cognitively connect the realm of DGE and the world of Euclidean Geometry.

Keywords: Dynamic geometry environments, Dragging, Variation, Abduction

Riding the Third Wave: Negotiating Teacher and Students' Value Preferences Relating to Effective Mathematics Lesson*

Chap Sam Lim
Universiti Sains Malaysia, Malaysia
cslim@usm.my

The "Third Wave" is an ongoing international collaborative mathematics education research project, involving 10 countries conducted over the years 2009–2011. Adopting the theoretical framework of social cultural perspective, the project aimed to explore the contextually-bound understanding and meaning of what counts as effective mathematics lesson from both the teachers and pupils' perspectives. This paper begins with a brief description of the Third Wave Study Project, the research framework and the general methodology used. Thereafter, it will concentrate on the main focus of the paper featuring a detailed discussion of the related findings from the Malaysian data. The data involved six mathematics teachers and 36 pupils from three types of primary schools. Multiple data sources were collected through classroom observations, photo-elicited focus group interviews with pupils and in-depth interviews with teachers. During each class lesson observation, the six selected pupils (as predetermined by their teacher) were given a digital camera to capture the moments or situations in the observed lesson that they perceived as effective. Pupils were then asked to elaborate what they meant by effective mathematics lesson based on the photographs that they have taken. Teachers were also interviewed individually immediately after each lesson observation and pupil's focus group interview. Findings of the study show that both teachers and pupils shared two co-values and two negotiated values in what they valued as an effective mathematics lesson. The two co-values are "board work" and "drill and practices" while the two negotiated values are "learning through mistakes" and "active student involvement". However, there are minor differences in teachers' and pupils' value preferences, for instance, pupils valued more of "clear explanation" from their teachers and active participation in classroom activities whereas teachers put emphasis on using different approaches to accommodate different types of pupils. More importantly, it was observed that an effective mathematics lesson is very much shaped by the continuous negotiation between teachers' and pupils' values and valuing. This paper ends with reflections on some possible implications and significant contributions of the study in mathematics education.

Keywords: Effective lesson, Expert/excellent teacher, Mathematics education, Primary School, Photo-elicited interview, Third wave, Values.

Learning Mathematics by Creative or Imitative Reasoning*

Johan Lithner
  Umea University, Sweden
  johan.lithner@umu.se
This paper presents (1a) a research framework for analysing learning difficulties related to rote learning and imitative reasoning, (1b) research insights based on that framework, (2a) a framework for research and design of more efficient learning opportunities through creative reasoning and (2b) some related ongoing research.

Keywords: Learning difficulties, Rote learning, Creative reasoning, Problem solving.

Features of Exemplary Lessons Under the Curriculum Reform in China: A Case Study on Thirteen Elementary Mathematics Lessons*

Yunpeng Ma
Northeast Normal University, China
mayp@nenu.edu.cn
Dongchen Zhao
Northeast Normal University, China
dongchenzhao@hotmail.com

Dramatic changes in mathematics education in China have taken place since the new mathematics curriculum standard was implemented in 2001. What do new features of exemplary lessons appear under the context of the curriculum reform? This paper will answer this question by presenting a case study on 13 elementary mathematics lessons that were evaluated as excellent exemplary lessons by mathematics educators in China. This study found that, consistent with the ideas advocated by the new curriculum, the selected lessons demonstrated the features of emphasizing on student's overall development, connecting mathematics to real-life, providing students the opportunities for inquiring and collaborating, and teachers' exploiting various resources for teaching. Meanwhile, the selected lessons also shared other common features in the lesson structure, the interaction between teacher and students, and the classroom discourse. The results reveal that the exemplary lessons have practiced the advocated ideas of the current reform, while they also embodied some elements that might be the stable characteristics of Chinese mathematics education.

Keywords: Chinese mathematics classroom, Teaching practice reform, Exemplary lesson, Elementary mathematics

Teachers, Students and Resources in Mathematics Laboratory*

Michela Maschietto
University of Modena e Reggio Emilia, Italy
michela.maschietto@unimore.it

This paper deals with the methodology of mathematics laboratory from two points of view: the first one concerns teacher education, the second one concerns teaching experiments in classes. Mathematics laboratory (described in the Italian national standards for mathematics for primary and secondary schools) can be considered as a productive "place" where constructing mathematics meanings, more a methodology than a physical place. It can be associated to inquiry based learning for students. An example of mathematics laboratory with cultural artefacts such as the mathematical machines (www.mmlab.unimore.it) is discussed.

Keywords: Mathematics laboratory, Instrumental genesis, Semiotic mediation, Mathematical machine, Teacher education.

The Common Core State Standards in Mathematics*

William McCallum
The University of Arizona, USA
wmc@math.arizona.edu
The US Common Core State Standards in Mathematics were released in 2009 and have been adopted by 45 states. We describe the background, process, and design principles of the standards.
Keywords: Standards, United States, Common Core.

From Practical Geometry to the Laboratory Method: The Search for an Alternative to Euclid in the History of Teaching Geometry*

Marta Menghini
Sapienza Università di Roma, Italy
marta.menghini@uniroma1.it
This paper wants to show how practical geometry, created to give a concrete help to people involved in trade, in land-surveying and even in astronomy, underwent a transformation that underlined its didactical value and turned it first into a way of teaching via problem solving and then into an experimental-intuitive teaching that could be an alternative to the deductive-rational teaching of geometry. This evolution will be highlighted using textbooks that proposed alternative presentations of geometry.
Keywords: Practical geometry, History of mathematics education, Textbooks.

Research on Mathematics Classroom Practice: An International Perspective*

Ida Ah Chee Mok
The University of Hong Kong, Hong Kong
iacmok@hku.hk
Research on Mathematics Classroom Practice encompasses very comprehensive themes and issues that may include any studies and scientific experiments happening inside the classroom, including consideration of the key agents in the

classroom (the teachers and the students), undertaken with diversified research objectives and theoretical backgrounds. To a certain extent, seeking an international perspective provides some delineation of the topic. Studies will then focus on those issues already prioritised as of interest by existing international comparative studies and those issues seen as significant within an educational system. This lecture will draw upon the work of an international project, the Learner's Perspective Study (LPS), an international collaboration of 16 countries with the aim of examining in an integrated and comprehensive fashion the patterns of participation in competently taught eighth grade mathematics classrooms.

Keywords: Mathematics classroom practice, Cross-cultural practice, Teaching strategies, Learning tasks, Student perspective

Mathematical Literacy for Living in the Highly Information-and-Technology-Oriented in the 21st Century: Mathematics Education from the Perspective of Human Life in Society*

Eizo Nagasaki
Shizuoka University, Japan
eenagas@ipc.shizuoka.ac.jp

This paper discusses mathematical literacy for living in our highly information-and- technology-oriented society in the 21st century. First, it inquires into the significance of thinking about mathematical literacy in terms of how it benefits modern individuals, as well as modern society. A summary of the past trends of mathematical literacy in Japan is given. This is followed by a consideration of a framework for thinking about mathematical literacy in the future. Here, the focus is on mathematical methods and the need to re-visit the meaning of studying mathematics. This is followed by a discussion of the design of school mathematics curricula that aim to nurture mathematical literacy. The discussion includes an examination of the general structure of school mathematics as it pertains to mathematical literacy, and the framework of school mathematics that addresses diversity. Concrete examples of the designs of school mathematics curricula based on research on mathematics education in Japan to date are given. Lastly, the maintenance and development of mathematical literacy outside school is touched upon.

Exploring the Nature of the Transition to Geometric Proof Through Design Experiments from the Holistic Perspective*

Masakazu Okazaki
  Okayama University, Japan
  masakazu@okayama-u.ac.jp

The gap between empirical and deductive reasoning is a global problem that has produced many students who have difficulties learning proofs. In this paper, we explore the conditions that aid students in entering into proof learning and how they can increase their ability before learning proofs through design experiments. First we discuss the theoretical backgrounds of the holistic perspective and didactical situation theory, and set our research framework as the transition from empirical to theoretical recognition consisting of the three aspects of inference, figure, and social influence. Next, we report our design experiments in plane geometry redesigned for the seventh grade, and examine how students may enter the world of proof by learning geometric transformation and construction as summarized in the three aspects of the framework. Finally, we suggest key ideas for designing lessons that promote transition.

Keywords: Transition to geometric proof, Holistic perspective, Empirical and deductive reasoning

Laying Foundations for Statistical Inference*

Maxine Pfannkuch
  The University of Auckland, New Zealand
  m.pfannkuch@auckland.ac.nz
  Chris J. Wild
  The University of Auckland, New Zealand
  c.wild@auckland.ac.nz

In this paper we give an overview of a five-year research project on the development of a conceptual pathway across the curriculum for learning inference. The rationale for why statistical inference should be part of students' learning experiences and some of our long deliberations on explicating the conceptual foundations necessary for a staged introduction to inference are described. Implementing such a pathway in classrooms required the development of new dynamic visualizations, verbalizations, ways of reasoning, learning trajectories and resource material, some of which will be elucidated. The trialing of the learning trajectories in many classrooms with students from age 13 to over 20, including some of the issues that arose, are briefly discussed. Questions arising from our approach to introducing students to inferential ideas are considered.

Keywords: Secondary-university students, Sampling variability, Visualizations, Verbalizations

Mathematics Education in Cambodia from 1980 to 2012: Challenges and Perspectives 2025*

Chan Roath
  Ministry of Education, Youth and Sport, Cambodia
  chan.roath@moeys.gov.kh

The Kingdom of Cambodia was a world leader in technology and scientific understanding from the ninth to the fifteen century as the Khmer Empire. Unfortunately the Pol Pot regime destroyed the education system in Cambodia between 1975 and 1979. The process of rebuilding the educational system of Cambodia was started by collecting the surviving educated people and by adapting the slogan: "The one who knows more teaches the one who knows less and the latter transfer's knowledge to illiterates". Mathematics education in Cambodia currently faces many problems such as a lack of well qualified teachers, a lack of knowledge in curriculum development, text book writing, methodology of teaching and use of ICT. Currently no quality assurance mechanism is available to ensure Cambodia's mathematics curriculum is up to international standards. The relatively low salary of teachers in the Kingdom remains an impediment to our educations system as it provides little motivation for people to become teachers.

The Cambodian Mathematical Society (CMS) was established on the 4th of March 2005 and recognized by the Royal Government of Cambodia to play a part in addressing the problems and improving the capacity of mathematical education in Cambodia. CMS is committed to promoting mathematics as a key "enabling" discipline that underlies other key disciplines and is at the heart of economic, environmental and social development in Cambodia.

A successful outcome for mathematical education in Cambodia depends on the creation and implementation of developmental goals that are appropriate for Cambodia.

The CMS has identified goals that will be made priorities in addressing the needs of mathematical education in Cambodia. These goals include improving the level of qualification of Cambodian mathematical teachers, upgrading the mathematical curriculum to a modern and internationally competitive level, improving the quality of teaching materials and textbooks available in the Khmer language, improving the pedagogical methods of teaching mathematics, promoting and supporting the use of information communication and technology (ICT) in mathematical instruction and encouraging participation in international mathematical programs and competitions as well as developing such competitions further in Cambodia.

Keywords: Cambodia, Mathematics education, Teaching skill, Information technology, Human resource.

The Challenges of Preparing a Mathematical Lecture for the Public*

Yvan Saint-Aubin
  Université de Montréal, Canada
  yvan.saint-aubin@umontreal.ca

As public curiosity and interest for science grow, mathematicians are invited more often to address a public that is not a classroom audience. Such a public talk should certainly convey "mathematical ideas", but it obviously differs from the classroom lesson. Preparing for such a talk offers therefore new challenges. I give examples from recent public lectures given by prominent mathematicians and by myself that try to tackle these challenges. I also reflect about how these efforts have changed my behavior in the classroom.

Keywords: Mathematical lecture, Public awareness, Public interest for mathematics, Science awareness, Mathematics communication.

Computer Aided Assessment of Mathematics Using Stack*

Christopher Sangwin
  University of Birmingham, United Kingdom
  C.J.Sangwin@bham.ac.uk

Assessment is a key component of all teaching and learning, and for many students is a key driver of their activity. This paper considers automatic computer aided assessment (CAA) of mathematics. With the rise of communications technology, this is a rapidly expanding field. Publishers are increasingly providing online support for textbooks with automated versions of exercises linked to the work in the book. There are an expanding range of purely online resources for students to use independently of formal instruction. There are a range of commercial and open source systems with varying levels of mathematical and pedagogic sophistication.

Numerical Analysis as a Topic in School Mathematics*

Shailesh A Shirali
  Rishi Valley School, India
  shailesh.shirali@gmail.com

Concerns about the divide between school mathematics and the discipline of mathematics are known in math education circles. At the heart of the debate is the sense that imperatives in school mathematics differ from those in the discipline of mathematics. In the former case, the focus is on remembering mathematical facts, mastering algorithms, and so on. In the latter case, the focus is on exploring, conjecturing, proving or disproving conjectures, generalizing, and evolving concepts that unify. It is clearly of value to find ways to bridge the divide. Certain topics offer greater scope at the school level for doing significant mathematics; one such is the estimation of irrational quantities using rational operations. This problem is ideal for experimentation, forming conjectures, heuristic reasoning, and seeing the power of calculus. The underlying logic is easy to comprehend. It would therefore be very worthwhile if we could make such topics available to students in high school.

Keywords: Numerical analysis, School mathematics, Discipline of mathematics, estimation, Irrational quantity, Rational operation.

Visualizing Mathematics at University? Examples from Theory and Practice of a Linear Algebra Course*

Blanca Souto-Rubio
Universidad Complutense de Madrid, Spain
blancasr@mat.ucm.es

With this communication, I will try to promote a discussion on visualization adapted to university level: how I understand it, why may it be important to understand advanced mathematics and, mainly, how it is currently taught. With this aim, five examples—obtained by the observation and my reflective practice in a Linear Algebra course—will be presented. The analysis of these episodes will enable deeper understanding of some issues of visualization in this particular context: relevant characteristics of visualization in Linear Algebra, some obstacles and opportunities of teaching visualization and some actions needed to improve the teaching of visualization at university level.

Keywords: Visualization, Linear algebra, Teaching at university level, Participant Observation.

On the Golden Ratio*

Michel Spira
Universidade Federal de Minas Gerais, Brazil
michelspira@gmail.com

The alleged appearances of the Golden Ratio Φ in natural phenomena, art, architecture and even literature have elevated it to a sort of mystical status inside and outside of Mathematics. In this article we present a skeptical view on this. We first discuss some of Φ's properties and then show that these properties are just particular cases of more general constructions. We then show how to find Φ when one believes it is there to be found. At the end, we discuss briefly one of the manifestations of the Φ cult.

Keywords: Golden ratio, Golden number, Geometry

The International Assessment of Mathematical Literacy: Pisa 2012 Framework and Items*

Kaye Stacey
    University of Melbourne, Australia
    k.stacey@unimelb.edu.au

The OECD PISA international survey of mathematical literacy for 2012 is based on a new Framework and has several new constructs. New features include an improved definition of mathematical literacy; the separate reporting of mathematical processes involved in using mathematics to solve real world problems; a computer-based component to assess mathematical literacy as it is likely to be encountered in modern workplaces; and new questionnaire items targeting mathematics. Procedures for quality assurance that arise in the preparation of an assessment for use in many countries around the world are illustrated with some items and results from the 2011 international field trial. The paper will provide background for the interpretation of the results of the PISA 2012 survey, which were published in December 2013.

Keywords: Mathematical literacy, Assessment, Comparative studies, Computer-based assessment, Achievement, Mathematical competencies.

Applications and Modelling Research in Secondary Classrooms: What Have We Learnt?*

Gloria Stillman
    Australian Catholic University, Australia
    gloria.stillman@acu.edu.au

This paper focuses on my 20 year program of research into the teaching and learning of applications and modelling in secondary classrooms. The focus areas include the impact of task context and prior knowledge of the task context during the solution of applications and modelling tasks, mathematical modelling in secondary school, and metacognition and modelling and applications. Some of the analysis tools used in this research are also presented.

Keywords: Modelling, Secondary school, Applications

Conflicting Perspectives of Power, Identity, Access and Language Choice in Multilingual Teachers' Voices*

Lyn Webb
    Nelson Mandela Metropolitan University, South Africa
    lyn.webb@nmmu.ac.za

Teachers in the Eastern Cape, South Africa teach mainly in English, which is not their home language. In order to elicit their inner voices about language conflicts and contradictions in their classrooms they were encouraged to write poetry about their perceptions of the impact of language in their lives. The most prevalent contradiction they expressed was the power and dominance of English juxtaposed against the subordination of their home languages. English gave them access to education and upward employment mobility, whereas they were excluded from various discourses when they used their home languages. Their home languages legitimised and defined their identities, but appeared to be negated in an educational and economic environment. Since the necessity for pupils to become fluent in English conflicted with the pupils' difficulties in understanding content knowledge expressed in English, the teachers faced a choice between teaching in English (for access to social goods) or their home language (for epistemological access), or both. The use of poetry evoked feelings and emotions that may not have been as obvious, or as evocative, if other data-gathering methods had been used. It appears that the self-reflection embodied in the poetry gave the teachers a sense of empowerment, self-realisation and solidarity.

Keywords: Multilingual teachers, Language, Power, Access, Identity, Language choice, Poetry

What Does It Mean to Understand Some Mathematics?*

Zalman Usiskin
The University of Chicago, USA
z-usiskin@uchicago.edu

Mathematical activity involves work with concepts and problems. Understanding mathematical activity in mathematics education is different for the policy maker, the mathematician, the teacher, and the student. This paper deals with the understanding of a concept in mathematics from the standpoint of the student learner. We make the case for the existence at least five dimensions to this understanding: the skill-algorithm dimension, the property-proof dimension, the use-application (modeling) dimension, the representation-metaphor dimension, and the history-culture dimension. We delineate these dimensions for two concepts: multiplication of fractions, and congruence in geometry.

Keywords: Curriculum, Mathematical understanding, Fractions, Congruence, Mathematical concepts

Mapping Mathematical Leaps of Insight*

Caroline Yoon
  The University of Auckland, New Zealand
  c.yoon@auckland.ac.nz

Mathematical leaps of insight—those Aha! moments that seem so unpredictable, magical even—are often the result of a change in perception. A stubborn problem can yield a surprisingly simple solution when one changes the way one looks at it. In mathematics, these changes in perception are usually structural: new insights develop as one notices new mathematical objects, attributes, relationships and operations that are relevant to the problem at hand. This paper describes a novel analytical approach for studying these insights visually using "mathematical SPOT diagrams" (SPOT: Structures Perceived Over Time), which display evidence of the mathematical structures students perceive as they work on problems. SPOT diagrams are used to compare the conceptual development of two pairs of participants, who investigate whether a gradient (derivative) graph yields information about the relative heights of points on its antiderivative; one participant pair experiences a leap of insight, whereas the other does not. Each pair's SPOT diagrams reveal key differences in the structural features they attend to, which can account for the disparate outcomes in their conceptual development.

Keywords: Mathematical insight; Calculus; SPOT diagrams; Mathematical structure

Mathematics Competition Questions: Their Pedagogical Values and an Alternative Approach of Classification*

Tin Lam Toh
  National Institute of Education, Nanyang Technological University, Singapore
  tinlam.toh@nie.edu.sg

In this paper, it is argued that the role of the various mathematics competitions could be expanded beyond helping the nation in identifying and developing the mathematically gifted students. Through an examination of some mathematics competition questions, it was identified that these competition questions could serve to help the general student population to (1) acquire mathematical problem solving processes through acquiring or developing a problem solving model; (2) learn mathematics beyond the constraint of the school mathematics curriculum; (3) deepen students' understanding of school mathematics; and (4) acquire mathematical techniques which are rendered obsolete by the evolving technology. With the availability of vast resource on competition questions, an alternative approach to classify the competition questions based on the function it could serve in the usual mathematics classroom is proposed.

Keywords: Mathematics competition, Problem solving, Classification

The Examination System in China: The Case of Zhongkao Mathematics*

Yingkang Wu

East China Normal University, China

ykwu@math.ecnu.edu.cn

Examination is a critical issue in education system in China. Zhongkao is a kind of graduation examination of junior high school, and at the same time, the entrance examination to senior high school. This paper describes the structure, features and changes in zhongkao mathematics papers in China based on a detailed analysis of 48 selected zhongkao mathematics papers from eight regions in recent six years. It is observed that the zhongkao mathematics papers stress computation, reasoning, and relations among different mathematics topics, but are less emphasized on applications of mathematics in real context. There are obvious region differences in zhongkao mathematics papers, with regions from west economic zone relatively less demanding and regions from east and central economic zones more demanding. Changes like more process-oriented questions and more real context questions are found. Examples of examination items are given to illustrate the identified features and changes.

Keywords: Zhongkao mathematics, Examination, Features, Changes, Junior high school graduates

Mathematics at University: The Anthropological Approach*

Carl Winsløw

University of Copenhagen, Denmark

winslow@ind.ku.dk

Mathematics is studied in universities by a large number of students. At the same time it is a field of research for a (smaller) number of university teachers. What relations, if any, exist between university research and teaching of mathematics? Can research "support" teaching? What research and what teaching?

In this presentation we propose a theoretical framework to study these questions more precisely, based on the anthropological theory of didactics. As a main application, the links between the practices of mathematical research and university mathematics teaching are examined, in particular in the light of the dynamics between "exploring milieus" and "studying media".

Keywords: University mathematics, Tertiary, Anthropological theory of the didactical.

Hidden Cultural Variables to Promote Mathematics and Mathematics Education—Are There Royal Roads?

Guenter Toerner

  University of Duisburg-Essen, Germany

  guenter.toerner@uni-due.de

Being a research mathematician as well as a researcher in mathematics education and teacher education the author is reflecting the history of mathematics and mathematics education in the last fifty years from the point of view as an executive committee member of a national mathematics society. Especially, he is interested in sociological aspects of mathematics learning and teaching.

It is self-evident for mathematics departments to recruit as many students as possible and to have as many as possible graduate successfully. However, the so-called success rate differs from country to country and is—in the country of the author—by no means pleasant. Are we as mathematicians aware of the various figures in our country, at our university?

The author also analysed the situation in his department as chair of the committee of education for which he is in charge for the European Mathematical Society. It is quite astonishing the cultural framework seems to influence the situation in the study of mathematics, e.g. the percentages of students studying mathematics in the group of all students differ enormously, they count 0.3 % in the Netherlands and 2.8 % in Germany. Of course, there are some obvious reasons and explanations and long-lasting strong traditions, but also there are some hidden variables. How do our learned societies reflect upon these parameters? Do they have master plans?

Mathematics at school needs friends, but who are the friends of mathematics? In some countries mathematicians and mathematics educators are fighting math wars, fortunately not in the author's country. There are very often quite rational explanations for the superfluous struggles, but how can these wars be ended and peace established. The author tries again to explain these issues from a European point of view and gives some recommendations. Shouldn't we internationalize these issues to develop joint initiatives?

Finally, we have to accept that teachers are the most influential people to promote mathematics. They are the stakeholders for mathematics education. Again, there are large differences between learned societies when it comes to caring for teachers and how to attract teachers. What is the percentage of teachers in our learned societies? How can we attract teachers to become members?

Since ICM Berlin in 1998, the German mathematical society has made large progress to acknowledge the work of teachers—at all grades—and to offer communication at eye level. Progress is small; however, we should be patient as we have to change attitudes on both sides. Recently we were successful to convince the Deutsche Telekom Stiftung for financing a national German institute of excellence for mathematics teacher education (DZLM) in which the author is involved. Again, some insights, ideas and recommendations are presented. What is the role mathematics at university might play to contribute to a continuous professional development of teachers for all grades?

Use of Student Mathematics Questioning to Promote Active Learning and Metacognition*

Khoon Yoong Wong
National Institute of Education, Nanyang Technological University, Singapore
khoonyoong.wong@nie.edu.sg

Asking questions is a critical step to advance one's learning. This lecture will cover two specific functions of training students to ask their own questions in order to promote active learning and metacognition. The first function is for students to ask themselves mathematical questions so that they learn to think like mathematicians who often advance knowledge by asking new questions and trying to solve them. This is also called problem posing, an important component of the "look back" step in the Polya's problem solving framework. The second function is for students to ask their teachers learning questions during lessons when they do not understand certain parts of the lessons. Students who are hesitant to ask learning questions need to be inducted into the habit of doing so, and a simple tool called Student Question Cards (SQC) can help to achieve this objective. These SQC cover four types of mathematics-related learning questions: meaning, method, reasoning, and applications. In a pilot study involving Grades 4 and 7 Singapore students, every student was given a set of these laminated cards. During lessons, the teacher paused two or three times and required the students to select questions from SQC to ask to clarify their doubts. This reverses the normal roles of teacher and students during classroom interactions. Teachers and students in this pilot study expressed mixed responses to the use of SQC. These two functions of student mathematics questioning have the potential to promote active learning of mathematics among school students through strengthening their metacognitive awareness and control. To realize this potential, teachers need to pay due attention to the science, technology, and art of student questioning.

Keywords: Student questions, Problem posing, Metacognition, Buddha, Confucius, Socratic dialogue

Open Access This chapter is distributed under the terms of the Creative Commons Attribution Noncommercial License, which permits any noncommercial use, distribution, and reproduction in any medium, provided the original author(s) and source are credited.

Part VIII
Topic Study Groups

Mathematics Education at Preschool Level

Tamsin Meaney

Introduction: The Aim and the Focal Topics

At the present time, research into mathematics education in the early years is receiving much attention internationally. There is much debate about whether mathematics teaching/learning in the early years should be about supporting children to develop their own interests or to prepare them for school. Alongside this debate is interesting research which shows young children's capabilities on working abstractly with a range of mathematical topics, previously considered too advanced. This topic study group of ICME 12 aimed to provide a forum for exchanging insights in early mathematical learning. While much research has focused on children's learning of number, a growing body of work examines the learning of geometry, measurement and other mathematical topics in preschool. TSG 1 provided a forum for sharing this work and exploring how the learning of these aspects of mathematics in pre-school can be strengthened. It also supported discussion of the preschool teacher education across different countries.

Participants took part in four sessions. Three sessions (1, 2, and 3) were devoted to research and project presentations and the discussions based on these presentations. Session 4 was organised as a discussion to outline a general research agenda. This session finished with a proposal for the group to write a book.

Organizers Co-chairs: Camilla Bjorklund (Finland), Malilyn Talor (New Zealand); Team Members: Haejyung Hong (Korea), Elin Reikeras (Norway); Laison IPC Member: K. (Ravi) Subramaniam (India).

T. Meaney (✉)
Bergen University College, Bergen, Norway
e-mail: tamsin.meaney@hib.no

© The Author(s) 2015
S.J. Cho (ed.), *The Proceedings of the 12th International Congress on Mathematical Education*, DOI 10.1007/978-3-319-12688-3_26

Presentations

Six presentations were made by researchers from 5 different countries. From Korea, Haekyung Hong discussed how after preschool mathematics classes had little academic gain for children even though many Korean children attended these classes. There were two presentations from Sweden where preschool children are expected to learn through play. In the first presentation by the research group Små Barn Matenatik, videos of children playing at one preschool were analysed using Alan Bishop's 6 mathematical activities. The second presentation by a Swedish graduate student, Laurence Delacour, discussed preschool teachers' adoption of a new curriculum from the perspective of the didactical contract. Oliver Thiel from Norway described a comparative study between Germany and Norway that looked at preschool teachers' competencies. Although most research on mathematics education in preschool tends to be about older children, Shiree Lee, New Zealand, presented research on very young children's exploration of space. The final presentation was by Brian Doig on a paper written with Connie Ompok on a cross-country investigation of games used to assess young children's mathematical knowledge.

Nosisi Feza, from South Africa, presented her poster about preschool teachers' knowledge of teaching mathematics and linked this to concerns about inequitable opportunities to learn.

Final Session

The final session was taken up with a general discussion about how mathematics education in preschools in different countries was conceptualised. In Sweden, 95 % of children attend preschool from the age of 12 months. They do not begin school until 7 years old and have an intermediate year, called preschool class, which still works with the preschool curriculum but acts as a bridge to school. However, in other countries a much smaller proportion of children attend preschools and school can begin as early as 4 years old (Ireland). Very few countries have a formal preschool curriculum and when they do there are differences between whether the focus is on the opportunities that preschools provide or on what children should learn. Preschool teacher education also differs with some countries requiring at least some staff at preschools to be university educated to other countries where staff have school qualification (Germany). Regardless of the education that preschool teachers have had, it seems that many governments are implementing professional development programs for teachers.

The active interest of governments in setting policies for early childhood sector makes this one of the most rapidly changing education fields in mathematics education. Consequently, one outcome of this topic group was the suggestion that a book should be written to document the current situation across the world.

Nosisi Feza from South Africa formulated the book proposal and our next step is to look for a publisher.

Framework for the Early Childhood Development of mathematics education across countries.

Introduction

- How is ECD mathematics education perceived in your country? What components of ECD are seen more important than others? What is the status of ECD provision generally and who is responsible for providing ECD facilities? What is the role of parents in the decision making? What drives ECD provision?

Historical Background

- The history of mathematics early childhood development in each country in terms of policies, national plans and challenges
- Reasons for change if any

Current Status

- Structural levels of ECD in your country e.g., ECD age range, beginning of formal schooling, preschool ages and structures etc.
- ECD policies
- Types of ECD facilities and their purpose
- Purpose of ECD provision generally
- Funding sources for ECD provisioning
- Departments that affiliate to ECD
- Access to ECD facilities in different settings
- ECD educator qualifications and training

Mathematics Education

- Visibility of mathematics education in Policies
- Curriculum with the focus to mathematics education
- Research and research funding on mathematics education in early years
- Monitoring systems for implementation of mathematics in ECD
- Regarding the current literature on early childhood education where is your country?
- What programmes and actions are taken towards ECD mathematics development in your country?
- What gaps do ECD mathematics practices have
- Diversity in the mathematics exposure from home to care and other ECD facilities
- What conclusions does the data make?
- What suggestions are conclusions making for policy, practice, and research of mathematics education in the ECD of your country

Open Access This chapter is distributed under the terms of the Creative Commons Attribution Noncommercial License, which permits any noncommercial use, distribution, and reproduction in any medium, provided the original author(s) and source are credited.

Mathematics Education at Tertiary Level and Access to Tertiary Level

Ansie Harding and Juha Oikkonen

Structure of TSG2 Sessions

TSG 2 had 4 sessions of 90 min each, themed as follows:

- Session 1: Teaching philosophies and professional development
- Session 2: Teaching practices
- Session 3: Student experiences/learning, also e-learning
- Session 4: Transition from school to university

The four sessions were all structured similarly. Presentations were classified as either long (15 min) or short (10 min). A session started with one long presentation followed by four or five short presentations (20 presentations in total). Each session closed with a discussion of 15 min. In addition three posters were discussed in the third session and displayed in the exhibition area.

Organizers Co-chairs: Ansie Harding (South Africa), Juha Oikkonen (Finland); Team Members Christopher Sangwin (UK), Sepideh Stewart (New Zealand), Miroslav Lovric (Canada), Sung-Ock Kim (Korea); Liaison IPC Member: Johann Engelbrecht (South Africa).

A. Harding (✉)
University of Pretoria, Pretoria, South Africa
e-mail: aharding@up.ac.za

J. Oikkonen
University of Helsinki, Turku, Finland
e-mail: Juha.Oikkonen@helsinki.fi

© The Author(s) 2015
S.J. Cho (ed.), *The Proceedings of the 12th International Congress on Mathematical Education*, DOI 10.1007/978-3-319-12688-3_27

General Comments

We are happy to report that TSG 2 ran smoothly and encountered no problems whatsoever. The team worked well together in organising the event before the time. Everyone stuck to deadlines and was forthcoming in suggestions and comments. During the conference itself the team members acted as chairs of the four sessions, respectively, and managed to create coherence amongst the attendees. The sessions were all well attended, drawing approximately 40 delegates per session. It was noticeable that many delegates seemed to develop a sense of belonging to TSG 2 and attended throughout. They were spontaneous in presenting questions and comments, especially during discussion sessions. Unfortunately co-chair Juha Oikkonen had to cancel attendance shortly before the conference on grounds of a medically related problem. He was extremely disappointed not to attend, having contributed in every respect to organizing TSG 2.

Comments Per Session

Session 1: Teaching Philosophies and Professional Development

This session kicked off with a presentation by the well-known twosome John and Annie Seldon, from the USA, a well-received presentation addressing the issue of student success in problem solving. This presentation was followed by four speakers giving an Iranian (Khakbaz Azimeh Sadat), Irish (David Wraith & Anne O'Shea) and Canadian (Miroslav Lovric) perspective, respectively, on related topics. The final presentation in this session was by Leigh Wood from Australia reporting on graduate skills necessary for successful transition from university to the professional environment.

- Annie Selden & John Selden: A Belief Affecting University Student Success in Mathematical Problem Solving and Proving
- Khakbaz Azimeh Sadat: How do Iranian Graduate Students Learn to Teach Collegiate Mathematics as Future Mathematics Professors?
- David Wraith & Anne O'Shea: The use of problem-solving techniques as a learning tool in university mathematics courses
- Miroslav Lovric: Learning Mathematics in an Interdisciplinary Science Program
- Leigh Wood: Preparing our graduates for the workforce

Session 2: Teaching Practices

This session started with a team of young but extremely competent educators from Finland describing an effective system introduced into tutorial sessions. Their enthusiasm added to the success of the session. The subsequent presentations described teaching practices from a variety of countries and a variety of perspectives, providing ample material for discussion.

- Terhi Hautala, Tiina Romu, Thomas Vikberg, Johanna Ramo: The Extreme Apprenticeship Method in Teaching Mathematics at University Level
- Olof Viirman: The Teaching of Functions as a Discursive Practice? University Mathematics Teaching from a Commognitive Standpoint
- Tolga Kabaca: Teaching the Cycloids by the use of Dynamic Software: Abstraction Process of Hypocycloid and Epicycloids Curves
- Liu Jiao & Yao Jing: The Application of Problem-based Learning in Higher Vocational Mathematics Teaching
- Rad Dimitric: Feedback from students' exams. A case study.

Session 3: Student Experiences/Learning, also E-Learning

The first presentation in this session was by Sepideh Stewart from New Zealand speaking on reactions of students to a particular approach to Linear Algebra. The presentation was informative and eloquently presented. Presentations focussed on how students learn and their experiences in doing so. Only one presentation was given on e-learning, perhaps surprisingly so as online learning is topical worldwide.

- Sepideh Stewart: Student Reactions to an Approach to Linear Algebra Emphasising Embodiment and Language
- Ann O'Shea, Sinead Breen, Kirsten Pfeiffer: An Evaluation of the Impact of Non-Standard Tasks on Undergraduate Learning
- Jeremy Zelkowski: Student Accountability & Instructor Variability: A research study in a terminal, required, applications focused calculus course.
- James Musyoka, Joyce Otieno, David Stern: Using e-learning to engage Mathematics and Statistics Students in a Kenyan University
- Ciriaco Ragual & Ester Ogena: Difficulties and Coping Mechanisms in Solving Mathematics Problems
- Diez-Palomar Javier: Family math education: New trends and possibilities for in the realm of mathematics at tertiary level (Poster)

- Haitham Solh: Strategies for Effective Teaching and Learning in Collegiate Mathematics Service Courses for Diverse Students (Poster)
- Ildar Safuanov: Design of a system of teaching elements of group theory (poster)

Session 4: Transition from School to University

Transition from school to university is a general problem as became apparent during this session. Ansie Harding gave the first presentation describing the problem faced in South Africa in this regard. Other presentations described transition problems experienced elsewhere in the world.

- Ansie Harding: On the horns of a dilemma: The transition from school to university in South African
- Randall Pyke: Initiatives at Simon Fraser University in First Year Mathematics and in the Transition from High School to University
- Lee Ji hyun: The Secondary-Tertiary Transition of the Axiomatic Method
- Hoda Ashjari: Recognising Texts in Undergraduate Mathematics Education

Conclusion

The four sessions were well-attended and enjoyed by all those who attended. New ties were established and collaboration possibilities were communicated.

Open Access This chapter is distributed under the terms of the Creative Commons Attribution Noncommercial License, which permits any noncommercial use, distribution, and reproduction in any medium, provided the original author(s) and source are credited.

Activities and Programs for Gifted Students

Peter Taylor and Roza Leikin

Introduction: The Aim and the Focal Topics

The aim of TSG-3 at ICME-12 was to gather educational researchers, research mathematicians, mathematics teachers, teacher educators, designers and other congress participants for the international exchange of ideas related to identifying and nourishing mathematically gifted students. The focal topics presented at the TSG-3 included but were not restricted to theoretical models of giftedness, the relationship between creativity and giftedness and the empirical research that will contribute to the development of our understanding in the field. Participants discussed effective research methodologies and research innovations (e.g., brain research) in the field of mathematical giftedness; the findings of qualitative and quantitative studies related to high mathematical promise, its realization, and the relationship between mathematical creativity and mathematical talent. Additional attention was given to the profiles of the gifted child: their range of interests, ambitions and motivations, social behaviour, how and at what age their giftedness is discovered or developed.

Educators who participated in TSG-3 discussed instructional design directed at teaching the gifted as well as development of appropriate didactical principles. The discussions were focused on the ways that lead students to discover and realize their

---

Organizers Co-chairs: Peter Taylor (Australia), Roza Leikin (Israel); Team members: Viktor Freiman (Canada), Linda Sheffield (USA), Mihaela Singer (Romania), Bo Mi Shin (Korea); Laison IPC Member: Shiqi Lee (China).

P. Taylor (✉)
University of Canberra, Canberra, Australia
e-mail: pjt013@gmail.com

R. Leikin
University of Haifa, Haifa, Israel
e-mail: rozal@edu.haifa.ac.il

© The Author(s) 2015
S.J. Cho (ed.), *The Proceedings of the 12th International Congress on Mathematical Education*, DOI 10.1007/978-3-319-12688-3_28

mathematical talents, and the ways of developing mathematical innovation at high level. The participants discussed mathematical activities that are challenging, free of routine, inquiry-based, and rich in authentic mathematical problem solving; types of mathematics suitable for challenging gifted students; creation of mathematics challenges; out-of-school ways of fostering giftedness, e.g., mathematics clubs, mathematical shows and competitions.

Last but not least we paid attention to teacher education aimed at mathematics teaching that encourages mathematical promise and promotes mathematical talents, including issues of the psychology of teaching talented students, socio-cultural and affective characteristics of the mathematically gifted, and the types of mathematics and pedagogy suitable for educating teachers of gifted students.

Participants took part in four sessions. Three sessions (1, 2, and 4) were devoted to research and project presentations and the discussions based on these presentations. Session 3 was organised with round table presentations. In what follows we present main topics of the sessions and some examples of the studies and projects presented at the TSG-3 at ICME-12.

Examples and Main Insights

Opening the Discussion

Session 1 was devoted to introduction to the central topics of the TSG. Three lectures, by Linda Sheffiled, Roza Leikin and Alexander Soifer, opened three main reviews of the TSG: international projects for realisations of students' mathematical potential with special emphasis on high mathematical potential (REF), systematic research on characterisation of mathematically gifted students, and mathematics for mathematically gifted.

Linda Shefield's talk "Mathematically Gifted, Talented, or Promising: What Difference Does It Make?" stressed the importance of the developmental perspective of mathematical abilities and the importance of providing each and every student with oportunities to realise these abilities. Based on the position that science, technology, engineering, and mathematics (STEM) are critical to the economy, security, and future of the world, Linda Sheffield argued that we need students who will become adults who understand the complexities of a technological world, who ask the essential questions to safeguard that world, and who will become the leaders, researchers and innovators in the STEM fields of the future. According to Sheffield, too often, in the United States, these students go unrecognized, unmotivated, and under-developed at a time when they are most vital. Sheffield discussed in her presentation whether the way we historically define these future STEM leaders and innovators has an effect upon their growth and development. This talk served as a starting point to the discussion of the international project devoted to the realization of students' intellectual potential related to STEM.

Roza Leikin stressed the importance of conducting systematic and well-designed research on the characteristics of mathematically gifted students. She presented large-scale Multidimensional Examination of Mathematical Giftedness that she conducts with colleagues from the research group in the University of Haifa (Mark Leikin, Ilana Waisman, Shelley Shaul). The presentation was devoted to brain activity (using ERP- Event-Related Potentials—methodology) associated with solving mathematical problems that require transition from a geometrical object to a symbolic representation of its property. Some 43 right-handed male students with varying levels of general giftedness (Gifted-G, Non-gifted-NG) and of mathematical expertise (Excelling-E, Non-excelling-NE) took part in the study. The researchers aimed to investigate the differences in brain activation among four groups of participants (G-E, G-NE, NG-E, and NG-NE). The findings demonstrated different patterns of brain activity associated with problem solving among the four experimental groups. In educational practice the results suggest that different groups of the study population need specific instructional approaches to realize fully their intellectual potential.

Alexander Soifer claimed that mathematics cannot be taught, it can only we learned by our students while doing it. According to Soifer, the classroom ought to be a laboratory where students actually touch the subject, overcome difficulties, which we sometimes call problem solving. "What kind of problems?"—asked the author, and answered: "here comes Combinatorial Geometry!" It offers an abundance of problems that sound like a "regular" school geometry, but require for their solutions synthesis of ideas from geometry, algebra number theory, and trigonometry and thus they are rich, challenging and insightful, and thus appropriate for the education of mathematically talented individuals.

When the three presenters finished their presentations it became clear that the contrast between the presentations enlightened the importance and openness of the following questions: Who are the mathematically gifted? Can giftedness be developed or rather is it realized? How do different perspectives on giftedness determine research and practice in the education of the mathematically gifted? and What kinds of mathematics problems are most appropriate to mathematically gifted?

International Experiences and Projects for Gifted

The second session was devoted to the projects of different kinds directed at educational activities with mathematically advanced students.

Mark Saul described activities of the Center for Mathematical Talent (CMT) at the Courant Institute of Mathematical Sciences (New York University) which was organized in the fall of 2010. Its mandate is to identify and support mathematically talented students in and around the New York City area—especially those from backgrounds where such services have traditionally been weak. The goal at the CMT is to create institutions, materials, and practices that will unlock and nurture these abilities in students, and will have an impact both on their lives as individuals and on the society in which they live.

Ildar Safuyanov reported on the experiences of fostering creativity of pupils in Russia. While the creative approach is understood by the authors and his colleagues as certain abilities and readiness of a person for creating something new, the purpose of educational process at school is the education of a person who would use a creativity approach for solving scientific or practical problems and for thinking independently. According to Safuyanov, differentiated teaching is an effective way of promoting creativity in conditions. Ildar Safuyanov discussed and compared different types of differentiated teaching and provided the audience with examples of internal differentiation by level of mathematical tasks.

Abraham Arcavi presented the Math-by-Mail project which is an online, interactive, extracurricular enrichment program in recreational mathematics conducted by mathematics educators from the Weizmann Institute of Science in Israel (leaders- Yossi Elran, Michal Elran, Naama Bar-On). Participants of the Math-by-Mail project are engaged in a multi-sense learning experience involving many skills such as comprehension, solving enquiry based problems and correspondence with mathematicians. The lecture demonstrated the scope of the program, its pedagogical and technological characteristics and its benefits for the talented math student.

Viktor Freiman from the University of Moncton, Canada, shared his innovative experience of designing and conducting professional learning communities with inclusive practices for students who "already know". In his project, mathematically gifted and talented students contribute to the virtual community. Same research findings demonstrated the effectiveness of the suggested approach as well as its complexity.

Duangnamol Tama reported on the project named "The Development and Promotion of Science and Technology Talented Project (DPST)". The project is supported by the the Thailand government. Thus national education focuses its efforts and policies on the national development of science, mathematics, and technology through the promotion of high caliber students in these areas.

At the end of this session the participants were exposed to the variety of approaches and variety of ideas directed at promotion of the mathematically gifted. Further discussion between the participants of the session was directed at answering the questions: Which features of the programs for mathematically gifted are culturally dependent and which of them are intercultural? Can successful projects from one country be applied in another country with a different cultural heritage? Do inclusive programs suit needs of the gifted?

Didactical Approaches and International Perspectives

At Session 3 participants of the TSG-3 were exposed to different didactical approaches and international perspectives on the education of mathematically advanced students. This was a round tables session. The authors were provided with an opportunity to present their papers several times to different people who were interested in their presentations. The groups changed each 10 min and each participant had an opportunity to learn about several works presented at this session. These works included:

- The program of making students create math problems: One of the methods of developing students' abilities to think and express by Nobuo Itoh from Japan,
- The role of student motivation in developing and assessing the acquisition of higher-order thinking skills, by Vincent Matsko, USA
- How the mathematically gifted and talented senior primary school students in Hong Kong understand mathematics, by Wai Lui Ka, Hong Kong
- The research on the mode of motivating the gifted students, by Wang He Nan, Beijing
- Enhancing mathematical research in high school, by Laura Morera, Spain
- Mathematical creativity and attachment theory: an interdisciplinary approach for studying the development of mathematical creativity of preschool children with a precarious childhood, by Melanie Münz, Germany.
- Problem modification as an indicator of deep understanding, by Mihaela Singer Florence, Romania
- Little University of Mathematics, by Laura Freija, Latvia
- Effects of Modified Moore Method on Elementary Number Theory for Gifted High School Students: An Exploratory Study, by Hee Kyoung Cho, Korea,
- Korean Middle School Student's Spatial Ability and Mathematical Performance: Comparison between Gifted Students and General Students by Sungsun Park, Korea

These presentations ended up with multiple questions about the research conducted by different participants and the practices implemented in different countries. The need for the better connections between theory and practice become more and more clear. Following this session we ask: What research approaches can inform us in the best way? How does research methodology depend on definition of gifted chosen in the study? How research and practice can be interwoven to advance theories of mathematical giftedness and advance effectiveness of the practical projects for mathematically gifted students.

Characteristics of Mathematically Gifted Students

The fourth session of the TSG focused on characterization of mathematically gifted students.

BoMi Shin from South Korea reported on a study that provided probability tasks to mathematically gifted students to investigate analogical reasoning as it emerges during the problem-solving process of students. Atsushi Tamura from Japan presented a case study about a gifted high school student in which he identified 5 prominent characteristics in thinking processes by investigating how he devised mathematical proof. Furthermore, this study found that sharing the thinking process of the gifted in the classroom had a good effect on both the class and the gifted himself.

Amaral Nuno and Susana Carreira from Portugal described analysis of creativity in the problem solving processes presented by eight students (from grades 5 and 6, aged between 10 and 11) who have participated in and reached the final phase of a Mathematical Competition. They suggested ways for evaluation of students' creativity in mathematical problem solving in a situation that includes a competitive factor and takes place beyond the mathematics classroom, which is often seen as restrictive for the development of mathematical creativity.

Brandl Matthias from Germany (in collaboration with Christian Barthel) suggested that there are two ways of selecting promising students for the purpose of fostering (in mathematics): whereas the standard procedure is to offer additional courses or material for volunteers or those chosen by the teacher, the other and perhaps more elitist—but with respect to quantitative aspects easier—way is to select the students with the best marks. Brandl argued that from a psychological perspective these ways represent two opposite sides of the causality between giftedness and assessment. One result of this investigation is the finding of strong correlations between the profiles of mathematical interests of specific subgroups that fulfill the characteristics which define mathematical giftedness.

The lecture by Marianne Nolte discussed relationships between "High IQ and High Mathematical Talent!". The findings followed from the long-term PriMa-Project in the University of Hamburg. This project is a research project and a project for fostering mathematically talented children. To detect among them mathematically especially talented children demands a highly comprehensive search for talents. Marianne Nolte stressed the complexity of the evaluation of mathematical talent and stressed that search for talent poses the risk that children may be classified wrongly as especially talented or that children's talents are not recognised.

In conclusion the following questions were raised by the group: Do we know more than Krutetskii after we perform studies on characteristics of students with high mathematical abilities? How do researchers choose their research paradigm? How do research methodologies correspond to the students' age or to a specific characteristic of giftedness that is examined? How studies on students thinking can/should inform educational practices?

The work of the group demonstrated how much is done in the field of the education of mathematically advanced students but moreover it stressed how much should be done in order to get a better understanding of the phenomena of mathematical giftedness and the effective ways of realization of mathematical potential in all students including mathematically talented ones.

Acknowledgments We would like to thank all the participants for their interest in the topic, all the contributors for their interesting presentations and hard work at this TSG, Team members—Viktor Freiman, Linda Sheffield, Mihaela Singer, Bo Mi Shin—for fruitful collaboration in preparation and conducting this TSG.

Open Access This chapter is distributed under the terms of the Creative Commons Attribution Noncommercial License, which permits any noncommercial use, distribution, and reproduction in any medium, provided the original author(s) and source are credited.

Activities and Programs for Students with Special Needs

Jean-Philippe Drouhard

Scope and Aims of the TSG 4

Around the world, a considerable number of primary and secondary teachers are involved in teaching mathematics to special educational needs learners ("SEN-L") and a fair proportion of teacher educators are involved in preparing these teachers. But both, teachers and educators, very often are working under somewhat isolated circumstances. They are isolated geographically—it is not always easy to identify others working with SEN-L regionally, let alone nationally or internationally. And they are also isolated in terms of particular focus—specialists working with blind students, for example, may have little professional contact, if any, with specialists in the education of deaf students and those of Down's syndrome. Professional groups tend to be based more on the nature of the special needs of the students rather than on the learning of mathematics. This means that in the dialogue amongst educators concerned with SEN-L, mathematics education is hardly ever at centre stage. On the other hand, mathematics education researchers and teachers seldom have the specific knowledge about SEN-L. Mathematics educators do consider what mathematics for all should be, but the "all" rarely include SEN-L. Issues related to the mathematics education of students with special educational needs are currently under represented in the research community. What seems to be lacking is a *community* of mathematics educators dedicated to exploring this domain. Hence, there is a need to create common references and shared resources (in particular in

Organizers Co-chairs: Jean-Philippe DROUHARD (France) and Sung-Kyu CHOI; Team Members: Heloiza BARBOSA, Petra SCHERER, Jacinthe GIROUX; Liaison IPC Member: Bernard HODGSON.

J.-P. Drouhard (✉)
Universidad de Buenos Aires, Buenos Aires, Argentina
e-mail: jpdrouhard@ccpems.exactas.uba.ar

© The Author(s) 2015
S.J. Cho (ed.), *The Proceedings of the 12th International Congress on Mathematical Education*, DOI 10.1007/978-3-319-12688-3_29

the case of inclusive education). In short, there is a strong need for a common culture of mathematics education for students with special educational needs.

At the end of the meeting it was agreed to create some kind of common Internet platform in order to communicate about how to start such a web of research and shared experience on special educational needs learners. A website (at the moment under construction) has been opened: https://sites.google.com/site/m4senl/ Contact: maths4senl@gmail.com.

What could Mathematics Education gain from the establishment of such common references and resources? First, mathematics education could become more significant in the lives of many students. There is a large number of young people and adult students for whom mathematics teaching may be "secondary" because the focus of their education is elsewhere. Second, insights developed in research with SEN-S could benefit mainstream mathematics teaching, through a re-analyses of assumptions about how mathematics is learned and what specific assessments tell us about students' abilities. Third, SEN-S may show unexpected dissociations between different aspects of mathematical knowledge. It is possible to find, for example, exceptional computational skills with little understanding of their conceptual basis in autistic. Finally, the discussion of different sorts of curricula with different resources appropriate for mathematics teaching while keeping mathematics as the focus of the discussion could lead to more diversified approaches to mathematics education.

Abstracts of the Communications and Posters Presented Within the TSG Meeting

Renato MARCONE, Miriam GODOY PENTEADO[1]: *A blind student at the university: Challenges for mathematics teachers.*
This presentation is based on the story of Mara, a student who became blind during a mathematics undergraduate course. The information for this case were obtained from interviews with Mara, her mother, university staff, colleagues and her teachers. As no blind student had ever before been at the mathematics faculty in question, the case of Mara took everyone by surprise. The first reaction from teachers was that Mara should take another subject—mathematics would be too difficult. However, given that Mara did not change her mind, the university staff had to define actions that would allow her to continue studying. In the article are presented more details of teachers' approaches. This case gives evidence of the challenge to be faced and possibilities that can be considered for teaching mathematics for students with special needs at the university.

---

[1] marcone.renato@gmail.com, mirgps@gmail.com

Solange FERNANDES[2], Lulu HEALY: *Representations of three-dimensional forms constructed by blind students: Relations between "seeing" and the "knowing".*

The aim of this paper is to analyse how blind learners manage the conflicts between "seeing" and "knowing" in relation to two-dimensional representations of two geometric solids (a cube and a square-based pyramid). It seeks to locate elements within their interactions which make up the repertoires of "knowing" of those who do not see with their eyes, treating the processes involved in such interactions as acts of perception, with their origins in the body, and which serve a mediating role between environment, culture and brain.

Juliane LEUDERS[3]: *Auditory representations for blind and sighted students.*

Research into special education teacher education and professional development is sparse. This study set out to investigate factors that support special education teachers' ability to teach students with special needs fraction ideas. Working with three teachers in high school settings, the year long investigation into teacher professional development identified a number of key factors that contribute to student misconceptions and what teachers can do to mediate their learning difficulties.

Teresa ASSUDE,[4] Jean-Philippe DROUHARD: *Mathematics teaching situations with deaf or hard of hearing pupils.*

This article aims to study some mathematics teaching situations which are proposed to the deaf or hard-of-hearing pupils in primary classroom for school inclusion (specialized classroom). We analyse some situations and identify some pupils' difficulties. Then we discuss the problem of the specificity or not of these teaching situations.

Rumiati RUMIATI, Robert J. WRIGHT[5]: *Research on number knowledge of students with Down syndrome: An experience from Indonesia.*

This chapter presents the results of a small scale research study on the number knowledge of students with Down syndrome in Indonesia. Five students with Down Syndrome and ages ranging from 7 to 19 years, from a special education school in Yogyakarta city were interviewed to document their abilities in identifying numerals, solving number problems involving the use of unscreened and screened collections of counters, and solving one-digit and two-digit number problems in horizontal format. The approach and the schedule of assessment tasks in the interview were adapted from that used in Mathematics Recovery. The interviews were conducted individually and videotaped in order to capture subtle clues related to students' abilities. The number knowledge of the five students with Down syndrome is described, compared and discussed.

---

[2] solangehf@gmail.com, lulu@baquara.com

[3] juliane.leuders@ph-freiburg.de.

[4] teresa.dos-reis-assude@univ-amu.fr.

[5] rumiati1@yahoo.co.id, bob.wright@scu.edu.au.

KOTAGIRI Tadato[6]: *Mathematical achievement and creativity inherent in children with special needs.*

The assessment of children's mathematical learning achievement entails recognition of the child's human rights to learn Basic Mathematics: (1) to be able to fulfil his/her potential, and more importantly, (2) to be prepared for creative participation in his/her community, both in work and in other activities. Nonetheless, because many children with Special Needs face severe difficulties in obtaining the Basic Mathematical understanding and skills which they both deserve and need, they are effectively being denied their basic educational rights. This paper, based on years of using a clinical approach to remedial education, provides evidence of such children's remarkable possibilities for the achievement of Basic Mathematics, in particular exposing instances of significant creative response.

Marjolijn PELTENBURG, Marja VAN DEN HEUVEL-PANHUIZEN, Alexander ROBITZSCH[7]: *Yes, I got them all? Special education students' ability to solve ICT-based combinatorics problems.*

This present study is aimed at revealing special education students' mathematical potential by means of a dynamic ICT-based assessment. The topic of investigation is elementary combinatorics, which is generally not taught in primary special education. Six combinatorics problems on finding all possible combinations of a number of different types of clothing items were presented on screen. Data were collected on students' performance in solving these items. The performances of students in regular education served as a reference. The total sample consisted of 84 students (8- to 13-year-olds) from special education and 76 students (7- to 11-year-olds) from regular education. Their mathematics ability ranged from halfway grade 2 to halfway grade 5. The results showed that special education students are able to solve combinatorics problems equally successful as regular education students.

Pamela PAEK[8]: *Longitudinal analyses of students with special education needs in the United States on high-stakes mathematics assessments.*

This paper analyzes one state's large-scale assessment (LSA) mathematics data over eight years in the United States, to identify patterns of progress and attrition rates for students with special education needs (SEN-S). A previous study (Paek and Domaleski 2011) showed that SEN-S tended to have slower growth and lower mathematics achievement compared to general education students (GE-S) across grades and years. However, the majority of SEN-S had missing data across years, indicating that any longitudinal reports of SEN-S' achievement and growth are not generalizable. Findings indicate that the majority of SEN-S do not have LSA data for a single year, change the types of assessment forms they take from year-to-year, and are not promoted to the next grade level as often as GE-S. These results reveal

---

[6] kotagiri@edu.u-ryukyu.ac.jp.

[7] M.Peltenburg@uu.nl, m.vandenheuvel@fi.uu.nl, robitzsch.alexander@googlemail.com.

[8] ppaek@nciea.org

why a significant amount of SEN-S' data is missing, and how assumptions about data to measure achievement and growth for SEN-S are currently not tenable.

Eugenie KESTEL, Helen FORGASZ[9]: *An investigation of a targeted intervention program delivered by personal Video-conferencing for primary and middle school students with mathematical learning difficulties.*
This paper describes an ongoing study investigating the effectiveness of an individual, conceptual instruction based, tuition program delivered by Personal VideoConferencing (PVC) for upper primary and middle school students with Mathematical Learning Difficulties (MLDs). The experimental intervention targets number sense and fluency with basic facts in mathematics. The effect of using a personal videoconferencing delivery modality on the mathematics anxiety levels experienced by students with MLDs is also investigated.

Rebecca SEAH: *Mathematics professional development for special educators: Lessons learned from the field.*[10]
Research into special education teacher education and professional development is sparse. This study set out to investigate factors that support special education teachers' ability to teach students with special needs fraction ideas. Working with three teachers in high school settings, the year long investigation into teacher professional development identified a number of key factors that contribute to student misconceptions and what teachers can do to mediate their learning difficulties.

Leticia Pardo[11] *Special Education in Xalapa, Mexico: A brief history.*
The main focus of this work is to discuss briefly the history of Special Education services in Xalapa, capital city of the Mexican state named Veracruz. After 31 years serving this government office has experienced three phases of evolution: Integrated groups, Complementary Aid and Educational Integration. We recall some of the main characteristics of every one of these periods of time to explain the way that children with special needs were detected and how they were helped. One conclusion is that the philosophical base of Special Education has evolved from a kind of medical point of view to one based in social aspects.

Open Access This chapter is distributed under the terms of the Creative Commons Attribution Noncommercial License, which permits any noncommercial use, distribution, and reproduction in any medium, provided the original author(s) and source are credited.

---

[9] eugenie.kestel@monash.edu, helen.forgasz@monash.edu

[10] rtkseah@gmail.com.

[11] rociopardo2000@yahoo.com.mx.

Mathematics Education in and for Work

Geoff Wake and Keiko Yasukawa

TSG Report

In considering the meaning of 'mathematics education in and for work', we viewed 'mathematics' as being inclusive of the formal academic discipline of mathematics as well as the range of practices in which mathematics is embedded. Thus we saw 'education' to be inclusive of formal, informal and non-formal learning, that is, in educational settings (e.g. adult community education, vocational and further education) as well as in the community and workplaces. Important to the work of our group is the consideration of learning as both an individual and collective endeavour. In addition we viewed 'work' to be inclusive of paid work and unpaid work such as work in the home, and activist work in community and social settings.

In the design of this Topic Study Group (TSG), focal topics chosen included empirical, theoretical and methodological issues related to questions such as:

- How is mathematics embedded in work practices; what is this mathematics like and how is it learned?
- What mathematics do people learn in preparation for work?
- How is mathematics/numeracy valued for and in employment in different societies?

Organizers Co-chairs: Geoff Wake (UK), Keiko Yasukawa (Australia); Team Members: Corinnes Hahn (France), Ok-Kyeoung Kim (Korea), Tine Wedege (Sweden), Rudolf Straesser (Germany); Liaison IPC Member: Morten Blomhøj (Denmark).

G. Wake (✉)
University of Nottingham, Nottingham, UK
e-mail: Geoffrey.wake@nottingham.ac.uk

K. Yasukawa
University of Technology, Sydney, Australia
e-mail: keiko.yasukawa@uts.edu.au

© The Author(s) 2015
S.J. Cho (ed.), *The Proceedings of the 12th International Congress on Mathematical Education*, DOI 10.1007/978-3-319-12688-3_30

- How does the mathematics taught and learned for work differ/match the mathematics used in work?
- How does the mathematics learning in and for work meet people's mathematical needs in other domains of their lives?

The presentations and discussions at the meetings of the TSG touched on these questions in intersecting ways. The number of papers formally submitted to the group was relatively low and raised concerns during our meetings about the evident lack of research and other activity associated with a fundamentally important aspect of mathematics education. We expand on the views of the group in relation to this at the end of this report.

Our common pattern of working in our meetings was to have a formal presentation of papers that had been submitted to stimulate discussion which after pursuing issues raised directly by the paper explored the themes and questions identified above.

The first paper presented was Ok-Kyeong Kim's 'Pharmacists and Mathematics'. Ok-Kyeong's study examined how two pharmacists recorded the mathematics that was embedded in their everyday practices as pharmacists. Although the pharmacists did not identify much mathematics in their work, when asked to keep a journal to record the use of mathematical thinking or skills, they began to notice their use of different mathematical concepts such as ratios, proportions, measurement and percentages. What was invisible to the pharmacists themselves at the commencement of the research project slowly emerged and gained visibility, stimulated by their recording of their everyday work practices. The paper raised important questions about the difference between invisibility and absence of mathematics in work, as well as the tensions in researching 'mathematics' in workplaces: is it mathematical practice, or is it pharmaceutical practice, and who has power in the naming of this practice?

Following Ok-Kyeong's presentation, TSG participants engaged in discussions about Jaime Carvalho e Silva's paper 'The Mathematics Teaching in Vocational Schools in Portugal'. Jaime reported on an initiative taken in Portugal of potential envy by mathematics educators in many other countries. The initiative has led to a nationally agreed set of mathematics modules for a wide range of vocational courses studied in the final three years of schooling. The modules cover a wide range of topic areas ensuring that there are suitable mathematical modules for each vocational course. Modelling and statistics feature strongly, and efforts are being made to incorporate 'realistic' examples and activities. Jaime reported that the focus now is on evaluating the efficacy of these modules from a range of perspectives including those of teachers, students and workers who have studied these them. The paper and ensuing discussions highlighted the ongoing question about how should we teach mathematics in vocational courses—as separate subjects or 'invisibly' as embedded content within the specialist vocational subjects.

Invisibility of mathematics in workplace practices featured again in the presentation of Keiko Yasukawa, Stephen Black and Tony Brown's paper, 'Mathematics Education for the Worker, for the Employer, and/or for the Global

Marketplace?—An Exploratory Study of a Complex Question'. The paper was based on a work in progress on the authors' investigation of what has been described as a 'crisis' of low levels of workers' literacy and numeracy levels in Australia that, according to policy makers and industry groups, are the cause of less than desirable productivity, especially in manufacturing. Keiko presented the researchers' preliminary findings from one factory where despite everyone (production workers and their managers) acknowledging that the workers' literacy and numeracy skills are very poor in relation to any normative measures, there is no impact on productivity or quality. As in Ok-Kyeung's study, the workers generally undervalued the mathematics involved in their work, arguably because so much of the mathematics was deeply embedded in the software systems they were using (for example, the computer aided design package used for modeling 3-dimensional objects). Their study did however point to an area of numeracy and literacy need that was (unsurprisingly) not identified by industry and employer groups: the literacy and numeracy practices required by workers, such as low-paid production workers, to critically interpret and negotiate to improve their working conditions.

The final paper presentation was Geoff Wake's paper, 'Seeking principles of design of general mathematics curricula informed by research of use of mathematics in workplace contexts'. Geoff's paper addresses the important question of how the mathematics curriculum can support students' transition from one mathematical (eg formal learning in school) context to another (eg informal learning in the workplace). Drawing on his previous studies of ways in which mathematics is often 'black-boxed', that is deeply embedded and invisible within workplace artifacts or procedures, and on learning as identity work among students in transition from school to work, Geoff articulated design principles for a general mathematics curriculum. These principles include viewing mathematics as not just an object of study, but as a practice that facilitates communication within, membership of, and transformation of a community of practice. Geoff's paper emphasized the value of using research on workplace practices to inform and transform general mathematics curriculum into one that affords students with authentic experiences of learning and becoming users and producers of mathematics.

A presentation of a poster by Minoru Ito based on his and his colleagues Tadashi Aoki and Akihiko Shimano on 'Partnership Program of Mathematics and Science Education in Japan' shifted the focus of the TSG members to a different kind of study. Minoru and his colleagues were involved in a partnership program between his university and a city in Japan to engage university academics and students to design and facilitate engaging mathematical experiences for students in the city's schools. This was an innovative and visionary project to address concerns both about growing disengagement of school students in mathematics and the expected demand of increased mathematical and technological knowledge that these same students are likely to face in their future to address the complex economic and environmental challenges in their society.

Lisa Bjorklund Boistrup and Marie Jacobson's poster presentation took a different but equally big picture view of mathematics education in and for work, in their discussion of the project led by Tine Wedege, 'Adults' mathematics: In work and for school'. Their project was still in its early stages, but aims to uncover the relationship between the mathematics containing competencies that adults encounter in their workplaces with the mathematics learning demands that students face in their vocational studies.

The presentations in this TSG represented studies being conducted in several European countries—the UK, Portugal and Sweden, the USA, Japan and Australia, about a range of workplace and educational contexts—pharmacies, factories, high schools, vocational schools, nursing and caring work, and transport and garages, with each raising salient issues. The value of understanding mathematics as a social practice was shared by many of the presenters and discussion participants. That there was a tension between learning mathematics as part of a workplace practice and learning mathematics more explicitly in order to be able to critique and perhaps transform existing practices was acknowledged, as well as its corollary, which is the question of who should teach mathematics in vocational preparation courses—the vocational specialist or a mathematics specialist?

The TSG presentations and discussions also highlighted the many theoretical resources that are informing research being undertaken to understand mathematics education in and for work. Along with the presenters' own prior research, the work of other colleagues in workplace mathematics research including Hoyles and Noss, Wedege and Zevenbergen were drawn upon by several presenters. Socio-cultural theories of learning including Vygotsky's/L'eontev's/Engestrom's activity theory, Lave's situated cognition theory and Wenger's ideas of community of practice featured in several of the papers, reflecting the need to account for the collective nature of mathematical practices in workplaces.

In the same way that Geoff Wake's paper highlighted the importance of workplace research informing general mathematics curriculum design, research in vocational and workplace mathematics education should perhaps be more strongly informing what happens in mathematics learning at earlier stages of schooling. A final discussion of the group focused on these and related issues. Members of the group expressed their concerns at the relative lack of interest of the ICME community in this area of research given the important role that mathematics education plays in preparing young people for future work and critical citizenship. It was resolved that the co-chairs would be pro-active in raising the profile of the issues that emerged during discussions of the group and would seek to explore the possibility of a future ICME survey group providing an overview of the state of play of mathematics education in and for work across a range of cultural settings around the world.

Final Timetable

Tuesday, July 10 Session 1—10.30–12.00
10.30–10.45 Introductions and opening remarks: Geoff Wake and Keiko Yasukawa
10.45–11.20 Presenter: Ok-Kyeong Kim—Pharmacists and Mathematics, Discussant: Jaime Silva

11.20–11.55 Presenter: Olda Covian-Chavez—Mathematics applications in Topography: What elements for the training? (*not presented*), Discussant: Geoff Wake

11.55–12.00 Closing remarks

Wednesday, July 11 Session 2—10.30–12.00

10.30–10.45 Introductions and recap of previous day: Geoff Wake and Keiko Yasukawa

10.45–11.20 Presenter: Jaime Silva—The mathematics teaching in Vocational Schools in Portugal, Discussant: Geoff Wake

11.20–11.55 Presenter: Keiko Yasukawa—Mathematics Education for the Worker, for the Employer, and/or for the Global Marketplace?—An Exploratory Study of a Complex Question, Discussant: Ok-Kyeong Kim

11:55–12:00 Closing Remarks

Friday, July 13 Session 3—11.00–12.30

10.30–10.45 Introductions and recap of previous day: Geoff Wake and Keiko Yasukawa

10.45–11.20 Presenter: Geoff Wake- Seeking principles of design of general mathematics curricula informed by research of use of mathematics in workplace contexts, Discussant: Keiko Yasukawa

11.20–11.55 Overall threads and observations: Rudolf Strasser

11.55–12.00 Closing remarks

Saturday, July 14 Session 4—10.30–12.00

Poster presentations

Presenter: Minoru Ito—Partnership Program of Mathematics and Science Education in Japan

Presenter: Lisa Bjorklund Boistrup—Adults' mathematics: In work and for school General discussions and future.

Open Access This chapter is distributed under the terms of the Creative Commons Attribution Noncommercial License, which permits any noncommercial use, distribution, and reproduction in any medium, provided the original author(s) and source are credited.

Mathematical Literacy

Mogens Niss

Introduction

The actual design and implementation of the structure and organisation of the four TSG sessions was carried out by *Mogens Niss*, with the assistance of *GwiSoo Na*, *Eduardo Mancera*, and *Michèle Artigue*. Unfortunately, Eduardo Mancera was eventually unable to attend the Congress and the TSG.

This TSG was included for the first time in the history of the ICMEs. Hence, there was no established ICME tradition to build on concerning this topic. Moreover, generally speaking, the very notion of mathematical literacy is not well-defined, especially as several related concepts, such as numeracy, quantitative literacy, mathematical proficiency, and mathematical competencies, are in general use as well. Against this background it was decided to devote a fair proportion of the session time to coming to grips with the notions of mathematical literacy and its "relatives".

The presentations given in the four sessions of TSG 6 were partly commissioned papers, partly contributed ones. As is often the case with TSGs, the attendance to this TSG was not completely stable, but varied across the four sessions, the average attendance being about twenty participants per session.

The themes of the four sessions were chosen as a reflection of perceived intellectual and scholarly needs, and of the papers contributed by participants. The main theme of the opening session was the *Notions and interpretations of mathematical literacy*, whilst *The role and impact of mathematical literacy in national and*

Organizers Co-chairs: Mogens Niss (Denmark), Hileni Magano-Kapenda (Namibia); Team Members: Eduardo Mancera (Mexico), GwiSoo Na (Korea), Jianming Wang (China); Liaison IPC Member: Michèle Artigue (France).

M. Niss (✉)
Roskilde University, Roskilde, Denmark

© The Author(s) 2015
S.J. Cho (ed.), *The Proceedings of the 12th International Congress on Mathematical Education*, DOI 10.1007/978-3-319-12688-3_31

international studies was the main theme chosen for Session 2. Session 3 was primarily devoted to the theme *The role, use and implementation of mathematical literacy in educational systems and institutions.* The main theme of the fourth and final session was *Mathematical Literacy and teachers.* In order to ensure discussion of the presentations, each session was concluded by questions or comments, in Sessions 1 and 4 assisted by a round-table.

Major Points from the Four Sessions

Session 1: Tuesday, 10th July, 10:30–12:00

To set the stage for the work of TSG 6, *Mogens Niss* gave a 30-minute introduction focusing on the notions of mathematical literacy and its relatives. He began by observing that mathematics educators have always insisted that knowledge, skills and insights pertaining to elementary mathematics go far beyond facts, rules and procedures. This was already the case with the First International Mathematics Study (FIMS), conducted by the IEA and published in 1967, which spoke about five "cognitive behaviours". Later on, organisations such as NCTM and OECD-PISA, and several individual researchers, made an effort to identify aspects of the "add-ons" involved, suggesting various terms for the enterprise. The term Mathematical Literacy was used at least as early as in 1944, but the first attempt at a definition seems to have been made in the first OECD-PISA framework in 1999, with minor modifications in subsequent frameworks. Other related terms are numeracy, quantitative literacy, mathematical proficiency, and mathematical competence (competencies). Niss asked whether these terms are just different names for the same thing, or each term stands for something independent. He concluded that when it comes to mathematical literacy, numeracy and quantitative literacy, many people use them interchangeably, even though it is actually possible to attach distinct specific meanings to these terms. Given that people tend not to stick to definitions, he proposed to use Mathematical Literacy as the overarching term for the common underlying idea of promoting mathematical empowerment by making mathematics functional in extra-mathematical contexts. In contrast, the terms mathematical proficiency and mathematical competencies refer to a much wider spectrum of mathematical mastery, pertaining also to intra-mathematical contexts.

Next, two 15-minute presentations on aspects of the range and scope of mathematical literacy were given. In the first one, *Steve Thornton* (with John Hogan) (Australia), suggested to utilise a notion of "slow mathematics"—inspired by the notion of "slow food" in contrast to fast food—as a metaphor for quality mathematics education and for mathematical literacy. Slow mathematics is meant to capture what working mathematically is actually about, going against the "one-size-fits-all" idea typical of traditional curricula. Thornton proposed that working mathematically should be made *the* curriculum, whereas content should be of secondary importance. He illustrated his ideas by two examples, one on sums occurring in the dice game of Yahtzee, and one on cyclones and tides. *Karen François* (Belgium) (co-author not present) discussed the relationship between mathematical literacy and statistical literacy from a theoretical perspective, based on a literature review. François concluded that whilst there are indeed clear similarities and links between mathematical and statistical literacy there are also significant differences (e.g. statistical literacy focuses on decision making under uncertainty), as reflected in the sociological fact that statistical literacy and statistics education have developed into independent notions and fields of study. In other words statistical literacy should not simply be perceived as a special sub-field of mathematical literacy.

The session ended with a round-table in which *Nitsa Movshovitz-Hadar* (Israel), the speakers and the members of the audience discussed the range and scope of the concept of mathematical literacy. Movshovitz-Hadar made the point that mathematical literacy should encompass insights into the reality of current mathematical developments and described a project in Israel in which secondary school students were exposed to contemporary "mathematical snapshots" once every two weeks.

Session 2: July 11th Wednesday, 10:30–12:00

The session opened by a 30-minute invited presentation by *Ross Turner* (Australia) (in charge of implementing the mathematics part of the OECD-PISA study for several cycles). After considering the genesis and meanings of the notions of mathematical and scientific literacy, numeracy and quantitative literacy in various reports, Turner zoomed in on the ways in which the notion of mathematical literacy was developed in different PISA cycles, right from the beginning. The concept of mathematical literacy in PISA has always given rise to some tension within the group of participating countries. The key tension can be phrased as one between seeing mathematics as a superset, having mathematical literacy is a smaller part, or seeing mathematical literacy as the overarching domain, with mathematics as a subset. The tension is both a conceptual one, reflected in the ways in which different versions of the PISA framework draws upon mathematical competencies and overarching content areas ("big ideas"), and a political one, reflected in the fear voiced in some quarters, that PISA, by focusing on contextualised mathematics, would not provide an adequate coverage of school mathematics curricula, as only relatively low level mathematics seems to be needed to solve PISA problems. Nevertheless, several PISA items could be solved by a tiny minority of students only. Another problem is that the word "literacy" does not exist in many languages, making translation difficult. These tensions gave rise to a strong pressure on the OECD, and then on those in charge of PISA, to change the focus of PISA 2012 towards a more traditional view of mathematics as being constituted by well-known content areas, without directly forbidding the use of the term mathematical literacy. Turner concluded by mentioning the promising work done by some of the PISA mathematics experts on the impact of mathematical competencies on the intrinsic demands of PISA items, and on these demands as predictors of observed item difficulty.

In the first of three 15-minute presentations, *Jeff Evans* (UK) offered a comparative analysis of the definition of numeracy in PIAAC (Project for International Assessment of Adult Competencies) and the definition of mathematical literacy in PISA, 2006. He found the PISA definition somewhat broader and more "humanistic" than that in PIAAC. Finally, Evans pointed to the criticism, raised by some, of the unidimensionality of the performance levels in both surveys.

Next, *Kees Hoogland* (The Netherlands) reported on a randomized, controlled, comparative study of 38,000 Dutch students solving image-rich, respectively word-based, numeracy problems. The aim of the study was to test the hypothesis that replacing word problems with image-rich problems would have a significant positive effect on students' result, and even more so with vocational students. In the study, 24 pairs of mathematically equivalent numeracy problems were constructed such that each pair contained a language-rich version and an image-rich version of the "same" problem. Each student was randomly given 12 problems of each type. The study was found to provide a fair degree of confirmation of the hypothesis stated.

The final presentation was given by *Yukihiko Namikawa* (Japan), who described a national project in Japan which first focused on scientific literacy and then moved on to mathematical literacy, focusing on citizenship. A key part of this project was the publication "Mathematical Literacy for All Japanese", containing chapters on the nature of mathematics, on the central objects and concepts of mathematics, on mathematical methods and mathematical competencies, on mathematical topics, and, finally, on the relationship of mathematics with humanity and science. Following a report published in 2008 by the Central Council for Education, a new comprehensive, national standards curriculum emphasising mathematical literacy for all is being phased in, challenging the education of teachers at all levels.

Session 3: Friday 13th July, 15:00–16:30

This session contained a variety of short presentations. *John Hogan* (with Steve Thornton) (Australia), after having proposed to define "being mathematically literate" as more or less the same as "being numerate", went on to suggest that this cannot be developed or observed in the mathematics classroom alone, it has to go across the curriculum. To illustrate how this can be pursued, Hogan briefly outlined some settings in the arts, English, health and physical education and science, corresponding to early, middle and later years, respectively. He finally sketched a numeracy framework developed for diagnostic, analytic and practical purposes.

Yelena Baishanski (USA, with co-author not present) spoke about achieving literacy through articulated reasoning in remedial mathematics courses for US community college students (i.e. La Guardia CC, New York). The project involved activities on simple applied arithmetico-algebraic problems, arising out of "current compelling issues" meant to be engaging and meaningful to students, on which they can develop and practice their own skills in reasoning and written communication about reasoning, so as to develop confidence in their own powers of deduction.

In the next presentation, *Jenna Tague* (USA, with co-authors not present), dealt with two linked topics: the so-called STEM (Science, Technology, Engineering and Mathematics education) reform in the USA and a related development project at Ohio State University, reconceptualising engineering courses by focusing on mathematical literacy. More specifically, Tague proposed to devise a mathematical literacy framework within a "STEM for engineering students" context, taking inspiration from the Danish KOM project on mathematical competencies.

Based on the observation that many interpretations of mathematical literacy give a crucial role to mathematical modelling, *Abolfazi Refiepour Garabi* (Iran) presented two related empirical studies, one of Iranian mathematics textbooks and one of teachers' views about application and modelling problems in their classrooms. Mathematical modelling and applications were introduced in Iranian textbooks in 2008/2009. Comparing with Australian textbooks, the author finds that measured by the number of real world modelling problems, these textbooks tend to have a larger emphasis on mathematical literacy than do Iranian textbooks. Iranian mathematics teachers experience difficulties in using applications and modelling problems in their classrooms, especially because they don't have access to adequate sources for modelling tasks.

The final presentation was given by *Luis Rico Romero* (Spain, with co-authors not present). He presented a study in progress on Spanish in-service secondary teachers' assessment of mathematical competences. In the Spanish curriculum of 2006, the notion of competence, including mathematical competence, is given a key role at all educational levels, and also in the related system of performance indicators. The study focuses on teachers' understanding of, and intended methods with regard to, competency assessment in mathematics. The components of a workshop on this topic for teachers were outlined.

Session 4: Saturday 14th July, 10:30–12:00

In the first presentation, *Cigdem Arslan* (with *Günes Yavuz*), Turkey, reported on a research study on the mathematical literacy self-efficacy of prospective mathematics teacher students (PTs) in different programmes in a Turkish university. The study was conducted by way of a 25 item questionnaire, where each item was to be answered in a five-point Likert scale format. The study found that PTs indicate an above-medium level of mathematical literacy self-efficacy, and that there were no significant differences between their mean scores with respect to their year in university, between male and female students, or with respect to their choice of programme.

The presentation by *Lyn Webb* (South Africa, with co-authors not present) was based on the fact that mathematical literacy was introduced in South Africa as a mathematics option for prospective teachers as an alternative to "usual" mathematics. This led to the establishment of mathematical literacy programmes at some higher education institutions. The degrees obtained are rather different from those of traditional mathematics programmes. Two mathematics programmes are offered at two universities in KwaZulu Natal. After comparing the programmes with respect to their overall design, Webb concluded from the study that a balanced mix of types of knowledge, particularly disciplinary, pedagogical, practical and situational learning, is essential for teacher training qualification, and that content knowledge is not sufficient.

The final presentation was delivered jointly by *Dave Tout* (Australia) and *Iddo Gal* (Israel). They set out by contrasting internal views of educational goals (learning the trade of the discipline) with external views of educational outcomes focusing on real-world functional demands ("literacies"/"competencies"). Different surveys of students (e.g. TIMSS, PISA) and adults (e.g. ALL, PIAAC) have been conducted to shape educational policies and to design interventions. Mathematical literacy and numeracy are of the same nature, but mathematical literacy sits (mainly) in student and school contexts and numeracy in adult world contexts. This is reflected in PIAAC's definition of numeracy, focusing on the mathematical demands of a range of situations in adult life and on associated facets of numerate behaviour. The presentation went on to highlight various results from PIAAC and other adult numeracy surveys, and concluded by calling attention to three kinds of challenges to mathematical literacy/numeracy: Conceptual challenges ("what is it?"), educational challenges ("how can we develop it?"), and systemic challenges ("where is it (to be) located?").

This conclusion provided a handy lead-on to the final part of the session, a combined round-table and discussion amongst participants. Members of the round-table, moderated by *Mogens Niss*, were *GwiSoo Na*, *Yukihiko Namikawa* and *Ross Turner*. The round-table and the audience focused on important points for future work on mathematical literacy, such as examining the relationship between mathematical literacy and mathematical knowledge and skills, and finding ways to develop teaching and learning of mathematical literacy so as to ensure that all students (and adults) get something out of their mathematical education of subjective and objective value.

Open Access This chapter is distributed under the terms of the Creative Commons Attribution Noncommercial License, which permits any noncommercial use, distribution, and reproduction in any medium, provided the original author(s) and source are credited.

Teaching and Learning of Number Systems and Arithmetic (Focusing Especially on Primary Education)

Joana Brocardo and Geoffrey B. Saxe

Aims, Themes and Organization

Aims and Themes

The group's focus is on individuals' elementary mathematical representations and understandings with a special interest in the way these aspects of cognition develop through activities in and out of school. The mathematical domains of concern include whole numbers, integers, and rational numbers as well as representations related to each of these domains.

A related interest of the group is socio-cultural analyses. These analyses would include the ways that mathematics (including mathematical argumentation, representations, problem solving, teaching-learning interactions) is constituted in everyday practices as well as the interplay between developing mathematical understanding and representations in and out of school.

The group encourages cross-disciplinary contributions, including (but not limited to) participation by educational researchers, mathematics educators, developmental psychologists, and cultural anthropologists.

Organizers Co-chairs: Joana Brocardo (Portugal), Geoffrey B. Saxe (USA); Team Members: Maria Lucia Faria Moro (Brazil), mlfmoro@sul.com.br, Minkyung Kim (Korea) mkkim@ewha.ac.kr; Liaison IPC member: K. Subramaniam.

J. Brocardo (✉)
Setúbal, Portugal
e-mail: joana.brocardo@ese.ips.pt

G.B. Saxe
Berkeley, USA
e-mail: saxe@berkeley.edu

© The Author(s) 2015   415
S.J. Cho (ed.), *The Proceedings of the 12th International Congress on Mathematical Education*, DOI 10.1007/978-3-319-12688-3_32

Organization

TSG 7 received 29 submissions. We decided to emphasize discussion, articulating oral presentation and its discussion with poster presentations.

Two members of the organizing team and one external reviewer reviewed each paper. From the reviews and interactions by email among the members of the Organizing Team, an agreement was reached on a final list of presentations and posters, leading to 10 oral presentations and 17 posters. Due to cancellations only 10 posters were presented in two slots with 5 in each one. This turned the poster sessions of the group into an interactive session, in which each poster was presented by the author(s) and then discussed with all the participants.

The participants in the group came from 15 different countries of North and South America, Asia, Africa and Europe.

Papers, Posters, and Discussion Topics that Emerged in the Sessions

The presentations and the discussion varied markedly, reflecting diverse orientations and focal interests in teaching and learning about number systems and operations. Though diverse, the papers and posters conformed to four general themes.

The first theme was formalization of mathematical ideas, mathematical contexts, and models. The presentations and the discussion highlighted potentialities and barriers to the learning and teaching of number system and operations.

The second theme engaged participants with elementary mathematical representations and understandings that individuals construct. The presentations included case studies that illustrate the development of representation and understandings through activities in and out of school.

A third concerned kinds of numbers that are the focus of teaching and learning. These papers focused on teaching and learning of whole, fractional and decimal numbers. Papers and posters presented and analyzed processes whereby students overcome their misunderstandings and difficulties.

Finally the group discussed examples of everyday practices in school that can promote understanding in the domain of number and operations as well as the interplay between developing mathematical understanding and representations in and out of school. This discussion included examples and ideas related with mathematical argumentation, representations, problem solving and teaching-learning interactions.

The schematic contained in Fig. 1 illustrates the principal focus of TSG7 on number systems and operations themes, the concern for understanding processes of teaching and learning related to the focus, and the paper presentations, posters, and discussion that emerged on the four themes.

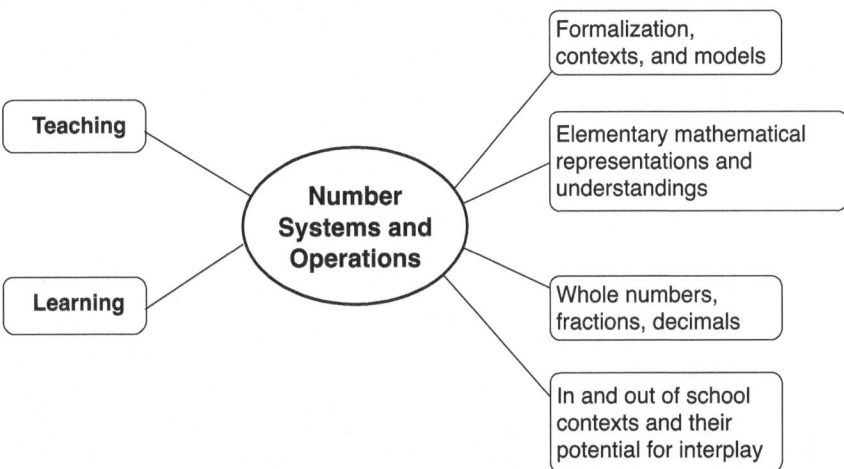

Fig. 1 Schematic of themes and presentation topics in TSG 7 for the 2012 meetings

Open Access This chapter is distributed under the terms of the Creative Commons Attribution Noncommercial License, which permits any noncommercial use, distribution, and reproduction in any medium, provided the original author(s) and source are credited.

Measurement—Focusing Especially on Primary Education

JeongSuk Pang and Kees Buijs

Preparation

Measurement, as well as related topics of geometry, forms an important mathematics domain on the level of both primary school and pre-vocational secondary school in many countries. At this level it relates primarily to quantifying certain aspects of real world physical objects such as the length, area, capacity, weight/mass, temperature or volume of objects, and to the reconstruction and application of the current measuring systems in a country (metrical or non-metrical). It also includes the use of measuring instruments such as the folding ruler and measuring tape, the measuring jug and the kitchen scale. Related geometrical topics include understanding of and working with the concept of scale, and the reconstruction and application of formulas for the area of a rectangle, triangle and other geometrical figures.

TSG-8 addressed researchers, curriculum developers, and reflective practitioners (teachers) working in the field of measurement and related geometry on the level of primary school. It aimed at providing a forum for generating discussion, exchanging insights, and establishing a state of the art sketch of the domain, including indications for the status of measurement as a foundation for advanced mathematics domains.

The TSG-8 organizing team called for papers dealing with various aspects of measurement such as theoretical perspectives on mathematical growth of students' thinking related to measurement, the development of measurement sense in students, connections between measurement and related domains such as number

Organizers Co-chairs: Jeong Suk Pang (Korea), Kees Buijs (Netherlands); Team members: Olimpia Figueras (Mexico), Silke Ruwisch (Germany), Andrea McDonough (Australia); Liaison IPC member: K. (Ravi) Subramaniam (India).

J. Pang (✉) · K. Buijs
Korea National University of Education, Cheongju, South Korea
e-mail: jeongsuk@knue.ac.kr

© The Author(s) 2015
S.J. Cho (ed.), *The Proceedings of the 12th International Congress on Mathematical Education*, DOI 10.1007/978-3-319-12688-3_33

sense and decimal numbers, curriculum development and implementation related to measurement, instructional approaches to foster students' development related to measurement, and culturally defined tools and practices for measurement and cultural supports for the learning and teaching of measurement.

Each of the 15 proposals which we had received was carefully and rigorously evaluated by three reviewers from the TSG-8 organizing team members with the support of K. Subramaniam. Having further discussed the initially accepted proposals amongst the TSG-8 team members, four papers were accepted for long oral presentation (30 min of presentation and 10 min of discussion) and eight papers for short presentation (15 min of presentation and 5 min of discussion). The remaining three papers were recommended for poster presentations during the general poster sessions of the ICME-12. Due to a cancellation, the final program of TSG-8 consisted of four long oral presentations and seven short ones.

We organized the accepted papers into four 90-munite sessions as follows:

- Session 1: Students' difficulties and teaching methods (July 10th),
- Session 2: Curricular materials and teaching methods (July 11th),
- Session 3: Delving into students' understanding (July 13th),
- Session 4: Measurement instrument and its use (July 14th)

Implementation

Session 1: Students' Difficulties and Teaching Methods

The first session was chaired by the co-chairs of TSG-8. At the beginning of the session, JeongSuk Pang from Korea welcomed all participants and introduced the organizing team members. Kees Buijs from Netherlands then delivered introductory remarks, showing a series of pictures taken in Seoul and related them to demonstrate measurement in a daily life.

Three papers were presented in this session (one long presentation and two short presentations) and vivid discussion was followed. First, Yah Hui Tan and Meng Hua Chua from Singapore investigated students' difficulties in learning the concepts of length and mass, and examined how teachers' use of an adapted version of the Kolb's *experiential learning cycle* was helpful to address their students' difficulties. They addressed the importance of using various measurement tools to assess students' understanding and misconceptions of measurement concepts.

Second, JeongSuk Pang, JeongWon Kim, and HyeJeong Kim from Korea identified key instructional elements in teaching measurement by comparing and contrasting two sets of measurement teaching practices which were recognized as good instruction in Korea. This presentation raised an issue on what counts as effective *measurement* instruction.

Third, Wayne Hawkins from Australia presented four primary teachers' pedagogical content knowledge in teaching measurement to students in Years 3 and 4. By exploring teachers' knowledge of mathematics along with knowledge of students and teaching, Wayne helped the audience understand the complex nature of pedagogical content knowledge and provoked a discussion on the dynamic nature of such knowledge.

Session 2: Curricular Materials and Teaching Methods

The second session was chaired by the TSG-8 organizing team member, Olimpia Figueras from Mexico. Three papers were presented in this session (one long presentation and two short presentations) and insightful issues were discussed afterwards. First, JeongSuk Pang, SuKyoung Kim, and InYoung Choi from Korea reported a comparative analysis of the statements in two Korean elementary mathematics textbook series in terms of two coding criteria: degree of guidance and key learning elements of the measurement domain. This presentation suggested the need of re-conceptualizing key learning elements of measurement as well as the possibility of developing a new coding system for textbook analysis. Several participants showed their interest in using this coding system in analyzing their textbooks.

Second, Silke Ruwisch from Germany presented third grade students' understanding of capacity and proposed the need for explicit comparison and measurement actions with many different containers before building up mental representation.

Third, Jeenath Rahaman from India presented different ways in which multiplicative thinking was involved in the measurement of area. She shared some tasks that had prompted students to use multiplicative thinking in finding the area of given figures. This also gave the participants an opportunity to reflect on the importance of designing tasks to explore the connection between multiplicative thinking and measurement of area.

Session 3: Delving into Students' Understanding

The third session was chaired by the TSG-8 organizing team member, Silke Ruwisch from Germany. Three papers were presented in this session (one long presentation and two short presentations) and thought-provoking issues were raised. First, Kees Buijs from Netherlands reported gaps between the informal and formal knowledge of 13–14 years old pre-vocational students, and suggested some ways to bridge such gaps. This presentation provided unique information mainly because of the characteristics of the students who had participated in this study. Despite their reasonable knowledge of measurement units and basic measurement sense, the difficulties that students had in solving more theoretical measurement problems were striking. As such, this presentation addressed a core issue in designing a measurement curriculum.

Second, Oyunaa Purevdorj from Mongolia presented second grade students' difficulties in understanding the given word problem, drawing a rectangle, and finding out the perimeter of a rectangle, and attributed the causes of such difficulties to the ways curriculum and textbook were designed, and the ways that teachers taught them in the country. This presentation helped participants understand the close relationship among curricular documents, teaching methods, and students' learning outcomes.

Third, Andrea McDonough from Australia reported on a design experiment to teach lower primary students about the measurement of mass. By illustrating multiple tasks and hands-on lessons in which students were expected to focus on the key measurement understandings of comparison and unit, Andrea prompted the audiences to grasp how to maximize the opportunity to learn the measurement of mass.

Session 4: Measurement Instrument and Its Use

The final session was chaired by the TSG-8 organizing team member, Andrea McDonough from Australia. Two papers were presented in this session (one long presentation and one short presentation) and general discussion was followed. First, K. Subramaniam from India presented measurement units and modes in the Indian context. He illustrated unique informal measurement units and multiple modes of quantification that are still being used in the Indian context. The presentation raised issues of how to design the school mathematics curriculum to incorporate students' practical knowledge of measurement and measurement sense.

Second, Bona Kang from USA reported four emerging sociomathematical norms regarding linear measurement and then the students' meaningful shift to use rigid tools. As such, she suggested the positive impact of social processes on the students' use of informal tools in measurement. This presentation raised an issue of a reflexive relationship between social and cognitive processes in measurement activity.

The final session was closed by two co-chairs. They appreciated all the participants who presented their studies, engaged in a rich discussion, and provided comments throughout the four sessions.

Reflection

The adequate number of papers presented in each session enabled TSG-8 to have an opportunity for participants to present their results, share ideas, and discuss issues within an affordable time frame. On the one hand, such an opportunity was effective in comparison to other TSGs because they had to run parallel sessions at the same time to provide more opportunities to present papers but had difficulties in sharing participants' ideas as a whole group. On the other hand, it was surprising that not many papers were submitted to TSG-8, even though measurement and related geometry are considered as an essential part of the mathematics curriculum especially at primary level in many countries. One reason might be a lack of attention to this domain. Another reason might be that a number of proposals were submitted to other TSGs by drawing more attention to the genre of research rather than the content domain of mathematics.

Generally speaking, TSG-8 had regular attendants who were ready to bring up rich discussion within a permissive atmosphere throughout the four sessions. Despite the relatively small number of papers presented in this group, a number of important issues came up and participants agreed the necessity of further international comparative studies in the domain of measurement. We hope that the topic study group dealing with measurement continues to serve a well-recognized group of the congress.

Open Access This chapter is distributed under the terms of the Creative Commons Attribution Noncommercial License, which permits any noncommercial use, distribution, and reproduction in any medium, provided the original author(s) and source are credited.

Teaching and Learning of Algebra

Rakhi Banerjee and Luis Puig

Overview

Topic Study Group 9 aimed to bring together researchers, developers and teachers who investigate and develop theoretical accounts of the teaching and learning of algebra. The group sought both empirically grounded contributions focussing on the learning and teaching of algebra in diverse classrooms settings, the evolution of algebraic reasoning from elementary through university schooling as well as theoretical contributions throwing light on the complexities involved in teaching and learning of algebra. Prospective contributors were requested to address one or more of the following themes: early algebra, use of ICT in algebra classrooms, proof and proving in algebra, problem solving, semiotics, designing of algebra curriculum.

Organization

We received 44 contributions for the TSG. Based on the review of these papers (each paper was reviewed by two members of the organizing team), 25 of these contributions were chosen for oral presentations and the rest were recommended for

Co-chairs: Rakhi Banerjee (India) rakhi.banerjee@gmail.com and Luis Puig (Spain) luis.puig@uv.es

Team Members: Swee Fong Ng (Singapore) sweefong.ng@nie.edu.sg, Armando Solares (Mexico) asolares@cinvestav.mx, Hwakyung Kim (Korea) hwakyung@gmail.com, Maria Blanton (USA)

R. Banerjee (✉)
Azim Premji University, Bangalore, India
e-mail: rakhi.banerjee@gmail.com

L. Puig
Universitat de València Estudi General, Valencia, Spain
e-mail: luis.puig@uv.es

© The Author(s) 2015
S.J. Cho (ed.), *The Proceedings of the 12th International Congress on Mathematical Education*, DOI 10.1007/978-3-319-12688-3_34

poster presentations. 16 of the oral presentations were short presentations (10 min for presentation and 5 min for discussion) and 9 were long presentations (20 min for presentation and 10 min discussion). For reasons of optimizing the available time and in order to fit in all the contributions, the group was divided into two subgroups and presentations were made simultaneously in the two sub-groups. The co-chairs of the team often helped in identifying the underlying theme in different presentations across the two sub-groups. Participants were requested to read up the articles to be presented in a session beforehand to be able to participate better. Some time was kept daily for the whole group to meet and discuss issues arising from the presentations or points which participants wanted to raise. More time was allotted for whole group activity on the first and the last day.

The participants were largely those who had contributed to the group and brought in perspectives from all over the world. The presentations touched upon students' understanding of different aspects of algebra, theoretical perspectives to make sense of students' work and help them learn better, teachers' understanding of the algebra they teach and professional development initiatives to help them focus on the important aspects of algebra. Pattern generalization and early algebraic thinking was an issue of discussion in various presentations. Problem solving and reasoning, proving, understanding of functions were explored in a few presentations. There were a couple of reports on algebra in particular culture/communities and curriculum/instruction status in a country. A few presentations focused on the use of computer aided tools for instruction or evaluation. An area which did not get any attention was how semiotics helps us understand students' developing knowledge of symbols, process of signification and communication.

Implementation

Session 1: July 10, Tuesday, 10:30–12:00 (Room no. 308a and 309)

On the first day, 45 min were kept for whole group discussion and only 4 presentations were scheduled for oral communication (2 long presentations and 2 short ones). The participants were reminded of the themes that the TSG would focus on and were given a general overview of the nature of the submissions received. They were further informed about the modalities of the conduct of the TSG.

The four presentations covered varied themes. One of the presentations focused on understanding of properties of operations with respect to fractions, operations on fractions, ability to think relationally and to perceive structure in expressions and their importance for learning algebra and developing algebraic thinking. Another one looked at the usefulness of variation theory as a means of improving teaching and learning and discussed how teachers went about designing lessons in the area of rational expressions using the theory. A third presentation looked at pre-service teachers' ability to engage in inductive reasoning and generalization in problem solving contexts. The fourth presentation focused on professional development of teachers in the area of functions that helped them develop and design activities that promote algebraic thinking among students.

Thus, we listened to an interesting set of ideas in terms of design of tasks, theoretical frameworks on the first day. They highlighted strengths and limitations of teachers' and students' thinking and reasoning while working on the tasks and provide valuable insights for designing of programmes with teachers and students in the future.

Session 2: July 11, Wednesday, 10:30–12:00, Room no. 308a and 309

Eight presentations were scheduled for the second day, four of them were long presentations and four short ones, divided equally between the two rooms.

All the presentations in one of the rooms dealt with pattern generalization. One of them discussed strategies used by students in secondary school for generalizing two patterns. Another one looked at difference in performance among students categorized by their abilities in mathematics on pattern generalization tasks as well as the strategies used for working on the tasks. One study compared competence of students in two countries: Hongkong and United States, in pattern generalization task. A last paper explored young Australian indigenous students' engagement with generalization of contextual growing patterns and ways in which cultural gestures help them in accomplishing the task.

The studies highlighted many categories of patterns and strategies for generalizing them and the participants discussed issues arising out of pattern generalizing tasks in their own countries and classrooms and made suggestions towards improving students' abilities to generalize, nature of tasks and instructions for students etc.

Presentations in the second room were not in a single topic. The long one dealt with algebraic proof in secondary education. In this presentation findings of a teaching experiment were reported on how the understanding of the generality of algebraic proof emerged when students study operative proofs. The study started from the knowledge from previous research that even students who are able to construct proofs using symbolic algebra rely on checking with numerical examples as a "proof". Students that followed the experimental teaching, that included the use of operative proofs, start noticing the generality of operation and then they start appreciating algebraic proofs.

The short ones dealt with very different issues. One presented a proposal to describe the structure of algebraic competence by using linear structure models. The second one dealt with algebraic reasoning in early algebra as generalized arithmetic, examining elementary school students' understanding of the properties of whole numbers operations. The reflection of students on the properties of operations with whole numbers is a way to teach and learn algebraic reasoning in early ages. In the study, it was found that students had capabilities in generalizing the properties of numbers and operations, but they had not developed such capabilities, because school practices have not provided enough opportunities and experience in order to develop them. This was showed by the fact that student were able to generalize the commutative law, but neither the associative nor the distributive laws.

The final one dealt with a research study on the ability of secondary students to translate statements between symbolic algebra and vernacular language and vice-versa. In this study, students performed better when translating from symbolic algebra expressions to vernacular language, and most errors when translating the other way round were attributed by the authors to "peculiar features of algebraic language".

Session 3: July 13, Friday, 15:00–16:30, Room no. 308a and 309

Eight more presentations were scheduled for this day, in a manner similar to Session 2.

Two presentations in one of the rooms highlighted students' capacities to reason algebraically in various situations. One of the presentations shared a teaching experiment aimed to promote the development of algebraic thinking among grade 4 students in the context of identifying numerical relations and patterns and thus deriving generalizations. Another one talked about an online game which focused on developing students' (grade 6) abilities to solve contextual problems dealing with covariation and functional relations and thus enter the domain of algebraic thinking. The other two focused on curricular issues. The third presentation analysed the differences in the treatment of the concept of function in two different kinds of middle school curricula used in the United States. The last presentation dealt with the status of algebra instruction, and in particular instruction of equations, in China where the author examined the textbooks, students' learning and teachers' instruction to come to understand the issue.

The four presentations in the other room focused on varied themes in algebra education. The first presentation briefed on a part of a larger study trying to understand the relationship between students' understanding of fractions as quantities and their abilities to form equations which require such multiplicative understanding. The second presentation reported on students' understanding of function concept among nursing students after they had worked in a context-based, collaborative instructional module. The third reported freshmen students' ability to use their algebra pre-requisite skills while working on calculus problems (Applied calculus optimization problem). The last presentation explored elementary school students' non-formal algebraic reasoning while solving word problems, especially by focusing on the mathematical structure or attending to the relation between quantities in the problem.

Session 4: July 14, Saturday, 10:30–12:00, Room no. 308a and 309

We had scheduled five presentations on the last day, however one of the presenters did not show up, so we ended with four presentations, two long and two short. The first paper in one of the rooms presented a theoretical framework to account for the difference in performance of students who have been taught problem solving through a particular heuristic of drawing a diagram and its impact on their ability to use letter-symbols later in learning algebra. The other presentation highlighted the use of geometrical method in a dynamic environment while solving quadratic equations.

The short presentation in the other room analyzed secondary school students' structure sense, while they had to reproduce rational expressions involving identities. The long one addressed the use of ICT for diagnostic and differentiation purposes, by presenting an online set of resources to diagnose students' knowledge on algebra, and to provide teachers with appropriate resources for managing a differentiated algebra curriculum to meet students' different needs.

A wide range of issues thus got addressed through the presentations and led to fruitful and engaging discussions. These highlighted the abilities and limitations of children's/students' understanding in various conditions; teachers' understanding and role in developing algebraic thinking; the effects of curriculum, textbooks, tasks and technology in promoting students' understanding and teachers' abilities to teach effectively. Participants often related their own experiences within their countries. An interesting pattern that emerged from the presentations and discussions was the way Asian countries emphasise and inculcate the use of symbols and symbolic writing from an early age, whereas this is a much delayed activity in US and many parts of Europe. Thus, the research studies also looked for evidences of having achieved this competence and/or ways to strengthen it. The western countries look for emerging symbols and idiosyncratic use of symbols among children to elucidate their reasoning and thinking processes. This eventually leads them to develop a better understanding of symbols and systematic use of them at a later stage.

Conclusion

We did of course face some difficulties in organizing the TSG. The organizing team worked quite well before the conference in giving inputs and reviewing the proposal submissions in time. However, the actual organization was not very simple. The breaking into subgroups although helped us logistically, we lost on listening to each speaker and had to satisfy with the summaries presented by them during the whole group discussion. This would have been more fruitful had everyone read the papers before the session, which was rarely the case. Unfortunately, often due to limited capacities of participants to express in the English language, summaries or discussions could not be taken beyond a point and those who had facility with this language were the ones who got heard more. Some more time or some other ways of organizing the TSG may prove to be more fruitful. Since all the presentations in every TSG were scheduled well in advance and all participants knew the exact listing of presentations by speakers, participants moved from one to another TSG. Thus, the group kept changing each day making it difficult to engage in themes, issues and concerns of a particular TSG.

Open Access This chapter is distributed under the terms of the Creative Commons Attribution Noncommercial License, which permits any noncommercial use, distribution, and reproduction in any medium, provided the original author(s) and source are credited.

Teaching and Learning Geometry

Colette Laborde

Aims, Themes and Organization of the Topic Study Group

Aims and Themes

This group provided a forum for discussion of the teaching and learning of geometry, with a focus especially on the middle and secondary school and university levels. The focus of the group was on theoretical, empirical, or developmental issues related to

- Curriculum studies of new curriculum implementation, challenges and issues, discussion of specific issues such as place and role of transformations
- An application of geometry on the real world and other subjects,
- The use of instrumentation such as computers in teaching and learning of geometry,
- Explanation, argumentation and proof in geometry education
- Spatial abilities and geometric reasoning
- Teacher preparation in geometry education.

The issues were addressed from the historical and epistemological, cognitive and semiotic, educational points of view related to students' difficulties and related to the design of teaching and curricula.

TSG 10 received 40 submissions. We decided to subdivide the group into 2 subgroups during 3 slots of the group and to organize a poster session during one slot.

Organizers Co-chairs: Colette Laborde (France), Linquan Wang (China); Team Members: Mathias Ludwig (Germany), Natalie Jakucyn (USA), Joong Kweon Lee (Korea); Liaison IPC Member: Hee Chan Lew hclew@knue.ac.kr.

C. Laborde (✉)
University of Grenoble and Cabrilog, Grenoble, France
e-mail: Colette.Laborde@cabri.com

© The Author(s) 2015
S.J. Cho (ed.), *The Proceedings of the 12th International Congress on Mathematical Education*, DOI 10.1007/978-3-319-12688-3_35

Organization

Each paper was reviewed by two members of the organizing team who gave an evaluation and suggestions for the writing of the full paper. From the reviews and interactions by email among the members of the Organizing Team, an agreement was reached on a final list of presentations and posters, leading to 3 long oral presentations, 17 shorter presentations and 20 posters. Finally, due to cancellations, 3 long oral presentations and 14 presentations took place. Only 4 posters were displayed at the poster session. Most of the poster presenters left their posters in the main poster session of the congress. This turned the poster session of the group into a very interactive and vivid session with a small number of papers, in which each poster was presented by the author(s) and then discussed with all the participants.

The presenters in the group came from 12 different countries of North and South America, Asia and Europe.

Content of the Group

Range of the Themes Addressed in the Group

Several themes dealing with various mathematical contents were addressed in the group (Table 1).

A Multifaceted Approach of Geometry

As visible in the previous table, geometry was approached from various points of view. It should be noted that these points of view are not independent but intertwined. For example, the notion of "geometric transformation" was addressed by several presentations focusing on various themes: curriculum design, students' learning or teachers' knowledge. Some key issues arose from the range of themes addressed by the group:

- the notion of shape and generally of representation in geometry teaching and learning with an extension to the use of Dynamic Geometry environments
- the link between geometry and the real world
- the notion of transformation
- teacher education

The notion of "shape" as a corner stone of school geometry was investigated by Usiskin in his long presentation: "(1) a "figure"—we study many different shapes in geometry; (2) a "type of figure", as in the declaration that an object is triangular-shaped; and (3) a "property of a set of similar figures", as in the statement that two

Table 1 The addressed themes and contents

Theme	Mathematical content	School level
Mathematical analysis of the domain	Shapes and relationships with functions, graphical representations	Secondary, University
Curriculum and textbooks	Plane geometry, transformations	Secondary
Problem solving	Combinatorial problems	Secondary, College, University
Reasoning and proving	3D and 2D configurations	Middle school
Modeling the real world	Mirror and line reflection, trigonometry	Elementary, Middle school, Secondary
Use of tools and technology	Centroids in 2D and 3D geometry, geometrical relationships, tessellations and transformations	Primary, Middle school, Secondary
Introduction to axiomatic system	Geometry of the sphere	College, University
Students' solving strategies	Area of trapezoids	Upper elementary, Early secondary, Secondary, College, University
Students' recognition of shapes	Solids	Primary, Middle school
Reading and writing	3D geometry	Upper secondary
Teacher education	Transformations, measurement	Pre and in-service teacher education

figures are congruent if they have the same size and shape, or two figures are similar if they have the same shape." Usiskin investigated how the notion of shape has been extended in school geometry with four components of present school geometry: coordinate geometry, transformations, applications of geometry, dynamic geometry software environments. An important claim of Usiskin is that whereas geometry is usually considered as studying abstractions of real objects, "geometry studies real figures as well as abstract ones".

This extension of the notion of shape can be linked to the notion of diagram or representation of geometric objects in 2D or 3D. The issue of representation was involved in several contributions.

In 3D, there is a larger variety of representations than in 2D: real models, 2D representations in various perspectives, computer representations. Ludwig and Steinwandel carried out an investigation on 242 10 to 15 year-old students who had to identify the shape of faces and to give the number of faces, edges and vertices of Platonic and Archimedean solids represented by either models, or computer animations or diagrams. In his long presentation, Ludwig showed that students benefit more from real models. The assistance by computer animations and by pictures was not so fruitful in tasks where the students need mental rotation to solve the task. Lavador used the Bruners' classification to design a teacher guide about measurement of solids, starting from enactive representations to move to images and iconic representations that lead then to symbolic representations.

The chosen representations in geometry problem solving (be it in 2D or 3D) may help or hinder a constructive reasoning for 12–15 year old students (Jones, Fujita and Kunimune); for the same problem depending on the diagram students may recognize or not the configuration for applying a known theorem. In his long presentation, Jones showed some examples in 2D and 3D and stressed the existence of prototypical representations that may turn into obstacles for recognizing the same property in other representations. Students' difficulties in interpreting diagrams seem to prevail across the world and are mentioned in contributions from Germany, Japan, and England. Jones concluded that "questions remain about how different mathematical representations influence students' decision making, conjecture production, and proof construction processes in the classroom, and how can such representations can be utilized by teachers to develop students' productive reasoning process." This is exactly the question also addressed in Kageyama's contribution that studies how students recognize analytical and logical properties of figures in construction tasks and use figural properties as justifying tools.

The link between geometry and the real world underlies several contributions and was even the focus of a few presentations. The issue seems to be more complex than expected. In some cases, referring to the real world can be very helpful for students (Ludwig). Whereas for Usiskin, although geometry is usually considered as studying abstractions of real objects, "geometry studies real figures as well as abstract ones", Boehm, Pospiech, Narciss and Körndle claimed that mathematics is an abstract world and they investigated what might be the potential confusions regarding a physical phenomenon after having experienced mathematics and physics lessons on this topic. Their study dealt with a very relevant phenomenon the mirror image in geometrical optics, as very often reflection is introduced in mathematics as modeling the mirror image. Their empirical data showed that we must pay attention to the fact that reality itself is not taught but a model of the reality and we must take into account the role of the used model in the teaching. It may happen that they do not go hand in hand as for reflection and mirror image and students may build inadequate knowledge. The results of the empirical study showed that students learn better when the scientific model is split into different science areas and when they are introduced to a multi-perspective modeling encompassing all model parts.

The link between real objects and theoretical objects of geometry was also viewed from the perspective of physical manipulations: real models for solid geometry (Ludwig, Suarez) but also strings, scissors, geoboard at elementary school (Faggiano). Faggiano stressed the fact that the manipulation by children contributes to the construction of meaning to geometric objects and relations only if they are involved in suitable tasks designed by the teacher.

Representations of geometric objects in Dynamic Geometry Environments are of a new nature and largely extending the range of manipulations and thought operations. Surprisingly a relative small number of contributions addressed this issue. Mammana (Ferrarelo and Pennisi) asked students to generalize properties from 2D to 3D by using two Dynamic Geometry environments (Cabri II plus and Cabri 3D). Their observations showed how the computer environments helped students not only to verify their conjectures but also to prove them. The same idea of combining exploring and generalizing was also investigated by Withney, Kartal and Zawojewsky with collegiate students using Lenart spheres for constructing an axiomatic system of spherical geometry. Faggiano combined the use of dynamic geometry and manipulatives at elementary school and concluded to the benefit of such combination. Lindamann carried out an investigation on the provocative question: "Which learning environment, DGE or traditional one produces a greater learning in a college geometry course?". No significant difference was found between the results of both kinds of learning environments. However as noted by Lindamann, students using technology gained other skills related to technology.

Transformations was a theme addressed by many contributions at least from two perspectives, a curricular perspective and from the perspective of pre- or in-service teacher education. La Ferla et al. compared the Common Core standards in the United States and the Turkish curricula and showed that the teaching of transformations is reinforced by the Common Core standards and becomes more aligned with the Turkish curriculum. Innovative teaching introducing pre-service or in-service teachers not only to transformations, but also to their use in solving geometry problems was reported by several contributions. Saego reported by means of very relevant examples about a professional development and its rich materials guiding teachers to move beyond conceptualizing similarity as a numerical relationship between two discrete figures to instead understand a precise conception of similar figures from a transformations-based perspective. Xhevdet Thaqi compared curricula of Spain and Kosovo and investigated "how do prospective teachers understand, learn and present each component of geometric transformations, if there is any differences between two different countries." The study concluded that of importance among student teachers is the concept image of transformation as displacement and change of place.

Teacher education was part of several presentations, be it the focus of the paper or joint to another issue such as the teaching and learning of transformations. As stressed by Somayajulu, teacher knowledge is especially fragile in geometry as a subject. This is certainly a major motivation for improving teacher education in geometry.

Geometry as a source of problems was illustrated by some contributions: Soifer presented geometry combinatorial problems for advanced students, Manizade and Mason carried out a thorough analysis of possible solving strategies of calculating the area of a trapezoid and showed how solving this task may be done at various Van Hiele levels. Hak Ping Tam and Hsin Han Wang concluded their study about the presentation of Pythagoras theorem in Taiwan textbooks by claiming that this theorem is a good opportunity for making students aware of the fact that multiple proofs can be given for the same theorem.

In conclusion, the various presentations of the group illustrated very well how rich the field of geometry teaching and learning is and how it can be investigated from various points of view with some emerging key issues, namely the nature and the role of representations.

Open Access This chapter is distributed under the terms of the Creative Commons Attribution Noncommercial License, which permits any noncommercial use, distribution, and reproduction in any medium, provided the original author(s) and source are credited.

Teaching and Learning of Probability

Per Nilsson and Jun Li

Aims and Focus

Probability has strong roots in the curricula of many countries but is relatively new in others. And although probability has been introduced into the mainstream school mathematics curricula in many countries, research does not necessarily support a rapid inclusion into the curriculum because many problems in teaching and learning probability are still unsolved. For example, should probability be taught to all students? When should students be introduced to probability? What is probability literacy? How is probability literacy developed? What kind of knowledge do teachers need in order to teach probability in more concrete, meaningful and effective ways? How do we facilitate the development of such teaching knowledge? How could investigating students' conceptions of probability from various perspectives further inform our teaching? At ICME 12 in Seoul, Topic Study Group 11 provided a forum for presentations and discussion from an international view about the current state and important new trends in research and practice related to the teaching and learning of probability.

Traditionally, the teaching of probability concerns two different interpretations of probability: (1) a classical conception, where probability is based on combinatorics or formal mathematics, and (2) a frequency conception, where probability is

Organizers Team Chairs: Per Nilsson (Sweden), Jun Li (China); Team Members: Enriqueta Reston,(Philippines), Egan Chernoff (Canada), Kyeong-Hwa Lee (Korea), Efi Paparistodemou (Cyprus); Liaison IPC Member: Gail Burrill (USA).

P. Nilsson (✉)
Växjö, Sweden
e-mail: per.nilsson@vxu.se

J. Li
Shanghai, China
e-mail: lijun@math.ecnu.cn

© The Author(s) 2015
S.J. Cho (ed.), *The Proceedings of the 12th International Congress on Mathematical Education*, DOI 10.1007/978-3-319-12688-3_36

based on empirical evidence and long-termed behaviour of random phenomena. The Topic Study Group (TSG) tried to look beyond these two interpretations and consider as the first focus how to teach probability concepts in ways that develop understanding and support the use of probability to make rational decisions in situations that affect peoples' lives and their work. It is important to note that the notion of probability as used in the Topic Study Group included aspects of chance, randomness, risk and its relationship to statistics.

The second focus was on teachers' knowledge for teaching probability. While teacher knowledge is critical for effective teaching of probability, very few studies deal with teacher knowledge and they (including the papers presented in the TSG) indicate that neither pre-service nor in-service teachers have enough knowledge for teaching probability. There is a growing global interest in learning what kind of knowledge teachers need to be able to effectively teach probability concepts and how to facilitate the development of such teacher knowledge. To promote more discussion and research in this area, the plenary panel discussion was narrowed to teacher knowledge for probability teaching.

The paper contributions were structured according to four general themes: *Curriculum Development and Policies, Research on Students' Thinking and Reasoning, Probability Literacy and Instructional Challenges, Teacher Knowledge in Probability Teaching*. They were presented in four sessions allotted to TSG 11.

The first three sessions began with an invited keynote speech: Ramesh Kapadia (United Kingdom), Manfred Borovcnik (Austria), Iddo Gal (Israel). The aim of these lectures was to sketch an overall picture of the TSG theme. A plenary panel was arranged for the last session that included all three keynote speakers and liaison, Gail Burrill, who were invited to reflect on the theme. Each session was closed by a summary by the session chair.

Session 1: Curriculum Development and Policies

Egan Chernoff, chaired the session., which began with opening remarks by co-chairs Per Nilsson and Jun Li, followed by an invited keynote speech by Ramesh Kapadia, and presentations by Jenny Gage (United Kingdom), Xianghui Wu (China).

Kapadia's address reviewed the main changes in the research related to probability education from the Piagetian-Fischbein era, the Kahneman_Tversky era to the current period. He summarized key research in the three eras and stressed the importance of developing new ideas from the past. He also provided an overview of curriculum development in England since the 1970s in the hope that some of the lessons can be applied elsewhere of the world. Based on the research and curriculum development, he suggested introducing probability at the elementary level, using a judicious mixture of subjective theory, a priori theory and frequentist theory of probability.

Gage presented an on-going project investigating mathematical modelling as a means for the learning of probability. She described school trials solving two problems by students between 10 and 14 years of age. The results suggested that the modelling approach and using values from the tally (natural frequencies), not probability, seemed to help students grasp the essence of the analysis of a problem and enabled them to use tree diagrams and 2-way contingency tables successfully.

Wu's paper was based on the belief that learning by game-playing should be central in children and adolescents' education as it stimulates the learning processes of flexibility, enjoyment, and adaptability. He shared with us his teaching experiences using three carefully designed games in his Grade 9 class.

In summarizing the session, Chernoff pointed to how the three talks highlighted that terms like misconceptions and subjective probability require serious discussion in future research. He raised the question of whether the frequency interpretation of probability should be emphasized with more care. He called on the need to address the teaching of risk and suggested we may benefit from research relevant to other TSGs, such as mathematical applications and modelling in the teaching and learning of mathematics.

Session 2: Research on Students' Thinking and Reasoning

Per Nilsson was the chair of Session 2. The session began with an invited address by Manfred Borovcnik, followed by presentations from Judith Stanja (Germany) and Theodosia Prodromou (Australia).

In his talk, "Conditional probability- a review of mathematical, philosophical, and educational perspectives", Borovcnik argued that conditional probability is a key concept in learning and accepting probability and that objective probability alone may not really help to change people's private criteria for dealing with conditional probability problems. He suggested the subjective approach is much closer to how people think and can thus much better explain conditional probabilities. He analyzed the need for teaching strategies to make plausible that conditional probabilities have nothing to do with time and causes, and showed various strategies for solving the Monty Hall problem. Borovcnik also reflected on translating probabilistic questions into *absolute (natural) frequencies*. His conclusion was that a wider conception of probability might be useful.

Stanja shared her attempt to characterize children's (age 8–9) elementary stochastic thinking by taking the role of semiotic means into account. Some theoretical ideas from Duval were outlined to serve as a basis for her description and analysis of interview data. She particularly stressed the complementarity of artefact and sign in learning probability and assessing child's understanding.

Prodromou addressed issues regarding the possibilities and challenges of using a computer-based modelling approach in the teaching of probability to 15 year-old students. In her investigations she particularly focuses on how the modelling approach can be used for building links between variation, theoretical models, simulations, and probability. Her results suggest that the way students express the relationship between signal and noise is of importance while building models from the observation of a real situation.

Summing up the session, Nilsson stressed the need to develop research methodologies in order to investigate the semiotic nature of teaching and learning probability. Approaching the teaching and learning through mathematical modelling seems timely. In Prodromou's study this was made in a computer-based learning environment. The session challenged research to develop real-world approaches for the teaching of probability through mathematical modelling.

Session 3: Probability Literacy and Instructional Challenges

Enriqueta Reston was the chair of Session 3. The session began with an invited address by Iddo Gal, followed by presentations from Hongshick Jang (Korea), Taek-Keun Oh and Kyeong-Hwa Lee (Korea).

Gal sketched an outline of probability literacy, its development, needs and connections to frameworks of adult competencies and mathematics curricula. He defined probability literacy by knowledge elements and dispositional elements and explained their relationships to both *internal* and *external* goals of probability education. To meet external demand better, he suggested teaching directly for probability literacy by increasing the use of tasks based on real-life problems in teaching and assessment, allowing time for subjective probability, and addressing dispositions and personal sentiments.

Jang suggested that empirical evidence involving the process of mathematical modelling in teaching is helpful to senior high school students' learning of probability. He presented his evidence both in terms of efficiency of teaching and motivation of students, but argued the necessity of mathematical formulation within the various types of uncertainty and the need to go beyond the conventional notion of mathematical modelling.

Oh and Lee addressed the teaching and learning of probability for gifted students. They found that learning through debate in solving probability tasks can be valuable for developing creativity of gifted Grade 11 students as the process stimulates flexibility, elaboration, and originality.

In summarizing the session, Reston reflected on whether there is any consensus on the meaning of probability literacy. Moreover, how does it relate to mathematics literacy? Statistical literacy? What are the overlaps? What are the gaps, if any? She also raised questions regarding what concrete actions and future directions will enable us to address instructional challenges in developing probability literacy among our students.

Session 4: Teacher Knowledge in Probability Teaching

Kyeong-Hwa Lee chaired the final session. After the presentations by Enriqueta Reston (Philippines), Per Nilsson (Sweden) and Egan Chernoff (Canada) the session ended with a panel debate on Teacher Knowledge in Probability Teaching.

Reston described a study exploring elementary mathematics teachers' conceptions of probability through inductive teaching and learning methods. As a background, she elaborated on the diversity of possible inductive teaching methods including, for instance, *inquiry teaching, problem-based teaching* and *investigations.*

Based on a survey study approach, Nilsson investigated correlations between Swedish teachers' content knowledge of probability and their level of education, teaching years and self-assessments of probability concepts. He found that the teachers have low confidence in understanding probability and have difficulties in applying the concepts in probability tasks.

Chernoff reported on research using the *attribute substitution model* to account for certain normatively incorrect responses of prospective teachers' understanding of random behaviour generated from a series of coin flips. His study considered individuals who, when presented a particular question, answer a different question instead. He argues that making connections between mathematics education and other domains of research will give mathematics education researchers new insights.

Before the Plenary Panel, Lee reviewed the meaning of knowledge for teaching given by Shulman in 1980s and Ball after 2000. During the panel time, Burrill, Kapadia and Borovcnik shared with all participants their insights on this topic.

Burrill choose teachers' pedagogical content knowledge for teaching probability as her main point. She indicated that having deep understanding of content knowledge is crucial for teaching. Teachers' knowledge of students and their ways of thinking about probability are essential as well. She recommended the Common Core State Standards for mathematical practices as a frame for engaging students in probability tasks and highlighted key points for teaching probability to teachers. Kapadia addressed teachers' content knowledge and pedagogical content knowledge as well. To develop probabilistic understanding, he appealed for investigations of teachers' knowledge across different countries with shared instruments. Borovcnik examined seven sources from which teachers could obtain their knowledge. He called for enhanced teaching of probability at the university level and connecting that closely to pedagogical issues, for example, to provide well-organized textbooks, which highlight modeling and other important ideas and to discuss the origins of students' misconceptions and how to use these in teaching to build understanding. He also listed and commented on several journals, websites of statistical associations and e-platforms he thought could be used to support teachers' development of probabilistic reasoning.

Several papers were presented in poster form: Haneet Gandhi, India; Zhengwu Long, China; Robyn Ruttenberg-Rozen, Canada; Narita, Masahiro, Japan; Tânia M.

M. Campos, Rosana Nogueira de Lima and Verônica Yumi Kataoka, Brazil; Natsumi Sekiya, Japan; Franziska Wandtner, Goetz Kersting, Reinhard Oldenburg, Germany; Michimasa Kobayashi, Japan. The posters elicited further discussion on the organizing themes of the sessions.

Time for formal presentations and discussions is always very limited at an international conference. But we are convinced that the work of the group initiated discussions on critical areas in probability education, such as teachers' knowledge for teaching, that will attract further investigations and support collaboration among people who are interested in the teaching and learning of probability.

Open Access This chapter is distributed under the terms of the Creative Commons Attribution Noncommercial License, which permits any noncommercial use, distribution, and reproduction in any medium, provided the original author(s) and source are credited.

Teaching and Learning of Statistics

Dani Ben-Zvi and Katie Makar

TSG-12 Rationale

Being able to provide sound evidence-based arguments and critically evaluate data-based claims are important skills that all citizens should have. It is not surprising therefore that the study of statistics at all educational levels is gaining more students and drawing more attention than it has in the past. The study of statistics provides students with tools, ideas and dispositions to use in order to react intelligently to information in the world around them. Reflecting this need to improve students' ability to think statistically, statistical literacy and reasoning are becoming part of the mainstream school and university curriculum in many countries.

As a consequence, statistics education is a growing and becoming an exciting field of research and development. Statistics at school level is usually taught in the mathematics classroom in connection with learning probability. Topic Study Group 12 (TSG-12) included probabilistic aspects in learning statistics, whereas research with a specific focus on learning probability was discussed in TSG-11 of ICME-12.

Organizers Co-chairs: Dani Ben-Zvi (Israel), Katie Makar (Australia); Team Members: Lisbeth Cordani (Brazil), Arthur Bakker (The Netherlands), Jangsun Paek (Korea); Liaison IPC Member: Gail Burrill (USA).

D. Ben-Zvi (✉)
University of Haifa, Haifa, Israel
e-mail: dbenzvi@univ.haifa.ac.il

K. Makar
University of Queensland, Queensland, Australia
e-mail: k.makar@uq.edu.au

© The Author(s) 2015
S.J. Cho (ed.), *The Proceedings of the 12th International Congress on Mathematical Education*, DOI 10.1007/978-3-319-12688-3_37

TSG-12 Meetings During ICME-12

This growing interest in statistics education was reflected in the popularity of this group and in the more than 40 papers accepted for presentation. The members of TSG-12 came from twenty different countries and varied significantly by experience, background and seniority. The presentations were divided into six themes of key issues in statistics education research: (a) theoretical issues in learning statistics; (b) integrating statistics with students' experiences; (c) the emergence of students' statistical reasoning; (d) teachers' statistical knowledge and learning and professional development of teachers; (e) learning technology in statistics education; and (f) learning statistics in school and at the tertiary level.

The four meetings of TSG-12 were organized to create a sense of community among all presenters and participants, who shared a common desire to improve statistics education by focusing on conceptual understanding rather than rote learning. To build and support this sense of community we asked participants to prepare for TSG-12 before they arrived in Seoul by reading all papers in advance, so we could discuss each other's work; the co-chairs kept informal correspondence with all participants before, during and after the conference; and finally, participants were asked to be involved every day of the program so we could get to know one another, develop collegial networks, welcome our emerging scholars and discuss the important work in statistics education research around the world.

Because of the large number of proposals we received, the time available only allowed for relatively short presentations by the authors. However, we felt it critical that all proposals be given time for presentation in some format. The four meetings were therefore organized to capitalize on community-building and discussions around our collective and individual research. Some of the sessions ran in parallel, some in roundtable format. While there was a poster session which is common for all TSGs, half of one TSG-12 session was dedicated to poster presentations so that the TSG-12 community could engage more directly with their authors and each other in a relaxed setting. Another highlight of the program was a panel of discussants on the final day to reflect as a community on the themes, presentations, issues raised and discussions over the four days.

The accepted papers were organized in the following ways:

- About twenty poster presentations to engage TSG-12 community discussions with diverse and thought-provoking studies;
- Eleven short presentations (5 + 5 min discussion) in a roundtable format organized into four themes to enrich understanding of the themes and allow for extended discussions around common interests;
- Twelve longer presentations and discussions (10 + 5 min discussion) to enhance the overarching themes of the short presentation and poster sessions;
- Four major long presentations (20 + 10 min discussion) to provoke initial discussions and stimulate final day reflections among the whole TSG-12 community. These papers were authored by Andreas Eichler and Markus Vogel

(Germany), Arthur Bakker, Xaviera van Mierlo and Sanne Akkerman (The Netherlands); Luis Saldanha and Michael McAllister (USA); and Dani Ben-Zvi and Keren Aridor-Berger (Israel).

TSG-12 Beyond the Conference

Informal feedback received after the conference was extremely positive. We felt at the end that much can be learned by integrating results from such a variety of research and practice in statistics education. This integration of theories, empirical evidence and instructional methods can eventually help students to develop their statistical thinking. These ongoing efforts to reform statistics instruction and content have the potential to both make the learning of statistics more engaging and prepare a generation of future citizens that deeply understand the rationale, perspective and key ideas of statistics. These are skills and knowledge that are crucial in the current age of information.

An informal set of proceedings was created to allow for immediate distribution of the TSG-12 papers among those within and beyond the TSG-12 members. The proceedings are available at: http://dbz.edtech.haifa.ac.il/publications/books. Many of the members of the community that came together for TSG-12 have remained in touch through a sharing of contact details and plans to meet again at the Ninth International Conference on Teaching Statistics (ICOTS-9) in 2014. Based on the TSG-12 papers, the book *Teaching and learning of statistics: International perspectives*, edited by Ben-Zvi and Makar, was published in 2014 by the Statistics Education Center, the University of Haifa, Israel.

TSG-12 Organizing Team

Co-chairs	Dani Ben-Zvi (Israel)	dbenzvi@univ.haifa.ac.il
	Katie Makar (Australia)	k.makar@uq.edu.au
Team members	Jangsun Baek (Korea)	jbaek@jnu.ac.kr
	Arthur Bakker (The Netherlands)	a.bakker4@uu.nl
	Lisbeth Cordani (Brazil)	lisbeth@ime.usp.br
Liaison IPC member	Gail Burrill (USA)	burrill@msu.edu

TSG-12 Resources

- TSG-12 Website: http://www.icme12.org/sub/tsg/tsgload.asp?tsgNo=12.
- TSG-12 Proceedings (eBook): http://dbz.edtech.haifa.ac.il/publications/books.

Open Access This chapter is distributed under the terms of the Creative Commons Attribution Noncommercial License, which permits any noncommercial use, distribution, and reproduction in any medium, provided the original author(s) and source are credited.

Teaching and Learning of Calculus

Victor Martinez-Luaces and Sunsook Noh

Aims

This Topic Study Group was a forum for discussions about the research and development in the teaching and learning of Calculus, both at upper secondary and tertiary level. Long and short presentations as well as the posters, showed advances, new trends, and an important work done in recent years on the teaching and learning processes of Calculus.

Organization

At ICME-12, TSG-13 had four one and a half hour timeslots and two general posters sessions. On the website of ICME-12 it is possible to access to all relevant documents including long presentations, short presentations and posters.

The accepted papers were organized as follows:

Organizers Co-chairs: Victor Martinez-Luaces (Uruguay), Sunsook Noh (Korea); Team Members: Margot Berger (South Africa), Francisco Cordero (Mexico), Greg Oates (New Zealand); Liaison IPC Member: Johann Engelbrecht (South Africa).

V. Martinez-Luaces (✉)
University of the Republic, Montevideo, Uruguay
e-mail: victoreml@gmail.com

S. Noh
Ewha Womans University, Seodaemun, Korea
e-mail: noh@ewha.ac.kr

© The Author(s) 2015
S.J. Cho (ed.), *The Proceedings of the 12th International Congress on Mathematical Education*, DOI 10.1007/978-3-319-12688-3_38

447

- 4 papers were accepted for long presentations.
- A group of 13 papers were accepted for short presentations. Finally, only 1 of these papers was not presented in the group sessions.
- In each session 3 short presentations and a long one were delivered.
- Each session was devoted to an important topic in Calculus teaching and learning.
- Posters were presented in the general poster session which was common for all TSGs.

The structure for each of the four 90-min session included some brief opening remarks by the co-chairs of the committee, followed by a long presentation (20 min) and 3 short paper presentations (10 min each). After the long and short presentations of each session, the whole group had at least half an hour for questions, comments and general discussion.

The following paragraph provides details on the 4 oral sessions and the 2 poster presentations related to TSG-13.

Long and Short Presentations Delivered

Tuesday, July 10

This morning session was devoted to an important topic in Calculus teaching: the derivative concept. The long presentation was delivered by William Crombie, from U.S.A, who proposed an alternative architecture of Calculus, in order to allow the access to advanced concepts from an elementary standpoint to a larger group of learners. An example of this approach is given by the idea of "transition line" that can be used even before developing limits and derivatives.

After that, the first short presentation was given by Jungeun Park, from U.S.A., who studied the student's discourses on the derivative using a communicational approach to cognition. Particularly, she focused on students' descriptions about the derivative and the relationships among a function, the derivative function, and the derivative at a point.

The next speaker, Miguel Diaz, from Mexico, documented the understanding of the derivative and its meaning on the part of 12 teachers, who teach Calculus in a high school in Mexico, using for this purpose several questionnaires specifically designed.

Finally, Hyang Im Kang from Korea reported how 11th grade students went through in reinventing derivatives on their own via a context problem involving the concept of velocity.

Wednesday, July 11

This second morning session was devoted to modelling, applications and other topics and it started with Victor Martinez-Luaces, from Uruguay, who described teaching experiences with inverse problems—of both causation and specification types—and modelling in Engineering Calculus courses.

The first short presentation was delivered by Mohammad Pourkazemi from Iran. He showed how by giving applied examples of Economics and Management in each section of Calculus, it is possible to increase the interest in Mathematics among students.

Next speaker, Anne D'Arcy-Warmington, from Australia suggested a reversal of the order, showing Calculus applications first and then the rules as a consequence in a semi-modelling style approach.

Finally, Greg Oates from New Zealand reported on 11 contemporary studies selected from the last Delta conference, which presented direct applications to, or important implications for, current practice in the teaching of undergraduate Calculus.

Friday, July 13

The third session was devoted to several important concepts in Calculus, like integrals, series, etc. The long presentation was delivered by Anatoly Kouropatov, from Israel. In his paper, he discussed the idea of accumulation as a core concept for a high school integral Calculus curriculum.

Short presentations started with Maria Teresa Gonzalez, from Spain. In her paper, she described the growth of mathematical understanding in university students, engaged in mathematics classroom tasks about the concept of numerical series.

The second short talk was given by Rafael Martinez-Planell, from Puerto Rico. His paper focused on student graphical understanding of two variable functions. His study—which applies APOS and Semiotic Representation theories—was based on semi-structured interviews with 15 students.

This Friday session finished with Jennifer Czocher, from U.S.A., who investigated about topics in introductory differential equations and their relation with the knowledge that students are expected to retain from their Calculus courses.

Saturday, July 14

The last session of TSG 13 was about pre-Calculus and first Calculus courses, and started with the long presentation delivered by Dong-Joong Kim, from Korea. In his paper, Kim investigates characteristics of the limit concept through the simultaneous use of historical and experimental analyses.

David Bressoud, from U.S.A. was in charge of the first short presentation. He showed the preliminary report of results from a large-scale survey of Calculus I students in the United States. The analysis highlights students' mathematical background as well as aspects of instruction that contribute to successful programs.

Another large scale survey—in this case, carried out in China—was the starting point of the following talk delivered by Xuefen Gao. Her study, involving 256 college-level Calculus students and 3 teachers, investigated the problems and misunderstanding of concepts in Calculus and designed concept-based instruction to help students to understand concepts.

Finally, Jose Antonio Fernandez, from Spain presented results of an exploratory study performed with students of ages 16–17. He investigated the different uses that these students make of terms such as "to approach", "to tend toward", "to reach" and "to exceed", terms that describe some properties of the concept of finite limit.

Poster Sessions

10 posters corresponding to TSG 13 were presented in 2 general poster sessions.

In poster 13-1, Young Gon Bae, from Korea studied how university students matched graphs and functions. In the next poster (13-2) Rie Mizukami, studied the main changes in the Calculus content at senior high schools in Japan. The third poster (13-3) explained by Jacinto Eloy Puig, from Colombia, analyses the important interconnections between infinity and infinitesimal quantities. In the next one—13-4—Youngcook Jun, from Korea, explored how to use CAS to develop a step-by-step solver for Calculus learning. In poster 13-5, Kazuki Chida, from Japan, proposed how to obtain laws about trigonometric functions from a very simple differential equation, without any reference to either an angle or a triangle. The next poster—13-6—showed by Kanna Shoji, from Japan, is aimed for the development of teaching materials, in order to make the students understand the relation between real-life and mathematics. In poster 13-7, Allan Tarp, from Denmark explored Calculus roots in primary and middle school. The next poster, i.e., the 13-9, expounded by Abolfazi Gatabi, shows how Iranian students participate in class-room discussion about infinite and infinitesimal concepts. Poster 13-11, presented by Mikie Takahashi, from Japan, focuses on approximate value calculation and its relation with practical high school mathematics. Finally, in the last poster (13-13), Misfer AlSalouli, from Saudi Arabia, investigates mathematics high school teachers' conceptual knowledge regarding the topics on Calculus.

At the end of the second poster session, the authors had the opportunity for oral presentation of their posters, having the benefit of an audience related to the TSG.

Conclusions

Several issues related to teaching and learning of Calculus regularly appeared in the general discussions located at the end of the oral sessions. The main themes in those discussions were: technology, visualisation, problem-solving, modelling and applications, and assessment, among others. TSG-13 papers also featured learning theories, construction of Calculus concepts and ideas (limits, integrals, derivatives, etc.), roots of Calculus concepts and other important topics in Calculus teaching and learning.

Most of the papers (long and short presentations and posters) showed an interest for innovative approaches to different topics, in order to help students to improve their knowledge and comprehension of Calculus. In several cases, these innovations were directly related to the use of technology, whereas in others, they were more involved in teaching approaches, courses materials, or specific tasks to be carried out by students of different educative levels and careers.

It is hoped that this interesting discussions and interaction between teachers and researchers of different countries will stimulate innovative ideas that will progress the advancement of mathematics education—particularly, in Calculus teaching and learning—into the following years of this new century.

Open Access This chapter is distributed under the terms of the Creative Commons Attribution Noncommercial License, which permits any noncommercial use, distribution, and reproduction in any medium, provided the original author(s) and source are credited.

Reasoning, Proof and Proving in Mathematics Education

Viviane Durand-Guerrier

Overview

The work of TSG 14 intended to serve a dual role: presentation of the current state of the art in the topic "Reasoning, proof and proving in mathematics education" and expositions of outstanding recent contributions to it. The topic will be considered at all levels of education: elementary, secondary, university (including pre-service teacher education), and in-service teacher education. The Organizing Team of the Study Group had invited theoretical, empirical or developmental papers that address one or more of the following themes: Historical/Epistemological/logical issues; Curriculum and textbook aspect; Cognitive aspect; Teaching and teacher education aspect, so that any paper of relevance to the overall focus of the Study Group.

The role and importance assigned to argumentation and proof in the last decade has led to an enormous variety of approaches to research in this area. Historical, epistemological and logical issues, related to the nature of mathematical argumentation and proof and their functions in mathematics, represent one focus of this wide-ranging research. Focus on mathematical aspects, concerning the didactical transposition of mathematical proof patterns into classrooms, is another established approach, which sometimes makes use of empirical research. Most empirical research focuses on cognitive aspects, concerning students' processes of production of conjectures and construction of proofs. Other research addresses implications for the design of curricula, sometimes based on the analysis of students' thinking in arguing and proving and concerns about didactical transposition. Recent empirical

Organizers Co-chairs: Maria Alessandra Mariotti (Italy), Stéphane Cyr (Canada); Team Members: Andreas Stylianides (UK), Viviane Durand-Guerrier (France), Youngmee Koh (Korea), Kirsti Hemmi (Sweden); liaison IPC member Hee Chan Lew (Corea).

V. Durand-Guerrier (✉)
University of Montpellier, Montpellier, France
e-mail: vdurand@math.univ-montp2.fr

© The Author(s) 2015
S.J. Cho (ed.), *The Proceedings of the 12th International Congress on Mathematical Education*, DOI 10.1007/978-3-319-12688-3_39

research has looked at proof teaching in classroom contexts and considered implications for the curriculum. The social-cultural aspects revealed in these studies motivate a current branch of research which is offering new insights. Comparative studies, trying to come to a better understanding of cultural differences in student's arguing and in the teaching of proof can be seen as part of this new branch of research. In this respect, papers presented at ICMI study 19 on "Argumentation and Proof" illustrate this diversity. Differences concern the focus researchers take in their approach, as well in the methodological choices they make. This leads not only to different perspectives, but also to different terminology when we are talking about phenomena. Differences are not always immediately clear, as we sometimes use the same words but assign different meanings to them. On the other hand, different categories that we build from empirical research in order to describe students' processes, understandings and needs are rarely discussed conceptually across the research field. Conceptual and terminological work is helpful in that it allows us to progress as a community operating with a wide range of research approaches.

Eleven papers and seven posters have been presented during the four sessions. There were thirty-five non-presenting participants who attended at least one session. The papers were from: Hong-Kong (1), Japan (2), Japan and UK (1), Turkey (1), UK (1), USA (5). The posters were from: Canada (1), Colombia (1), France (1), Japan (1), Peru (1), USA (2). The non presenting participants came from: Denmark (1), France (2), Germany (2), Hong-Kong (1), Japan (5), Korea (10), Norway (1), Portugal (1), South Africa (1), Sweden (3), Thailand (2), UK (1), USA (4).

For each session the attendee ranged from forty to fifty participants. The composition of the attendee was representative of the diversity of the participants in the congress: mathematicians, didacticians, expert researchers as well as young researchers, teachers form primary school to university.

According to the topics addressed by the papers accepted we identified four main themes to which each paper and poster may be related:

- Theme 1: Conception of proof from different theoretical perspectives.
- Theme 2: Proof in the classroom: the role of the teacher.
- Theme 3: Evaluation of proofs.
- Theme 4: Curriculum and materials.

Each of the four 90-min sessions (July 2013 10th, 11th, 13th, and 14th) were devoted to one of these themes. The structure for each 90-min session included some brief opening remarks by the chair of the session; the presentations, 10-min for long presentation and 5 min for short presentations. The general discussion on the papers and posters took place at the end of each session.

Theme 1: Conception of Proof from Different Theoretical Perspective (10th July 2013)

In this session, three papers were presented, so that two related posters, presenting a variety of theoretical backgrounds.

Kotaro Komatsu (Japan), in line with a long tradition of considering Lakatos epistemology as relevant for mathematics education, proposed to consider *Lakatos' Heuristic Rules as A Framework for Proofs and Refutations in Mathematical Learning: Local Counterexample and Modification of Proof.* Ysuke Tsujiyama (Japan) paid interest to characteristization of proving process in school mathematics based on Toulmin's concept of field, while Michelle Zandieh, Kyeong Hah Roh, Jessica Knapp (USA) explore *Student Proving through the Lens of Conceptual Blending.*

In their posters, Paul Dawkins, Kyeong Hah Roh (USA) emphasized *the Roles of Metaphors for Developing Students' Logical Control in Proof-oriented Mathematics,* while Shiv Karunakaran (USA) considered *Examining the Structure of Proving of Experienced Mathematics Doctoral Students.*

The final discussion enlightened the diversity of the theoretical perspectives; questions were addressed from teachers to researchers on the relevance of their theoretical backgrounds for designing tasks aiming to develop reasoning, proof and proving in class.

Theme 2: Proof in the Classroom: the Role of the Teacher (11th July 2013)

In this session three papers and one poster were presented; various aspects of the delicate role of teachers in classroom concerning proof have been enlightened.

Annie and John Selden presented the paper from Milos Savic (USA) who considers the controversial question *Where is the Logic in Student-Constructed Proofs?* Andreas J. Stylianides and Gabriel J. Stylianides (U.K.) focused on *"The big hurdle we have to overcome is getting students out of the mode of thinking that math is just plug-in-and-move-on kind of thing": Challenges in beginning to teach reasoning-and-proving.* Anna Marie Conner (USA) considered *Warrants as Indications of Reasoning Patterns in Secondary Mathematics Classes.*

In his poster, Medhat H. Rahim (Canada) proposed to consider *Description and Interpretation of Student-Teachers' Attempts to Construct Convincing Arguments and conjectures through Spatial Problem Solving Tasks.*

The discussion in the session, along with the content of presentation, put light on the difficulties for teachers to engage students in mathematical activity involving proof and proving; a main issue concerns the possibility of making students aware of the necessity for proof and proving. Taking in consideration that Geometry was the most represented mathematical domain in the papers and posters presented in the group, a question raised in the discussion: is this matter of fact unavoidable, or is it possible to work on proof in class in other mathematical domains. Finally participants agreed that although geometry is a relevant traditional domain for teaching reasoning, proof and proving in secondary school in many countries, there are also other relevant domains such as arithmetic, linear algebra, analysis etc., depending on the level.

Theme 3: Evaluation of Proofs (13th July 2013)

Two papers and two posters were devoted to evaluation of proofs or arguments. A third poster related to the theme 2 was also presented.

Yeşim İmamoğlu, Ayşenur Yontar Toğrol (Turkey) have presented *An Investigation of Senior Mathematics and Teaching Mathematics Students' Proof Evaluation Practices*. Yating Liu, Azita Manouchehri (USA) focused on means for *Nurturing High School Students' Understanding of Proof as a Convincing Way of Reasoning* and look for a theoretical framework.

In their posters, Shintaro Otsuka (Japan) paid interest on *Reasoning in Explaining False Statements: Focusing on Learner's Interpreting Propositions*, while Viviane Durand-Guerrier, Thomas Barrier, Faiza Chellougui, Rahim Kouki (France, Tunisia) provided *An Insight on University Mathematics Teaching Practices about Proofs involving Multiple Quantifiers*. Maria Nubia Soler Alavarez (Colombia) presented *Types of Rasoning used by Training Mathematics Teacher in a Class about Rational Numbers*.

Questions concerning validity were at the core of this session. The papers showed the variety of practice related to this question, opening a discussion on the distance between requirement addressed to students concerning rigor and ordinary teachers practices which generally do not fulfill these requirements. Finding an adequate balance between these two aspects in class is not easy.

Theme 4: Curriculum and Materials (14th July 2013)

In this session, three papers and one poster were presented, providing a variety of landscapes.

Mikio Miyazaki, Taro Fujita, Keith Jones (Japan, U.K.) presented material for *Introducing Proof in Lower Secondary School Geometry: A Learning Progression Based on Flow-chart Proving*. Yip-Cheung Chan (Hong-Kong) aim *Rebuilding The*

Harmony Between Figural and Conceptual Aspects For Reasoning, Proof and Proving in Dynamic Geometry Software. Ruthmae Sears (USA) investigates *The Impact of Subject-specific Curriculum Materials on the Teaching of Proof and Proof Schemes in High School Geometry Classrooms.* Estela Vallejo and Uldarico Malaspina (Peru) offered *A Look at the Justifications in the Basic Education in Peru: the National Curricular Design and some Texts used in the 1st Grade of Secondary level.*

The discussion on the presentations concerned the diversity of approach in curriculum and material, enlightening the interest of comparative studies on reasoning, proof and proving.

As closing remarks, the participants agreed that the discussion which took place at the end of each session were rich and concerned as well the implication for teaching, the theoretical assumptions, the role of logic, the specificity of geometry, the need for proof or proofs without words.

A common feeling was that, although *Reasoning Proof and Proving* have been studied for a long time, further international researches are needed.

Open Access This chapter is distributed under the terms of the Creative Commons Attribution Noncommercial License, which permits any noncommercial use, distribution, and reproduction in any medium, provided the original author(s) and source are credited.

Mathematical Problem Solving

Manuel Santos-Trigo and Zahra Gooya

Introduction

The program was designed to set up to organize, structure, and discuss the academic agenda of mathematical problem solving and its developments. The program included an open invitation to the mathematics education community to contribute and reflect on research and practicing issues that involve: (a) Addressing the origin, characterization, and foundation of mathematical problem solving, (b) discussing problem solving frameworks used to support research and curricula reforms in mathematical problem solving; (c) analyzing local and international research programs in mathematical problem solving; (d) discussing curriculum proposals that support the development of mathematical problem solving; (e) analyzing different ways to assess mathematical problem solving performances; (f) discussing the role played by the use of different digital tools in students' development of mathematical problem solving proficiency; (g) addressing programs that foster learners' development of problem solving approaches beyond school; and (h) identifying future developments of the field.

The international problem solving community responded to the invitation and sent more than 30 proposals, of those 18 were selected for presentation during the

Organizers Co-chairs: Manuel Santos-Trigo (Mexico), Zahra Gooya (Iran); Team Members: Jiang Chunlian (China), Mangoo Park (Korea), Dindyal Jaguthsing, Singapore; Liaison IPC Member: Yuriko Baldin (Brazil).

M. Santos-Trigo (✉)
Mathematics Education Department, Centre for Research and Advanced Studies, Cinvestav-IPN, Tsukuba 07360, Mexico
e-mail: msantos@cinvestav.mx

Z. Gooya
Shahid Beheshty University, Tehran, Iran
e-mail: zahra.gooya@yahoo.com

© The Author(s) 2015
S.J. Cho (ed.), *The Proceedings of the 12th International Congress on Mathematical Education*, DOI 10.1007/978-3-319-12688-3_40

sessions, and 10 were assigned to the poster session. In this report, we inform about the subjects and themes that authors addressed in their written proposals, and the results and discussions that emerged during the authors' oral presentations held during the development of the sessions at the ICME conference. A pdf file that includes all authors' contributions can be retrieved from: http://www.matedu. cinvestav.mx/~santos/icme12/ICME12TSG15book.pdf.

An Overview

The authors' contributions addressed and discussed several issues that were identified in the open invitation letter they received and was available through the congress web-page. Here, we highlight common issues addressed in the contributions that include mathematical reflections on what problem solving entails, the variety of studies and methodologies used to frame research studies, the range of participants in those studies that involves elementary, secondary, high school students, in-service and practicing teachers, and university students, and a variety of theories used to support and develop problem solving research.

(a) Two contributions reviewed issues regarding what types of problems are relevant to discuss with students, and the importance for instructors to create an instructional environment in which students can actively be engaged in problem solving experiences. One example used to illustrate problem solving strategies and conjectures that emerged during the solution process was a variant of a task discussed by Polya (1954, pp. 43–52): *Into how many parts is space divided by 5 planes?* The discussion became important to identify ways to formulate and pursue conjectures in which a set of heuristics appears important during the entire solution process. The same theme "heuristic methods" is also addressed in another contribution to discuss examples where students have an opportunity to rely on strategies such as pattern recognition, working backwards, guessing and testing, looking for simpler problems, etc. to solve tasks set in different contexts. Both contributions offer ways to analyze tasks that can be useful to construct instructional paths to foster students' mathematical problem solving experiences.

(b) Eight contributions recognized the importance for learners to work on small groups to discuss and defend their ideas, listen to others, and communicate results. Two contributions emphasized students' social interactions as a way to enhance cognitive experiences. One proposes a teaching module to guide university students to comprehend and develop conceptual knowledge associated with a first differential equation course. In general, authors used a bricolage perspective that relies on several conceptual frameworks to support the study; another contribution builds up a local conceptual framework to guide practicing elementary teachers to develop problem-solving experiences through social interactions.

(c) Four contributions rely on statistical analyses to compare students' problem solving performances. For example, three studies emphasize the use of pre and post-tests to analyze and compare groups of students' problem solving achievements as a result of receiving differential problem solving instruction. For instance, one group explicitly addressed the importance of using analogical thinking in their approaches versus a group that followed a regular teaching approach. Other studies relied on the use of Case Study methodology in which the participants' problem solving behaviors are analyzed in detail. It is common in this process or use of task-based interviews, groups or class videos, or a combination of qualitative tools to gather data and to foster the development of problem solving approaches. In general, a tendency in six contributions was to rely on both the use of quantitative and qualitative tools to analyze learners' problem solving behaviours.

(d) It was observed that five contributions have explicitly relied on frameworks that extend problem-solving approaches such as models-and-modeling perspectives. The analyses of problem solving performance of students that consistently have shown high achievement in international assessments was also addressed in seven of the contributions. For example, a study focused on analyzing the extent to which some Korean students epistemological beliefs about mathematics are related to their problem solving behaviours. Similarly, another study analyzes how a problem-based learning (PBL) was implemented in China.

It must be noted that the use of mathematical competitions to promote learners' development of problem solving skills has been encouraged in different countries. For example, one study analyzes how a web-based mathematical problem competition became important for 13–14 years to engage in problem solving experiences that go beyond those that appear in regular classroom contexts. Yet, another contribution analyses how a set of didactic techniques based on the problem centred Japanese tradition is implemented in Swedish. In this particular study the author relies on the use of Anthropological Theory of Didactics which is a framework commonly used in the French mathematics education tradition.

(e) Problem solving activities also play an important role in teachers professional development programs and the education of prospective teachers. A contribution focuses on fostering both prospective and practicing teachers' competence to pose, formulate, and pursue questions or problems. The framework that authors used to support the problem posing experiences involves epistemic, cognitive, and mediation analysis of tasks and learners interaction and is called an Onto-Semiotic approach. Likewise, the implementation of problem solving activities has taken different directions and aims. For instance, one contribution emphasizes the second Polyas' proposed stage of problem solving "designing a plan or planning the solution" to improve colleges students abilities to solve arithmetic problems.

Remarks and Future Directions

Learning, constructing, or developing mathematical knowledge via problem solving activities continues to be an important goal in curriculum proposals and a central theme in research programs around the world. However, a salient feature of the group contributions is that there are multiple ways and a variety of interpretations of what a problem solving approach to learn mathematics entails, and ways to frame and implement curriculum proposals. To analyze and reflect on common aspects around problem solving approaches we must construct and activate an international community that continuously shares research programs and discusses problem-solving developments. This community must include active researchers whose academic agenda involves both theoretical and practicing themes in problem solving. And teachers who show clear interest in implementing problem solving approaches in their classrooms are key elements since they look for ideas to consistently frame their practices around problem solving activities. In particular, teachers' discussions focus on demanding actions and directions that will help reduce efficiently a long list of contents and to concentrate on problem solving activities to study key concepts deeply. What fundamental mathematical ideas and processes should be central in curriculum proposals that promote problem-solving approaches?

Another teachers' interest is to address the role of students' international assessments (PISA, TIMMS) in problem solving approaches. That is, to discuss the extent to which the mathematics and ways of reasoning involved in those international assessments is consistent with problem solving approaches. Another important issue that emerged during the group session is the role played by the use of different forms of digital technology in fostering learners' development of mathematical problem solving experiences. It was recognized that there is little information on the type of mathematical reasoning that students construct as a result of using several technologies, and how that reasoning expand or complement paper and pencil approaches. There was a consensus that it is urgent to include in the research and practicing agenda the extent to which theoretical and conceptual frameworks used in problem solving needs to be adjusted in order to explain and foster the students' development of mathematical learning in problem solving scenarios that enhance the systematic use of digital technology.

Open Access This chapter is distributed under the terms of the Creative Commons Attribution Noncommercial License, which permits any noncommercial use, distribution, and reproduction in any medium, provided the original author(s) and source are credited.

Reference

Polya, G. (1954). *Induction and analogy in mathematics education. Volume 1 of mathematics and plausible reasoning.* NJ: Princeton University Press.

Visualization in the Teaching and Learning of Mathematics

Gert Kadunz and Michal Yerushalmy

Report

The history of visualization within mathematics education is a long one. Since the beginning of the 1980s mathematics educators are interested in the practical challenges of teaching visualization, in visualization of mathematics as exhibits in school or aligned with educational psychology and are looking for theoretical frameworks.[1] Recall the earlier text of Norma Presmeg (cf. Presmeg 1986, 1994, 1997), Theodore Eisenberg's widely recognized paper "On the understanding the reluctance to visualize" (Eisenberg, 1994) and more recent analysis of visualization in mathematics education can be found in Arcavi (2003) or David (2012). Regardless of their focus these papers nearly all offer a common picture for which a mathematician's success owes a considerable amount to visualization skills (Heintz 2001). On the other hand the history of mathematics shows visualization to have been cut back and even avoided to a certain extent. In the time of Leonhard Euler the visual was also used as a means for proving or establishing the existence of a mathematical object, whereas the mathematicians of the 19th and 20th century

---

Organizers Co-chairs: Gert Kadunz (Austria), Michal Yerushalmy (Israel); Team Members: Mathias Hattermann (Germany), Michel Hoffmann (USA), Kyoko Kakihana (Japan), Jaehong Shin (Korea); Liaison IPC Member: Bernard Hodgson (Canada).

---

[1] E.g. Jerome Bruner and his view on the use of images, Jean Piaget's learning theory or George Lakoff and his view on metaphors.

G. Kadunz (✉)
University of Klagenfurt, Klagenfurt, Austria
e-mail: gert.kadunz@uni-klu.ac.kr

M. Yerushalmy
University of Haifa, Haifa, Israel
e-mail: michalyr@construct.haifa.ac.il

© The Author(s) 2015
S.J. Cho (ed.), *The Proceedings of the 12th International Congress on Mathematical Education*, DOI 10.1007/978-3-319-12688-3_41

reduced the use of visualization for gaining new ideas when solving problems. Heuristics was the task of visualization. We suspect that this gap between the two trends was one reason why dealing with visualization became a significant topic for researchers in mathematics education.

Beyond our specific domain, for the last two decades we have seen a growing interest in the use of images as a general cultural change. It was Thomas Mitchel's dictum that the linguistic turn is followed now by a "pictorial turn" (Mitchel 1994) or Gottfried Boehms (Boehm 1994) "iconic turn". Their concentration on visualization in cultural sciences is based on their interest in the field of visual arts and it is still increasing (Bachmann-Medick 2009). Other technology-enabled visualization developments such as medical imaging, which have introduced sophisticated methods for reconstructing and manipulating images, changed the public and scientific conventions in regard to what formerly was invisible. As happened with modern telescopes which allow us to see nearly infinite distant objects or microscopes which bring the infinitely small to our eye structures become visible and with this kind of visibility they become a part of the scientific debate. Visualization technology causes new paradigms to be developed as structures that could only speculated about are now subject of scientific debate. We may say that their ontological status has changed and in that regard images became a major epistemological factor.

Such new developments, caused substantial endeavour within cultural science into investigating the use of images from different perspectives. Mitchell (1987), Arnheim (1969) or Hessler and Mersch (2009) are examples. The introduction to "Logik des Bildlichen" (Hessler 2009), which we can translate as "The Logic of the Pictorial", focusses on the meaning of visual thinking. In this chapter they formulate several relevant questions on visualization which should be answered by a science of images. Among these questions we read: epistemology and images, the order of demonstrating or how to make thinking visible.

When we consider these short deliberations then we can recognize two positions. We have a long tradition of visualization within mathematics education which is based and supported by practical and theoretical practices. At the same time there are several recent developments within cultural science concerning visualization. Hence there is a need to find means of transmission and terms that would support the exchange of ideas and research questions between cultural science and mathematics education. A theory-based example of such means of transmission is relevant to a topic that our group explored in regard to the relevancy of the semiotic system. Here we mention the semiotics of Charles S. Peirce and more precisely, his idea of diagrammatic thinking which became a tool for investigating mathematical activities (Dörfler 2005; Hoffmann 2005).

The presentation of the visualization group at ICME12 can now be seen as a realization of the above mentioned views on visualization that reflect the diversity of challenges of visualization within mathematics education. Among these presentations we find theoretical deliberations concentrating on visual semiotics, presentations central to mathematics education visualization and curriculum attempting to use technology to bridge the gap between mathematicians and mathematics education views, presentations concentrated on the use of new software and newer hardware to enhance visualization and on what might develop into new paradigm of the visualization science using brain imaging technology attempting to make the invisible visible. In the next few paragraphs we attempt to sketch the group work with illustrations from the many papers[2] presented.

As the first example we refer to Christoph Schreiber presenting his view on Peirce's semiotics "Semiotic Analysis of Collective Chat-Based Problem-Solving Processes". Schreiber illustrated the development of 'Semiotic Process Cards' based upon Charles Sanders Peirce's triadic sign relation. These cards were used as instruments for analyzing mathematical chat sessions. Within a certain teaching situation called 'Math Chat', students were asked to solve mathematics' problems while being restricted to the use of visible inscriptions only. The characteristics of this experimental setting was that pupils were required to document all their attempts at solving mathematical problems as visual inscriptions in written and graphical form. To develop a suitable instrument Schreiber combined an interactionist approach together with Perice's semiotic perspective. As a result Schreiber was able to describe the

Mathias Hattermann's text "Visualization—the Key Element for Expanding Geometrical Ideas to the 3D-Case" is an example of the group discussion in regard to the visual qualities of design of learning with technological tools. In his Hattermann described the activities of students at university level when using software for 3D-geometry (Cabri 3D). To do so he started with the presentation of two geometric constructions from plane geometry. Hattermann asked how do basic ideas in the context of plane geometry can foster or hinder similar constructions of 3D-geometry? It is the intimate relation between the tool used and the visible geometric diagrams or in other words the *instrumental genesis* of the software and the process of geometrical construction which is in the core of Hattermann's answer to his question. In this respect an experimental approach using the drag mode in 3D can help to find answers to describe the finding of a correct solution. The instrumental genesis of the utilized tool must be accomplished so that mental schemes can be used to extend basic ideas to the 3D-case.

The design and qualities of software was one component of the "Visual Math" curriculum design story that Michal Yerushalmy presented. The challenge was to establish technology-based setting that would motivate algebra students to argue, refute, and revise conjectures, and to study whether prominent visualization habits of mathematical reasoning can become part of the routine pedagogy of school mathematics. Beyond software Yerushalmy described why did the design of an organizational map was a major challenge in finding out how known algebra tasks may be redesigned into a sequence emphasizing quasi-empirical process of reasoning. The museum view was a leading image in the design of the VisualMath interactive eBooks in algebra, functions and calculus. Based on theoretical

---

[2] All papers presented within TSG 16 can be found at http://www.icme12.org/sub/tsg/tsg_last_view.asp?tsg_param=16.

framework of interactive diagrams that is based on visual-semiotic analysis, Yerushalmy design Interactive Diagrams that provide opportunities for the interactive text to present the curriculum's ideas to be the subject of the reader's inquiry.

Roza Leikin's "From a Visual to Symbolic Object in Algebra and Geometry: ERP[3] Study with Mathematically Excelling Male Adolescents" is in a sense the literal realization of our aforementioned hint "how to make the invisible visible". Leikin and coauthors performed a comparative analysis of brain activity associated with transition from visual objects to symbolic objects in algebra and geometry. The goal of this study was to examine differences in ERPs between gifted and non-gifted excelling in mathematics adolescents while solving mathematical tasks in algebra and geometry. One finding regarding the giftedness effect was that, relative to gifted participants, non-gifted participants produced greater brain activity. This finding is consistent with the neural efficiency hypothesis of intelligence, stating that brighter individuals display lower brain activation while performing cognitive tasks. Another finding indicates a significantly higher brain activity connected to geometry test compared to algebra test. Hence Leikin and assumes that geometric tasks increase the participants' working memory load by keeping the visual geometric object in working memory until the problem is solved.

In addition to the aforementioned view on the relation of the visual and mathematics these examples reflect a fruitful diversity of visualization too. In this respect visualization appears to be a vivid part of research within mathematics education.

Open Access This chapter is distributed under the terms of the Creative Commons Attribution Noncommercial License, which permits any noncommercial use, distribution, and reproduction in any medium, provided the original author(s) and source are credited.

References

Arcavi, A. (2003). "The role of visual representations in the learning of mathematics." *Educational Studies in Mathematics*, 52, 215–241.

Arnheim, R. (1969). *Visual thinking*, The Regents of the University of California.

Boehm, G. (1994). Die Wiederkehr der Bilder. *Was ist ein Bild?* G. Boehm. München, Wilhelm Fink, 11–38.

David, M. M., & V. S. Tomaz (2012). "The role of visual representations for structuring classroom mathematical activity." *Education Studies in Mathematics*, 80(3), 413–431.

Dörfler, W. (2005). Diagrammatic Thinking. Affordance and Constraints. *Activity and sign: grounding mathematics education*. M. Hoffmann, J. Lenhard and F. Seeger. New York, Springer: 57–66.

Eisenberg, T. (1994). "On understanding the reluctance to visualize." *Zentralblatt fuer Didaktik der Mathematik*, 26(4), 109–113.

Heintz, B., & J. Huber, Eds. (2001). *Mit dem Auge denken*. Wien, New York, Springer.

Hoffmann, M. H. G., J. Lenhard, et al., Eds. (2005). *Activity and sign*. New York, Springer.

---

[3] An event-related potential (ERP) is the measured brain response that is the direct result of a specific sensory, cognitive, or motor event.

Hessler, M., & D. Mersch, Eds. (2009). *Logik des Bildlichen: Zur Kritik der ikonischen Vernunft.* Bielefeld, transcript Verlag.

Mitchell, W. J. T. (1987). *Iconology: image, text, ideology.* Chicago, The University of Chicago Press.

Mitchell, W. J. T. (1994). *Picture theory.* Chicago, The University of Chicago Press.

Presmeg, N. C. (1986). "Visualisation in high school mathematics." *For the Learning of Mathematics,* 6(3), 42–46.

Presmeg, N. C. (1994). "The role of visually mediated processes in classroom mathematics." *Zentralblatt fuer Didaktik der Mathematik,* 26(4), 114–117.

Presmeg, N. C. (1997). Generalization Using Imagery in Mathematics. *Mathematical reasoning: analogies, metaphors and images.* L. English, D. London, Lawrence Erlbaum Associates, Inc.: 299–312.

Mathematical Applications and Modelling in the Teaching and Learning of Mathematics

Jill Brown and Toshikazu Ikeda

Introduction

Applications and modelling have been an important theme in mathematics education during the last 40 years; in particular, through ICMEs regular working/topic groups and lectures on applications and modelling, and the series of International Community on the Teaching of Mathematical Modelling and Applications (ICTMA) conferences, held biennially since 1983. Relations between the real world and mathematics are particularly topical. One reason for learning mathematics is to understand and make sense of the world. The mathematics education community was invited to submit proposals addressing one of six themes and related issues. The focus could be at any level of education including teacher education and the work of mathematicians in the field. It is not surprising therefore that this TSG attracted much attention, with 44 papers submitted. Papers were reviewed by two reviewers. Thirty-six papers were accepted for presentation, from 17 countries (Australia, Austria, Brazil, China, Cyprus, Germany, Israel, Japan, Korea, Mexico, Portugal, Singapore, South Africa, Sweden, Switzerland, UK, USA) and included several teacher authors. Authors received feedback from the co-chairs, and were given time to revise papers in response to this. Also 24 posters related to this TSG (from 10 countries) were presented. Accepted papers were assembled into groups

Organizers Co-chairs: Jill Brown and Toshikazu Ikeda; Team Members: Sung Sook Kim, (Korea), Nicholas Mousoulides (Cyprus), Jussara de Loiola Araújo (Brazil); Liaison IPC member: Morten Blomhoej (Denmark).

J. Brown (✉)
Australian Catholic University, Melbourne, Australia
e-mail: jill.brown@acu.edu.au

T. Ikeda
Yokohama National University, Yokohama, Japan
e-mail: toshi@ynu.ac.jp

© The Author(s) 2015
S.J. Cho (ed.), *The Proceedings of the 12th International Congress on Mathematical Education*, DOI 10.1007/978-3-319-12688-3_42

for summary, presentation, and discussion. Given the large number of papers, two concurrent sessions were held with participants together initially and for the final discussion. Given space constraints, only presenting authors are named.

Goals and Curriculum

Paraic Treacy presented *The role of mathematical applications in the integration of mathematics and science*, using the authentic integration triangle to argue how secondary students in Ireland can be supported to apply their mathematical knowledge to authentic tasks particularly in science contexts. Also looking at secondary school curriculum, Karen Norwood discussed *Mathematics instruction using decision science and engineering tools (MINDSET): A multi-step problem solving and modelling course for high school students.* She reported on the development and implementations of a year long US curriculum using a problem solving modelling approach. Xiaoli Lu presented a *Comparative study on mathematics applications in mathematics textbooks* where selected current texts from China and the US were scrutinised for mathematical applications. They report, disappointingly 'most examples in textbooks are traditional mathematical problems without real world contexts'. Jean-Luc Dorier's report on *Modelling: a federating theme in the new curriculum for mathematics and sciences in Geneva compulsory education (age 4-15)* outlined a new curriculum with modelling as a central theme. However, the definition of modelling was modified from that of Niss, Blum and Galbraith (2007); so rather than modelling involving the extra-mathematical and mathematical domains, although two domains are required, the real world is not an essential one of these.

Jussara Araujo presented *Critical construction of mathematical models: An experience on the division of financial resources,* reporting on graduate mathematics education students engagement in a critical mathematical modelling task where 'fair criteria' had to be determined to allocate money. The task raised awareness amongst the participants of the importance of modelling. Jung-Ha An reported on *Developing mathematical modelling curriculum using difference equations.* Examples were shown to demonstrate the use of difference equations in the modelling process in a general mathematics education course. Also at tertiary, *Mathematical experiments: A new-designed course for non-mathematical undergraduates in Chinese universities* was the focus of Jinxing Xie who shared experiences in designing and teaching courses, for non-mathematics students, on applied mathematics through experiments, modelling and software use.

Teaching Material, Pedagogy, and Technology

At the primary level, Nicholas Mousoulides presented *Modelling as a bridge between real world problems and school mathematics*. He argued for a modelling approach, using engineering MEAs, as a rich source of situations that build on and extend students' existing mathematical learning. Takashi Kawakami presented *Necessity for modelling teaching corresponding to diversities: Experimental lessons based on dual modelling cycle framework for the 5th grade pupils*. He reported on a teaching experiment with students working on two related tasks. Focussed on teachers of Year 8–9 students, Janeen Lamb presented *Planning for building models of situations: What is involved?* Data from 8 participants in a project aimed at enhancing teachers' instructional practices were analysed. After completing a modelling activity using an applet, teachers planned how to implement the task in their classrooms. Two studies focused on Year 12 Japanese students: Masahiro Takizawa presented *Colors and Mathematics*, illustrating how the colour of an image can be used to teach functions and transformations, by adopting a modelling approach. The paper presents a teaching experiment with Year 12 students, using the 'Colors' software. Tetsushi Kawasaki presented *A study of mathematical modelling on Year 12 students' function education*, reporting the use of modelling in promoting the teaching and learning of two variable functions. The author reports results of a teaching experiment with 15 students.

Issic Leung presented *The effect of changing dimensions in illustrative examples in enhancing the modelling process*, arguing for a greater emphasis on illustrative examples (e.g., a sketch or diagram). Making greater sense of what is represented should subsequently support mathematical modelling. Also taking a theoretical stance, Vince Geiger presented, *On considering alternative frameworks for examining modelling and application activity: The role of texts and digital tools in the process of mathematical modelling*, discussing several modelling cycles and frameworks used in either engaging in modelling or by researchers in the field. He argued that models for teaching and learning can be applied to modelling situations. His focus is on the interplay between task, teacher, students and tools.

Experimental Research

Irit Peled presented *More than modelling skills: a task sequence that also promotes children's meta knowledge of modelling*, reporting on the development of meta-knowledge of modelling by Year 5-6 students as they worked on 10 tasks. Meta-knowledge included different ways of mathematising a given problem and hence different models for a single situation can be used appropriately. Focused on Year 5 students Maike Hagena and Rita Borromeo Ferri presented, *How do measurement sense and modelling competency influence each other? An intervention study about German middle class students dealing with length and weight*. Susanne Grunewald presented *Acquirement of modelling competencies: First results of an empirical comparison of the effectiveness of two approaches to the development of (metacognitive) modelling competencies of students,* reporting use of modelling activities in measurement contexts with Year 8 students. Stanislaw Schukajlow and Andre Krug presented *Treating multiple solutions in the classroom and their influence on students' achievements and the affect–The preliminary results of the quasi-empirical study,* comparing Year 9 students' work on 'Pythagoras tasks' where no assumptions were required to those where different assumptions and hence multiple solutions were possible, hypothesising the latter leads to better achievement (modelling and intra-mathematical).

Jin Hyeong Park reported on *Conceptual understanding of mathematical knowledge through mathematical modelling in a spreadsheet environment.* Park sees modelling as representing real phenomena mathematically in order to understand the real world reporting a case study of 15 gifted Year 8 students engaged in an Iced Coffee Task. Findings included development of conceptual calculus understanding and ability to mathematise from their models back to the real world. Also focussing on spreadsheet use, Manfred Borovcnik reported *Applications of probability: The Limerick experiments* that is, responses of probability workshops participants (inservice secondary teachers), arguing that probability is best taught from a modelling and applications perspective, particularly where technology is used. Here any situation in a classroom is considered as being 'real world'.

Xueying Ji presented *A quasi-experimental study of high school students' mathematics modelling competence,* reporting modelling competence of Year 10–11 students in China. She found students did not realise the importance of validating their results or critically assessing their models. Milton Rosa presented *Ethnomodelling: A research concept on mathematical modelling,* arguing the application of techniques in ethnomathematics along with the tools of modelling allows us to see a different reality. Further, research should be from an etic and an emic perspective.

Assessment, Teacher Education, and Obstacles

Peter Frejd presented *Alternative modes of modelling assessment: A literature review,* reporting different assessment methods (i.e., written tests, projects) and viewpoints (atomic or holistic). Xenia-Rosemarie Reit and Matthias Ludwig's paper, *A cross-section study about modelling task solutions,* reported a study where 337 solutions to the *Restringing a tennis racket task* were analysed. Four main solution approaches were identified. Differences were found in terms of approach taken and progress on the solution path. Kaino Luckson presented *The nature of modelling activities and abilities of undergraduate students: some reflections on students' mathematics portfolios,* focusing on modelling tasks undertaken by pre-service teacher education students via distance education. Michael Besser reported on *Competency-oriented written feedback in every-day mathematics teaching: How to report on students' solutions of modelling tasks and how to assess the quality of these reports?* This study looked at teacher feedback in the context of technical and modelling tasks, considering strengths/weaknesses of specialized written competency based feedback.

In the field of pre-service teacher education, Thomas Lingefjärd presented *Learning mathematics through mathematical modelling,* arguing that by developing modelling tasks and then engaging in teaching scenarios conceptual understanding occurs. In addition, students came to understand that technology changes what is possible in developing modelling tasks. Dawn Ng presented, *Activating teacher critical moments through reflection on mathematical modelling facilitation* where the focus was on the teacher's role and in particular, on the teacher interpretation of student ideas and interventions. The interplay between listening and questioning was critical. Also focused on the role of the teacher was Peter Stender on *Modelling in mathematics education development of forms of intervention and their placement in the teacher education* and Dominik Leiss *Adaptive teacher interventions in mathematical modelling.* Both report studies where the balance between student autonomy and teacher interventions was critical.

A Final Word

There are many interpretations of the terms *mathematical modelling* and *applications.* Whilst diversity is desirable, it is helpful to have a common basis for our interpretations. TSG discussion contributes to a shared understanding and the majority of teachers and researchers, see the *real world* as a critical and essential component of modelling and applications. Following Niss et al. (2007), both mathematical modelling and applications are seen as connecting the mathematical world and the real world. These two worlds are distinct, with the later "describing the world outside mathematics" (p. 3). It is also important to distinguish between modelling and applications. The former begins in the real world and requires a modeller to mathematise the situation, that is, to translate the problem situation into a mathematical situation. In an application, this mathematising has already been done for the solver who works in the mathematical world.

Open Access This chapter is distributed under the terms of the Creative Commons Attribution Noncommercial License, which permits any noncommercial use, distribution, and reproduction in any medium, provided the original author(s) and source are credited.

Reference

Niss, M., Blum, W. & Galbraith, P. (2007). Introduction. In W. Blum, P. Galbraith, P., H.-W. Henn & M. Niss (Eds.), *Modelling and applications in mathematics education–The 14th ICMI study* (pp. 4-32). New York: Springer.

Analysis of Uses of Technology in the Teaching of Mathematics

Morten Misfeldt and Wei-Chi Yang

Overviews

This Topic Study Group aimed at providing a forum to discuss the current state of art of the presence of technology in diverse aspects of teaching mathematics conveying a deep analysis of its implications to the future. Technology was understood in a broad sense, encompassing the computers of all types including the hand-held technology, the software of all types, and the technology of communication that includes the electronic board and the Internet. The discussions served as opportunity for all interested in the use of technology in education environment, to understand its diverse aspects and to share the creative and outstanding contributions, with critical analysis of the different uses.

The Topic Study Group had 42 contributions and more than 80 participants. The topics addressed were diverse but evolve around the use of technology in the classroom practice, design and use of digital teaching materials, Technology in teacher education, Distance education and the use of learning management systems. The use of technology in the teaching of mathematics is an expanding and diverse field, and in the following we will summarize the status and consensus that became apparent through the work in the Topic Study Group. One way to gain an overview of the image of the field presented in the topic study group is to look at the different

Organizers Co-chairs: Morten Misfeldt (Denmark), Wei-Chi Yang (USA); Team Members: Erol Karakirik (Turkey), Ngan Hoe Lee (Singapore), Cheong Soo Cho (Korea), Matte Andersen (Norway); Liaison IPC Menber: Yuriko Baldin (Brazil).

M. Misfeldt (✉)
Aarhus University, Aarhus, Denmark
e-mail: mmi@dpu.dk

W.-C. Yang
Radford University, Radford, USA
e-mail: wyang@radford.edu

© The Author(s) 2015
S.J. Cho (ed.), *The Proceedings of the 12th International Congress on Mathematical Education*, DOI 10.1007/978-3-319-12688-3_43

technologies involved. The technologies adopted and described in the TSG did mainly fall into 6 categories (1) handheld and pc based computer algebra systems, (2) dynamic geometry systems, (3) learning management systems and internet access, (4) domain specific visualizations and manipulatives, (5) video streaming, and (6) touch technology such as ipads and smartboards.

Another way to gain such overview is to present the discussions and concerns that were prevalent in the discussion and contributions. These concerns relate to (a) an increase in efficiency of mathematics instruction with the aid of technology—including technological support for development of specific mathematical concepts and competencies, (b) teacher training and teacher practice with technology, (c) the use of technology to support motivation and recruitment to mathematics, and (d) technological support for teaching processes—such as digital task assignment and marking.

Technologies Used for Mathematics Instruction

Computer Algebra Systems, Dynamic Geometry Systems and spreadsheets has been a part of mathematics instruction for decades, yet the mediations of the technologies as well as the research problems addressed by the community is still developing. The presentations in the TSG showed that these technologies are to some extend adopted in the mathematics education practices. The contributions relating to these technologies hence addressed issues relating to teachers adoption, the possibility to deploy such technologies for supporting low achievers, the teaching of specific mathematical concepts in new ways with technology, and the integration of these technologies into learning management systems. Learning management systems signifies a class of systems that is used to support and augment teaching practices. In relation to these systems a number of initiatives to augment their mathematical capabilities were presented. Apart from the integration of Computer Algebra Systems and Dynamic Geometry Software into learning management systems, the work related to the use of such systems related to the construction of multimodal learning environments including video and interactive manipulatives, within learning management systems. Technology that allows for the development of interactive visualizations and for sharing content were presented for several topics and educational levels ranging from primary school to university. Online task environments for students to train their skills with mathematical tasks were also presented. Online streaming of video was presented both as stand-alone and as a part of an online environment for teaching of mathematics. One project applying tablet pcs and interactive whiteboards was also presented. Hence a wide range of the applicable educational technologies where present in the Topic Study Group.

Problems Addressed in the TSG

The main problem addressed in the contributions was the potentials of using technology to enhance teaching of mathematics to become a more efficient enterprise. This problem was addressed in a multitude of ways in the topic study group. Interventions aiming at using technology (typically CAS and DGS) to teach mathematical topics in new ways where presented in the groups. It is difficult to summarize the role and influence of technology across these interventions since many factors other than the use of technology influences such interventions. Teacher training and teacher practice with technology conducted was addressed in several of the contributions. One motivation for a specific attendance to this area is that the teachers' choices and practices are, in many ways determining for the success of technology integration in the teaching of mathematics. Motivation and recruitment is an important theme underlying several of the contributions to the topic study group. New interactive illustrations or video presentations might not only make it possible for more students to grasp the abstract mathematical concepts, it might also make mathematics more appealing to larger groups of students.

Apart from addressing concerns, some of the reports in the Topic Study Group also demonstrated new technological developments, addressing the aim of improving mathematics instruction. Technology can automate aspects of the process of teaching mathematics, such as assigning and marking tasks. This is a development with many possible advantages and represents an area were the technological development currently is quite rapid. Another area where new technological developments were presented was the integration of mathematical tools such as CAS and DGS, into web 2.0 internet technology, in a way that supports collaboration and distance education.

A special issue of the electronic Journal of Mathematics and Technology is under preparation. This issue will include papers from the Topic Study Group.

Open Access This chapter is distributed under the terms of the Creative Commons Attribution Noncommercial License, which permits any noncommercial use, distribution, and reproduction in any medium, provided the original author(s) and source are credited.

Analysis of Uses of Technology in the Learning of Mathematics

Marcelo C. Borba and Hans-Georg Weigand

Introduction

In ICME12, the role of technology in mathematics education was divided into two distinct study groups: Analysis of uses of technology in the teaching (TSG 18) and learning (TSG 19) of mathematics. Of course, these two aspects of mathematics education are closely intertwined, but we tried to concentrate the TSG 19 discussions around the aspect of LEARNING with ICT (Information and Communications Technology).

The TSG 19 especially addressed the following issues in the learning of mathematics:

- the design of digital technology
- the design of learning environments
- large-scale and long-standing digital technology implementation projects
- assessing mathematics learning with and through digital technologies
- the interaction between ICT and learners of mathematics
- connectivity of ICT
- theoretical and empirical models for learning with ICT
- the implementation of curricula

Organizers Co-chairs: Marcelo C. Borba (Brazil), Hans-Georg Weigand (Germany); Team Members: Ornella Robutti (Italy), Mónica Villarreal (Argentina), Tom Dick (USA), Youngcook Jun (Korea); Liaison IPC Member: Yuriko Baldin (Brazil).

M.C. Borba (✉)
UNESP, São Paulo, Brazil
e-mail: mborba@rc.unesp.br

H.-G. Weigand
University of Wuerzburg, Wuerzburg, Germany
e-mail: weigand@dmuw.de

© The Author(s) 2015
S.J. Cho (ed.), *The Proceedings of the 12th International Congress on Mathematical Education*, DOI 10.1007/978-3-319-12688-3_44

Outline of Contributions

All submitted papers were reviewed by three reviewers and 33 papers and one poster were finally accepted. For presentation, papers were grouped into four groups:

- Group A: E-learning, Interactive Textbooks, Games, Mobile Applications
- Group B: Theoretical Aspects
- Group C: Dynamic Geometry Systems (DGS), Calculators, CAS
- Group D: Topics in Mathematics

Each of the four 90-min sessions was devoted to one of these four groups of papers. The time available did not allow for formal presentations of every paper by their authors. Two papers from each group were selected for presentations by the authors. The remaining papers in that group were summarized by a member of the organizing committee, with opportunities for comments by the authors and for discussion of the papers by all participants. The structure for each 90-min session included some brief opening remarks by the co-chairs of the committee, followed by a 30-min period for summary and discussion of those papers not presented later in the session. Following this summary discussion, each of the two selected papers were presented by their authors (15 min each, with 10 min for presentation and 5 min for discussion). After the individual paper presentations, participants engaged in 15 min of roundtable discussions focused around questions of emergent issues raised by the papers considered in that session. At the conclusion of each session, the TSG 19 co-chairs had made some brief closing remarks.

Group A: E-learning, Interactive Textbooks, Games, Mobile Applications

- Gerry Stahl (College of Information Science, Drexel University, Philadelphia, USA): Designing a Learning Environment to Promote Math Discourse
- Robyn Jorgensen (Griffith University—Australia), Tim Lowrie (Charles Sturt University—Australia): Digital Games and Mathematical Learning: A summary paper

Gerry Stahl emphasized the fact that more and more teachers and students were learning online—with distance education, online masters programs, home schooling, online high schools, etc.—which makes the incorporation of virtual collaborative learning environments a natural trend. He presented a virtual GeoGebra learning environment that integrates synchronous and asynchronous media with an innovative multi-user version of a dynamic math visualization and exploration toolbox.

Jorgensen and Lowrie presented a summary of a three-year project that explored the possibilities of digital games to enhance mathematical learning. They especially found that using games in classrooms might have much more benefits than just learning mathematics.

Group B: Theoretical Aspects

- Abramovich Sergei (*State University of New York at Potsdam, USA*), Eun Kyeong Cho (*University of New Hampshire, USA*): Pre-teachers' learning of mathematics through technology-enabled problem posing
- Barbara Schmidt-Thieme (University of Hildesheim Germany), Hans-Georg Weigand (University of Wuerzburg, Germany): Choosing adequate Digital Representations,

Abramovich and Cho considered the potential of new technologies to turn a routine arithmetical problem into a challenging mathematical investigation. The authors suggested that an important didactic task for teachers will be to decide if technology-enabled problem posing results in a contextually, numerically, and pedagogically coherent problem. This influences the choice of the adequate software.

Schmidt-Thieme and Weigand presented examples of students' working with representations and posed some main future research questions concerning the use of representations in a technology-based environment, e.g.: Which criteria characterize an adequate representation of a problem's solution? Which different levels of argumentation, reasoning and proof are related to a special representation? Which criteria characterize a good (in the sense of giving some feedback about learners' competencies) documentation of a solution of a problem?

Group C: Dynamic Geometry Systems (DGS), Calculators, CAS

- Arthur B. Powell, Loretta Dicker (Rutgers University, USA): Toward Collaborative Learning with Dynamic Geometry Environments
- Thomas Lingefjärd, Jonaki Ghosh, Aaloka Kanhere (Technology Working Group of the Indo Swedish Initiative in Mathematics Education): Students Solving Investigatory Problems with GeoGebra—A Study of Students' Work in India and Sweden,

Powell and Dicker presented a model of collaborative, online learning with a dynamic geometry environment that supports collaboration around mathematical problem solving and development of significant mathematical discourse. The authors especially intend to motivate in-service secondary teachers in designing curricular units that develop students' significant mathematical discourse as they develop geometric ideas.

Lingefjärd, Ghosh and Kanhere started with the hypothesis that the use of technology in mathematics instruction might lead from an experimental mathematics, that is, verification and conjecturing, to theoretical mathematics, that is, formal abstract concepts and proofs. The authors had done a parallel experimental study in Sweden and India using a dynamical geometry environment and getting quite similar results concerning the working styles of students in these two countries.

Group D: Topics in Mathematics

- Christian Bokhove (St. Michaël College, Zaandam, the Netherlands/Freudenthal Institute, Utrecht University, Utrecht, the Netherlands), Paul Drijvers (Freudenthal Institute, Utrecht University, Utrecht, the Netherlands): Effects Of A Digital Intervention On The Development Of Algebraic Expertise
- Jens Jesberg, Matthias Ludwig (Goethe University Frankfurt, Germany): MathCityMap—Make mathematical experiences in out-of-school activities using mobile technology

Bokhove and Drijvers especially wanted to answer the question about the effect of an intervention, consisting especially of diagnostic digital modules, on the development of algebraic expertise, including both procedural skills and symbol sense. They observed "a large effect on improving algebraic expertise" after an intervention of just 5 h.

Jesberg and Ludwig presented a "MathCityMap-project", which is based on a GPS technology. High school students experienced mathematics at real locations and in real situations within out-of-school activities, with the help of GPS-enabled smartphones and special math problems.

Conclusions

More than thirty years have passed since the first ICMI study group on technology. Papers presented in this TSG show that the work with technologies can present new trends even though one can no longer refer to digital technologies as "new technologies". Digital tablets and devices that increasingly enhance the possible interactions between humans and technology were presented as means for transforming the way students can know. Many of these devices imply changes in curriculum and challenge the structure of time in school. In other words, if they are to be used in school, students will either have to be outside class using mobile technology, or in class using them for longer periods of time. TSG 19 was diverse enough that many papers also proposed how technology can be used now, without many changes in the way school is organized. "Geogebra" is one of those key applications used at this conference. The free software seems to have found many different followers in different countries and it has been used in different manners. Some have incorporated it into online learning environments, while others are developing ways of annotating the screen of Geogebra.

Last but not least, findings of new technological developments and of research results were discussed in small groups, overcoming language barriers. The situation is the same in mathematics classrooms all over the world. Apart from special and valuable cultural divergence and distinctions new technologies reveal the same or at least similar problems in mathematics learning all over the world and they may be a catalyst to forward important developments in mathematics classroom activity.

Open Access This chapter is distributed under the terms of the Creative Commons Attribution Noncommercial License, which permits any noncommercial use, distribution, and reproduction in any medium, provided the original author(s) and source are credited.

The Role of History of Mathematics in Mathematics Education

Renaud Chorlay and Wann-Sheng Horng

Report

At the ICME 2012 Conference, *history of mathematics* (HM) *in maths education* was specifically discussed in several contexts: one of the 37 *Topic Study Groups* (chaired by W.S. Horng and R. Chorlay); one discussion group on "Uses of History of Mathematics in School (Pupils aged 6–13)" (organised by B. Smestad); one regular lecture on "History, Application, and Philosophy of Mathematics in Mathematics Education: Accessing and Assessing Student's Overview & Judgment" (by U. Jankvist); and one general presentation of The HPM international study group, among the organizations affiliated with the ICMI. A parallel TSG dealt with *the history of mathematics teaching and learning* (chaired by K. Bjarnadottir and F. Furinghetti).

Eleven talks and fourteen posters were presented in the context of TSG 20, with participants from (nearly) all continents; unfortunately, the African continent was not represented. The TSG was a great success if success is to be measured by attendance.

Being of a multi-faceted nature, the topic was addressed from a great variety of viewpoints, which testifies to the richness of our field. Our goal here is not to

Organizers Co-chairs: Renaud Chorlay (France), Wann-Sheng Horng (Taiwan); Team Members: Hyewon Chang (Korea), Kathy Clark (USA), Abdellah El Idrissi (Moroco), Manfred Kronfellner (Austria); Liaison IPC member: Evelyne Barbin.

R. Chorlay (✉)
ESPE, Paris, France
e-mail: renaud.chorlay@espe-paris.fr

W.-S. Horng
National Taiwan Normal University, Taipei, Taiwan
e-mail: horng@math.ntnu.edu.tw

© The Author(s) 2015
S.J. Cho (ed.), *The Proceedings of the 12th International Congress on Mathematical Education*, DOI 10.1007/978-3-319-12688-3_45

summarize the talks (which are still available on-line at http://www.icme12.org/) but to stress this variety of viewpoints and research perspectives.

Research in the history of mathematics was represented by A. Cauty's talk on Aztec calendars, providing the rest of the community with fresh material for future, more teaching-oriented work. At the other end of the spectrum, several innovative teaching or training experiments were presented and discussed: a course for undergraduate students, with a focus on the role of mathematics in European culture (J. Wanko); an undergraduate course on propositional logic and the meaning of "if-then" statements, emphasizing student work on original sources (J. Lodder); a course designed for newly qualified teachers, with an emphasis on the role of HM as a means to foster mathematical content knowledge (S. Lawrence); a course on the history of mathematics for pre-service teachers in Norway, with a focus on the interactions between historical content knowledge, image of mathematics, and attitude toward the inclusion of HM in teaching (B. Smestad).

Finding the right tools (be they conceptual, or quantitative) to describe, analyze and assess teaching practices is another endeavour that calls for further research. These questions are by no means specific to the HPM community, and it is well-worth investigating the extent to which shared tools are relevant in an HPM context. Along this line of research, M. Alpaslan presented his on-going doctoral work on the assessment of a pre-service teacher-training course in HM in Turkey, with a view to improving its design in a context of institutional reform. U. Jankvist presented a joint work (with R. Mosvold, J. Fauskanger, and A. Jakobsen) on the MKT framework (Mathematical Knowledge for Teaching), and argued for its usefulness both as an analytical tool and as a means of communication with the math-education community at large.

Four case-studies were presented, which used specific historical texts to address didactical/epistemological research questions. The role of visualization in proofs was studied on the base of Archimedes' "mechanical proof" of the theorem on the volume of the sphere (M. del Carmen Bonilla); CABRI 3D was used as a visualization tool. S. Xuhua argued that several justifications for algorithms in the multiplicative theory of fractions that can be found in the Chinese classic *The Nine Chapters* could improve students' understanding of the standard rules, and help fight well-known systematic errors. T. Kjeldsen reported on an experiment conducted at high-school level, in which students were asked to make sense and compare two historical texts bearing on the notion of function (Euler, Dirichlet); among other effects, this unusual task was shown to help make "meta discursive rules" more explicit. Finally, A. Michel-Pajus presented a collection of algorithmic texts—some well-known, some excitingly new—and studied them from an epistemological and comparative perspective; the algorithms were studied both in terms of expression (algorithmic *texts*, in a semiotic and instrumental context), and justification.

It should be stressed that in the ICME context, the TSG on HM in maths-education attracts many newcomers to the field of HPM, thus challenging members of the HPM community to make their "common culture" and their quality requirements more explicit. For instance, the fact that most of us stress the importance of the use of *original* sources may have come as a surprise to some; even without considering use in the classroom, the fact that original sources *are* available (availability being highly dependent on language) is not always so well known. When original sources are considered, working with them does require some know-how. We hope this TSG was instrumental in raising awareness on these aspects; we were pleased to see that many participants, including newcomers, could attend the HPM meeting in Daejeon (16–20 July 2012).

The chairpersons would very much like to thank all those who helped organize this TSG, in particular the members of the "team": Hyewon Chang, Kathy Clark, Abdellah El Idrissi, and Manfred Kronfellner; and, Evelyne Barbin, who acted as liaison with the IPC.

Open Access This chapter is distributed under the terms of the Creative Commons Attribution Noncommercial License, which permits any noncommercial use, distribution, and reproduction in any medium, provided the original author(s) and source are credited.

Research on Classroom Practice

Yeping Li and Hélia Oliveira

Introduction

Classroom practice, as a process, involves multiple agents and their interactions within the classroom as a system. The process can be manifested in diverse formats and structures, and its effectiveness can be influenced by numerous factors both internal and external to the classroom. Research on (mathematics) classroom practice can thus take different perspectives, and much remains to be examined and understood as we all try to improve mathematics teaching and learning through classroom practice.

Although it has long been recognized that research on classroom practice is important, large-scale systematic research on classroom practice in school mathematics is a relatively new endeavour. In fact, this Topic Study Group is only the second time in the ICME history to take a primary focus on classroom practice. As the quality of classroom instruction is a key to students' mathematics learning, this Topic Study Group focuses on finding ways for understanding, assessing, and improving the quality of classroom practice.

The entire organizing team worked together before the congress in planning and organizing TSG 21. The TSG 21 was well attended in all four 90-min sessions,

Organizers Team Chairs: Yeping Li (USA), Hélia Oliveira (Portugal); Team Members: Merrilyn Goos (Australia), Kwangho Lee (Korea), Raimundo Olfos (Chile); Liaison IPC Member: Fredrick Leung (Hong Kong).

Y. Li (✉)
Texas A&M University, College Station, USA
e-mail: yepingli@tamu.edu

H. Oliveira
University of Lisbon, Lisbon, Portugal
e-mail: hmoliveira@ie.ul.pt

© The Author(s) 2015

S.J. Cho (ed.), *The Proceedings of the 12th International Congress on Mathematical Education*, DOI 10.1007/978-3-319-12688-3_46

which indicates strong interest in this topic by congress delegates. This report provides an overview of the aim and focus of TSG21 and a summary of the discussion that occurred throughout the sessions.

Aims, Focuses, and Themes

As set by the organization team, the general aim of TSG 21 was, in the international mathematics education community, to elevate people's understanding of the importance, specific nature, and challenges in research on classroom practice, to promote exchanges and collaborations in identifying and examining high-quality practices in classroom instruction across different education systems, and to enhance the quality of research and classroom practice. More specifically, through its official program during the congress and other activities (including those before and after the congress), TSG 21 was intended to provide an international platform for all interested parties (e.g., mathematics educators, mathematics teachers, educational researchers, etc.) to disseminate findings from their research on classroom practice with the use of various theoretical perspectives and methodologies, and to exchange ideas about mathematics classroom research, development, and evaluation.

The main focus of TSG 21 was a discussion of research related to mathematics classroom practice, which includes activities of learning and teaching processes located within the classroom as a system. This requires a study of the interactions among the mathematical content to be taught and learned, the instructional practices of the teacher, and the work and experiences of the students. In the interaction processes, mathematical content is contextualized through situations, the teacher plays an important instructional role drawing on his/her knowledge, and the students involve themselves in the learning processes. It is important to understand through research the nature and extent of these interactions, the complexity of the didactic system, the roles of the teacher and students in the interaction processes when the mathematical content is taught and learned, and the complexity of the activities in the classroom.

The 39 accepted papers were assembled into the following eight themes for presentation and discussion during the congress:

- Theme 1: Theoretical and methodological considerations
- Theme 2: Instructional context, reflection, and improvement
- Theme 3: High-quality instructional practices
- Theme 4: Students' perception, class work, and learning
- Theme 5: Teaching and learning elementary mathematics
- Theme 6: Teachers' questioning and response in classroom instruction
- Theme 7: Instructional design and practice
- Theme 8: Curriculum/task implementation

In addition, there were nine proposals accepted for poster presentations in a separate session organized by the Congress.

Each of the four 90-min sessions (July 10, 11, 13, and 14 in 2012) was devoted to two of these eight themes (4–5 papers for each theme), which were carried out simultaneously in two separate rooms. In the following sections, we briefly summarize the paper presentations and discussions during these sessions.

Session 1 (Theme 1: Theoretical and Methodological Considerations)

Gade adopted a theory/practice approach based on Vygostky for researching classroom practice, with the potential of informing practitioner's inquiry in ongoing classrooms. Morera and Fortuny illustrated the use of an analytical method of classroom episodes as a proposal to develop systematic research on whole-group discussions. Mesa, Lande and Whittemore argued for the need to attend to two dimensions of classroom interaction when describing it, by one study where they, simultaneously, analyzed the complexity of mathematical questions and the interactional moves that the teachers use to encourage student involvement in the lesson. Canavarro, Oliveira and Menezes illustrated the use of an analytical tool for lessons driven by an inquiry-based perspective in the case of one teacher who adopted a four phase model for the lesson structure. Xolo reported one expanded coding scheme that focuses on learning outcomes and teachers' didactic strategies from video recordings of sequences of lessons, intended to capture a greater degree of nuance in classroom practice.

In synthesis, these papers propose new analytical tools to investigate the classroom practice that contribute to having a better picture of what is happening in the classroom, showing a deep concern for acknowledging the teachers' work.

Session 1 (Theme 2: Instructional Context, Reflection, and Improvement)

Andersson presented a study of disengaged students' identity narratives in the senior secondary years. The instructional context was defined by tasks, situations (tools, activities, participants), school structures, the socio-political context, and the societal context. Olfos and Estrella described the use of a short video rich in potential problem situations to help primary school teachers initiate a lesson on fractions via problem posing. The lesson study approach resulted in the lesson being successively improved as each teacher implemented it. Oliveira, Menezes, and Canavarro reported on a project that created multimedia cases to stimulate reflective analysis of lessons in teacher education. Lee and Kim analysed one teacher's discourse during lessons involving small group work and used the lesson video to stimulate the teacher's reflective thinking towards improvement. Vanegas,

Giménez, and Font i Moll illustrated the use of a two dimensional grid for identifying nine types of democratic mathematical practices in the classroom.

In synthesis, each of the papers presented in this theme reported on attempts to support democratic, equitable and critical classroom practices. The authors also investigated processes of teacher change by supporting teachers' systematic reflection and iterative improvement of their lessons.

Session 2 (Theme 3: High-Quality Instructional Practices)

Zhao and Ma found that lessons taught decades apart had similar content and teacher-student interaction but different types of tasks. Lee reported on the classroom practice of a teacher with high levels of mathematical knowledge for teaching. This study highlights a need for sensitivity in building respectful relationships between the researcher and teacher when classroom observation is also used for teacher evaluation. Zhao examined secondary school teaching practices in China where few teachers have a background in statistics. Focusing on teachers' interpretation of statistical graphs, the study found that teachers had limited understanding of key statistical concepts and gave more emphasis to procedures than conceptual understanding. Lewis, Corey, and Leong compared research from Japan, Singapore, and the US and found similarities in the categories used to define high quality practice. Li asked what could be learned from culturally valued classroom practices in China, and proposed a framework comprising macro pedagogy and micro pedagogy perspectives for understanding classroom instruction.

In synthesis, the papers in this theme proposed a variety of analytical frameworks for observing lessons and evaluating the quality of instruction. But each was concerned with the question of what counts as "high quality" instruction, and whether there are common or different criteria across countries and cultures.

Session 2 (Theme 4: Students' Perception, Class Work, and Learning)

Olteanu presented some results from a longitudinal study whose aim was to provide and develop a repertoire of reliable practices and tools to solve immediate problems in teachers' daily professional lives; namely, to improve students' learning in mathematics. Gao and Tian concluded that the students in the class where an open inquiry to problem solving was adopted were more accurate and succinct, quicker, and more fluent in language than the students in the class that followed a guided inquiry. Yang and Leung found that secondary students generally do not perceive their mathematics classroom environment very favorably. Gender differences were also found. Yau and Mok reported five consecutive lessons that showed that most students imitated the teacher's examples completely or partly. The authors argued that the strong direct role of teacher might help the students master their mathematical content. Araya, Varas, Giaconi and Foltz analyzed pupil's perceptions about mathematics, math learning and teaching in Chile and Finland. Considering the significant difference between these two countries, results showed surprising similarities connected to prototypical ideas.

Session 3 (Theme 5: Teaching and Learning Elementary Mathematics)

Silvestre and Ponte showed that the teaching/learning experience supports the conjecture that proportional reasoning develops when students explore, solve problems, and work with different representations. Yong, Zanzali, and Jiar showed that by developing a favorable learning environment and through scaffolding the students (low achievers) could progressively adapt themselves to a child-centered approach and begin to think more autonomously. Goos, Geiger, and Dole presented a model of numeracy whose elements comprise mathematical knowledge, dispositions, tools, contexts, and a critical orientation to the use of mathematics, and applied it to analyze changes in one teacher's planning, classroom practice, and personal conceptions of numeracy. Kwon and Thames showed that despite variations in the use of the task and the collective work with students, the work of teaching involves several core features: hearing mathematical reasoning, mathematical needs, and key mathematical concepts; and comparing different solutions and making alternative solutions reasonable. Pinto studied the development of the meaning of multiplication and division of non-negative rational numbers, arguing that problem solving helps students to overcome some difficulties and to understand and to formalize mathematical concepts.

In synthesis, these papers illustrate good practices that draw on the use of powerful mathematical tasks alongside with approaches that promote students' autonomy and critical orientation in solving problems.

Session 3 (Theme 6: Teachers' Questioning and Response in Classroom Instruction)

Lee analyzed the changes in one pre-service teacher's questioning practices, as she starts to give her students the opportunity for explaining and justifying their mathematical ideas. Subramanian illustrated several forms of questioning by one Indian teacher, which she argues is a culture-influenced pedagogy in that country and thereby widely practiced in the classrooms. Fox reported how two teachers who were observed throughout one unit of instruction were able to handle unanticipated questions by posing counter examples or simpler related questions. Sun compared the questioning practices of two teachers (Chinese and Czech) by observing the video of one lesson from each teacher. The questions posed by the Chinese teacher tended to require only a short answer, in a short period of time, and without the teacher's help. On the contrary, the questions by the Czech teacher were more cognitively demanding, but he provided no scaffolding. Aizikovitsh-Udi, Star, and Clarke presented two case studies demonstrating that good teacher questioning involves more than just good questions.

In synthesis, these papers show the growing interest in the teachers' questioning practices as a consequence of the recognition of its pedagogical value for the students' learning. Some professional cultures seem to value the power of questioning for a long time, but the nature and objectives of the questions the teachers pose differ substantially from setting to setting.

Session 4 (Theme 7: Instructional Design and Practice)

Mogensen shared recent efforts in Denmark to focus on mathematical pedagogical goals and mathematical points in mathematics teaching. Choquet analyzed the practice changes of a primary school teacher resulted from using '*problème souverts*' (open problems). Sekiguchi examined how Japanese mathematics teachers handle multi-dimensions of coherence and coordinate coherence and variation. Japanese mathematics teachers seemed to achieve multidimensional coherence by utilizing a double-anchored process schema, and their deliberate use of variation seemed to facilitate students' reflection. Lin described a general procedure of conceptual variation via either diagram form (more on perceptual knowledge), or verbal/symbolic form (more on rational knowledge). A lesson plan of conceptual variation on the topic of elliptical definition was also given to illustrate how to use the general procedure to design conceptual variation. Varas, Martínez, Fuentealba, Näveri, Ahtee, and Pehkonen presented results from a three-year follow-up Finland–Chile research project that introduced open-ended problem solving activities in third grade classes.

In synthesis, these papers present different perspectives and approaches used in developing and designing classroom instruction, with particular focuses on the use and organization of mathematical ideas/points, open-ended problems, instructional coherence and variation.

Session 4 (Theme 8: Curriculum/Task Implementation)

Huang, Li and Yang reported one study with three primary teachers in the context of the implementation of a new mathematics curriculum, in which the notion of variable was taught. All teachers promoted students' use of numbers and letters to describe realistic problems and explain conclusions, but they provided few opportunities for students to experience the problem-solving process. Moreira and

Campelos discussed the implications of the implementation of a new mathematics curriculum on teachers' practices, focusing on the balance between the collective and individual component of the practice. Grow-Maienza presented the results of one program that promoted the teachers' integration of principles abstracted from a Korean curriculum into the curriculum in use in one elementary school in the USA. Bingolbali and Bingolbali analyzed one teachers' practice concerning the implementation of one task in the classroom, arguing that a low fidelity to the task plan may be an expression of the teacher's flexibility to attend to students' needs.

In synthesis, these papers show that curriculum reforms are fruitful contexts to research the classroom practice, that may provide good opportunities to rethink the professional development of teachers, but that it is also necessary to understand how the intended innovations relate with the collectively and individually established teachers' practices.

Closing Remarks

Among the main points discussed across the four sessions we highlight the following ones:

- The search for what characterizes "high-quality" practices and the frameworks used to evaluate these practices taking into account the cultural and national diversity;
- The evolving classroom practices in many countries that reflect a move from the traditional instruction to innovative ways of teaching, and the demanding teacher's role associated with that transformation;
- The practices of questioning and inquiry-based approaches in different countries and their commonalities and differences;
- The teachers' practices concerning the work with mathematical tasks, namely their concern about the role played by the contexts, and the students' dispositions and perspectives concerning mathematics;
- The "Chinese paradox" and other countries' paradoxes concerning the relationship between students' achievement and classroom practice;
- The development of new analytical tools to do research on classroom practice.

Naturally, in such a broad topic as classroom practice, many questions remain to be addressed. The diversity of themes and focuses presented suggests many different perspectives that contributors took on what constitutes "classroom practice", which aspects of classroom practice are to be focused, and how "practice" is conceived using different analytical frameworks. The participants shared a strong interest in continuing the TSG's dynamics, and proposed the possibility of exploring joint projects in different countries and new publications focusing on some of the main themes discussed, and of gathering at other international conferences to do informal meetings to continue to do some work together.

Open Access This chapter is distributed under the terms of the Creative Commons Attribution Noncommercial License, which permits any noncommercial use, distribution, and reproduction in any medium, provided the original author(s) and source are credited.

Learning and Cognition in Mathematics

Gaye Williams and Hsin Mei Huang

Aims

Learning and cognition is a classical and very vital area in research on mathematics education. Researchers have published many valuable research findings that have contributed to significant development in this area. The continued efforts of researchers now and in the future will, we hope, lead to extensive 'pay-offs'. Different to many other special and related TSGs, such as teaching and learning of algebra, geometry, measurement, statistics, calculus, reasoning, proving and problem solving, to mention a few, TSG22's participants will contribute a more general focus on learning and cognitive activity, and insights into students' characteristics; their strengths and weaknesses in the process of mathematics learning. The TSG focus can include any teaching and learning contexts: from kindergarten to tertiary level, adult education, and teacher professional development. TSG22 discussions should be balanced between theories and their practical applications in mathematics teaching and learning.

Organizers Co-chairs: Gaye Williams (Australia), Hsin-Mei Huang (Taiwan), Team Members: Sungsun Park (Korea), Mariana Saiz (Mexico), Jerry Becker (USA); Liaison IPC Member: Shiqi Li (China).

G. Williams (✉)
Deakin University, Geelong, VIC, Australia
e-mail: gaye.williams@deakin.edu.au

H.M. Huang
Taipei Municipal University of Education, Taipei, Taiwan
e-mail: hhuang22@gmail.com

© The Author(s) 2015

497

S.J. Cho (ed.), *The Proceedings of the 12th International Congress on Mathematical Education*, DOI 10.1007/978-3-319-12688-3_47

Focus and Themes

Psychological characteristics of students that influence their inclination to think creatively in mathematics

- Effects of psychological characteristics on students' test performances
- The role of optimism (resilience) in mathematical problem solving

Cognitive processing associated with the creative constructing of knowledge

- What aspects of curriculum development/materials contribute to developing learners' mathematical thinking, mathematical inquiry or mathematical creativity?
- What cognitive processes are associated with autonomous student development of new knowledge and what 'teacher moves' can promote such activity?

Mathematical thinking accompanied by affective elements

- In what ways are cognitive, social, and affective elements connected during the development of new knowledge?
- The nature of affective elements that can accompany creative mathematical thinking.

Social interactions associated with creative mathematical thinking

- What aspects of teaching mathematics (teaching behaviors) contribute to developing mathematical thinking, mathematical inquiry or creativity in mathematics?
- What characteristics of classroom interaction or discourse (students-students; teacher-students) facilitate or contribute to knowing mathematics or developing thinking or inquiry abilities?
- What mathematical problems are there that have good use in the classroom by teachers that contribute towards developing cognition in mathematics?

The nature of mathematical understanding

- Children's interpretation of and performance on national and international math tests
- The rationale behind selecting a wrong answer in multiple-choice items in mathematics assessments.
- Contexts for developing mathematical understanding

Number of Submissions and Attendants

31 papers were reviewed and the following decisions were made: 5 long presentation (16 %), 8 short presentation (26 %), 11 posters (35 %), Overall acceptance rate was 77 %.

The number of attendants at each session was between 30 and 45. Each of the four TSG22 sessions attracted a large audience and this added to the stimulating nature of the discussions. With so many thought provoking contributions, and the differences in perspectives communicated, there was insufficient time to pursue all of the interesting questions and issues that arose. The panel's post-ICME communications with participants and others visiting the ICME TSG22 site illustrate ongoing interest and reflections arising from the work of TSG22.

Schedule of TSG22

Session 1, Tues 10, 10:30-12
Welcome, Overview
Luis Radford (Invited Plenary), Sensuous Cognition: Mathematical thinking as a Body- and Artifact-based Social Practice (30 min)
Round Table 1: In school and out of school mathematics learning (12 min)

- Paper 1, Michaela Regecova & Maria Slavickova, How Students' Everyday Experiences Influence Their Mathematical Thinking
- Paper 2, Rankin Graham, Homework: Pre-calculus Algebra Class
- Paper 3, Kadian M. Callahan, Prospective middle School teachers' generalizing actions (reasoning about algebraic and geometric representations)
- Question/Discussion

Poster Session (Parallel to Round Table 1)
Jorge Soto-Andrade & Pamela Reyes-Santander, Mathematical cognition in young offenders
Shin-Yi Lee (Invited Early Career Researcher), Analysis of "look back" strategies in mathematical problem solving
Hsin-Mei Huang, Children's thinking about measuring areas
Plenary Discussion (4 min)

Session 2: Wednesday, July 11, 10:30-12
Introducing Session
Lianghuo Fan (Invited Plenary), Learning of Algorithms: A Theoretical model with focus on cognitive development (30 min)
Rosa Ma. Garcia & Mariana Saiz (Electronic), Listening to children explain wrong answers
Terezinha Nunes & Peter Bryant, Children's' Understanding of Probabilities
Yasufumi Kuroda & Naoko Okamoto, How can brain activity contribute to understanding of mathematical learning process
Plenasry Discussion (10 min)

Session 3: Friday, July 13, 11-12:30
Introducing Session
Rina Hershkowitz, Tommy Dreyfus, Michal Tabach, Chris Rasmussen, Megan Wawro (Invited Plenary Team) (55 min)

- Hershkowitz, Dreyfus, & Tabach, Exponential growth: Constructing knowledge in the classroom
- Chris Rasmussen, Megan Wawro Documenting collective activity in the classroom
- Michal Tabach, Rina Hershkowitz, Chris Rasmussen, & Tommy Dreyfus, Exponential Growth: Co-ordinating Construction of Knowledge and Documenting Collective Activity in the Classroom
- Question/Discussion

Hong Seek Eng, Lee Ngan Hoe, & Darren Yeo Jian Sheng, Metacognitive approach: Kick- starting problem solving activity
Gaye Williams (Co-chair) Linking confidence, persistence, and optimistic problem solving activity
Plenary Discussion (13 Mins)

Session 4: Saturday, July 14, 10:30-12
Introducing Session
Alan Schoenfeld (Invited) (30 Mins) Social dynamics for supporting creative mathematical thinking and problem solving
RT2a: Promoting creative thinking: international perspectives (12 min)

- Paper 1, Xianwei Yuan Van Harpen: Creativity and problem posing in US and China
- Paper 2, Yeojoo Jin: Problem solving in Korea
- Paper 3, Cristina Frade, Steve Lerman, Luciano Meira, Peter Winbourne: Working with the ZPD to Identify Learning as Participation in Mathematical Practices
- Question/Discussion

RT2b: Developing understandings of complex mathematical ideas (Parallel to RT 2a) (12 min)

- Paper 1, Revathy Parameswaran: Expert mathematicians approach to understanding definitions
- Paper 2, Megan Wawro: Student reasoning about invertible matrix theorem in linear algebra
- Paper 3, Jun Mun Kyeong Semantic and syntactic reasoning on the learning of algebra
- Question/Discussion

Yuka Koizumi & Keiko Hino Social interactions of competent teacher: Stimulating creative thinking
Plenary Discussion (11 Mins)
Where to Now? (20 Mins)

Brief Summary of Outcomes

The ICME TSG22: Learning and Cognition co-chairs Hsin-Mei Huang and Gaye Williams provide a brief overview of what occurred in preparing for and participating in the ICME-12 TSG22 Learning and Cognition. The TSG22 Panel invited five researchers/research team presentations (Luis Radford; Alan Schoenfeld; The Tommy Dreyfus, Rina Hershkowitz, Michal Tabach, Chris Rasmussen, Megan Wawro Team; The Terezinha Nunes, Peter Bryant Team; and Lianghuo Fan), and two Early Career Researchers (Michal Tabach and Shin-Yi Lee) to highlight cutting edge research in this TSG. The announcement of the 2011 ICMI Awards (Hans Freudenthal Award: Luis Radford; Felix Klein Award: Alan Schoenfeld) contributed further to the interest already shown in this TSG. The quality and number of papers submitted through the reviewing process created dilemmas: how could we enable the sharing of the rich contributions proposed? We decided upon round tables presented simultaneously with many short presentations. Researchers rose to the challenge of showcasing their studies succinctly but with sufficient depth to allow others to follow up on their work. The TSG22 Poster Sessions were well attended and contributed further to TSG22 research.

Some of the connections identified between various presentations are now identified. For example, Radford, and Koizumi and Hino, and Schoenfeld focused differently on 'culture'. Radford on cognition as a 'culturally and historically constituted form of creative responding' with 'sensation considered as a substrate of the mind …', Koizumi and Hino on the learning culture set up by the teacher to 'stimulate[s] children's creative mathematical thinking', and Schoenfeld on the development of classroom cultures in which 'the students had internalized the relevant mathematical standards' to become 'accountable to the discipline (as opposed to, or in addition to, accountable to the teacher)' and 'able to speak with mathematical authority'. The Hershkowitz Team adapted existing methodological tools to network theories in studying 'the role played by individuals and groups in the class as well as by the class as a whole, in the knowledge constructing process'. Williams examined psychological influences on processes associated with creative construction of new knowledge during problem solving, and Ngan Hoe Lee's Team, and Shin-Yi Lee examined metacognitive processes associated with problem solving. Mathematical understanding and how it develops was explored in probability (by Nunes and Bryant), and in children's developing understandings of area formulae (by Huang). A 'theoretical model for the learning of algorithm with focus on students' cognitive development' was presented by Fan. Soto-Andrade and Reyes-Santander illustrated creative mathematical activity amongst young offenders thus identifying a fruitful area for further research, and Yasufumi Kuroda and Naoko Okamoto's research on brain activity provided a reminder of an expanding area of research in learning and cognition.

The 2014 ZDM Special Edition 'New Perspective on Learning and Cognition in Mathematics Education' (presently under construction) extends many invited presentations and long presentations within TSG22 along four broad themes:

- Contributions of 'Culture' to Cognition;
- Cognitive, Social, and Psychological Elements of Knowledge Construction;
- Influences of the Mathematics as 'Taught' on Mathematical Thinking and Mathematical Understandings; and
- Focusing Students on Learning Processes Including Problem Solving Processes.

Open Access This chapter is distributed under the terms of the Creative Commons Attribution Noncommercial License, which permits any noncommercial use, distribution, and reproduction in any medium, provided the original author(s) and source are credited.

Mathematical Knowledge for Teaching at Primary Level

Len Sparrow

Overview

The group generated considerable interest with 30 papers and abstracts being submitted. A review system was established by the TSG 23 Chair Christoph Selter whereby each paper was read and reviewed by one of the Co-Chairs and a Team Member. From this process 19 papers were accepted for presentation in Seoul.

The presentations were given over four days with each day being allocated 90 min in the main program. These sessions were chaired by Len Sparrow with help from Pi-Jen Lin on Day 2. Due to the high number of papers, and a wish of the organising team for as many colleagues as possible to experience presenting at the Congress, paper presentations were short (15 min). Each presentation had an allowance for questions and comments by the TSG participants. Papers were grouped under similar themes so that there was an element of coherence each day. The Chair summarised the issues and questions for each day and presented these to the TSG members for comment at the next session. They are copied below. Attendance at the presentations was typical of such groups with a group of stalwarts attending every presentation and every day while others attended only for their presentation. The group attracted a range of participants from early researchers to highly experienced professors and was enriched by this diversity.

Organizers Co-chairs: Christoph Selter (Germany), Suck Yoon Paik (Korea); Team Members : Catherine Taveau (France), Pi-Jen Lin (Taiwan), Len Sparrow (Australia); Liaison IPC member: Mercy Kazima (Malawi).

L. Sparrow (✉)
Curtin University, Perth, WA, Australia
e-mail: l.sparrow@curtin.edu.au

© The Author(s) 2015
S.J. Cho (ed.), *The Proceedings of the 12th International Congress on Mathematical Education*, DOI 10.1007/978-3-319-12688-3_48

Schedule

Session 1: Tuesday, 10th July, Teachers' mathematical knowledge

10:35 Christine Browning, Understanding Prospective Elementary Teacher Content Knowledge: Common Themes from the Past Decade.

10:50 Siew Yin Ho, Pre-service teachers' specialised content knowledge on multiplication of decimals.

11:05 Pi-Jen Lin, Future teachers' proof of universal and existential elements.

11:20 Di Liu, A comparative study of Chinese and US pre-service teachers' mathematical knowledge of teaching in planning and evaluating instruction.

11:35 Cheng-Yao Lin, Enhancing pre-service teachers' computational skills through open approach instruction.

11:50 Eva Thanheiser, Preservice elementary teachers' understanding of multi-digit whole numbers: Conceptions and development of conceptions.

Session 2: Wednesday, 11th July, Teachers' knowledge about children's mathematical thinking and reasoning.

10:40 Jeong Suk Pang, Novice Elementary Teachers' Knowledge of Student Errors.

10:55 Yusuke Shinno, Issues on prospective teachers' argumentation for teaching and evaluating at primary level: Focussing on a problem related to discrete mathematics.

11:10 Mi Sun Pak, Teachers' knowledge and math teaching in a reform curriculum.

11:25 Mustafa Alpaslan, Preservice mathematics teachers' conceptions regarding elementary students' difficulties in fractions.

Day 3 Friday 13th July—Teachers' beliefs, attitudes and orientations

15:10 Audrey Cooke, Anxiety, awareness and action: Mathematical knowledge for teaching.

15:25 Ronald Keijzer, Mathematical knowledge for teaching in the Netherlands.

15:40 Sharyn Livy, Foundation and connected mathematical content knowledge for second year primary pre-service teachers developed in practice.

15:55 Hyun Mi Hwang, Korean elementary teachers' orientations and use of manipulative materials in mathematics textbooks.

Session 4 Saturday, 14th July, Theoretical conceptualisation of teachers' knowledge

10:40 Minsung Kwon, Mathematical knowledge for teaching in the different phases of the teaching profession.

10:55 Tibor Marcinek, Learning to interpret the mathematical thinking of others in preservice mathematics courses: Potential and limitations.

11:10 Miguel Ribeiro, Teachers' mathematical knowledge for teaching and its role on practice.

11:25 Arne Jakobsen, Using practice to define and distinguish horizon content knowledge.

Summary of Issues Raised in Topic Study Group

Session 1:

- We already know a lot about the content knowledge of preservice primary/ elementary teachers in USA.
- Similar information is available from Non-USA countries.
- Generally, they lack deeper forms of conceptual knowledge especially in number related areas.
- What causes these limitations? Procedural teaching? Other?
- What are the consequences of this? Why is it a problem?
- Results in procedural teaching and a continuation of the cycle of procedural teaching?
- What are strategies to overcome this limited knowledge?
- Is it important to overcome these limitations?
- What has already been done? National testing of pre-service teachers in UK —Evidence that it is effective? Teaching primary mathematics content in University programs/courses/units.
- Is this phenomenon in all countries? If not, how are they different? Singapore? China? Finland? Korea?
- What mathematics should pre-service teachers know?
- Should there be an entry standard in mathematics for pre-service primary teachers? If so, what should it be? Higher level mathematics?

Session 2:

- What mathematics should primary teachers know? Pre-service/In-service?
- How will they come to know this?
- How will others know they know?
- Should we employ mathematics specialists?
- How does better teacher mathematics knowledge impact the classroom/ children's mathematics learning?
- How will they come to gain knowledge of children's errors, thinking, misconceptions?
- Is it important that primary teachers know about and undertake investigations, proof, explanations in mathematics?

Session 3 and 4:

- Is it possible to teach sufficient mathematics content while teaching about mathematics pedagogy?
- How can you motivate pre-service/in-service teachers to learn the mathematics needed for primary teaching?
- Do teachers need knowledge of how to use materials for teaching mathematics?
- Should we develop teachers' numeracy or mathematical knowledge?

- How do you find out what mathematics pre-service/in-service teachers know/understand?
- Is developing teacher confidence in mathematics the key?
- How can you tell which teachers are in denial or are just unaware of their limited mathematical knowledge?
- What are situations that help pre-service/in-service teachers identify gaps in their knowledge?
- How do you help when you/they spot gaps in knowledge?
- What knowledge do teachers need to make practice 'mathematically demanding' and 'pedagogically exciting'?
- How can one help develop horizon content knowledge?

Open Access This chapter is distributed under the terms of the Creative Commons Attribution Noncommercial License, which permits any noncommercial use, distribution, and reproduction in any medium, provided the original author(s) and source are credited.

Mathematical Knowledge for Teaching at the Secondary Level

Aihui Peng and Hikma Smida

Overview

TSG 24 at ICME-12 aimed to especially examine current scholarship and research on mathematical knowledge for teaching at the secondary level by collecting, comparing and discussing research experiences in this area, through the following three questions: What mathematical knowledge is needed for teaching at secondary level? What are the status quo of knowing and using mathematical knowledge for teaching at secondary level? How should we move forward (or what we have done) towards better equipped with mathematical knowledge for teaching at secondary level? In ICME 12, TSG 24 gathered 23 oral presentations from Canada, China, Finland, France, India, Ireland, Korea, Norway, South Africa, Spain, Sweden, and Turkey. They were presented in terms of four subtopics.

Organizers Co-chairs: Aihui Peng (China), Hikma Smida (Tunisia); Team Members: Hakan Sollervall (Sweden), Dongwon Kim (Korea), Karin Brodie (South Africa); Liaison IPC Member: Mercy Kazima (Malawi).

A. Peng (✉)
Southwest University, Chongqing, China
e-mail: Aihuipeng@gmail.com

H. Smida
Université de Carthage, Tunis, Tunisia
e-mail: hikma.smida@ipest.inu.tn

© The Author(s) 2015
S.J. Cho (ed.), *The Proceedings of the 12th International Congress on Mathematical Education*, DOI 10.1007/978-3-319-12688-3_49

Theoretical Perspective and Conceptual Framework for Mathematical Knowledge for Teaching at Secondary Level

The first presentation entitled "secondary school teachers' mathematical problem-solving knowledge for teaching" was presented by Olive Chapman (Canada). The study identified the nature of mathematical problem-solving knowledge for teaching and how this knowledge could support students' development of proficiency in problem-solving, which has significant implications for teacher education. In particular, the author discussed what should teachers know to teach for problem-solving proficiency and what knowledge should teachers hold to help students to become proficient in problem solving. These questions were addressed from a theoretical perspective and from a study that investigated secondary school teachers' knowledge in terms of their conceptions and teaching of problem solving in relation to contextual problems.

The second presentation "the ladder of knowledge: A model of knowledge for second level mathematics teachers" was presented by Niamh O'Meara (Ireland). In this study, the authors developed a new model of knowledge to meet the needs of curricula with a strong focus on mathematical applications.

The third presentation "competence in didactic analysis in the pre-service training of secondary school mathematics teachers in Spain" was presented by Vincent Font (Spain). The study illustrated how one of the components of the broad competence in didactic analysis (identifying potential improvements to be implemented in future classes) was developed within the context of the University of Barcelona.

The fourth presentation "coordinating theories to analyze the relationship between teachers' actions and teachers' knowledge—a presentation of a methodological approach" was presented by Erika Stadler (Sweden). The study presented a tentative methodological framework to analyze what kind of mathematical knowledge for teaching, MKT, novice mathematics teachers use when teaching. The main idea of the framework is to coordinate three different theoretical frameworks, which provide a methodological tool for analyzing the relationship between teachers' teaching actions and mathematical knowledge.

The fifth presentation "the structure of knowledge of teaching of student teachers on the topic of distance formula" was presented by Lin Ding (China). The presentation provided a new approach of interpreting knowledge of teaching (KOT) of secondary mathematics student teachers by examining its structure (i.e. mathematics, student and pedagogy). A brief analysis on two examples regarding the structure of KOT was provided in order to illustrate how this approach works.

The sixth presentation "A pre-analysis of the creation of teacher's resources for developing instruction in basic logic in French high schools" was presented by Zoe Mesnil (France). The author presented studies on the role of logic in mathematics education in order to show how it can help students to improve their skills in language and expression. Through the analysis of curricula and textbooks, the study presented an overview of the process of didactic transposition for teaching the concepts of logic.

Pre-service Mathematics Teachers' Knowledge

This subtopic consists of five presentations from USA, Ireland and Turkey. The first presentation "secondary teacher candidates' mathematical knowledge for teaching as demonstrated in their portfolios" was presented by Hari Koirala (USA). Their study focused on prospective secondary school teachers' mathematical knowledge and their ability to demonstrate how their learning of mathematics from their university courses applies to the teaching of secondary school mathematics.

The second presentation "Chinese and US pre-service mathematics teachers' knowledge for teaching algebra with a focus on representational flexibility" was presented by Rongjin Huang (USA). Their study examined Chinese and U.S prospective middle grade teachers' knowledge of algebra for teaching with a focus on representational flexibility. It was found that the Chinese participants not only demonstrated sound knowledge needed for teaching the concept of function, but also had the flexibility in using representations appropriately. In contrast, the U.S. counterparts showed their weakness of using these concepts to solve problems and using appropriate representations.

The third presentation "whose fault is it anyway? The truth about the mathematical knowledge of prospective secondary school teachers and the role of mathematics teacher educators" was presented by Miriam Liston (Ireland). The author presented an empirical research study which aims to contribute to the understanding of prospective secondary level mathematics teachers' mathematical knowledge for teaching. The findings suggest that prospective mathematics teachers may not have sufficient subject matter knowledge to alter their teaching strategies and ultimately teach for understanding.

The fourth presentation is "pre-service secondary school mathematics teachers' specialized content knowledge of complex numbers" presented by Fatma Aslan (Turkey). The author reported the findings of a study of pre-service secondary school mathematics teachers' learning of complex numbers during a content course. According to the author's findings, participants were able to build connections between their mathematical understanding as teachers with their teaching practice and students' mathematical ideas.

The fifth presentation "a comparative analysis of the content knowledge for secondary pre-service mathematics teachers" was presented by Wei Sun (USA). In his study, it focused on the knowledge that the secondary pre-service teachers gain during their study in the teacher education program. Two mathematics teacher preparation programs were examined, one from China and the other from the US, with the intent to shed light on this important issue and help mathematics educations understand mathematics teacher education from a broader (international) perspective.

In-service Mathematics Teachers' Knowledge

The first presentation "seeing mathematics through processes and actions: investigating teachers' mathematical knowledge and secondary school classroom opportunities for students" was presented by Rose Mary Zbiek (USA). The study described the processes and actions approach. The authors proposed a more general way to characterize MKT than is typically used.

The second presentation "what is pre-service and in-service Teachers' MKT in concept of vector" presented by Hyunkyoung Yoon (Korea) was to investigate the mathematical knowledge for teaching (MKT) of pre-service and in-service mathematics teachers on the concept of vector. 80 pre-service and 124 in-service mathematics teachers were asked to perform three questions based on MKT's subdomain. The results show that pre-service teachers have stronger common content knowledge. On the other hand, in-service teachers have stronger specialized content knowledge, knowledge of content and teaching.

The third presentation "pedagogical knowledge, pedagogical content knowledge, and content knowledge for teaching mathematics: how do they shape teaching practices?" was presented by Hee-Jeong Kim (USA). This empirical study offered a case of a proficient middle school mathematics teacher, well known as a highly skilled teacher in her district, and explored the teacher's decision-making in different teaching contexts. The author discussed what the contributions of different kinds of knowledge were and implied how we can support teachers with regard to knowledge for better mathematics teaching.

The fourth presentation "hypothetical teaching trajectories (HTT): analysing contingency events in secondary mathematics teachers' practice" was presented by Jordi Deulofeu (Spain). This paper showed through the work done by a future secondary mathematics teacher called Gabriel in his initial training at the university, how analyzing HTT can serve a double role: giving information about the prospective teacher's mathematical knowledge and helping to validate an instrument that serves teachers to reflect on their own mathematical knowledge in practice.

The fifth presentation "developing craft knowledge in mathematics teaching" was presented by Inger Nergaard (Norway). Her study focused on teachers' opportunities to develop craft knowledge through their engagement with students. Using video recordings of mathematics lessons and following up conversation with the teachers, two episodes of teaching were considered. In the first episode the teacher appears to close down opportunities for discussion of the unanticipated situations that arose and thus she denied herself opportunities to learn from the situation, while the second episode concerned a teacher who invites students into her teaching and thus enable further development of existing knowledge.

The sixth presentation "understanding teachers' knowledge of and responses to students' mathematical thinking" was presented by Shikha Takker (India). She reported a case study which aimed at understanding teachers' knowledge about students' mathematical thinking in situ. Teacher's response to students' mathematical thinking was characterized based on classroom observations, task-based interviews, complemented with the anticipation and reflection of students' responses to 'proportion' problems. It was found that such a framework helps in creation of conflict in the teacher and is a potential source of teacher reflection.

Methodology Issues on Mathematics Teachers' Knowledge

The first presentation "using scenarios validated as measures to explore subject matter knowledge (SMK) in an interview setting" was presented by Sitti Patahuddin (South Africa). The presentation focused on one scenario adapted from LMT (e.g. from the Learning Mathematics for Teaching—LMT—project) in order to explore how teachers interviewed engage with each of the responses offered.

The second presentation "instruments for improving teachers' use of artifacts for the learning of mathematics" was presented by Håkan Sollervall (Sweden). The author argued that teachers' mathematical knowledge has to include instruments for controlling how the artifact used become involved when students engage in solving mathematical tasks. The authors proposed to meet this demand by coordinating the matching notions of affordances (planning) and objects of activity (evaluation). They briefly illustrated how these notions can be used as analytical instruments in a fashion that connects to what teachers already do in their daily work.

The third presentation "exploring the influence of teachers' use of representation on students' learning of mathematics" was presented by Emmanuel Bofah (Finland). The aim of the study was to examine how teachers' use of different mathematics representations, in the domain of functions, affects students' behavior in the process of doing and learning mathematics.

The fourth presentation "consensuating the best profile of a mathematics teacher in the transition to secondary school; a discussion of experts using the Delphi method" was presented by Sainza Fernandez (Spain). This on-going investigation, embedded in a larger project that targets primary-secondary transition in mathematics, explored the knowledge of a group of expert mathematics teachers and experts involved in teachers' education using the Delphi method. The results arisen point at secondary teachers as more responsible for the success or failure of the process and their sensitivity as professionals of mathematics education as particularly determinant.

The fifth presentation is "a case study on the status quo of the development of Tibetan mathematics teacher's pedagogical content knowledge (PCK) in Lasha" (China). A case study was used to analyze the development of Tibetan mathematics teacher's PCK in secondary school in Lasha.

The sixth presentation "The project: collaborating to advance secondary teachers' mathematics proficiency for teaching" was presented by Pier Junior Clark (USA). Using the *Provisional Framework for Proficiency in Teaching Mathematics* as a guideline, the author examined the changes in the secondary teachers' mathematics proficiency and efficacy for teaching data analysis and statistics over a year-long professional development project.

Summary

TSG24 included presentations from many points of view:

Conceptual frameworks for mathematical knowledge for teaching at secondary level, e.g., what is the nature of mathematical knowledge for teaching at secondary level? What mathematical knowledge needs to know and how to use it from an advanced perspective for a secondary school teacher? What are the approaches, from the practice point of view, that could support teachers developing their mathematical knowledge that they need to know and know how to use it?

Empirical researches that aim to contribute our understanding of what mathematical knowledge is needed or how it is assessed in different scenarios, e.g., teachers' mathematical knowledge for teaching in specific activities, teachers' mathematical knowledge for teaching in specific domain, teachers' mathematical knowledge for teaching in special situations, such as information and communication technology environment, innovative and creative approaches of developing mathematical knowledge and the instruments for assessing these approaches specifically.

Empirical researches to explore relationships between teachers' learning of teaching (both pre-service and in-service) and students' learning of mathematics, e.g., the effect of mathematics knowledge for teaching on student achievement, the innovative and creative approaches of developing the effect of mathematics knowledge for teaching on students' learning and achievement.

Open Access This chapter is distributed under the terms of the Creative Commons Attribution Noncommercial License, which permits any noncommercial use, distribution, and reproduction in any medium, provided the original author(s) and source are credited.

In-Service Education, Professional Development of Mathematics Teachers

Shuhua An and Andrea Peter-Koop

The aim of TSG 25 at ICME-12 was to discuss the experiences and approaches developed in different countries to support the professional development of teachers for practice, in practice and from practice. The study group 25 received 74 paper submissions from scholars, graduates, and practitioners in various countries and regions, and accepted 69 papers. A total of 63 papers were presented at 10 sessions at ICME 12 conference. Participants discussed research based practices and state-of-the-art approaches to the in-service education and professional development of teachers from a multi-national and globe perspectives. This report will address some key ideas in the following topics from TSG 25:

- Research studies and projects in professional development of primary and secondary school teachers
- Research studies and projects in in-service education and teacher education programs
- Classroom teaching research and lesson study in professional development of primary and secondary school teachers
- In-service education in STEM field in secondary school settings—Research studies and projects
- Mentor and coaching programs in professional development of primary and secondary school teachers

Organizers Co-chairs: Shuhua An (USA), Andrea Peter-Koop (Germany); Team Members: Barbara Clarke (Australia), Yimin Cao (China), Gooyeon Kim (Korea); Liaison IPC Member: Gabriele Kaiser (Germany).

S. An (✉)
California State University, Long Beach, Long Beach, USA
e-mail: san@csulb.edu

A. Peter-Koop
Institut für Didaktik der Mathematik, Universität Bielefeld, Bielefeld, Germany
e-mail: andrea.peter-koop@uni-bielefeld.de

© The Author(s) 2015
S.J. Cho (ed.), *The Proceedings of the 12th International Congress on Mathematical Education*, DOI 10.1007/978-3-319-12688-3_50

Professional Development of Primary and Secondary School Teachers

One of the challenges in teacher professional development is the nature of the research and the differing agendas of stakeholders. Much of the research takes the form of evaluation of teacher development projects and while they build on a growing body of research, the contexts in which they occur are complex. As a result it is difficult to synthesize the findings in ways that can inform future planning. How can our small pieces of research contribute to our understanding of the whole picture?

The role of teacher attitudes within the context of professional development is important but can be overemphasized at the expense of actions. A number of the papers helped focus on the role of practice in teacher development. The value of ensuring that participants have a voice was a common theme.

The important discussion focused on content of professional development and measurement of effects of professional development. A number of papers indicated the needs of paying attention to specific knowledge, such as error analysis, and measuring teachers' knowledge and teacher learning from error analysis and engaging learners in avoiding the errors.

Participants discussed the forms of professional development. Presentations shared different forms of lesson studies, such as Teacher Research Group in China, an important form of school based professional development.

In-Service Education and Teacher Education Programs

The presentations shared their effective approaches in in-service education and teacher education programs. However, the discussions indicated the challenges in in-service education and teacher education programs. The examples of the challenges: (1) How can we best prepare math teachers? (2) How to measure teachers' pedagogical content knowledge, (3) How to support new teachers and teacher retention issue, (4) Design different models of professional development that support teachers in new initiatives, (5) Relationship between professional development and classroom teaching, (6) Teaching work load and time to plan lesson in US, (7) Tools for reflection, and (8) Leadership roles.

Classroom Teaching Research and Lesson Study in Professional Development and Teacher Education Programs

Classroom teaching research and lesson studies have various forms in different countries. The following focused questions regarding classroom teaching research and lesson study in professional development of primary and secondary school teachers were asked during the discussions:

- What is effective classroom teaching?
- What math teacher educators should know about effective classroom teaching?
- How can we best prepare math teachers to teach math effectively?
- How do we enhance the effectiveness of professional learning communities for math teachers?

In-Service Education in STEM Field in Secondary School Settings

In-service education in secondary schools with a focus on integrating science and technology is an interesting topic of TSG 25 sessions. A range of contexts and countries were represented both in the papers and the discussions and there was considerable overlap in the issues of concern. The role of technology provides an added challenge as both software and hardware is constantly being updated. The comfort zone of teachers was a common issue and the acknowledgement that in-service education and professional development often requires teacher to move out of their comfort zone. This is particularly relevant in technology rich or cross discipline environments.

Mentor and Coaching Programs in Professional Development of Primary and Secondary School Teachers

There were a range of papers focusing on leading teacher change through a variety of models. Mentoring and coaching models are increasingly being used in many countries. One model that was particularly promising was "teacher researchers" in China. They are a form of master teacher with considerable expertise who supports teacher development. This systemic approach also provides for teacher progression within the profession that is not available in many countries.

Whole Group Discussion

The whole group discussion focused on key issues, major findings, insights, international trends in research and development in professional development and in-service education, and indicated open questions to be addressed in the future.

Questions to be addressed in the future

- How do we support new teachers in the new initiatives?
- What are the different models of professional development? Especially, what are good models for new initiatives?
- What is the relationship between professional development and effective classroom teaching?
- What are the common strategies in professional development and classroom teaching in different countries? Diverse issue is needed to address also.

Discussion on future planning: Publications arising from TSG 25

- Publication of selected papers in an edited volume to be published by Springer Mathematics Teacher Education and Development series (Research based papers)
- Publication of selected papers in a special issue in Journal of Mathematics Education (USA) (Research based papers)

More opportunities:

- Routledge Education, Taylor & Francis expressed their interest in publishing TSG 25 papers
- A journal editor from Singapore also expressed her interest in publishing TSG 25 papers in a special issue

Joint project

- Participants supported the idea to work together for a joint project that compares in-service education and professional development of mathematics teachers in different countries.

Open Access This chapter is distributed under the terms of the Creative Commons Attribution Noncommercial License, which permits any noncommercial use, distribution, and reproduction in any medium, provided the original author(s) and source are credited.

Pre-service Mathematical Education of Teachers

Sylvie Coppé and Ngai-Ying Wong

Overview

The topic study group on pre-service mathematical education of teachers is dedicated to sharing and discussing of significant new trends and development in research and practice about the various kinds of education of pre-service mathematics teachers and of pre-service primary teachers who teach mathematics and are trained as generalists. It aimed to provide both an overview of the current state-of-the-art as well as outstanding recent research reports from an international perspective. The group discussed research experiences with different practices of pre-service mathematical education of (mathematics) teachers throughout the world, i.e. similarities and differences concerning the formal mathematical education of teachers, types and routes of teacher education, curricula of (mathematics) teacher education, facets of knowledge and differences in their achievements and beliefs about the nature of their training, and a variety of factors that influence these differences.

Organizers Co-chairs: Sylvie Coppé (France), Ngai-Ying Wong (Hong Kong); Team Members: Lucie De Blois (Canada), Björn Schwarz (Germany), Insun Shin (Korea), Khoon Yoong Wong (Singapore); Liaison IPCMember: Gabriele Kaiser (Germany).

S. Coppé (✉)
University of Lyon, Lyon, France
e-mail: sylvie.coppe@univ-lyon2.fr

N.-Y. Wong
The Chinese University of Hong Kong, Hong Kong, China
e-mail: nywong@cuhk.edu.hk

© The Author(s) 2015

S.J. Cho (ed.), *The Proceedings of the 12th International Congress on Mathematical Education*, DOI 10.1007/978-3-319-12688-3_51

Session Schedule

We received 51 proposals from different countries, 6 were rejected and at last we had 40 papers and only 37 presentations. As we had four 90-min sessions (July 10, 11, 13, and 14), two groups ran parallel in order to let 10 min to each presentation.

Each session was devoted to different issues in affect research in mathematics education.

Session 1: Tuesday, July 10, 10:30–12:00

Group A:

Buchholtz Nils, Studies on the effectiveness of university mathematics teacher training in Germany

Francis-Poscente Krista, Preparing elementary pre-service teachers to teach mathematics with math fair

Jennifer Suh, 'Situated learning' for teaching mathematics with pre-service teachers in a math lesson study course

GwiSoo Nah, A constructivist teaching experiment for elementary pre-service teachers

Qiaoping Zhang, Pre-service teachers' reflections on their teaching practice

Group B

Liora Hoch, Miriam Amit, When math meets pedagogy: the case of student evaluation

Hugo Diniz, Math Clubs: space of mathematical experimentation and teacher formation

Huk Yuen Law, Becoming professional mathematics teachers through action research

Levi Elipane, Integrating the elements of lesson study in pre-service mathematics teacher education

Müjgan Baki, Investigating prospective primary teachers' knowledge in teaching through lesson study

Session 2: Wednesday, July 11, 10:30–12:00

Group A

Zhiqiang Yuan, Developing prospective mathematics teachers' technological pedagogical content knowledge (TPACK): a case of normal distribution

Roslinda Rosli, Elementary pre-service teachers' pedagogical content knowledge of place value: A mixed analysis

Steve Thornton, Saileigh Page, Julie Clark, Linking the mathematics pedagogical content knowledge of pre-service primary teachers with teacher education courses

Rachael Kenney, Writing and Reflection: Tools for developing pedagogical content knowledge with mathematics pre-service teachers

Group B

Jan Sunderlik, Soetkova, Identification of learning situations during prospective teachers' student teaching in two countries

Yali Pang, Using a Video-based Approach to Develop Prospective Teachers' Mathematical Knowledge for Teaching and Ability to Analyze Mathematics Teaching

Xiong Wang, The Video Analysis of the Authentic Classroom as an Approach to Support Pre-service Teachers' Professional Learning: A Case from Shanghai Normal University, China

Namukasa Immaculate, Measuring teacher candidate's conceptual, procedural and pedagogical content knowledge

Session 3: Friday, July 13, 15:00–16:30

Group A

Hyun Young Kang, Korean Secondary Mathematics Teachers' Perspectives on Competencies for Good Teaching

Rongjin Huang, Pre-service secondary mathematics teachers' knowledge of algebra for teaching in China

Björn Schwarz, Relations between future mathematics teachers´ beliefs and knowledge with regard to modelling in mathematics teaching

Yeon Kim, Challenges to teach mathematical knowledge for teaching in mathematics teacher Education

Group B

Yuki Seo, Enhancing mathematics thinking for training mathematics teachers: a case at the department of engineering

Kiril Bankov, Curriculum for preparation of mathematics teachers: a perspective from TEDS-M

Lin Ding, A comparison of pre-service secondary mathematics teacher education in Hanover (a city in Germany) and Hangzhou (a city in China)

Khaled Ben-Motreb, Pre-service teachers' teaching practices and mathematics conceptions

Ildar Safuanov, Master programs for future mathematics teachers in Russian federation

Session 4: Saturday, July 14, 10:30–12:00

Group A

Claire Berg, Barbro Grevholm, Use of an inquiry-based model in pre-service teacher education: Investigating the gap between theory and practice in mathematics education

Loretta Diane Miller, Brandon Banes, Teaching pre-service elementary teachers mathematics through problem-based learning and problem solving

Ji-Eun Lee, Towards a holistic view: analysis of pre-service teachers' professional vision in field experiences

Diana Cheng, Discourse- based instruction in small groups of pre-service elementary teachers

Kwang Ho Lee, Eun-Ha Jang, The research on PBL Application in mathematics method course

Group B

Ceneida Fernandez, Julia Valis, Salvador Linares, An approach for the development of pre-service mathematics teachers' professional noticing of students' mathematical thinking

Erika Löfström, Tuomas Pursianen, "I knew that sine and cosine are periodic… but I was thinking how I could validate this": A case study on mathematics student teachers' ersonal epistemologies

Ju Hong Woo, The change of mathematics teaching efficacy beliefs by student teaching

Mi Yeon Lee, Preservnrique Galindo, Pre-service Teachers' Ability to Understand Children's Thinking

Ravi Somayajulu, Manjula Joseph, Candace Joswick, Characterizing secondary pre-Service mathematics teachers' growth in understanding of student mathematical thinking over a three-course methods series

Main Questions Discussed

Main questions were discussed such as:

- What are fundamental concepts to study the field of pre-service teacher in comparison of in-service teacher? What are special challenges for respective studies arising from the particular characteristics of pre-service teacher education and how to face them?
- What knowledge contribute to the development of the pre-service teacher? Which actions push the pre-service teacher to lost their initial experience of pupil to integrate new epistemological posture?
- What are the contribution of the different tolls (technology, writing, reflection, video) during the teacher training? How can a common core of the concept of "pedagogical content knowledge" be described against the background of its different conceptualizations?
- Are the challenge different in function of countries? What is the influence of the curriculum on practice of pre-service teacher?
- What kind of mathematic could contribute to the development of pre-service teacher? And how can it be taught adequately?
- Why do we teach mathematics and why this answer influence the teacher training?

Issues and Findings

Quite a number of issues on pre-service teacher education were identified, which includes considerable drop out rate, lack of knowledge and even lack of interest in mathematics among potential teachers in some countries. There also exists disagreement between goal and reality. For instance, while constructivism is advocated in the school curriculum, teacher education programmes did not provide such experience to student-teachers.

A number of means were introduced to address the above, arriving at promising results. The use of math fair, lesson studies, situated learning, ICT, writing, enquiry/ problem based learning and reflections are some of them. We observed the influence of the cultural context concerning education or mathematics teaching/learning from different countries or different parts of the world.

A salient focus among the presentations is teacher's knowledge, ranging from subject content knowledge, pedagogical content knowledge to belief. There were discussions on how teacher education programme can strike a balance between the mathematics component and the pedagogical component and how these two can be linked together.

Probably, the use of video in teacher training sessions is revealed as an important tool which could create or contribute to create these links. But we concluded that using video in pre service teacher training is not easy. We need to elaborate research programs to study how it could be possible to develop video based training. There were discussions on the different kinds of video (for example, showing expert or novice teachers, ordinary lessons or experimental), on the different goals (to show, to analyze, to observe the teacher or the students) on the different points of view (the teacher or the students) on the different conditions and on the limits. These remarks led to another issue: how could the teacher trainer introduce and use video to help the pre-service teacher to develop different kinds of knowledge or skill for mathematics teaching? How could the video give some informations on the student learning…

As for the recurrent issue of PCK, it was realised that it is cultural/context and student dependent. In other words, for a single subject matter, it depends on the 'target audience' for searching for the best way to have it presented. Rather than instoring potential teachers with a bundle of PCK (corresponding to a single SK), it might be more realistic and effective to equip them with the ability to adjust the presentation (of SK) spontaneously according to the subtle variations of their students. Again reflection comes into play.

How to build a path from fun to formal mathematics, from elementary mathematics to advanced mathematics is another issue of concern. All these involve all the parties: the student-teacher, the teacher trainer, the mentor and the pupils (during field experience). All these would not only result in reflections among student-teachers, professors and even teacher education curriculum developers should have their reflections too.

Summary

There were a lot fruitful discussions in this topic study group. We appreciated the different topics of the papers. We observed that there were a lot of very interesting issues which are very similar from a country to another and we hope our discussion will continue to bear fruits and impacts on our future programme for pre-service mathematics education. We learned from the different points of view and the cultural contexts.

Open Access This chapter is distributed under the terms of the Creative Commons Attribution Noncommercial License, which permits any noncommercial use, distribution, and reproduction in any medium, provided the original author(s) and source are credited.

Motivation, Beliefs, and Attitudes Towards Mathematics and Its Teaching

Birgit Pepin and Ji-Won Son

Report

Affect has been a topic of interest in mathematics education research for more than 30 years. More recently, and as emphasized in the last ICME 11 report, beliefs has turned from a 'hidden' to a more 'visible' variable. Today we know that affective variables can be regarded as explicit factors which influence mathematics learning outcomes as well as instructional practice. The different research perspectives used in studies of affect include psychological, social, philosophical, and linguistic. Those various views were represented in the ICME 12 research presentations. It also became clear during the conference, and this was expected, that the construct of 'affect' encompasses related constructs such as 'motivation', 'beliefs', 'values' and 'attitudes', to name but a few. We invited, and received, presentation proposals on all areas of affect in mathematics learning and teaching.

The organizing committee organized the accepted papers and posters for TSG 27 in the following ways:

Organizers Co-chairs: Birgit Pepin (Norway), Ji-Won Son (USA); Team Members: Bettina Roesken (Germany), Inés Mª Gómez-Chacón (Spain), Nayoung Kwon (Korea); Liaison IPC Member: Bill Barton.

B. Pepin (✉)
Soer-Troendelag University College, Trondheim, Norway
e-mail: birgit.pepin@hist.no

J.-W. Son
University of Tennessee, Knoxville, USA
e-mail: sonjwon@utk.edu

© The Author(s) 2015
S.J. Cho (ed.), *The Proceedings of the 12th International Congress on Mathematical Education*, DOI 10.1007/978-3-319-12688-3_52

- One 'elicited' Roundtable on 'Methodological issues in Affect Research';
- Six groups of short paper presentations and discussions (15 min);
- Three long paper presentations (30 min);
- Posters in the general poster session.

We had a large number of proposals and rigorously reviewed them, each proposal being evaluated by three reviewers (members of the TG27 team) according to a common set of criteria (agreed review scheme). At the advice of the ICME organizing committee we accepted most, only rejecting about six proposals, and arranged the accepted proposals in sessions. As we had four 90-min sessions (July 10, 11, 13, and 14) available, we decided to run parallel sessions, allocating 20–30 min for long and 10–15 min for short presentations. Each session was chaired by one of the co-chairing team members (unfortunately Inés Mª Gómez-Chacón could not attend ICME 2012). Posters were allocated to the poster session, which was common for all TSGs. One of the highlights of the TSG 27's sessions was the 'elicited' Roundtable on methodological issues, which had a 60-min time allocation.

The following will provide a 'taste' of the presentations and issues discussed.

On the 10th July the co-chairs opened up the first of four one and half hour sessions. Subsequently, Jill Cochran presented her research asking questions concerning values and ideals in mathematics education. She argued that teachers, policy makers, curriculum developers, and other professionals often held ideals that were in opposition to each other, and that this created conflicts of interest, in particular for classroom teachers. The following two sessions ran parallel, and each parallel session included three short presentations on the following topic areas: 'Students' views of mathematics'; and 'Mathematics teacher knowledge and efficacy'. Each series of presentations was followed by a discussion of the presentations.

On the 11th July the (elicited) Methodology Roundtable and one short presentation were scheduled. The panel members of the Roundtable were all well-known researchers in the field of affect in mathematics education: Markku Hannula; Gilah Leder; Ilana Horn; and Guenter Toerner. Each outlined their insights concerning methodological issues, and Markku Hannula presented a theoretical framework for the inclusion of the different 'lenses'. Then questions about the framework and relevant issues were discussed.

The 12th July session started with a (long) presentation by Mac an Bhairs Ciaran and colleagues on the 'effect of fear on engagement with mathematics'. They reported on a comparative study of first year undergraduate mathematics students: one group had failed their first year examinations; the second had successfully completed the first year. It was argued that whilst both groups named 'fear' as a factor for engagement (or not) with mathematics, for one group it emerged as a positive motivation, in the sense that it formed part of their coping mechanisms when dealing with the various obstacles that they encountered. The subsequent parallel sessions included six (short) presentations, under the headings of 'Motivation and conditions for pupil learning' and 'Teacher beliefs concerning curriculum and tracking'. Again, each series of presentations was followed by a discussion of the presentations.

The last TGS 27 session had a similar structure, albeit more time was allocated for discussion of the whole TGS, insights gained and implications for future research (as this was the last session of the TSG). In an opening (long) presentation Birgit Pepin reported on a study of 'Affective systems of Norwegian mathematics students/teachers in relation to 'unusual' problem solving'. She argued that results from the three different groups (each at different stages of their educational and professional development) showed that positive engagement structures were linked to working together in a group and previous (positive) experiences, whereas 'giving up' was connected to 'working alone' and the 'unusual' problem-solving situation. The subsequent two parallel sessions (including altogether four (short) presentations) were in the two themes of 'Teacher beliefs and practices' and 'Teachers' views on mathematical tasks'.

In a final discussion the following issues were raised:

1. Five minutes for (short) presentations is not sufficient. Hence, either a different mode of running the TGS should be found, or (fewer) presenters should be given more time, also for discussion. This has implications for acceptance of future proposals: this ICME the TGS 27 had a very large number of proposals, and approximately half were accepted as short or long presentations (19), and approximately half accepted as posters (with a small number of rejections). Hence, questions arise: should the reviewing process (TGS 27 had three reviewers and developed its own evaluation schedule) be more rigorous, and more papers be rejected? Or should the TSG be 'inclusive' and find another mode of running the group?

2. The question of 'publication' was raised: presentations were 'published' in the ICME 12 pre-proceedings, but how does this count/is acknowledged in terms of publications?

3. It was suggested to be more selective about the accepted papers and support, and perhaps elicit, more 'novelty' topic areas: e.g. affect and mathematical thinking (including suitable theoretical frameworks and measurement instruments/ methodological tools for this field of research); affect as a dynamic system (including affective systems and 'collectives' in social contexts); intervention studies/design-based research on 'affect and cognition'.

4. TGS 27 was provided with two rooms close to each other (and this was beneficial for participants to be able to attend sessions). However, it was difficult for the group to 'merge' as a whole, as many discussions took place in separate sessions, and some participants wanted to share their ideas in a whole group discussion.

5. Overall, it was emphasized that this ICME's TGS on affect went well (as did previous groups) and that this group is now an established and well-recognized part of ICME.

List of Groups, Presentations and Presenters of Long and Short Presentations

Tuesday, 10th July
  Jill Cochran, Does a balanced philosophy in mathematics education exist?
  Student views of mathematics:

- **Mario Sanchez Aguilar**, Alejandro Rosas and Juan Gabriel Molina Zavaleta,
  Mexican students' images of mathematicians
- **Sally Hobden**, After graduation? The beliefs of alumni bachelor of education
  students reading mathematics and the formation of mathematical knowledge
- **Veronica Vargas Alejo**, Cesar Cristobal Escalante and Jamal Hussain, Beliefs
  and attitudes toward mathematics at university Level, development of mathematical knowledge

  Teacher knowledge and efficacy

- **Janne Fauskanger**, Teachers' epistemic beliefs about HCK
- **Giang-Nguyen Nguyen**, Diagnosing student motivation to learn mathematics:
  A form of teacher knowledge
- **Ayse Sarac** and Fatma Aslan-Tutak, The relation of teacher efficacy to students'
  trigonometry achievement

Wednesday, 11th July
  Methodological issues
  Dohyoung Ryang, The viability of the mathematics teaching efficacy beliefs
  instrument for Korean secondary pre-service teachers
  Roundtable (Co-chairs: Bettina Roesken and Birgit Pepin; Panel members:
  Markku Hannula, Gilah Leder, Ilana Horn and Guenter Toerner)

  Methodological issues in Affect Research: distinguishing between 'state' and 'trait' in
  mathematics education research.

Friday, 13th July
  Mac an Bhaird Ciaran, The effect of fear on engagement with mathematics
  Motivation and conditions for pupil learning

- **Chonghee Lee**, Sun Hee Kim, Bumi Kim, Soojin Kim and Kiyeon Kim,
- **Denival Biotto Filho** and Ole Skovsmose, Researching foregrounds: About
  motives and conditions for learning
- **Nelia Amado**[1] and Silvia Reis, A young student's emotions when solving a
  mathematical challenge
- **Suela Kacerja**, "Cultural products are girls' things!" Interests Albanian students
  retain for real-life situations that can be used in mathematics

  Teacher beliefs concerning curriculum and tracking

- **Qian Chen**, Teachers' beliefs and mathematics curriculum reform: A comparative study of Hong Kong and Chongging

- **Benjamin Hedrick** and Erin Baldinger, Beliefs about tracking: Comparing American and Finnish prospective teachers

Saturday, 14th July

Birgit Pepin: "Exploring affective systems of Norwegian mathematics student/ teachers in relation to 'unusual' problem solving"

Teacher beliefs and practices

- **Dionne Cross** and Ji Hong, "I'm not sitting here doing worksheets all day!": A longitudinal case study exploring perceived discrepancies between teachers' beliefs and practices
- **Ralf Erens**[1] and Andreas Eichler[1], Teachers' curricula beliefs referring to calculus

Teacher views on mathematics tasks

- **Esther Levenson**, Affective issues associated with multiple-solution tasks: Elementary school teachers speak out
- **Anika Dreher** and Sebastian Kuntze, Pre-service teachers'views on pictorial representations in tasks

Open Access This chapter is distributed under the terms of the Creative Commons Attribution Noncommercial License, which permits any noncommercial use, distribution, and reproduction in any medium, provided the original author(s) and source are credited.

Language and Communication in Mathematics Education

Tracy Craig and Candia Morgan

Introduction

The topic of "Language and Communication in Mathematics Education" covers a wide range of areas of interest, ranging from the question of what constitutes "language" in mathematics, through investigations of communicative interactions in mathematics classrooms and study of issues involved in teaching and learning mathematics in multilingual settings. This breadth was well represented in the papers accepted for presentation in the Topic Study Group at ICME12. In order to facilitate discussion, the paper presentations in each session were divided into two sets, with participants choosing which set to attend. This allowed the discussion to focus in greater depth on common themes. In addition, one session of the TSG was devoted to a panel discussion on the topic of "Theoretical and methodological issues in studying language in mathematics education" and a final plenary meeting enabled participants to reflect on the TSG as a whole, the common issues addressed, the lessons learnt and aspirations for future work on the topic. In this report, we present an overview of the major themes arising in the papers presented and in the discussions during the congress.

Organizers Co-chairs: Tracy Craig (South Africa), Candia Morgan (UK); Team Members: Marcus Schuette (Germany), Rae Young Kim (Korea), David Wagner (Canada); Liaison IPC Member: Oh Nam Kwon (Korea).

T. Craig (✉)
University of Cape Town, Cape Town, South Africa
e-mail: tracy.craig@uct.ac.za

C. Morgan
Institute of Education, University of London, London, UK
e-mail: c.morgan@ioe.ac.uk

© The Author(s) 2015
S.J. Cho (ed.), *The Proceedings of the 12th International Congress on Mathematical Education*, DOI 10.1007/978-3-319-12688-3_53

Classroom Interactions

The nature of classroom interactions and their relationship to the doing and learning of mathematics is a major area of research, forming the focus of many of the papers presented in the TSG. The majority of these papers were concerned with the construction of mathematics and mathematical thinking and, in particular, the ways that teachers and teaching methods shape the possibilities for students' mathematical thinking and the ways in which mathematical knowledge is developed in interactions between teacher and students and among groups of students.

Drageset characterised different ways in which teachers respond to student contributions, offering a framework for analysing how different practices may have potential to help student thinking to progress. Milani also discussed how different forms of interaction may relate to learning, identifying dialogic questioning as a form that involves students as active participants in the learning process. Focusing on the development of spatial perception in young children, Schuette's study investigated the different ways in which this domain is talked about in the three contexts of primary school, infant school and in the home. Park used a semiotic approach to analyse and describe students' proportional reasoning, finding that multiplicative strategies were more successful than either additive or formal strategies.

Lee et al. looked at the effects of using "story-telling" instead of formal proof when teaching about transformation of functions, suggesting that students have similar success with both methods but that the story-telling approach has affective benefits. Investigating students' ability to present their solution methods and explanations in writing, Misono and Takeda identified a need for teachers to work with students to develop their use of mathematical language and their communication skills. Another approach to thinking about teaching methods was provided by O'Keefe and O'Donoghue, who offered a linguistic analysis of textbooks, using this to characterise how the nature of mathematics is portrayed.

Looking in detail at a teacher working with a small group of children, Gellert analysed an episode in which a disagreement arises, identifying the epistemological development and how the teacher and students negotiate mathematically. In Barcelona, a group of researchers is investigating classroom interaction from the point of view of studying the social construction of mathematical knowledge. This group presented two papers looking deeply at the mathematical activity of students when working in pairs (Badillo, Planas, Goizueta and Manrique) and in whole group discussion (Chico, Planas and Goizueta).

Language is not only used for communicating knowledge but is also a means for establishing our identities and relationships. This function of language was addressed by Heyd-Metzuyanim, whose paper presented an analysis of the "identifying" and "mathematizing" interactions in two small groups of students while they were engaged in problem solving. She suggested that, for the lower attaining group, the struggles over identification may have hindered their progress in learning.

Multilingualism in Mathematics Education

There has been a longstanding interest in the issues involved in teaching and learning mathematics in different languages. This originated to a large extent in the context of post-colonialism at a time when many countries with a legacy of education in the language of the ex-colonial power were struggling to value their own national and local languages and to develop the use of these languages in education. Political struggles over choice of language of instruction continue, while research is adding to our understanding of how characteristics of specific languages may affect the nature of the mathematics that is done using the language as well as how they may affect student learning. Two papers by Edmonds-Wathen and by Russell and Chernoff both addressed the differences between Aboriginal Englishes, spoken in indigenous communities in Australia and Canada respectively, and the standard forms of English spoken by the majority of their teachers and used in the classroom. While appearing similar in some respects, these languages carry different cultural and conceptual underpinnings with consequent possibilities for meaning making that teachers need to be aware of.

With increased mobility of populations as well as national decisions to offer mathematics education in a range of languages, mathematics educators across the world are increasingly needing to deal with classrooms in which students speak more than one language and have varying levels of competence in the main language of instruction. While this is often portrayed as being a 'problem', the papers presented in the TSG demonstrate that mathematics educators are dealing in subtle and important ways with the complex issues involved. Indeed, the research reported by Ní Ríordáin and McClusky from Ireland indicates that bilingual students with good competence in both languages (Irish and English) outperformed those for whom one language was dominant. Investigation of the students' language use while problem solving suggested that bilingualism was associated with enhanced metacognitive ability. The benefits of bilingualism are one of the motivations behind the introduction of Content and Language Integrated Learning (CLIL), a policy supported by the European Commission, involving teaching curriculum content through the medium of a foreign language. Maffei, Favilli and Peroni reported on the introduction of CLIL in Italy, teaching mathematics through the medium of English in secondary schools.

Whereas the students investigated by Ní Ríordáin and McClusky and by Maffei et al. experienced teaching and learning in both languages, Craig's study looked at the experience of university students in South Africa, studying mathematics through the medium of English only, in spite of the fact that for some of them this was not their main language. She introduced writing activities into the classroom as a means of developing students' understanding of mathematical concepts and found that both English and non-English main language students grappled similarly with the mathematical content but that language was a source of difficulty and a potential obstacle for less well-prepared students. The question of how pedagogic methods may have differential effects for students from different linguistic and cultural backgrounds was also addressed by the study proposed by Björklund Boistrop and Norén. Their concern was to investigate teachers' assessment practices in interactions with students in multilingual classrooms in Sweden.

Theory and Methodology

A wide range of theoretical perspectives and methodologies was apparent in the papers presented and this was a focus of much discussion during the TSG sessions as participants sought to understand the basis for analyses and conclusions and to interrogate and develop the rigour of the methods used to study language and communication. Two presentations took as their main topic the use and development of theory and methodology. Nachlielli and Tabach addressed the combination of two theories: the social semiotics and Systemic Functional Linguistics of Halliday (1974), a general semiotic and linguistic theory, and Sfard's theory of commognition (2008), which addresses the nature of mathematical discourse specifically. They used these theories to develop a framework for analysing classroom interaction. Similarly, Tang, Morgan and Sfard drew on the same two theories to present the development of an analytical framework for studying examination papers and the nature of the mathematical activity that students taking these examinations are expected to engage in.

Given the widespread interest in theory and methodology among those attending the TSG, a plenary panel discussion on this topic was organised. Three presenters, Einat Heyd-Metzuyanim, Candia Morgan and Máire Ní Ríordáin were asked to identify and reflect upon the theoretical and methodological issues that had arisen for them in their research programmes, the choices they had made and the ways these choices may have affected the outcomes of the study. The presenters also questioned each other and responded to these questions and to those raised by other members of the TSG. Issues raised included the definition and operationalization of constructs, use of quantitative and qualitative methods, and the effects of language used by a researcher on the nature of data collected.

Final Reflections

Underpinning many of the presentations were the intertwined themes of politics and culture. It was repeatedly observed that language in education is inherently political, in more than one way. National or cultural politics can influence the choice of language and teaching methods, the roles language plays in the classroom, researcher access to classrooms and the uses to which research findings are put. Language is similarly influenced by culture and is an indicator of cultural identity. Politics, culture, language and teaching and learning are interrelated. Additionally, culture can influence research methodology.

Language, from the point of view of the learner, both gives and limits access to mathematics. Communicative activities in and outside the classroom shape mathematical thinking and thus language mediates access to mathematics. From the point of view of the researcher, language is both a research tool and a focus for research into mathematics teaching and learning. There is a relationship between language and learning, but also one between language and pedagogy. Analysis of communicative activities in the context of mathematics teaching and learning allows us to understand both. For successful learning to occur the teacher needs to effectively communicate mathematics, bringing issues such as open and closed discourses, specialised and everyday registers, multimodality and multilingualism to the attention of the researcher of language.

The practical topics of data collection, processing and analysis were of particular interest. Analysing language issues in the mathematics classroom can be difficult, there are methodological dilemmas and challenges. The logistics of gathering and analysing language data can benefit from further investigation, addressing issues such as how to analyse large corpuses of data when the method of analysis calls for detailed attention to small amounts of text. Large bodies of language data could benefit from being made accessible to large groups of people to work collaboratively, but that in itself brings in complications of ethics and multiple languages. Also, context is key to understanding and, in data sharing, the context of the data collection could be obscured. The role of language in the communication of mathematics is complex; in trying to capture that complexity we tend to reduce it for ease of understanding. This introduces a tension for researchers as something is inevitably lost in that reduction. Analysis of language as communication of mathematics benefits from the insights offered by cross-disciplinary perspectives, such as from linguistics.

The Topic Study Group closed with an appreciation of the small community which had formed at ICME, a hope to collaborate (and data share?) in future and a call to pool our skills and knowledge with one another.

Open Access This chapter is distributed under the terms of the Creative Commons Attribution Noncommercial License, which permits any noncommercial use, distribution, and reproduction in any medium, provided the original author(s) and source are credited.

References

Halliday, M. A. K. (1974). Some aspects of sociolinguistics *Interactions between linguistics and mathematical education symposium*. Paris: UNESCO.

Sfard, A. (2008). *Thinking as Communicating: Human Development, the Growth of Discourses, and Mathematizing*. Cambridge: Cambridge University Press.

Gender and Education

Olof Steinthorsdottir and Veronique Lizan

Report

While mathematics are universal, it appears that delicate process in the classroom, but not only there, lead boys and girls to perceive things differently. And from this perception at school depends the future of the jobs. If the teacher, male or female, is conscious of this, what can he/she do to provide to each pupil or student, boy or girl, the opportunity of understanding, participating and finally appreciating mathematics at best?

The subject is not new: it merges explicitly at ICME3 in Karlsruhe (Germany) in 1976. «[...] Moreover, it is recommended that the theme 'Women ans Mathematics' be an explicit theme of ICME 1980.»: this ends the third and last resolution of the Congress. This recommandation became realised at ICME4 in Berkeley in 1980 and goes on since.

From the proposals received for ICME12 from all over the world, the reflection at Topic Study Group «Gender and Education» was organised along four themes: gender issues in research and learning environmental; student's achievement, assessment and classroom activities; self-efficacy and attitudes; gendered views of mathematics.

Organizers Co-chairs: Olof Steinthorsdottir (USA), Veronique Lizan (France); Team Members: Collen Vale (Australia), Laura Martigon (Germany), Sun Hee Kim (Korea); Liaison IPC Member: Cheryl Praeger (Australia).

O. Steinthorsdottir (✉)
University of Northern Iowa, Cedar Falls, USA
e-mail: olly.steintho@uni.edu

V. Lizan
Institut de Mathématiques de Toulouse, Toulouse, France
e-mail: veronique.lizan@math.uni-toulouse.fr

© The Author(s) 2015
S.J. Cho (ed.), *The Proceedings of the 12th International Congress on Mathematical Education*, DOI 10.1007/978-3-319-12688-3_54

The subject deals with the notion of «gender», that has merged in sociology studies during 70s and it took time to work out a definition since gender doesn't reduce to sex. The term appeared in ICMI history first in 1992; it was introduced at ICME7 in Québec by IOWME.

Indeed, gender and mathematics is at the crossing of different subjects (sociology, psychology, biology or anthropology for example) what is not surprising since teaching mathematics to pupils or students generates interactions between the teacher and the classroom but also between classroom members. So it is at the same time a complex but also a completely natural subject, so natural that it can sound unrelevant.

What Do We Learn on «Gender and Mathematics» at ICME12?

The aim of the first session was to establish some basis: precisely define vocabulary, revisit the term « gender » for maths classrooms and develop a methodology to study what happens in a math class when considered from a gendered viewpoint. Indeed, crossing gender with mathematics stakes very delicate process and it is essential to circumscribe the studied objects and the way they'll be studied in any research on the subject.

The second session pointed that different social parameters impact pupils achievement to international tests or national selection process, especially those that concern family background. The type of tests or criteria of selection can also introduce unsuspected bias into selection process. Gender interfers with mathematics achievement not only in the classroom but everywhere from the moment there are human relationships, and more acutely when mathematics are assigned a role of selection, quite a social selection role.

The third session enlighted how important is the way of teaching to catch the interest of pupils—the girls of the study appreciate to be responsibilized and active —and also how important is the involvement of parents for maths studies or topics in pupils' interest for maths and their success, especially concerning girls. In maths teaching process, the content is important of course, but the manner also is of importance as well as the environment knowledge to try to equally imply most if not all pupils or students of a classroom and make them feel concerned by the maths class. Reading ability of course is also a technical factor of success for students in mathematics through their self-assessment—the best they read and the more acurately self-assessment is perceived to perform—Self-efficacy that boys and girls don't live in he same way especially during problem-solving tasks is also a parameter of importance in mathematical activity environment. It is precisely when the maths activity perturbates the pupil, the pupil's security in some way (difficult question or open problem for example) that some aspects of each pupil's personality built since childhood stake. In that sense maths activity actively participates in the personal construction of each pupil.

The fourth session pointed that children at pre-school are already submitted to gendered stereotypes during mathematics activities, and also that gender and mathematics are related to cultural parameters even if statistics show differences between boys and girls achievements in the same sense everywhere: mathematics are abstract and universal but the question is the same everywhere independantly of cultures.

Perspectives for the Future

Different gender activities were disseminated in ICME12 program and one could concoct a quite full time «Gender and mathematics» program during the congress : part of Gilah Leder's talk since gender is one of her interests; an overview «Gender and Mathematics education (revisited)»; 2 IOWME (International Organization of Women in Mathematical Education) meetings; a Girls' day organised by KWMS (Korean Women in Mathematical Sciences) and WISET (Korea Advanced Institute of Women in Science, Engineering and Technology); and of course the topic study group «Gender and Mathematics» and its four sessions. The Girls' day mentoring activity was of special interest because it involved about 110 girls and also mentors, women maths researchers or scientific engineers; it was related to the WISET stand at the Mathematical Carnival and also to activities especially for girls. Analogous days also exist in Australia, France or USA for example, and they constitute a first step to an active treatment of gender and mathematics, or more generally science, topic.

Anyway the public at the topic study group was essentially constituted by people already conscious that mathematical activity at school has not the same social meaning or psychological impact for boys as for girls. But, are all of us that teach mathematics to both female and male conscious (or convinced?) that both publics don't deal with mathematics in the same way? And how to make a math course equally attractive for boys and for girls?

Of course, ICME takes the subject of gender and mathematics into account since its very beginning. Anyway it is not a timeworn leitmotiv since the corpus on gender and mathematical education constitutes along years. On the contrary, it is necessary to wake up that the question is of importance and to become aware that it is closely related to the future of mathematics and science that lack of students for both research, engineering and technology.

Scientists are already active on the subject (Cf. Girls' day and also the work of the devoted associations). When will teachers be systematically trained to consider their pupils also as boys and girls and then when will teachers take into account in their practice gender angle to tackle their classes? And what contents for training teachers on the subject? Perhaps subjects at a plenary talk in a future ICME.

Open Access This chapter is distributed under the terms of the Creative Commons Attribution Noncommercial License, which permits any noncommercial use, distribution, and reproduction in any medium, provided the original author(s) and source are credited.

Mathematics Education in a Multilingual and Multicultural Environment

Anjum Halai and Richard Barwell

Introduction

For this topic study group, 35 papers were accepted from a range of different cultural, linguistic and country contexts. The papers were discussed under specific thematic questions. These themes provide an organizing framework for this report that draws its content from the papers and the discussion in the TSG 30 sessions. The submissions illustrated the rich diversity in the kinds of issues that arise in mathematics education in multilingual and multicultural environments. These include challenges for teaching, learning, curriculum, pedagogy, teacher education and use of technology in and for multilingual and multicultural settings. Issues were at the level of policy (e.g. language of instruction) and at the level of classrooms (e.g. teaching methods, curriculum) and teacher education (e.g. models of pre-service and teacher professional development). Diversity was also seen in terms of the geographical spread of the contexts from where papers were presented. The diversity of contexts reflects technologically advanced countries with increasingly large immigrant populations (e.g. Australia, Canada, Germany, Sweden, USA, UK), postcolonial countries with concomitant colonial languages as the medium of instruction (e.g. Ghana, Pakistan, Malaysia, South Africa, Tanzania) and countries with varied indigenous and official languages (e.g. China, India, Indonesia, Mexico,

Organizers Co-chairs: Anjum Halai (Pakistan), Clement Dlamini (Swaziland); Team Members: Richard Barwell (Canada), Nancy Chitera (Malawi), Dong Joong Kim (Korea); Liasion IPCMember: Frederick Leung (Hong Kong).

A. Halai (✉)
Aga Khan University, Dar es Salaam, Tanzania
e-mail: anjum.halai@aku.edu

R. Barwell
Univeristy of Ottawa, Ottawa, Canada
e-mail: richard.barwell@uottawa.ca

© The Author(s) 2015
S.J. Cho (ed.), *The Proceedings of the 12th International Congress on Mathematical Education*, DOI 10.1007/978-3-319-12688-3_55

New Zealand). The overwhelming prevalence of issues related to quality of mathematics education in multilingual and multilingual contexts illustrates its significance.

Theme One: What Is Distinctive About Learning and Teaching of Mathematics in Multicultural and Multilingual Settings?

Presenters and participants identified several teaching strategies and distinctive elements of multilingual classrooms, highlighting potential for improving learners' mathematical skills. These included the use of group work, judicious questioning, implementation of second language teaching techniques in mathematics classrooms, promoting a positive climate in the classroom, enabling "translanguaging" i.e. to switch between the linguistic resources and cultures that learners have at their disposal (e.g. Farasani's work with British Iranian learners), and "exploratory talk" (e.g. the work of Webb and Webb in South Africa) as a vehicle to promote dialogue to enhance learners' reasoning skills in mathematics. An enduring concern for mathematics learning was students' lack of competence in the language of instruction. It was also noted that the discussion of papers in this theme emphasized issues arising specifically from multilingualism, as compared to multiculturalism.

Theme Two: What Is the Experience of Education Systems that Have Changed the Medium of Instruction in Mathematics?

Experiences were shared of learners and teachers from different country contexts where the medium of instruction was changed or different from the first language of the learners (e.g. Kasmer's and Kajoro's work in Tanzania) and multilingual classrooms with immigrant learners from several different first language backgrounds (e.g. Meyer's work with immigrant learners in Germany). For learners in multilingual postcolonial classrooms, presenters discussed several linguistically and culturally responsive teaching strategies such as the use of pictorial and other representations of mathematical ideas, situating the mathematics tasks in a familiar context, and code switching to facilitate learning. However, it was noted that there were tensions in classroom dynamics where a position of power and prestige was given to the language of instruction while learners' first language was not seen as a language of choice (e.g. Ampah-Mensah's work in Ghana).

In the case of classrooms where learners, often from immigrants communities, came from multiple language backgrounds not shared by the teacher and often not by other learners, it was concluded that an official language of the classroom was necessary to enable communication in the whole class. However, this necessity need not preclude strategies such as small group work where learners could use their home languages. Empowering the learners to take responsibility for their learning in small groups, and looking at the outcomes of the group work, could be strategies that teachers could employ in such multilingual settings. It was agreed in the discussion that the range of strategies and methods being employed by teachers and learners in the multilingual classrooms needed to be evaluated for their efficiency and effectiveness.

Theme Three: How Can Mathematics Teaching Respond to the Oppression of Cultural and Linguistic Minorities?

Studies in this theme reported different models (e.g. the "bi-cultural curriculum model" in New Zealand presented by Jorgensen), and teaching methods (e.g. Matematika GASING Method in Indonesia by Surya and Moss) for responding to the needs of learners from cultural and linguistic minorities. While there were subtle differences in the orientation and motives of these methods and models, they were mainly premised on the view that all children can learn mathematics provided they have opportunity to do so, and that the opportunity should be to access culturally and linguistically relevant mathematics teaching and learning. It was also recognized by these proponents that language, culture and mathematics pedagogy are integrally bound in a complex relationship. The models and methods proposed certain key elements of teaching that could be employed in mathematics classrooms for learners from culturally and linguistically marginalized or minority groups. For example, exposing learners to multicultural visual representation and conceptual tools before abstract mathematics notation; ensuring "respect" for learners in multiethnic classrooms by creating ample space to listen to them and guide their thinking (e.g. Averill and Clark's work in New Zealand); and taking a "bi-cultural focus" in the curriculum that legitimizes the culture of the school and of the community. However, in the discussion an issue was raised that culture was a broad and potentially nebulous term and needed further clarity in terms of its application to mathematics education.

Theme Four: How Does/Should Teacher Education Take Account of Cultural and Linguistic Diversity?

In this strand, it was pointed out that pre-service teacher education must take account of multilingual classrooms and recognized that a vast majority of learners learn mathematics in a second or third language. Exemplars of teacher education programmes included the presentation by Prediger and team, on the notion of an inter-disciplinary teacher education course proposing that mathematics teachers need to have didactic and linguistic knowledge and cultural sensitivity to understand the challenges that might be faced by the learners from diverse settings. Likewise interventions in teacher education provided a range of strategies and techniques that could be employed with teachers and students. These included, dialogic strategies and "exploratory talk" to promote mathematical reasoning among students, extended wait time for second language learners of mathematics, need for clarity and avoidance of slang in use of language in multilingual classrooms, utilizing learners' fluency in their main language as well as to garner the aid of a more able peer. The few studies that harnessed the potential of technology to enhance the cultural understanding and experience of learning mathematics in a second or third language included the use of video-conferencing, social media and Skype as a medium to provide experience of teaching in a multilingual setting and enhance cultural understanding (e.g. the work of Moss and Boutwell with pre-service teachers in USA, Singapore and Haiti). A conclusion was that technology provided a relatively easy opportunity for teaching mathematics within a multicultural and multilingual environment. With creativity, connections, and technology, pre-service mathematics teachers could learn about mathematics, teaching, and culture in other countries without leaving their own.

Theme Five: How Do Curricula and Policy Take Account (or not) of Cultural and Linguistic Diversity?

In this theme the focus was more on curricular processes (not necessarily curricular content) embedded in instructional sequence, pedagogy and teaching strategies for improved teaching and learning in diverse contexts. For example a teaching sequence was presented by Xaab Vasquez, based on the philosophy of "Wejën Kajën" in Oaxaca in Mexico, which encourages reflection on the prevailing education processes and the need to make explicit that learners are not isolated but are situated in a wider social and cultural context. Cooperative learning strategies were presented as an approach to create space for marginalized learners to improve achievement in mathematics. Similarly, presentations proposed differentiated instruction sensitive to the needs of minority students and "equitable strategies" that encourage collaborative knowledge production, student authority and ownership of knowledge, and mutual respect (e.g. the work of Manjula and Erchick in USA). Such strategies should be guided by the principle of reducing discontinuities between the lives of students by drawing on their cultural heritage to create an egalitarian context for supporting the learning of all students (e.g. the work of Ryoon Jin Song and team in South Korea). Use of mathematics investigations, films, print literature and internet websites were also seen as ways to accommodate cultural diversity in the classroom. The case was also presented of the International Baccalaureate Diploma Program, IB, which operates in three languages (English,

French and Spanish). It was pointed out that the IB curriculum is integrally concerned with the international dimensions of mathematics and the multiplicity of its cultural and historical perspectives, which in turn helps to discover new perspectives and horizons in international mathematical education.

Theme Six: What Theoretical Perspectives on Cultural and Linguistic Diversity Are Most Helpful in Investigating The Teaching and Learning of Mathematics?

Several theoretical frameworks and conceptual models were presented in this theme to provide tools for understanding and analyses of issues related to teaching and learning of mathematics in contexts of cultural and linguistic diversity. For example these included the presentation by Essien and team on an extension of Wenger's work on "communities of practice" for application to pre-service teacher education for multilingual mathematics classrooms. Likewise an integrated model was presented that integrates three hitherto disparate registers: those of code switching, transitions between informal and academic (mathematical) forms of language within a given language, and transitions between different mathematical representations. However, it was pointed out that further research was required to establish the efficacy of this model. Sevensson's presentation raised issues related to research methodology in ensuring that "students' voices" are heard. Barwell and team presented work that extended Bakhtin's (1981) theory of language and claimed that the theory provides a framework for looking at the tensions in mathematics classrooms in diverse language contexts but go on to state that more research is needed in this area.

Concluding Remarks

Certain key overarching questions or concerns were raised for further deliberation about the quality of mathematics education in diverse linguistic and cultural settings. These include: "Where is the mathematics in talking about the methodological, political and equity issues in multilingual and multicultural classrooms?" It was reiterated that meetings like ICME are primarily about mathematics education and therefore mathematics should be in the foreground. A concern was that meta-concepts like "culture" and "language" were employed in the discussion as if there existed a shared understanding of these concepts. However, there needs to be discussion and debate to problematize these notions and clarify their usage in mathematics education. Also it was noted that even though the title of the TSG 30 and the themes included "multilingualism" and "multiculturalism" the papers and discussion tended to focus on issues related to multilingualism.

Acknowledgments The contribution of the committee members especially Clement Dlamini, authors and participants in TSG 30 are sincerely acknowledged.

Open Access This chapter is distributed under the terms of the Creative Commons Attribution Noncommercial License, which permits any noncommercial use, distribution, and reproduction in any medium, provided the original author(s) and source are credited.

Reference

Bakhtin, M. M. (1981). *The Dialogic Imagination: Four Essays.* (Ed., M. Holquist; Trans, C. Emerson and M. Holquist). Austin, TX: University of Texas Press.

Tasks Design and Analysis

Xuhua Sun and Lalina Coulange

Aims

A critical topic in mathematics education is the design and analysis of open-ended, realistic, and exemplary tasks. Task design and analysis is a relatively new field, appearing for the first time as a topic of study (TSG 34) at ICME-11 in Monterrey, Mexico. It is developing quickly and dynamically as an area of international attention and active research.

Topic Study Group 31 will bring together researchers, developers and teachers who systematically investigate and develop theoretical and practical accounts of task design and analysis. We welcome proposals from both researchers and practitioners and encourage contributions from all countries. Presentations and discussions will target new trends, new understanding, and new developments in research and practice.

We have a particular interest in empirically grounded contributions that underline design principles and theoretical approaches, and give examples of tasks

Organizers Co-chiars: Xuhua Sun (China), Lalina Coulange (France); Team Members: Eddie Chi-keung Leung (Hong Kong), Nguyen Chi Tanh (Vietnam), Hea-Jin Lee (Korea/USA); Liaison IPC Member : Masataka Koyama (Japan).

X. Sun (✉)
University of Macau, Macau, China
e-mail: xhsun@umac.mo

L. Coulange
Universite de Bordeaux, Bordeaux, France
e-mail: lalina.coulange@gmail.com

© The Author(s) 2015

S.J. Cho (ed.), *The Proceedings of the 12th International Congress on Mathematical Education*, DOI 10.1007/978-3-319-12688-3_56

designed for promoting mathematical development. We plan to discuss (but are not limited to) the following themes:

- Theoretical and practical development that guides task design and analysis
- Diverse theoretical approaches or principles that guide task design and analysis
- Diverse practical traditions/approaches that guide task design/analysis and their theoretical accounts
- Examples of task analysis for studying the relations between tasks, psychological development, and mathematical development
- Critical literature studies or meta-analysis of task design and analysis

The group will welcome contributions that focus on primary or secondary education. Research and development in task design and analysis presented at ICME-11 is retrievable at (http://tsg.icme11.org/tsg/show/35).

Organizations

On the website of ICME-12 it was possible to follow the planning process and eventually access all relevant documents including the timetable for TSG sessions. Each Session has four 90 min timeslots (on Tuesday, Wednesday, Friday and Saturday mornings). This made TSGs the prime forum for participation. We expected that participants engage in the review process prior to the conference, and we nominated respondents to all presentations in order to enable deeper levels of critical discussion during the conference. The presenters worked in pairs and made short comments or elaborated on each other's work after every presentation. In this way, TSG 31 was an active study group.

Submissions and Theme

The organizing committee received 12 submissions with 100 % acceptance rate (11 short oral presentations and 1 poster). The organizing committee assembled the accepted papers for TSG 31 into four groups for summary, presentation, and discussion:

- *Dynamic Geometry Environments and the Role of Representations*
- *Categorizations of Tasks and Textbooks*
- *Tasks Enacted by the Teacher and Students*
- *Discoveries and Justifications*

Schedule

Session 1 Tuesday, 10th July, 10:30–12:00, Dynamic Geometry Environments and the Role of Representations (Number of attendants: 24).

Opening remarks: Sun Xuhua susanna and Lalina Coulange (20 min).

Mickael Edwards, Task Design and Analysis using the Measure-Trace-Algebratize Approach (25 min).

Teresa B. Neto, Xuhua Sun, Task design and analysis of on-to semiotic approach (25 min).

Eddie Chi-keung Leung, Hea-Jin Lee, Sun Xuhua (Main discussant speakers): Round-table discussion with the whole group on the 2 contributions (20 min).

Session 2 Wednesday, 11th July, 10:30–12:00, Categorizations of Tasks and Textbooks (Number of attendants: 32).

Regina Bruder, Eight target structure types of Tasks as background for learning surroundings (25 min).

Hyungmi Cho, Jaehoon Jung, Ami Kim and Oh Nam Kwon, An analysis of the mathematical tasks in the Korean 7th grade mathematics textbooks and workbook (25 min).

Lianzhong Fan, Jiali Yan, Xuhua Sun, The Changes of Task Design for Development "Two-Bases" in China after Ten-year Curriculum Reform (25 min).

Hea-Jin Lee, Nguyen Chi Tanh, Lalina Coulange (Main discussant speakers), Round-table discussion with the whole group on the 3 contributions (15 min.)

Session 3 Friday, 13th July11:00–12:30 Tasks Enacted by the Teacher and Students (Number of attendants: 35).

Rina Namiki and Yoshinori Shimizu, On the Nature of Mathematical Tasks in the Sequence of Lessons (25 min).

Julie Horoks, Analysing tasks to describe teachers' practices and link them to pupils' learning in mathematics (25 min).

Marita Barabash, Raisa Guberman, Multiple informal classifications of geometrical óbjects as an ongoing process of developing young students' geometric insight (25 min).

Eddie Chi-keung Leung, Nguyen Chi Tanh, Lalina Coulange Main discussant speakers: Round-table discussion with the whole group on the 3 contributions (15 min).

Session 4 Saturday, 14th July10:30–12:00, Discoveries and Justifications (Number of attendants: 25).

Michael Meyer Forming concepts through discoveries and justifications (25 min).

Celine Constantin, Lalina Coulange In search for a specific algebraic task design or how to elaborate a situation highlighting algebraic techniques in second grade (25 min).

Eddie Chi-keung Leung, Nguyen Chi Tanh, Hea-Jin Lee Main discussant speakers Round-table discussions of the session papers (15 min).

Sun Xuhua, Lalina Coulange Closing remarks: Whole group discussion on the work of the group and conclusion (25 min).

On-line Discussion Notes

https://docs.google.com/document/d/1Bll7r2tN7ha8PQrr2J3xJu5gEQ05wPveDFV
83EZdpvc/edit.

Open Access This chapter is distributed under the terms of the Creative Commons Attribution Noncommercial License, which permits any noncommercial use, distribution, and reproduction in any medium, provided the original author(s) and source are credited.

Mathematics Curriculum Development

Koeno Gravemeijer and Anita Rampal

Introduction

The purpose of TSG 32 was to gather congress participants who are interested in research, policy or design that focuses on mathematics curriculum development. The TSG aimed at including presentations and discussions of the state-of-the-art in this topic area and new trends and developments in research and practice in mathematics education. Curriculum was perceived at two levels. On a national or state level, where the focus is on content and goals for the primary or secondary school mathematics curriculum. And on a more specific level of curriculum design which concerns the developmental trajectories of mathematics content and the best ways to represent them. In relation to this theme, we especially solicited papers that might foster the deliberation on the varied aims of the curriculum and bring concerns and experiences from different contexts.

The papers that were submitted could be arranged in four categories, which were used to structure the sessions:

- Authenticity and Inquiry
- Implementation

Organizers Co-chairs: Anita Rampal (India), Koeno Gravemeijer (the Netherlands); Team Members: HyeJeong Hwang (Korea), Margaret Brown (United Kingdom), Cyril Julie (South Africa); Liaison IPC Member: Cheryl Praeger (Australia).

K. Gravemeijer (✉)
Eindhoven School of Education, Eindhoven, The Netherlands
e-mail: koeno@gravemeijer.nl

A. Rampal
Department of Education, University of Delhi, Delhi, India
e-mail: anita.rampal@gmail.com

© The Author(s) 2015

S.J. Cho (ed.), *The Proceedings of the 12th International Congress on Mathematical Education*, DOI 10.1007/978-3-319-12688-3_57

- The Syllabus
- Math Topics

Each session consisted of one long paper and a number of short paper presentations.

Authenticity and Inquiry

The session on authenticity and inquiry started with a presentation by *Anita Rampal* (India) (with *Katie Makar* (Australia)) of a paper on the topic of embedding authenticity and cultural relevance in primary mathematics. She observed that there is an increasing need for a more democratic and universal participation in elementary school, better numeracy among citizens and mathematical competence and expertise in the workforce, but that accountability systems have often worked in opposition to these elements to further suppress authentic problems in favor of those that can be easily tested. In their paper they highlighted approaches to tackle this problem using innovative curriculum materials in two diverse contexts—India and Australia. These materials were designed with the specific intent of increasing students' opportunities for learning mathematics in ways that are relevant to their familiar and local contexts and cultures. Specifically, to increase the use of culturally relevant thematic units in Indian primary school textbooks, and to embed inquiry-based learning using authentic problems in the Australian curriculum. As half of India's children do not complete elementary education owing to the alienation they face in school, a social constructivist approach has been adopted to ensure more inclusive and democratic participation of all children. This has led to the development of new textbooks which, especially at the primary level, attempt to locate mathematics in the diverse socio-cultural contexts of children's lives. A new national curriculum in Australia has sought to align the curricula across the states and territories and to reflect a stronger focus on disciplinary knowledge and proficiencies, general capabilities and cross-curricular priorities. A seven year longitudinal study has been researching teachers' experiences and pedagogical practices as they adopt and adapt inquiry-based teaching in their classrooms, by engaging students in addressing ill-structured problems that required students to continually re-negotiate their understandings of mathematics within a rich context.

This presentation was followed by three short paper presentations:

Shelley Dole (with *Katie Makar*, and *Gillies Robyn*) (Australia) presented a paper on how the inquiry pedagogy of the intended curriculum was enacted in Australian classrooms. To answer this question, they assembled video data, classroom observations, and interviews with teachers involved in a design research-project. This concerned 40 teachers (of Grades Prep to 7) who attended three professional development meetings per year, and taught 3–4 inquiry-mathematics units per year. The teacher meetings provided the teachers with an opportunity to discuss their thoughts about and experiences with inquiry. It showed that during these meetings teachers identified the benefits of inquiry. The classroom observations showed that the teachers were keen to undertake inquiry in their classrooms, but it showed also that inquiry is difficult for both teachers and students.

Danrong Ying (China) presented a study in which a comparison was made between inquiry tasks in three high school mathematics series in China. Two textbooks were based on the "Obligatory High School Standards", issued by the Chinese Ministry of Education, the other one was based on the "Shanghai Primary and Middle School Mathematics Curriculum Standards". The results reveal that mathematics inquiry tasks in three series mainly focused on "Number and Algebra". And even though the textbooks based on the "Obligatory High School Standards", gave various names to mathematical inquiry tasks, the actual presentation was mainly in the form of pure mathematical problems. In all three selected textbook series, the tasks labeled "experiment" all focused on using information technology to solve mathematical problems, while clear procedures were given.

Yamei Zhu (with *Yun Gan*, and *Yaping Yang*) (China) presented a paper on a comparative study of mathematics textbooks in Shanghai, Singapore and America. Some differences could be traced to the different cultural backgrounds. The Shanghai and the Singapore textbooks reflected a typical "eastern culture" and the American textbooks a typical "western culture". In the former the teacher is dominant, the textbooks offer structured and coherent knowledge, and there is an emphasis on pure mathematics which leads to "the multi-steps, logic-based and knowledge-rich mathematics problems". The American textbooks focus on what the authors qualify as "isolated and incoherent knowledge". At the same time, the USA textbooks use context problems which convey the meaning of mathematics study.

Implementation

The session on implementation started off with a presentation by *Margaret Brown* (with *Jeremy Hodgen*, and *Dietmar Kuchemann*) (United Kingdom) of a paper on changing the grade 7 curriculum in algebra and multiplicative thinking at classroom level in response to assessment data. In this presentation, the methods and results of the project were reported. Phase 1 of the project took the form of assessment of attitude and understanding in the areas of algebra and multiplicative thinking of a nationally representative sample of students in Grades 6–8 in England. The results revealed that the majority of students in Grade 8 had an understanding of ratio, which did not extend beyond scaling up by multiplication by a small whole number, while 40 % had an understanding of algebra, which did not extend beyond that of treating letters as objects or direct evaluation. Phase 2 of the project involved working with eight teacher researchers to research the understandings of their own Grade 7 students in these areas and to explore ways of improving their students' understanding. The understanding of many students was 'patchy'. To some extent this reflected a lack of connections in the understanding of the teacher researchers. This in turn limited the possibility of formative assessment. Analysis of the recommended schemes of work and of the most popular textbooks showed that each new topic was covered rapidly and superficially, with teachers often reducing the content to routine procedures to enable students to do the class work exercises. There was no time for deep treatment of topics, discussing the power of different models/representations, relating them to connected ideas, or discussing how they could be applied to more complex problems. In Phase 3 the project is extended to more teachers using interlinked sequences of 40 outline lessons designed by the research team.

This presentation was followed by two short paper presentations and a chat with presenters.

Ji-Won Son (United States) presented a short report on a comparative study on inquiry tasks in three senior high school mathematics textbook series in China. The purpose of the paper was to examine teachers' transformation of cognitive demand of textbook problems. A survey was carried out among 183 teachers teaching from 1st to 6th Grades, of whom eight teachers were observed. It showed that the cognitive demand of the textbook problems plays an important role in deciding the cognitive demand of the problems used by the teachers, but the teachers used lower level teacher questions. An in depth and broad analysis with respect to teachers' textbook use showed that a wide variety of factors influenced the quality of instruction.

LV Shi-hu (with *YE Bei-bei* and *CAO Chun-yan*) (China) presented a study on the implementation of the new mathematics curriculum for compulsory education in Chinese mainland. Surveys were carried out in the Gansu province, among 300 primary and middle school teachers, and 1,360 students in Grades 7–9. The surveys used both questionnaires and interviews. A comparison with earlier surveys showed that the application of the Standards had increased, and that the teachers had acquired a better understanding of the Standards, even though only 20 % said to "completely understand" the Standards. The student questionnaires revealed that different teaching methods, especially cooperative learning, exploratory learning and independent learning were used by the teachers.

The Syllabus

In the third session the syllabus was the central theme. *Tamsin Meaney* (Sweden) (with *Colleen McMurchy* and *Tony Trinick*, New Zealand) presented a paper on the contested space of Maori mathematics curriculum development in Aotearoa-New Zealand. This concerned the development of the first mathematics curriculum in te reo Māori, the Māori language, in New Zealand in the 1990s and its revision in the mid-2000s. They argued that the development of national mathematics curricula in te reo Māori involved contestation, not just around indigenous knowledge and epistemology, but also around language. The authors stressed the power relationships that existed between the various actors involved in the curriculum development process. They argued that the power embedded in the Ministry of Education allowed it to keep a firm grip on the curriculum development process, although the process was contested and in some cases subverted by Māori because of their expectations about the use of te reo Māori. There has been a strong movement amongst some Māori communities for language revitalization and growth since the 1970s. The revision of curricula was thus done with an expectation that it would be less proscriptive, supporting a more community-developed approach to the mathematics that would actually be taught in schools. This supported Māori parents' aspirations for greater fluency in their children's Māori language and opportunities to strengthen their children's tribal identities. More of the specialist mathematical terms and grammatical structures were developed so that mathematics could be taught more easily at higher levels in Maori. The 2008 curriculum minimized the linguistic confusion that arose from the introduction of many new Māori terms in the 1990s. The revised curriculum has an emphasis on mathematical communication that has clearly been indicated by the inclusion of a Māori language strand. So this process like the earlier one has contributed to the teaching of mathematics in te reo Māori.

This presentation was followed by a series of short paper presentations, and a chat with presenters.

Anette Jahnke (Sweden) presented a paper on the process of developing a syllabus, in which she presented critical reflections from a syllabus developer. She had been involved in writing the new (2011) Swedish national syllabuses for kindergarten, elementary and upper secondary school. She observed that every tenth year politicians initiate a reform, often only in one part of the school system. Usually a small number of teachers and/or teacher educators are hired to write a draft during a very short period of time, which is then sent out a number of times for reactions. Often reforms did not result in coherent syllabus from K–12. One of the reasons of failure of syllabus reform was that teachers did not understand or even mis-understood the syllabus. This resulted in very restrictive instructions to the syllabus writers.

Tomas Hojgaard (Denmark) presented a paper on what he called "The fighting of syllabusitis". He coined the term syllabusitis as a name for a disease consisting of focusing on the mastering of individual subjects. As an alternative he suggested using a set of mathematical competencies, while using a matrix structure of the relation between subject specific competencies and subject matter. He argued that such a matrix structure has proven to be a crucial element when attempting to put the competence idea into educational practice, not least because it makes it possible for teachers to take an active part in such a project and welcome it as a developmental tool.

Math Topics

In the last session we gave attention to specific math topics. *Tomoko Yanagimoto* (with *Yuichi Hayano*) (Japan) presented a paper on the teaching and learning of knot theory in school mathematics. Knot theory is studied actively world-wide, since, even though the basis is simple, it has many unsolved problems. Furthermore, it can be related to scientific research fields, such as Genome DNA. The members of the project have written up teaching contents for pupils from elementary school to high school as a book, "Teaching and Learning of Knot Theory in School Mathematics". Experimental teaching—based on the results of the study on teaching knot theory in elementary school and junior high school—started in public junior high school in 2009 and in elementary school in 2011. It showed that knot theory was effective in helping students improve their spatial visualization, in elementary school pupils in particular. In junior and senior high schools, knot theory led students to become more engaged in their mathematical activities. Typical for this project was that mathematicians, researchers of mathematics education and professional practitioners of education cooperated, respecting and trying to understand researches of others' professional fields while creating materials for education. An expert in knot theory could indicate the value of each teaching material from the view point of knot theory. Researchers of mathematics education could indicate the value of each teaching material from the view point of mathematics education. School teachers could realize the teaching in their classrooms based on their pupils' cognition. Officials of mathematics education society could help the teachers carry out the experimental teaching in their public school by asking the director of the school.

This presentation was followed by short paper presentations.

Qintong Hu (with *Ji-Won Son*) (United States) presented a comparative study of the initial treatment of the concept of function in selected math textbooks in the US and China. They analyzed the initial treatment of the concept of function in three curricula: a US traditional text, a US Standards-based text, and a Chinese reform text. The textbook problems were analyzed on three dimensions, contextual feature, response type, and cognitive expectation. It showed that both the US traditional textbook and the Chinese textbook were designed for teacher-centered instruction. While the reform-oriented US textbook was designed for student-centered instruction. However, the US reform-oriented textbook was more similar to the Chinese textbook in putting problems in illustrative contexts, emphasizing connections, reasoning and proof.

Linda Arnold (with *Ji-Won Son*) (United States) presented a paper on a content and problem analysis on learning opportunities related to linear relationships in USA textbooks. The textbook analysis methods included both problem and content analysis. They examined examples of four types of mathematics textbooks: (1) two different commercial texts; (2) a so-called "back-to-basics" text; (2) a reform-oriented U. S. National Science Foundation (NSF) funded text; and (4) a once commonly-used historic text, published a generation ago. There were numerous similarities between the historic textbook and present day commercial texts, suggesting that little had changed over 50 years. All of the problems of the NSF-funded text involved real world context and were geared toward extended thinking, in contrast to the back-to-basics and historic texts that showed a high degree of procedural presentation. It further showed that students using commercial texts were expected to master an especially broad array of objectives.

Sunyoung Han (Korea) presented a study on the effect of science, technology, engineering and mathematics (STEM) project based learning (PBL) on students' achievement. Even though Science, Technology, Engineering and Mathematics (STEM) are critical fields to ensure a financially sound national economy, students have been under-enrolling in STEM courses. To address this problem, STEM Project Based Learning (PBL) developed an instructional method using "ill-defined tasks". The purpose of the study was to examine how STEM PBL lessons affect students' achievement in terms of four mathematical topic areas (i.e. algebra, geometry, probability and problem solving). The participants were diverse students enrolled in a small, urban, and low socio-economic high school. The study showed STEM PBL positively influenced most mathematical topics.

The Topic Study Group meeting ended with some closing remarks by Anita Rampal and Koeno Gravemeijer. It was noted that the special time for chat with presenters' gave greater opportunity for small group intensive discussions that brought out specific issues from different cultural and country contexts.

Endnote

The design and organization of the four TSG sessions was carried out collaboratively by the organizing team. Unfortunately, Cyril Julie was eventually unable to attend the Congress and the TSG.

Open Access This chapter is distributed under the terms of the Creative Commons Attribution Noncommercial License, which permits any noncommercial use, distribution, and reproduction in any medium, provided the original author(s) and source are credited.

Assessment and Testing in Mathematics Education

Christine Suurtamm and Michael Neubrand

Introduction

The purpose of Topic Study Group 33 was to address issues related to assessment in mathematics at all levels and in a variety of forms. Assessment and evaluation play an important role in mathematics education as they often define the mathematics that is valued and worth knowing. Furthermore, sound assessment provides important feedback about students' mathematical thinking that prompts student and teacher actions to improve student learning.

Our Topic Study Group sought contributions of research in and new perspectives on assessment in mathematics education that address issues in current assessment practices. Initially we saw these issues as falling into two main strands, large-scale assessment and classroom assessment. Our original call suggested that papers might address one or more of the following topics:

Organizers Co-chairs: Christine Suurtamm (Canada), Michael Neubrand (Denmark); Team Members: Belinda Huntley (South Africa), Liv Sissel Grønmo (Norway), David C. Webb (USA), Martha Koch (Canada), Heidi Krzywacki (Finland); Liaison IPC Member: Johann Engelbrecht (South Africa).

C. Suurtamm (✉)
University of Ottawa, Ottawa, Canada
e-mail: suurtamm@uottawa.ca

M. Neubrand
Universität Oldenburg, Oldenburg, Germany
e-mail: michael.neubrand@uni-oldenburg.de

© The Author(s) 2015
S.J. Cho (ed.), *The Proceedings of the 12th International Congress on Mathematical Education*, DOI 10.1007/978-3-319-12688-3_58

557

Large-Scale Assessment

- Issues related to the development of large-scale assessments, which might include such areas as the conceptual foundations of such assessments, designing tasks that value the complexity of mathematical thinking, etc.
- Issues related to the purposes and use of large-scale assessment in mathematics.
- Issues related to the development of large-scale assessment of mathematics teachers' mathematical and pedagogical content knowledge.

Classroom Assessment

- Issues connected to the development of teachers' professional knowledge of assessment and their use of assessment in the mathematics classroom.
- Issues and examples related to the enactment of classroom practices that reflect current thinking in assessment and mathematics education (e.g. the use of assessment for learning, as learning, and of learning in mathematics classrooms)

Broad Issues

- The development of assessment tasks that reflect the complexity of mathematical thinking, problem solving and other important competencies.
- The design of alternative modes of assessment in mathematics (e.g. online, investigations, various forms of formative assessment, etc.).

We received over 50 papers from a range of countries and continents and needed to solicit assistance from committee members and others to review the papers. All papers were reviewed by at least two reviewers. The papers presented a wide variety of issues in assessment and testing and most of the papers were accepted for plenary presentations, small group presentations or poster presentations. The difficult task for the co-chairs was to create a meaningful schedule so that all of these issues could be presented and discussed within the time frame allotted for the Topic Study Group Sessions.

The work was organized into three strands:

- Strand 1: Large-scale assessment and the implications for the development of teaching and learning
- Strand 2: Classroom assessment and developing students' and teachers' knowledge
- Strand 3: Task and test design: Various perspectives

Papers were then categorized according to these three strands and after our initial meeting to introduce the topics and structure of the group, each day consisted of the presentation of plenary papers or posters that are connected to these three strands, and then a division into three subgroups with one subgroup focused on each strand. We also had a poster session with open discussion within the TSG program, as well as posters shown only in the general poster exhibition. The following presents a summary of the main themes presented and discussed in each of the strands. We have also included the ideas from the plenary papers which typically stretched over several strands.

Strand 1: Large-Scale Assessment and the Implications for the Development of Teaching and Learning

There were over 15 papers and several plenary papers presented in this strand over several days. Numerous issues emerged through the discussion of the papers. The range of papers demonstrates that mathematics education researchers are using large-scale assessment results for many different purposes and to investigate a range of complex aspects of mathematics teaching and learning in various contexts. For instance, there are comparative studies (e.g. Wo, Sha, Wei, Li and An) and studies analyzing issues in special regions (e.g. Cheung; Fengbo; Leung; Mizumarchi), studies concentrating on specific topics (e.g. Hodgen, Brown, Coe, and Küchemann), and studies of a more experimental nature (e.g. Li). Several papers that were presented illustrate the challenge of looking for broader trends or patterns across schools and districts while being careful to acknowledge and investigate local contextual factors.

Other papers discuss the use of assessments to investigate a range of factors such as students' higher order thinking skills at different levels of schooling (e.g. Bai; Zhang), gaps in knowledge (e.g. Gersten and Woodward), teacher knowledge (e.g. Shalem, Sapire and Huntley), or teaching approaches (e.g. Thompson), and to improve connections of instruction, assessment, and learning (e.g. Paek). These papers remind us that great care must be taken to ensure that the interpretations being made from the test scores are appropriate. Paper presentations and discussion suggest that using a range of methodological approaches may help to better address the complex questions being investigated in assessment in mathematics education research. For instance, cluster analysis of large-scale data can be used to find patterns in scores but methods such as case study, think aloud protocols while students respond to test items, student and/or teacher focus groups and interviews would enrich our understanding of the patterns observed. The use of a variety of methods is helpful in making sound assertions from data from large-scale assessments.

Strand 2: Classroom Assessment and Developing Students' and Teachers' Knowledge

The papers in this strand were organized into several different categories for presentation: classroom assessment in primary grades (e.g. Makar, and Fry; Hunsader, Thompson, and Zorin), assessing conceptual thinking in the classroom, and teachers' knowledge (Grønmo, Kaarstein, and Ernest). Specific topics that arose in presentation and through discussion included:

- The teachers' role and conceptions of assessment and mathematics (e.g. Esen, Cakiroglu, and Capa-Aydin; Hoch and Amit)
- Task design to elicit students' thinking (e.g. Kim, Kim, Lee, Joen, and Park)
- The students' role and responses to open, though provoking questions (e.g. Mangulabnam)
- Development of students' self-reflection, self-assessment, and self-regulation (e.g. Teong, and Cheng)
- Developing transparency, for students in particular, in classroom assessment (e.g. Semana and Santos)
- Teachers' experiences in implementing formative assessment (e.g. Koch and Suurtamm; Krzywacki, Koistinin, and Lavonen)
- Assessing conceptual understanding through alternative assessments (e.g. Türegün)

Across all of these categories was a strong emphasis on formative assessment and at the heart of most, if not all of these papers, was the desire by either researcher and/or teachers to make sense of what students are thinking and learning. The presentations attended to various ways that students' mathematical reasoning is elicited and interpreted by teachers through classroom assessment.

There was also a great deal of discussion about initiatives in various jurisdictions to improve classroom assessment and to support teachers' use of formative assessment. These initiatives included assessment resources, collaboration, professional development, and support from Ministries of Education. It was noted that this support coupled with valuing teachers' autonomy and professional judgment seemed to provide fertile ground for sound classroom assessment practices. It was noted however that this is not occurring in all jurisdictions and we discussed the differences in teacher autonomy in different countries. International forums such as ICME provide a rich setting where these comparative discussions can occur and may prompt other jurisdictions to develop new initiatives that support strong classroom assessment.

Strand 3: Task and Test Design, Various Perspectives

The core of test design is the creation of appropriate tasks. However that is a business that requires consciousness about the various purposes tests are constructed for. Therefore, tasks and test design has aspects of conceptual and practical nature, and implementation issues are also to be considered.

Thus, the sessions in this Strand addressed a rich bundle of aspects. We started with reports on studies on teachers' knowledge (e.g. Webb). One question addressed is, how knowledge and behavior come together, and how that interplay can be measured. Studying teachers' knowledge has also internationally comparative aspects, insofar as pedagogical content knowledge for teaching has to be effectively operationalized (e.g. Kaarstein).

How specific items for goals of assessment can be constructed appropriately—closed or open as well—was the topic of the next session, with contributions of Hong and Choi; Toe and de la Torre; Kwong and Ming; Kang and Lee; Hong, Kim, Lee, and Joo. Also this question has various perspectives from elementary mathematics classrooms to college-bound students; various mathematical topics have to be attended from big ideas about measurement to the issues of learning to prove; dealing with the answers of the students is decisive and ranges from descriptions to the analysis of the competencies which can be detected in the student responses by appropriate models. All these aspects require also the discussion of methodological issues.

Finally, we also discussed some broader aspects of using tests. One topic was how teachers view and use an on-line, formative assessment system and what conclusions they can draw for their teaching (e.g. Stacey and Steinle). And even broader, was the general question as to whether entrance tests to universities are necessary (e.g. Kohanova).

Concluding Remarks

There was discussion within this topic study group as to whether it should have been two topic study groups—one for large-scale assessment and another for issues in classroom assessment. The discussion concluded by recognizing that these should not be separated as it is critically important that these two groups share their issues, ideas and practices if there is to an alignment between assessment that is ongoing, such as in a classroom and assessment that is an event, such as in large-scale assessment. The participants also found that discussions across countries pointed up many similarities in issues such as teacher professional development in assessment, transparency to students, task design to elicit student thinking, and meanings given to assessment results.

Acknowledgment The contribution of the committee members, authors and participants in TSG 33 are sincerely acknowledged.

Open Access This chapter is distributed under the terms of the Creative Commons Attribution Noncommercial License, which permits any noncommercial use, distribution, and reproduction in any medium, provided the original author(s) and source are credited.

The Role of Mathematical Competitions and Other Challenging Contexts in the Teaching and Learning of Mathematics

Mariade Losada and Ali Rejali

Statement of Purpose

The organizing group and the participants come to challenging mathematics from many different perspectives, but all firmly believe that mathematics education for the twenty-first century requires all teachers, schools and extra-curricular experiences to provide structure and support that allow and entice each student and citizen to strive to reach his or her personal best in mathematics. In the words of the discussion document for ICMI Study 16, "Mathematics is engaging, useful, and creative. What can we do to make it (engaging, useful and creative mathematics) accessible to more people?"

Aims

1. To gather teachers, mathematicians, mathematics educators, researchers and other congress participants who are interested in mathematical competitions and other challenging contexts in the teaching and learning of mathematics at all levels.

Organizers Co-chairs: Maria de Losada (Colombia), Ali Rejali (Iran); Team Members: Andy Liu (Canada), John Webb (South Africa), Jaroslav Svrcek (Czech Republic), Kyung Mi Choi (Korea); Liaison IPC Member: Petar Kenderov (Bulgaria).

M. Losada (✉)
Universidad Antonio Nariño, Bogotá, Colombia
e-mail: mariadelosada@gmail.com

A. Rejali
Isfahan University of Technology, Isfahan, Iran
e-mail: a.rejali@yahoo.com

© The Author(s) 2015
S.J. Cho (ed.), *The Proceedings of the 12th International Congress on Mathematical Education*, DOI 10.1007/978-3-319-12688-3_59

2. To present research results and reports on activities that will allow the group to make an updated sketch of the state of the art, thus further developing the aims of the 16th ICMI Study, and colouring it in by addressing new trends and developments in research and practice in mathematics competitions and other challenging contexts and their effect on mathematics teaching and learning and in pinpointing research problems of special interest to the group.

In summary, the organizing team welcomed all contributions related to mathematical challenges, the state of the art, follow-up studies and the results of studies on the impact of these activities on mathematics education. The organizing team has asked those wishing to join the study group to submit a paper of between 1500 and 2500 words in length addressing issues highlighted or others that make a significant contribution to the aims and focus of the group, and they have also invited speakers to submit their papers to the WFNMC journal (http://www.amt.edu.au/wfnmc/journal.html) for possible publication in a special issue.

Questions that Could Have Been Addressed Were

Do mathematical challenges better reflect the nature, the beauty and other characteristics of the corpus of elementary mathematics, as well as the experience of doing mathematics, than ordinary school mathematics? Does this make the mathematics involved more likely to engage the learners?

Does the widespread use of calculators and computers—marvelous tools that they are—imply that mathematics education can only justify itself (aside: in as much as it prepares the learner to use a calculator or computer in an intelligent fashion, or) in as much as it is challenging, non-routine and cannot trivially be done on a calculator or computer, that is, in as much as it provides opportunities for all learners to be engaged in challenging mathematics?

How does this last question apply to in-service and future teachers? What are the needs and characteristics of teacher education with regard to challenging mathematics?

What are the implications for more challenging assessment in mathematics—both in and beyond the classroom?

Does the involvement of teachers in challenging contexts in and outside the classroom affect their behavior in their teaching mathematics?

Does the engagement of the learners in challenging contexts affect their learning ability in mathematics?

Realizations at ICME-12

The organizing team recieved 25 contributions. Each contributed paper was reviewed by at least two referees. Finally 13 papers were accepted for presentation.

Contributions from participants from all continents addressed challenging mathematical experiences in many contexts. Unfortunately one of the contributors from India and one from Iran could not participate due to lack of funding and some contributors as well as members of the organizing team were unable to attend the Congress due to programming conflicts with IMO in Argentina, and their joint interest with other TSG's especially TSG 3 (Activities and programs for gifted students).

A joint paper, given by Emily Hobbs, Kings College London, (with David Stern and Michael Obiero Maseno University, Kenya, Zachariah Mbasu, Makhokho School, Kenya, Jeff Goodman, Lycee Francais Charles de Gaulle, UK, Tom Denton, York University, Canada) focused on the motivation challenging mathematics gives to students in Kenya to continue their studies on the university level. The talk was titled" Report on the 1st Maseno Maths Camp: a mathematics popularisation event in Kenya". It introduced a mathematics camp in Kenya, developed from the need to create a forum where mathematics could be discussed and explored at the secondary level in such a way as to show that there is more to mathematics than calculations and correct answers, mentioning that the aim of the camp was to expose young minds to new ideas in mathematics relevant to the world they live in. They reported a one-week programme which was developed for school students focusing on problem solving, promoting play, experimentation, and using computers to explore mathematical ideas. Their goal was to spark a life-long love for mathematics in students, which will both improve their performance in school and increase the chances that they will pursue mathematics and science in the longer term. They mentioned that participation of school teachers was also encouraged in order to expose them to innovative teaching methods and computer resources. The structure and content of the 1st Maseno Maths Camp 2011, the future of this camp and plans for Mini Maths Camps around schools in Kenya were explained in this presentation.

From India, we learned of challenging mathematics for students from deprived backgrounds through the paper titled "Turning Tension into Thrill (of joy), Tournament as a Tool—a case study" prepared by Arundhati Mukherjee, The New Horizon School, India. The speaker was unable to attend ICME12, but her contribution remains part of the scene sketched there.

The use of the Internet in reaching out to students with mathematical challenges was highlighted by Susana Carreira, Sciences and Technology, University of Algarve, Portugal in conjunction with Nelia Amado, of the same university and Rosa Antonia Tomas Ferreira, from the University of Coimbra, Portugal, as well as Jaime Silva, also from the University of Coimbra in their paper titled "A web-based mathematical problem solving competition in Portugal: Strategies and approaches". After each problem is posted students have two weeks to submit their answers either through their personal e-mail or on the webpage platform. As speaker Susana Carreira mentioned some more details of the competition and discussed the results. Mark Applebaum of Kaye Academic College of Education, Israel, (in a paper prepared jointly with Margo Kondratieva, Memorial University, Canada and Viktor Freiman, University de Moncton, Canada,) gave results from the Virtual Mathematical Marathon, addressing student and family participation, and the general enthusiasm it has raised especially in Israel's Jewish community. The study, although not the presentation itself, concentrated on the following questions: what are participation patterns in an online problem-solving competition by boys and girls and how successful were participants according to the gender, and how their intermediate result related to their further participation in the event.

In a related theme, Yahya Tabesh from the Sharif University of Technology in Iran (along with Abbas Mousavi of the same institution) explained an internet resource that allows students to search and learn from ingenious solutions published on the Internet He introduced a new tool called "How to iSolve It!" which develops problem solving over the net as a smart system. He mentioned that iSolve is a reference system for problem solving which through a wiki on a social network would develop skills. He explained that the system could be referred on the net as a retrieval information system with smart algorithms which lead to more advanced results. Finally he introduced the iSolve system properly and some pilot results were also presented in his talk.

"Developing a much more challenging curriculum for all" was the theme treated by María de Losada of the Universidad Antonio Nariño, Bogotá, Colombia, a proposal to make challenging mathematics an integral part of the mathematics curriculum. In her presentation she reported on research regarding the construction of a much more challenging curriculum for students of grade six, based on her and her colleagues' own research as well as that of many other mathematics educators who have analyzed basic research, the panorama of failure and the spectrum of success.

Alexander Soifer from the University of Colorado, USA, brought the group's attention to the relationship between mathematical challenges and research in mathematics itself in his talk titled "The goal of mathematics education, including competitions, is to let students touch "real mathematics"; We ought to build that bridge". Professor Soifer maintained that as in "real" mathematics, Olympiad problems ought to include not just deductive reasoning, but also experimentation, construction of examples, synthesis in a single problem of ideas from various branches of mathematics, open ended problems, and even open problems. Olympiad problems should merit such epithets as beautiful and counter-intuitive. He explained problem creation and research subjects drawn from problems in competitions and connections between the following: mathematics olympiads, open problems, synthesis, construction, example, and mathematics, research.

From the Russian Federation the group had the experience of listening to the paper "South Mathematical Tournament: Tasks and Organization Hints" read by Daud Mamiy from Adyghe State University written jointly with Nazar Agakhanov, Moscow Institute of Physics and Technology. As the title implies an innovative tournament nevertheless true to the traditional roots of original and challenging problem solving gave many new ideas to the participants of TSG 34. The South Mathematical Tournament has been held in Orlyonok Children's Recreational Center on the Caucasus Black Sea coast. The authors mentioned that the tournament has been administered and organized by Adyghe State University. The report showed that this tournament is a team mathematical contest structured as a series of mathematical battles between secondary school students representing various regions of Russia, that the team members are usually students who are well prepared and possess the experience of participation in competitions at various levels. In this presentation the authors claim that every year the tournament includes some of the winners and awardees of National and International Mathematics Olympiads and candidates to Russia's national mathematics team. The scheme and some of its problems were explained in detail in the presentation.

The experience of organizing a competition simulating investment strategies for university students of the administrative sciences was recounted by Yahya Tabesh in a paper titled PitGame and prepared in collaboration with Mohammad H. Ghaffari Anjadani and Farzan Masrour,also of the Sharif University of Technology, Iran. He reported that PitGame minimizes the downsides and obstacles of contests through a sort of double creativity which seems to cause a fresh environment stressing the joy of problem solving. He mentioned that contests are mainly a competitive learning activity and it is usually on an extracurricular level that creativity and problem solving skills are developed. He claimed that competitive learning could assist educators in discovering students' abilities and creativity as well as improve students' skills, and that it would support improvement of the educational system too. He explained that this problem solving contest is based on competitive learning, game theory, and role playing.

Typical of research relating into the impact of participation in high-level mathematical problem-solving competitions, Kyung-Mi Choi of Korea and the University of Iowa, USA, in a paper prepared with Laurentius Susadya also of the University of Iowa informed the TSG34 group of "Impacts of Competition Experiences on Five IMO Winners from Korea".

Conclusions

Ali Rejali as one of the group's co-chairs opened the first session and María de Losada the second co-chair made the closing remarks. One theme running throughout the wide variety of experiences and research presented is the motivational quality of challenging mathematics on all levels, allowing each student to contribute his own ideas, benefit from the ideas of his peers and "own" the mathematics being developed through the solving of original and non-routine problems. A more challenging mathematics curriculum for all is being developed. Use of the Internet is becoming more prevalent, reaching out to students everywhere and sometimes getting families involved. Much research focuses on results analyzed by groups distinguished by gender, by level of competitivity, or by social and economic background. The heart of mathematical competitions are the problems created and posed, intimately related to the driving force in the creation of mathematics itself., but it is the students and teachers, and the transformation of their relationship to mathematics that drives the activity of this study group.

Open Access This chapter is distributed under the terms of the Creative Commons Attribution Noncommercial License, which permits any noncommercial use, distribution, and reproduction in any medium, provided the original author(s) and source are credited.

The History of the Teaching and Learning of Mathematics

Fulvia Furinghetti

History of Mathematics Teaching and Learning

History of mathematics teaching and learning is a subject that concerns two domains of research and may generate fruitful synergies between them. In 2000, during the International Symposium celebrating the centenary of the first international journal on mathematics teaching (*L'Enseignement Mathématique*), the interplay between the present educational problems in mathematics and their historical evolution through the twentieth century brought to the fore the potentialities of the field of research, "History of mathematics teaching and learning," not only for historians, but also for educators, see Coray et al. (2003). This field of research became particularly visible at ICME-10 in 2004 at Copenhagen, where a Topic Study Group (TSG 29) was dedicated to it, see Special issue (2006), Schubring and Sekiguchi (2008). History of mathematics education then became a subject of talks and workshops in various international meetings, for instance at the European Summer Universities (ESU-4 in Uppsala in 2004, ESU-5 in Prague in 2007, ESU-6 in Vienna in 2010), and at the Congresses of European Research in Mathematics Education (CERMEs). During the TSG 38 at ICME-11 in 2008 in Monterrey, research into this topic again proved its productivity, with papers presented on the history of the reform movements, on the analysis of classical textbooks, and on historical practice (inside and outside institutions), see Special issue (2009). In 2008 the celebration of the centenary of International Commission on Mathematical Instruction (ICMI), also emphasized the importance of the dialogue

Organizers Kristín Bjarnadóttir (Iceland), Fulvia Furinghetti (Italy); Team Members: Amy K. Ackerberg-Hastings (USA), Alexander Karp (USA), Snezana Lawrence (UK), Young Ok Kim (Korea); Liaison IPC Member: Evelyne Barbin (France).

F. Furinghetti (✉)
Department of Matematics, University of Genova, via Dodecaneso 35, 16146, Genova, Italy
e-mail: furinghetti@dima.unige.it

© The Author(s) 2015
S.J. Cho (ed.), *The Proceedings of the 12th International Congress on Mathematical Education*, DOI 10.1007/978-3-319-12688-3_60

between the present and the past in mathematics education, see Menghini et al. (2008). In 2006 the first international journal devoted to this field of study, the *International Journal for the History of Mathematics Education*, was launched. Recently, specialized international research symposia took place in Iceland (2009) and in Portugal (2011), see Bjarnadóttir et al. (2009, 2012).

On the occasion of ICME-10, a first international bibliography of research in the field was prepared. The bibliography is now retrievable at the following address: http://www.icme-organisers.dk/tsg29/BiblTSG.pdf.

This bibliography outlined streams in research: transmission and socio-cultural reform movements; aspects of teaching practice (textbooks, methods, teacher professional development); cultural, social and political functions of mathematics instruction; and comparative studies.

History of Mathematics Teaching and Learning at ICME-12

Following the already established tradition of research in history of mathematics education, the International Program Committee of ICME-12 included in the scientific program a TSG 35 entitled "The history of the teaching and learning of mathematics". In the announcement of the conference the following possible themes were proposed:

- changes and roles of teachers' associations
- changes of curricula in the various countries
- changes of mathematics education as a professional independent discipline
- general trends in the organizing of the lesson
- interdisciplinarity and contexts
- methods
- policies in teacher education
- reforms movements
- the cultural and social role of mathematics
- the overall impact of digital technologies in the learning and teaching of mathematics
- the role of textbooks in the teaching and learning of mathematics
- the situation of journals on mathematics education
- treatment of particular topics (geometry, algebra, etc.)

Four timeslots of one and one-half hour each were allowed to the TSG 35. Among the submitted papers the following were selected for the presentation at ICME-12, see *ICME-12 Final Program* (2012). The full texts are reported in *ICME-12 Pre-Proceedings* (2012):

- Amy Ackerberg-Hastings (UMUC and NMAH, US). Teaching Mathematics with Objects: The Case of Protractors
- Senthil Babu (French Institute of Pondicherry, India). Learning of Mathematics in Nineteenth Century South India

- Kristín Bjarnadóttir (University of Iceland, Iceland). The Implementation of the 'New Math' and its Consequences in Iceland. Comparison to its Neighbouring Countries
- McKenzie (Ken) A. Clements, and Nerida F. Ellerton, (Illinois State University, US). Early History of School Mathematics in North America, 1607–1861
- Gregg DeYoung (The American University in Cairo, Egypt). Evangelism, Empire, Empowerment: Uses of Geometry Textbooks in 19th Century Asia
- Viktor Freiman (Université de Moncton, Canada) and Alexei Volkov (National Tsing Hua University, Taiwan). Common Fractions in L.F. Magnitskii's *Arithmetic* (1703): Interplay of Tradition and Didactical Innovations
- María Teresa González (University of Salamanca, Spain). Notebooks as a Teaching Methodology: A Glance through the Practice of Professor Cuesta (1907–1989)
- Alexander Karp (Teachers College, US). Russian Mathematics Teachers: Beginnings
- Kongxiu Kuang (Southwest University, China), Yimin Xie (Jinan University, China), Qinqiong Zhang (Wenzhou University, China), Naiqing Song (Southwest University, China) Development, Problems and Thoughts of New China (PRC)'s Mathematics Education
- Snezana Lawrence (Bath Spa University, UK). The Fortunes—Development of Mathematics Education in the Balkan Societies in the 19th Century (Distributed paper)
- Lucieli M. Trivizoli (Universidade Estadual de Maringa, Brazil). Some Aspects of Scientific Exchanges in Mathematics between USA and Brazil
- Alexei Volkov (National Tsing Hua University, Taiwan). Scholarly Treatises or School Textbooks? Mathematical Didactics in Traditional China and Vietnam

Alexander Karp presented the *Handbook on the History of Mathematics Education*, edited together with Gert Schubring (University of Bielefeld, Germany and U.F.R.J., Brazil). About 40 distinguished scholars from all over the world have agreed to participate in this major project. The publisher of the book is Springer-Verlag. This *Handbook* is a real landmark in the development of the theme in question.

It is worth mentioning other activities related to the theme of TSG 35 that enriched the panorama of the themes treated.

Regular Lecture

RL5–9, Marta Menghini (University of Rome La Sapienza, Italy). From Practical Geometry to the Laboratory Method: The Search for an Alternative to Euclid in the History of Teaching Geometry. See the text in *ICME-12 Pre-Proceedings*.

Posters and Oral Presentations

* Tanja Hamann and Barbara Schmidt-Thieme (Germany). "Macht Mengenlehre Krank?": New Math at German Primary Schools
* Sanae Fujii (Japan). Mathematics Teaching Using "Sanpou shojyo (Algorithm Girl)" for Junior High School Students
* Sung Sook Kim (Korea). Seok-Jeong Choi and Magic Squares
* Shinya Itoh (Japan). Structure of Didactical Principles in Hans Freudenthal's Didactics of Mathematics, Oral Presentation.

The abstracts are in *ICME-12 Pre-Proceedings*. The contributions cover important subjects of mathematics education:

* physical devices for teaching mathematics
* teacher professional development
* systems of instruction
* exchanges between countries
* reforms
* textbooks
* treatment of parts of mathematics
* eminent people in mathematics education.

Both specificity of national contexts and internationality of themes inherent in mathematics education were treated in the presentations and the discussions.

Final Remarks

We know that the vision and mission that inspired the journal *L'Enseignement Mathématique* and afterwards ICMI enhanced internationalization and communication in the world of mathematics education, see Furinghetti (2003). These goals were pursued throughout the ICMI's existence and, in particular, ICME conferences have been a powerful means for realizing them, see Furinghetti and Giacardi (2008). TSG 35 and the related activities are an example of internationalization and communication among researchers. All five inhabited continents have presented contributions to the history of mathematics education: Africa (Egypt), Asia (China, India, Japan, Korea, Taiwan), Europe (Germany, Iceland, Italy, Spain, UK), North America (Canada, US), and South America (Brazil).

In spite of the limitation of the scheduled time, the contributions at ICME-12 on the history of mathematics teaching and learning have allowed reflection on the double aspect of this topic. On the historical side, they showed that the present situation of mathematical education does not come out of the blue but has old roots and accompanied the growth of civilizations and societies. On the educational side, history offers to educators a different point of view for looking at educational problems and provides insights into possible solutions. Then, really, we may see the history of mathematics education as a bridge between the past and the future.

Open Access This chapter is distributed under the terms of the Creative Commons Attribution Noncommercial License, which permits any noncommercial use, distribution, and reproduction in any medium, provided the original author(s) and source are credited.

References

Bjarnadóttir, K., Furinghetti, F., & Schubring, G. (Eds.) (2009). *"Dig where you stand".* *Proceedings of the conference on On-going research in the History of Mathematics Education.* Reykjavik: University of Iceland – School of Education.

Bjarnadóttir, K., Furinghetti, F., Matos, J. M., & Schubring, G. (Eds.) (2012). *Proceedings of the second International Conference on the History of Mathematics Education.* Caparica, Portugal: UIED.

Coray, D., Furinghetti, F., Gispert, H., Hodgson, B. R., & Schubring, G. (Eds.) (2003). *One Hundred Years of L'Enseignement Mathématique*, Monographie n. 39 de *L'Enseignement Mathématique.*

Furinghetti, F. (2003). Mathematical instruction in an international perspective: the contribution of the journal *L'Enseignement mathématique.* In D. Coray, F. Furinghetti, H. Gispert, B. R. Hodgson, & G. Schubring (Eds.), *One hundred years of L'Enseignement Mathématique*, Monographie n. 39 de *L'Enseignement Mathématique*, 19-46.

Furinghetti, F., & Giacardi, L. (Eds.) (2008). *The first century of the International Commission on Mathematical Instruction (1908-2008).* http://www.icmihistory.unito.it/

ICME-12 Pre-Proceedings (2012). http://www.icme12.org/ Retrieved 30 November 2014.

ICME-12 Final Program (2012). http://www.icme12.org/ Retrieved 30 November 2014.

Menghini, M., Furinghetti, F., Giacardi, L., & Arzarello, F. (Eds.) (2008). *The first century of the International Commission on Mathematical Instruction (1908-2008). Reflecting and shaping the world of mathematics education.* Rome: Istituto della Enciclopedia Italiana.

Schubring, G., & Sekiguchi, Y. (2008). Report on TSG 29: The history of the teaching and the learning of mathematics. In M. Niss (Ed.), *Proceedings of ICME-10 2004 (10th International Congress on Mathematical Education* (pp. 422-425). Roskilde: IMFUFA, Roskilde University.

Special issue (2006). Schubring, G. (Ed.). History of teaching and learning mathematics. *Paedagogica Historica, 42*(4&5). [Proceedings of TSG 29 at ICME-10].

Special issue (2009). d'Enfert, R., & Ruiz, A. (Eds.). *International Journal for the History of Mathematics Education, 4*(1). [Proceedings of TSG 38 at ICME-11].

The Role of Ethnomathematics in Mathematics Education

Pedro Palhares and Lawrence Shirley

Report

Kay Owens (Charles Stuart University, Dubbo, NSW, Australia) presented a paper illustrating how schools can change when funds are available to assist schools and communities to implement appropriate and effective professional development, to establish partnerships between school and community, to revise teaching approaches and curriculum, to overcome disadvantage, and to value family and Aboriginal cultural heritage. She stressed that the people involved and their planning are critical for transformation. The schools were in a Smarter Stronger Learning Community so they supported each other across schools but other programs in the various schools were also important in achieving change.

Zhou Chang-jun, Shen Yu-hong, Yang Qi-xiang (Dehong Teachers' College) presented a paper about Dai ethnic mathematical culture, which is an important part of Dai ethnic culture. Mathematical elements show in their daily life. Through a research project of the Yunnan Dehong Dai people in southwest China, they collected the first-hand information, tried to do a small investigative study, and collected mathematics teaching resources that are useful to primary and secondary schools students on mathematics learning in this minority areas.

Organizers Co-chairs: Pedro Palhares, Lawrence Shirley; Team Members: Willy Alangui (Philippines), Kay Owens (Australia), Paulus Gerdes (Mozambique), Ho Kyung Ko (Korea); Liaison IPC Member: Bill Barton (New Zealand).

P. Palhares (✉)
University of Minho, Braga, Portugal
e-mail: palhares@ie.uminho.pt

L. Shirley
Towson University, Towson, USA
e-mail: LShirley@towson.edu

© The Author(s) 2015
S.J. Cho (ed.), *The Proceedings of the 12th International Congress on Mathematical Education*, DOI 10.1007/978-3-319-12688-3_61

575

Annie Savard (McGill University) discussed problems of bridging the Inuit culture of northern Canada with the official and cultural requirements of Canada's school mathematics curriculum, especially when goals seem to clash.

Igor Verner, Khayriah Massarwe and Daoud Bshouty (Technion – Israel Institute of Technology) presented a paper discussing pathways of creativity and focusing on the one going through practice in creation and analysis of useful and mathematically meaningful artifacts. They propose to involve prospective teachers in practice of construction and analysis of geometric ornaments from different cultures as well as in teaching geometry. They considered perceptions and attitudes that triggered students' creative learning behavior in this context.

Milton Rosa and Daniel Clark Orey [Universidade Federal de Ouro Preto (UFOP)] think that the application of ethnomathematical techniques and tools of modeling allows us to examine systems taken from reality and offers us an insight into forms of mathematics done in a holistic way. According to them, the pedagogical approach that connects a diversity of cultural forms of mathematics is best represented through ethnomodeling, a process of translation and elaboration of problems and the questions taken from academic systems. Seen in this context, they attempted to broaden the discussion of possibilities for the inclusion of ethnomathematics and associated ethnomodeling perspectives that respect the social diversity of distinct cultural groups with guarantees for the development of understanding different ways of doing mathematics through dialogue and respect.

Karen Francois [University Brussels (Vrije Universiteit Brussel)] and Rik Pinxten (University Ghent) started by the statement from the Vygotsky and the Cultural psychology approach (M. Cole) that 'learning is situated, socioculturally contextualized'. Learning happens in the space of background/foreground (of the learner) in his or her particular environment of experience. Math learning implies an implicit understanding, categorizing and conceptualization of reality. e.g., set theory implies intrinsically a part-whole framing of reality. They think that the tremendous dropout from math classes and the structural gap between good and bad performers (PISA) is caused by disregarding the linguistic and socioculturally formatted background/foreground of the learners. They want to use anthropological study in the classroom to know/map the child's background/foreground and adapt the entry into mathematics courses accordingly, hence their option for multimathemacy.

Maria do Carmo S. Domite (Faculty of Education, University of São Paulo) (electronically) presented an attempt to make possible an approach between ethnomathematics and the mathematics learning processes in the scholar context—however it does this from an ethnomathematician's point of view, not that of a Cognitive Psychology studious. She therefore focused on two notions of the mathematics education processes: the notion understood as the student's "prerequisite" and the notion of the teacher's "listening". She brought to the centre of discussion that the teacher should know to understand the students' initial mathematics knowledge—how he/she uses them-, as well as know how to listen to what the students have to say—respecting the cultural and social differences in order to help them build a more critical and elaborate thinking about mathematics ideas.

Andrea V. Rohrer (Universidade Estadual Paulista) and Gert Schubring (Universidade Federal do Rio de Janeiro) started to remember that since the creation of the International Study Group on Ethnomathematics, several researchers have debated on how could or should a theory of ethnomathematics exist, and, if so, how it is to be conceptualized. So far, there exists no consensus on how this theory should be defined. During the last International Conference on Ethnomathematics (ICEM-4) in Towson, Maryland (July, 2010), Rik Pinxten emphasized on the necessity of reopening this debate. Ethnomathematics will only be acknowledged by other scientific communities if we, as ethnomathematicians, are able to establish a proper conceptualization of this field of study. They presented one possible approach to a conceptualization of a theory of ethnomathematics a theory that needs to be regarded as an interdisciplinary discipline that covers theories from both the exact and social sciences.

Alexandre Pais (Aalborg University) and Mônica Mesquita (University of Lisbon) consider that the push to marry off local and school knowledge has been a growing concern within educational sciences, particularly in mathematics education where a field of studies by the name of ethnomathematics has been producing research around the uses people do of mathematics outside school's walls. Notwithstanding the good will of educational agents in bringing to schools local knowledges, criticisms have been made on the sometimes naive way in which such a bridge is theorized and implemented. After a brief description of these criticisms, they presented the Urban Boundaries Project as an attempt to avoid the inconsistencies of schooling, and the promotion of a non-scholarized ethnomathematics.

Joana Latas (EBI/JI de Aljezur, CIEP-U. Évora) and Darlinda Moreira (Universidade Aberta) claim that the integration of cultural aspects in curricula is a means of legitimizing students' experiences and of answering to the cultural diversity in favor of a meaningful mathematical learning. (e.g. Bishop 2005; Gerdes 2007; Moreira 2008). They attempted to highlight the role of cultural mathematics in the development of the predisposition to establish mathematical connections. Such an objective was framed in a broader investigation (Latas 2011) in which a curricular project was developed, whose conceptualization followed an ethno-mathematical approach. The results suggest that students: (i) appropriated cultural distinct practices through the relation that they established with their previous knowledge; (ii) gradually revealed a greater predisposition to establishing mathe-matical connections; (iii) deepened local and global mathematical knowledge in the interaction between both dimensions.

Roger Miarka (Universidade do Estado de Santa Catarina, Brazil) and Maria Aparecida Viggiani Bicudo (Universidade Estadual Paulista, Brazil) presented a paper, based on a PhD research, aiming to discuss the conception of mathematics, and its developments in terms of methodology, of five preeminent ethnomathe-matics researchers: Bill Barton (University of Auckland, New Zealand), Eduardo Sebastiani Ferreira (Universidade Estadual de Campinas, Brazil), Gelsa Knijnik (Universidade do Vale do Rio Sinos, Brazil), Paulus Gerdes (Centro Moçambicano de Investigação Etnomatemática, Mozambique) and Ubiratan D'Ambrosio (Universidade Bandeirante de São Paulo, Brazil). The research was carried out under a phenomenological perspective, and its methodology involved an interview with each of the above-mentioned researchers. Theses interviews were analyzed hermeneutically, and through phenomenological reductions, thematic categories were articulated. In this presentation they brought the category about the presence of mathematics within ethnomathematics.

Also, there were several posters and short presentations, including reports from Portugal (Pedro Palhares), Tibet (Xiawu Cai Rang), Nepal (Bal Luitel and Amril Poudel), Philippines (Rhett Latorio), China (Xueying Ji), Mozambique (Marcos Cherinda) and Zambia (Mitsuhiro Kimura), especially on details of local mathematics and applications of local culture in school mathematics.

Marcos Cherinda made a special presentation, inviting participants (and all interested in ethnomathematics) to attend the Fifth International Conference on Ethnomathematics (ICEM-5), to be held in July 2014 (specific date to be announced), in Chidenguele, Gaza, Mozambique.

Participants

There were thirty-five participants (from twenty-two countries) in the TGS-36 sessions: Maria Aparecida Bicudo (Brazil), Bill Barton (New Zealand), Marcos Cherinda (Mozambique), Sandy Dawson (USA), Tournés Dominique (France), Cris Edmonds-Wathen (Australia), Karen François (Belgium), Kgomotso Garegae (Botswana), Kangu Hyun Jin (Korea), Jason Johnson (United Arab Emirates), Traore Kalifa (Burkina Faso), Jiyeon Kim (Korea), So Yoang Kim (Korea), Mitashiro Kimura (Japan), Ho Kyung Ko (Korea), Rhett Latonio (Philippines), Joana Latas (Portugal), Chan Gyu Lee (Korea), Bal Luitel (Nepal), Danilo Mamangon (Micronesia), Roger Miarka (Brazil), Epi Moses (Palau), Kay Owens (Australia), Alexandre Pais (Denmark), Pedro Palhares (Portugal), Amrit Poudel (Nepal), Andrea Rohrer (Brazil), Annie Savard (Canada), Lawrence Shirley (USA), Edmir Terra (Brazil), Koichi Tomita (Japan/Malaysia), Rhoda Velasques (Philippines), Igor Verner (Israel), Lim Byong Yang (Korea), Hossein Zand (United Kingdom).

Open Access This chapter is distributed under the terms of the Creative Commons Attribution Noncommercial License, which permits any noncommercial use, distribution, and reproduction in any medium, provided the original author(s) and source are credited.

Theoretical Issues in Mathematics Education: An Introduction to the Presentations

Angelika Bikner-Ahsbahs and David Clarke

Report

In the last decade, the issue of theories has been raised more than once in international conferences on mathematics education (e.g. PME 2005, 2010; CERME since 2005; ICME 11). Since 2006, a European group for the networking of theories has researched the question how mathematics education can deal with different theories (Bikner-Ahsbahs et al. 2010). All these events have shown the diversity of theories to be inherent in mathematics education. TSG 37 of ICME 12 has gathered an up-to date version of the state of the reflection on theories with respect to the theoretical questions and underpinnings of the international field, at the same time stimulating insightful exchanges and discussions crossing theoretical cultures within mathematics education. Group discussions addressed the following issues: What theories do we need in mathematics education? What do they have to cover according to the conditions, roles and functions of theory use and development, and how can we deal with the diversity of theories in a scientific fruitful way?

Radford (2008) provided a meta-theoretical frame for theories. Referring to Lotman (1990), Radford characterized the space of theory cultures as a semiosphere: a dynamically evolving space. According to Radford, theory is a dynamic way of understanding, provided and performed on the basis of a triplet (P, M, Q):

Organizers Co-chairs: Angelika Bikner-Ahsbahs (Germany), David Clarke (Australia); Team Members: Cristina Sabena (Italy), Minoru Ohtani, (Japan), Gelsa Knijnik (Brazil), Jin Young Nam (Korea).

A. Bikner-Ahsbahs (✉)
University of Bremen, Bremen, Germany
e-mail: bikner@math.uni-bremen.de

D. Clarke
University of Melbourne, Melbourne, Australia
e-mail: d.clarke@unimelb.edu.au

© The Author(s) 2015

S.J. Cho (ed.), *The Proceedings of the 12th International Congress on Mathematical Education*, DOI 10.1007/978-3-319-12688-3_62

the set P of principles of the theoretical culture, the set M of methodologies that refer to P including methods and the set Q of paradigmatic questions in the core of the theory, its P and M. The developmental dynamic is constituted by research and the exchange of research results R referring back to the triplet (P, M, Q) (Radford 2012). This understanding of the terms *theory* and *semiosphere* provides a framework for understanding the connection of theories as exchanges and dialogues between their parts. Connections can be created among the theories parts [(P, M, Q), R] and structured according to their degree of integration in the landscape of networking strategies: *understanding and making understandable, comparing and contrasting, combining and coordinating, and locally integrating and synthesizing* (Bikner-Ahsbahs and Prediger 2010). The first two pairs can always be done but the third and the fourth pair can only be executed if the principles of the theories are close enough. TSG 37 offered an introduction about the networking of theories, eight long presentations, five short presentations and short statements about the posters within four sessions. The first session involved questions concerning how theories from outside mathematics education might fruitfully inform the dynamics of research within mathematics education; in particular, it addressed the challenge of identifying theories suitable for use in mathematics education, contrasting the treatment of particular constructs relevant to mathematics education within two or more theories, suggesting inadequacies in the capacity of currently available theories to meet the needs of mathematics education, and recommending what developments are required. It was established early in the discussion that no single theory can claim to be comprehensive and so all theories are consequently partial and selective in their focus and the phenomena they describe and attempt to explain (Clarke 2011). Two presentations were discussed. Knijnik positions culture at the heart of teaching and learning mathematics and addresses the issue of Ethno-mathematics as an offer to think of cultural differences in grammar and logic. Her approach can be regarded as a coordination of two background theories rooted in the work of Wittgenstein and his language games and of Foucault and his work on how discourse establishes truth in the culture. Pais and Valero point to the demand of mathematics education for all and recommend the inclusion of economic considerations in the theories employed in mathematics education. According to Pais and Valero, current theories do not accommodate these concerns. The two contributions offer different perspectives: according to Pais and Valero, we have to be more open in the direction of political and economic value of mathematics, and, according to Knijnik, we must see philosophies of teaching and learning as parts of the distinct cultures from which they have developed and in which they are applied.

The second session investigated the role and function of theories in mathematics education (and mathematics education research), their capacity to provide insight into one or more different contexts or issues in mathematics education, and the methodological entailments of selecting particular theories in the process of research and design. Drawing on Vygoskian theory, Albert positioned learning mathematics as a cultural-historical activity mediated by a sign system and applied this to serve teaching practice by the use of algebra tiles. Hatfield asked how a theory could be built to capture lived mathematical experience in order to investigate this phenomenon. He included two views, the phenomenological and the constructivist view, to start building a theory. Trninic and Kim adopt a radical position on embodied cognition. They regard learning mathematics to be cognitively embodied and employ this in the design of computer-based environments. To do so, theory and design have to co-inscribe, e.g. they mutually inform and entail each other. These three presentations accorded theory different roles and functions: (1) theory as a source of models to be applied to the practice of teaching and learning, (2) theory constructed to understand a specific phenomenon, and (3) theory as informing instruction to be co-developed with design towards a specific goal. The researcher's perspective on learning mathematics and on the aim and function of research determines what kind of theory is considered suitable. Hatfield grappled with the new idea of lived mathematical experience within learning suggesting that this focus has to be intensively theorized before it brings a theory to the fore. Trninic and Kim reconceptualised embodied cognition by situating it in design experiments and by looking at the theory in a new way. Albert used research results from a background theory with a long tradition (Vygotski's social psychology) employed in mathematics education to inform practice.

Session three discussed the question of how to deal with different theories in a scientific way, addressing the challenge of utilising the results of research studies in mathematics education undertaken using different theories. The generic term "networking" was employed to include strategies such as connecting, comparing, contrasting, combining, coordinating, integrating, and synthesising. Such strategies are intended to provide heightened insight into a complex setting. The session involved reporting examples of the networking of theories, their limits and their potential for advancing mathematics education. Three presentations addressed these issues. Even investigated the same data set from two philosophical traditions: constructivist theory, investigating learning by looking at cognitive development; and activity theory, used to investigate the teacher's participation. She showed that the object of investigation and the research questions were different according to the particular theory and therefore the results of the analysis and the answers to the research questions were also different, but complementary and mutually informative. She inferred from this that the use of more than one theory demands parallel lines of research and meta-theoretical exchanges. Trigueros et al. undertook a theoretical study to investigate the different meanings of *mathematical object* in Action-process-object-schema (APOS) theory and in the onto-semiotic approach. She showed that some concepts within one theory could be interpreted by the other, suggesting that these concepts could be associated with measures or results amenable to comparison, that is, they were commensurable, whereas results associated with other concepts might be incommensurable but compatible because they were not mutually contradictory; but could be seen as disjoint or complementary. The relationship between different theories cannot be simplistically categorised as either commensurable/incommensurable or as compatible/incompatible. Theories may partly overlap and may be mutually informative to some extent. The issue of limits was raised by the third presentation from Kidron and Monaghan, which discussed the complexity of dialogue between theories. In this presentation, Kidron showed that dialogues as exchanges between theoretical cultures can be regarded on two levels: (i) the cultural level of theories participant in research processes—a possible mechanism to network theories; but also (ii) the individual level, where researchers work with different theories within one project and must forge connections between theories in the process of constructing their findings. Both data and results are constructed within research through methodologies that reflect the choice of theories. In this way, the networking (combination or juxtaposition) of theories might lead to uncover blind spots in the making of data and in the analyses, clarify the theories' boundaries but also advance research through enriching results.

The use of multiple theories (and associated parallel analyses) in a single research project can serve several purposes (Clarke 2011):

- By addressing different facets of the setting/s and providing a richer, more complex, more multi-perspectival portrayal of actors and actions, situations and settings;
- By offering differently-predicated explanations and differently-situated propositions;
- By increasing the authority of claims (and instructional advocacy), where findings (both explanations and emergent propositions) were coincident across analyses;
- By qualifying the nature of claims, where findings of the parallel analyses were inconsistent or contradictory (cf. Even's analyses of "the same data");
- By providing a critical perspective on the capacity of each particular theory to accommodate and/or explain data related to the same events in the same setting.

In session four, the evolving discourse could be used to discuss short presentations, reporting studies with multiple theoretical perspectives on mathematical imagination (Aralas), on mathematical visuality (Flores), and on concept formation (Rembowski). Rosa and Aparecida addressed philosophical considerations regarding mathematical technology and Kusznirczuk suggested according theories the status of organising principles for coordinating the objects that populate our discourses and our methods.

In summary: Since research questions are intimately connected with the theoretical frameworks in which they are elaborated, it may appear problematic to use different theories to answer the same research question. However, different theories may usefully address different questions about the same setting (e.g. the mathematics classroom) or even the same issue (e.g. the instructional use of representations). Researchers should draw on the expertise of the various theoretical cultures to enrich the general discourse of the mathematics education community and respond to society's major questions at an appropriate level of complexity. The discussion raised the question up to what point researchers might be able to consciously choose a theory or a theoretical paradigm for their research and it brought to the fore the criteria under which theories might be evaluated: the dichotomy true/ wrong was contrasted with the useful/useless one, and the concepts of validity and viability were considered. The work carried out in this TSG has constituted a small but solid step in fighting back this danger for mathematics education.

Open Access This chapter is distributed under the terms of the Creative Commons Attribution Noncommercial License, which permits any noncommercial use, distribution, and reproduction in any medium, provided the original author(s) and source are credited.

References

Bikner-Ahsbahs, A., & Prediger, S. (2010). Networking of Theories—An Approach for Exploiting the Diversity of Theoretical Approaches. In B. Sriraman & L. English (Eds.), *Theories of mathematics education: seeking new frontiers* (pp. 479-512). New York: Springer, Advances in Mathematics Education series, Vol. 1.

Bikner-Ahsbahs, A., Dreyfus, T., Kidron, I., Arzarello, F., Radford, L., Artigue, M., & Sabena, C. (2010). Networking of theories in mathematics education. In M. M. Pinto, & T. F. Kawasaki (Eds.), *Proc. 34th Conf. of the Int. Group for the Psychology of Mathematics Education*, vol. 1 (pp. 145-175). Belo Horizonte, Brazil: PME.

Clarke, D. J. (2011). A Less Partial Vision: Theoretical Inclusivity and Critical Synthesis in Mathematics Classroom Research. In J. Clark, B. Kissane, J. Mousley, T. Spencer & S. Thornton (Eds.) *Mathematics: Traditions and [New] Practices. Proceedings of the 2011 AAMT–MERGA conference (pp. 192-200).* Adelaide: AAMT/MERGA.

Lotman, Y. (1990). *Universe of the mind. A semiotic theory of culture.* London: I. B. Taurus.

Radford, L. (2008). Connecting theories in mathematics education: Challenges and possibilities. *ZDM Mathematics Education*, 40, 317-327.

Radford, L. (2012). *On the growth and transformation of mathematics education theories.* Paper presented at the International Colloquium: The Didactics of Mathematics: Approaches and Issues. A homage to Michèle Artigue. http://www.laurentian.ca/educ/lradford. Accessed 29 October 2013.

Part IX
Discussion Groups

Current Problems and Challenges in Non-university Tertiary Mathematics Education (NTME)

James Roznowsk and Huei Wuan Low-Ee

Report

The countries represented during the discussion group's meetings included: Australia, Canada, China, Iran, Israel, Philippines, Singapore, the United Arab Emirates, and the United States. The types of institutions varied among the countries and included: two and four-year vocational institutions, community colleges, programs related to retraining adults, and a one-year preparatory program in Israel for individuals coming out of military service and interested in attending university.

The discussion group team suggested five questions that were developed from the proposal. These dealt with:

- Student placement;
- Student learning of mathematics;
- Use of technology in teaching, learning, and assessment of mathematics;
- Classroom research; and
- Faculty development.

The attendees were asked to select the topics they were most interested in discussing. The topics with the most interest were technology, classroom research, and faculty development.

Organizers Co-chairs: James Roznowsk (USA), Low-Ee Huei Wuan (Singapore); Team Members : Vilma Mesa (USA), Steve Krevisky (USA), Auxencia Limjap (Philippines); Liaison IPC Members: Johann Engelbrecht (South Africa).

J. Roznowsk (✉)
Harper College, Palatine, USA
e-mail: jroznowski@harpercollege.edu

H.W. Low-Ee
Singapore Polytechnic, Singapore, Singapore
e-mail: lowhw@sp.edu.sp

© The Author(s) 2015

S.J. Cho (ed.), *The Proceedings of the 12th International Congress on Mathematical Education*, DOI 10.1007/978-3-319-12688-3_63

To facilitate the discussion of issues related to technology, attendees were asked to answer the following question: What types of technology are available to teach, learn and assess mathematics at your institution? The notes from this discussion follow.

Notes on Uses of Technologies at Non-university Tertiary Institutions

Types of technologies:

- Graphing calculator
- Computer software (e.g. Excel™)
- CAS—good for learning math
- Internet resources
- Projectors and presentation devices
- Online learning platforms (homework problems)—there can be differences between the way in which a topic is explained in online tutoring software and how the instructor teaches, especially when the online tutoring comes with the textbook. In Singapore, instructors typically develop their own materials.
- Course management software (e.g. Blackboard™, WebCT™, Angel™)

Graphing calculators are handled with a variety of approaches around the world —either not allowed, allowed, or required. Singapore—graphing calculators are compulsory for mathematics at A-levels, and also adopted by high schools offering integrated programmes, the calculators are reset before the start of the A-levels mathematics papers. US—graphing calculator may be required (may be rented) in community college, but then may not be allowed at universities. Canada—required at high school, not permitted at most post-secondaries. Philippines—had assessments both with and without graphing calculators within a single course. Used among pre-service teachers in a course on technology. Worksheets are given to guide the students to gain further understanding from the work done with the hand-held calculators.

The discussion related to technology also led to a discussion of developmental mathematics and different terminologies used in different countries or regions. At some institutions "intermediate" algebra and "college" algebra are considered separate topics. Some states in the US are not having developmental math taught in lecture format, and students are being placed in front of computers instead. There is some concern about how well weak students will do under these conditions. There are similar issues in the Philippines. Universities are phasing out foundations mathematics. In Singapore, many of the students who opt to join the five polytechnics are weaker in math. At Singapore Polytechnic, weaker students are given customised CD-ROMs and are required to complete exercises before starting at the polytechnic. They are given assessment test upon arrival, and that determines whether or not they are required to attend face-to-face remedial math sessions. Recorded lectures given by selected lecturers to supplement the regular lectures are also provided in Blackboard.

For student teachers in Australia the problem is that when teachers first start, they usually reproduce what they saw when they were in school, so in training teachers, lecturers try to introduce more and different tools that are available.

The second session of DG1 focused on classroom research. Individuals did 5-min presentations on the different aspects of classroom research. These included research projects on student/teacher interaction, assessment of student learning, and potential research topic areas. After the presentations, the group discussed the expectations of instructors at their varied institutions with regard to conducting classroom research. At many institutions in the United States and Canada, research is not an expectation of faculty at community or vocational colleges. They are often not given the time or resources needed to do classroom research. In fact, at some institutions, instructors who conduct classroom research may be thought of as taking time away from their assigned responsibilities.

An attendee from Singapore shared information about a new requirement of faculty at her institution to participate in research projects. Questions by others involved faculty reaction to this expectation and professional development for faculty who are new to such research. Information about classroom research in Singapore was distributed to the discussion group attendees after ICME. It was offered as a model that can be adapted to encourage classroom research by instructors at institutions in a variety of countries.

Before the end of the session, participants reviewed ways to continue the discussion beyond ICME 12. The session closed with agreement among attendees that the discussion was of great value and arrangements were developed to make sure a proposal for the continuation of the discussion group would be submitted for ICME 13.

Open Access This chapter is distributed under the terms of the Creative Commons Attribution Noncommercial License, which permits any noncommercial use, distribution, and reproduction in any medium, provided the original author(s) and source are credited.

Creativity in Mathematics Education

Hartwig Meissner

Report

The Discussion Group 2 on Creativity in Mathematics Education at ICME-12 in Seoul, affiliated with MCG, was a great success. Co-chaired by Board members Emily Velikova and Vince Matsko, the group focused on issues relating to creativity in the classroom as well as training both pre-service and in-service teachers regarding effective ways of fostering creativity.

DG 2 faced difficulties in communicating with ICME organizers, and as a result, participants were not able to see the proposals before the conference. As a result, the chairs decided to begin each of the two ninety-minute sessions with brief summaries of the papers delivered by their authors. These were both lively and informative, and gave the 100+ participants in DG 2 a chance to be brought up to speed.

The sessions focused on the following questions, developed by the Board earlier in the year:

- What does creativity mean in the process of teaching and learning mathematics?
- How can we develop or stimulate creative activities in and beyond the mathematics classroom?
- How might we balance mathematical skill training and mathematical creativity?
- What should be done in teacher training programs at the pre-service and in-service levels to foster creativity in the classroom?

Organizers Chair: Hartwig Meissner (Germany); Team Members: Emiliya Velkova (Bulgaria), Jong Sul Choi (Korea), Vince Matsco (USA), Mark Applebaum (Israel), Ban Har Ywap (Singapore); Liaison IPC Member: Bernard Hodgson (Canada).

H. Meissner (✉)
University of Münster, Münster, Germany
e-mail: meissner@uni-muenster.de

© The Author(s) 2015
S.J. Cho (ed.), *The Proceedings of the 12th International Congress on Mathematical Education*, DOI 10.1007/978-3-319-12688-3_64

(Note: a link to the DG 2 website, including a full description of the Discussion Group as well as all the submitted papers, is accessible from the MCG website.)

The first session addressed the first two questions, while the second centered around the last two. The chairs subdivided the first two questions into six, so that smaller discussion groups could be formed. DG 2 participants selected the question they wished to discuss, and elected a representative who summarized the discussion at the end of the session. Many participants remarked on quality of the discussions, and all were stimulated by the sharing of ideas.

The second session was rather smaller than the first, occurring on the last full day of ICME-12. As a result participants voted to have a large group discussion rather than breaking into smaller groups. This proved to be effective—and even though participants were tired after a hectic week, it proved difficult to make sure everyone had a chance to speak. Those involved had a real passion for creating engaging activities in the mathematics classroom, and there was no shortage of ideas to share.

Of course, in discussions like these, more questions are raised than are answered. These questions came up as a response to concerned teachers truly wanting to be more creative in the classroom. Among the questions raised by the DG 2 participants were:

- How do we decrease pressure on students so that they are more free to be motivated and involved in mathematics?
- How can we use technology to allow students to demonstrate originality, flexibility, and fluency of thought?
- How can we develop creativity within a pre-service teacher's university experience?
- Given we believe that *all* students can be creative, how can we create opportunities for students to do so?
- How can we deliberately foster creative thinking to encourage innovation?
- How can we provide accessible resources for teachers so that they may more easily bring creative activities into their classrooms?
- How can we change the climate of university education departments so that developing creativity in teachers is valued and addressed in the curriculum?
- How can creativity in mathematics education be made a priority at a regional or national level?

Of course none of these questions has an easy answer. But one or more of them might be suitable for a discussion forum or a special session of a conference on education. We welcome contributions to this newsletter from mathematics educators who have successfully answered one of these questions either in their classroom, or who made an impact regarding one of these questions on a local, regional, or national level.

Open Access This chapter is distributed under the terms of the Creative Commons Attribution Noncommercial License, which permits any noncommercial use, distribution, and reproduction in any medium, provided the original author(s) and source are credited.

Issues Surrounding Teaching Linear Algebra

Avi Berman and Kazuyoshi Okubo

Report

Linear Algebra is one of the most important courses in the education of mathematicians, scientists, engineers and economists. DG 3 was organized by The Education Committee of International Linear Algebra Society (ILAS) in order to give mathematicians and mathematics educators the opportunity to discuss several issues on teaching and learning Linear Algebra including motivation, challenging problems, visualization, learning technology, preparation in high school, history of Linear algebra and research topics at different levels. Some of these problems were discussed. Around 50 participants participated in the discussion.

Motivation

The interest in learning linear algebra can be motivated by real life (high tech) applications and by challenging problems. The following examples were mentioned:

Organizers Co-chairs: Avi Berman (Israel), Kazuyoshi Okubo (Japan); Team Members: Steven Leon (USA), Sepideh Stewart (New Zealand), Sang-Gu Lee (Korea), David Strong (USA); Liaison IPC Member: K. (Ravi) Subramaniam.

A. Berman (✉)
Israel Institute of Technology, Haifa, Israel
e-mail: berman@technion.ac.il

K. Okubo
Hokkaido University of Education, Sapporo, Japan
e-mail: okubo.kazuyoshi@s.hokkyodai.ac.jp

© The Author(s) 2015
S.J. Cho (ed.), *The Proceedings of the 12th International Congress on Mathematical Education*, DOI 10.1007/978-3-319-12688-3_65

Google's Page Rank; Edge detection methods; Neural networks; Problems in graph theory; Properties of the Fibonacci sequence; Computer games—"Fiver" (see http://www.math.com/students/puzzles/fiver/fiver.html), Another popular game that can be solved by linear algebra (see http://matrix.skku.ac.kr/bljava-v1/Test.html).

Technology

Sang-Gu Lee from Korea described the work on Sage. He talked about what can be done in Mobile learning Environment and mentioned his coming regular lecturer on Mobile LA. He mentioned Bill Barton' words on Monday "We know that Education using ICT will improve the quality of Math Education. But it is clear that we are not THERE yet" (ICT Revolution in Math Education). http://www. sciencetimes.co.kr/article.do?todo=view&atidx=0000063949. Avi Berman from Israel described the use of clickers to promote the students' involvement and the communication between them and the professors.

J. L. Dorrier from France raised a question "Is it a good idea to use computers?" Sang-Gu said "If we do it properly, there is no reason for not using technology for education. I am used to teach and discuss in a traditional way, but I encourage our students to use whatever they can use for better understanding of Linear Algebra." Dorrier said that he prefers to wait with the use of computers to a later stage. Other participants said that they use Maple from the beginning of the course. Ludwig Paditz from Germany pointed out that technology sometimes gives incorrect results. "Modern calculators (CAS) resolve some problems but sometimes incorrect". It is important that students check if the computed results make sense.

Understanding

Megan Wawro from the USA described her research with Chris Rasmussen on how students engage with eigenvalue-eigenvector system making connections with functions.

Sepideh Stewart (New Zealand, USA) described her PhD thesis on teaching and learning Linear Algebra. She made a framework using Tall's embodied symbolic and formal words of mathematical thinking in conjunction with Ed Dubinsky's APOS theory. She found that the majority of the students were comfortable in the symbolic world but struggled with formal definitions and theorems. She also found that embodied (giving body to an abstract idea) thinking helped some students to have a better grasp of Linear Algebra. Her thesis is available on the web.

Teaching Mathematics and Engineering Students

Saeja Kim at U Mass, Dartmouth said she is not in favor of introducing Linear Algebra in an abstract way. She suggested to start with linear equations pointing out that most students are struggling with even the basics ideas. Avi said he also starts with linear equations but quickly moves into more abstract theory. Dorrier said that Linear Algebra is not about solving systems. The main thing is having mental views. He believes that abstraction ability of math majors should be developed. Avi said that at the Technion this is done also for students of electrical engineering and computer science.

Michelle Zandieh (Arizona) and many other participants emphasized the importance of Geometry and Visualization.

Chris Rasmussen (San Diego) asked how the differential equations and the linear algebra courses can be combined. The question was answered by a presentation by Karsten Schmidt from Denmark titled: "Revising the first semester math course for engineering students".

The names of the participants and photos from the two sessions can be found in http://matrix.skku.ac.kr/2014-Album/ICME12-DG3-report-v1.htm.

Open Access This chapter is distributed under the terms of the Creative Commons Attribution Noncommercial License, which permits any noncommercial use, distribution, and reproduction in any medium, provided the original author(s) and source are credited.

The Evolution of Mathematics-Teachers' Community-of-Practice

Nitsa Movshovitz-Hadar and Atara Shriki

Aim and Rationale

A successful implementation of educational change depends on teachers' professional development, and their ability to translate innovative ideas into practice. Although teaching, by its very nature, is a complex practice, most teachers work in isolation, making their own planning and decisions, and solve pedagogical problems having limited consultation with and feedback from their colleagues. The past decade has seen increasing demand to improve school mathematics, which, as a result, generated a need for teachers to join forces and share individual knowledge and experience with the community. Thus, the need to nurture mathematics teachers' communities of practice became a primary goal.

Wenger (1998), who coined the term "community of practice" (CoP), maintains that in order for a community to be recognized as a CoP, a combination of three characteristics, cultivated in parallel, is necessary: (i) The domain: A CoP is identified by a common domain of interest; (ii) The community: A CoP consists of members who are engaged in joint activities and discussions, help each other, share information, and build relationships that enable them to learn from one other; (iii) The practice: Members of a CoP are practitioners. They develop a shared repertoire of resources, such as experiences, stories, tools, and ways of addressing recurring problems, thus learn with and from each other. In general, national communities of mathematics teachers conform to Wegner's first two characteristics: they definitely share an interest in mathematics, its teaching and learning, meet in professional

Organizers Co-chair: Nitsa Movshovitz-Hadar (Israel), Atara Shriki, (Israel); Team Members: Diane Resek (USA), Barbara Clarke (Australia), Jiansheng Bao (China); Liaison IPC Member: Gabriele Kaiser (Germany).

N. Movshovitz-Hadar (✉) · A. Shriki
Technion Israel Institute of Technology, Haifa, Israel
e-mail: nitsa@technion.ac.il

© The Author(s) 2015
S.J. Cho (ed.), *The Proceedings of the 12th International Congress on Mathematical Education*, DOI 10.1007/978-3-319-12688-3_66

conferences, read professional journals, and share a professional jargon enabling them to learn from one another. However, the third characteristic, to a large extent, is still missing in many communities of mathematics teachers, as only few develop a shared repertoire of resources. Even those communities of mathematics teachers who do develop such resources usually count on leaders of the community to put them together for the benefit of the entire community.

In light of the above, DG4 focused on issues related to the formation of a mathematics teachers' CoP (MTCoP) and their on-going handling from both theoretical and practical points of view.

Session 1: Triggers and Needs for CoPs to Be Formed—Theory and Practice

Following a short introduction that presented views from three continents (Barbara Clarke, Australia; Jiansheng Bao, China; Diane Resek, USA) participants were asked to share experiences and promising practices, and to consider the following questions in small groups:

- What triggers and needs for CoPs to be formed, can you identify based upon your own experiences/beliefs/research?
- Who are the initiators and what are their drivers?

In as much as possible, please anchor your perceptions in a theoretical framework.

The following are some of the issues and challenges identified during the discussion:

- Arriving at shared goals for the purpose of teaching, defining problems of teaching, and agreeing on problem definitions/boundaries is not a simple process, but no doubt challenging;
- Sometimes groups are dysfunctional and there are some features to be wary of in groups: For example, blaming the student rather than taking personal responsibility;
- It can be challenging to develop a genuine CoP due to norms of privacy being evident in many schooling cultures. For example: reluctance to 'open the classroom door' to other teachers;
- Getting teachers to focus on results of their change of practice versus just doing activities should be at the heart of working with MTCoPs;
- Leadership, trust, sustainability, and quality of relationships are required for an effective community of practice. These issues raise questions regarding who should lead and run a CoP (School teachers? University professors? Researchers? Consultants?), and how the nature of leadership effects the commitment and sustainability of the group;

- The community needs to continue learning, which may require redirection. Systemic support can be effective when it establishes a culture of professional collaboration with appropriate expertise.

Session 2: Forming, Running and Sustaining an Effective MTCoP

Following a short introductory presentation (Atara Shriki and Nitsa Movshowitz-Hadar, Israel), three subgroups were formed focusing on three themes that emerged from the first session:

- Forming and running of MTCoP: Bottom up vesus top-down models;
- Collective efficacy: How do we build mutual trust, sense of belonging and ownership;
- Sustainability of MTCoP.

In relating to these questions, participants were asked to provide concrete examples from their previous experience. Since it turned out that participants observed reciprocal connections between forming and running of MTCoP and its sustainability, we present these concerns together.

The following is a brief summary of the issues discussed:

Forming, running and sustaining MTCoPs. The design of professional development programs is mostly 'top down', done by teacher educators who are not necessarily members of the MTCoP to whom the program is targeted. The designers of such programs hardly ever ask teachers for their urgent needs and spend time responding to them. This might be one of the reasons for the unsustainability of most MTCoPs. Therefore, the question is what should be done in order to nurture these CoPs as independent groups that keep developing professionally without external assistance. It is also assumed that sustainability is dependent on the initial motivation for the group and whether it was internally or externally initiated. Namely, the sustainability of a MTCoP is directly affected by the driving force of the community. There has to be a desire (whether intrinsic or external) to change, to learn, and to transform. Some further related questions are: How to bring teachers to acknowledge the need to change their practice? What would teachers consider as change? How can teachers develop their ability to reflect on their change of practice?

Trust and Efficacy. Tensions exist in a functioning MTCoP. Although these are not bad, they need to be managed productively to move the group forward. One needs conflict to make changes, but also needs to build a rapport. There can be tensions between leadership and the ownership of participants, and tensions between making meetings compulsory versus having voluntary participation. Thus, it is necessary to be aware of these possible tensions, and discuss them openly with teachers.

Conclusion

DG4 provided an opportunity for productive dialogue and sharing of experiences from a range of contexts and countries. There are many positive experiences and experiential knowledge that need to be shared. We hope to continue these conversations into the future.

Open Access This chapter is distributed under the terms of the Creative Commons Attribution Noncommercial License, which permits any noncommercial use, distribution, and reproduction in any medium, provided the original author(s) and source are credited.

Reference

Wenger, E. (1998). *Communities of practice: Learning, meaning, and identity.* Cambridge, UK: Cambridge University Press.

Uses of History of Mathematics in School (Pupils Aged 6–13)

Bjørn Smestad

Report

Activities concerning history of mathematics have been a part of ICMEs since ICME2 in Exeter in 1972. They are now regular features of ICMEs, organized by the HPM (International Study Group on Relations between History and Pedagogy of Mathematics). The premise of this discussion group was that research on history of mathematics in education tends to have older pupils and students in mind, and that there is a lack of both research and resources on how to include a historical perspective when teaching younger pupils.

Three key questions were pointed out in the invitation to the discussion group:

- Which ideas from HPM can be used with children (aged 6–13) in such a way that produces good results (e.g. improved student engagement, positively impacted student learning)?
- What would be criteria for finding, developing and selecting materials to be used with children (aged 6–13)?
- How does the HPM community in particular (and mathematics education community more broadly) assure that high-quality material that cover a variety of topic are produced and shared?

Question 1 was discussed in the first session and questions 2 and 3 were discussed in the second session.

Organizers Co-chairs: Bjørn Smestad (Norway), Funda Gonulates (USA); Team Members: Narges Assarzadegan (Iran), Kathy Clark (USA), Konstantinos Nikolantonakis (Greece); Liaison IPC Member: Evelyne Barbin (France).

B. Smestad (✉)
Oslo and Akershus University College of Applied Sciences, Oslo, Norway
e-mail: bjorsme@hioa.no

© The Author(s) 2015

S.J. Cho (ed.), *The Proceedings of the 12th International Congress on Mathematical Education*, DOI 10.1007/978-3-319-12688-3_67

In the first session, Kathy Clark gave a short introduction of ideas from the vast literature on how and why to include history of mathematics in teaching. Thereafter, Narges Assarzadegan gave a short talk on how she has been working with her students in Iran on the topic. Kathy Clark subdivided question 1 into further sub-questions: What *are* the ideas for which HPM contributes meaningfully to the mathematical experience of pupils aged 6–13? What are the forms of *good results* we wish to happen? *How do we know* when good results occur? What are some of the *obstacles* that teachers using HPM with pupils of this age may encounter—and what are ways to address or minimize the obstacles?

Group discussions on the first question brought forward a wealth of ideas: the use of historical instruments, finding good problems from history to engage children of this age range, using concrete materials to visualize mathematics, working with words instead of symbols, exploring cross-curricular themes (for instance historical measuring units), using source material from the middle ages, studying materials from the cultures of children's parents and grandparents, and studying positive/negative numbers through history, to mention a few. More generally, it was pointed out that although "storytelling" was in our introduction described as just one of many ways of working with history of mathematics to kids, storytelling is indeed particularly important at this age level and should not be disparaged. Teachers that are able to fascinate their pupils with great (and meaningful) stories from the history of mathematics have a wonderful gift.

The good results participants wish to happen at this age level mostly has to do with the attitudes of the children: we want them to see mathematics as a fascinating cultural and human activity and make them connect to it in new ways. We will probably never be able to prove beyond doubt that using history of mathematics with children do have positive effects, as history of mathematics will always be just one of several elements a teacher uses simultaneously to engage his students. For the teacher, however, such proofs are not necessary—just seeing the pupils engaged is good enough.

Of course, there are obstacles—both in terms of resources and in teachers' opinion that history of mathematics will take time from mathematics. Moreover, as work on history of mathematics is not mandated in curricula in most countries, there is the ever-present need to justify it to colleagues who are not interested. This can also be lonely work. Some of these issues can partly be remedied by working on what we discussed in session 2, however.

In the second session, as an introduction to discussions on question 2 and 3, Bjørn Smestad and Kathy Clark gave some good examples of use of history of mathematics in teaching, including some from online sources.

The group came up with a long list of criteria for materials, noting that not every resource need to fit every criterion. The resource should:

- Include significant mathematics (and be curriculum-related)
- Include activity/task/problem/something for pupils to "do"
- Fire-up the imagination; inspire pupils to do mathematics
- Tell a story

- Have multiple representations (pictures, text, sound, video, interactivity)
- Show mathematics as a human endeavor (e.g., have a cultural aspect)
- Be doable in a "reasonable amount of time"
- Generate discussion, debate among the pupils
- Be authoritative and accurate

There are lots of materials on the internet, and at first you feel lost as it is difficult to see what is of good quality. After a while, you start being able to determine what "makes sense", but still you need to sort through a lot of bad stuff while looking for the gems. (But to get even there, you will probably need experience in using the materials—and where do you get that?) Thus, there is a need of a "clearing house" for keeping valuable materials in one location. This idea was developed further later in the discussion: what we need is a "Kantor project" (named after Moritz Kantor), mimicking the "Klein Project" in providing high-quality resources to teachers, for instance with comments both from historians of mathematics and from teachers who have used the resources with pupils (including information on how it was used and the perceived outcomes). In addition, the need for History of Mathematics courses and better resources at libraries, were mentioned.

The discussion group consisted of about 25 people from around the world, with a good mix of well-known faces in the HPM community and newcomers. This lead to good discussions where everybody took part. In that respect, we view the discussion group as a successful experience, and hope that the discussions here will inspire further work on teaching with history of mathematics for young pupils. We do hope there will be increased dissemination of ideas for this purpose in the years to come.

This report was written by Bjørn Smestad. He is happy to be contacted at bjorsme@hioa.no for further information on the works of this DG.

Open Access This chapter is distributed under the terms of the Creative Commons Attribution Noncommercial License, which permits any noncommercial use, distribution, and reproduction in any medium, provided the original author(s) and source are credited.

Postmodern Mathematics

Paul Ernest

Aim and Rationale

The goal was to elucidate the nature of postmodern mathematics and mathematics education, including the multiplicity of the subject of mathematics; and to explore and share ideas of how postmodern perspectives offer new ways of seeing mathematics, teachers, learners, mathematics education theories, research and practice. The two key themes originally used to organizer the sessions are as follows.

Theme 1 What does postmodern mathematics mean? Part of this means exploring perspectives of mathematics as a plurality, as having multiple forms, identities, locations. Even the name mathematics is plural, although modern(ist) usage overlooks this anomaly in treating it/them as a single entity.

Theme 2 What does postmodern mathematics education mean? This involves considering multiple-self perspectives of the human subject as teacher, learner or researcher.

Organizers Co-chairs: Paul Ernest (UK), Regina Möller (Germany); Team Members: Ubiratan D'Ambrosio (Brazil), Allan Tarp (Denmark), Sencer Corlu (USA), Gelsa Knijnik (Brazil), Maria Nikolakaki (Greece); Liaison IPC Member: Michèle Artigue (France).

P. Ernest (✉)
University of Exeter, Exeter, UK
e-mail: p.ernest@ex.ac.uk

© The Author(s) 2015
S.J. Cho (ed.), *The Proceedings of the 12th International Congress on Mathematical Education*, DOI 10.1007/978-3-319-12688-3_68

Background

Postmodernism rejects a single authoritative way of seeing mathematics, teachers and learners, for each can be seen and interpreted in multiple ways. Mathematics can be seen as axiomatic and logical leading to indubitable conclusions, but it can also be seen as intuitive and playful, open-ended, with surprises and humour, as evidenced in popular mathematical images and cartoons. Additionally it can be seen in its applications in science, information and communication technologies, as well as in everyday life and ethnomathematics. All of these dimensions are part of what makes up mathematics and they all co-exist successfully.

This perspective asserts that there is no such thing as mathematics. There is no unique object named by the term. There is no unique or fixed identity for the various knowledge realms, activities, practices, forms, identities or locations connoted by the term. However, mathematics (in the plural) do exist—evidently there is a multiplicity connoted by this term that varies according to the times, places, and purposes for which they are invoked. These different mathematics bear what Wittgenstein terms a family resemblance. There is no essential defining character shared by them all, but to a greater or lesser extent, they are recognizably related.

It is also important to recognize that all human subjects have multiple selves and that we all (mathematicians, teachers and learners—and we are always all three) have access to different selves: authoritative knowers, researchers, learners, appreciators and consumers of popular and other cultures, as well as having non-academic selves. Thus mathematics teachers can be seen as epistemological authorities in the classroom as well as co-explorers of unfamiliar realms both mathematical and cultural; and as ring-masters in the mathematical circus. Students can be seen as receivers of mathematical knowledge, but also as explorers, interpreters and sometimes even creators of mathematics and cultural realms that can be related to mathematics. All of these perspectives and selves are resources we can use to enhance the teaching and learning of mathematics, but many are currently overlooked or excluded.

Discussion Group aimed to provoke discussion on these and other issues, to raise and discuss these ideas and explore and generate examples relevant to classroom practices.

Papers, resources and discussions were shared on-line before the conference both via the official ICME-12 site and via Allan Tarp's MATHeCADEMY.net website. The purpose was to begin discussions before the conference and to share planned presentations in advance so that participants could prepare themselves and presentations could be kept short and much of the group time devoted to discussion and contributions from participants. As judged by informal feedback, the discussion group was very successful in this respect.

Key Questions

- Is there such a thing as postmodern mathematics?
- What is postmodern thinking in mathematics and mathematics education? What is new or different about it and what are the implications for research in mathematics education?
- Given a postmodern multiple-perspectives view of mathematics what illuminations and surprises can be found for mathematics and its teaching and learning in multidisciplinary sources including: history of mathematics, ethnomathematics, science, information and communication technology, art works, stories, cartoons, films, jokes, songs, puzzles, etc.?
- How might the new emphases and differences foregrounded by postmodern perspectives impact in the primary and secondary mathematics classrooms? What concrete examples serve to illustrate these differences?
- How can a multiple-selves view of the human subject be reflected in the mathematics classroom and in mathematics teacher education? How can a multiple-selves view of the teacher facilitate teacher education?
- To what extent are the theories and presentations offered at the conference and elsewhere in publications actually post-modern?

The discussion group opened with the showing of an animated movie discussion on Postmodern Mathematics Education between avatars for Paul Ernest and Allan Tarp. The movie is accessible at: http://www.youtube.com/watch?v=ArKY2y_ve_U and the script for the animated movie has been published in The Philosophy of Mathematics Education Journal no. 27, retrievable via http://people.exeter.ac.uk/PErnest/.

Among the key distinctions made in this dialogue were those between philosophical and cultural postmodernism. The former is concerned with multiple epistemologies and representations of knowledge, whereas the latter is more frivolous concerning the eclectic conjoining of different styles, with pastiche, bricolage and irony. The second distinction is between Anglo postmodernism (e.g., Rorty) with its focus on knowledge and uncertainty, and continental postmodernism with its emphasis on power (e.g., Foucault).

The remaining presentations in the group were as follows:

- Bill Atweh, Is the Good a Desire or an Obligation? The Possibility of Ethics for Mathematics Education
- Bal Chandra Luitel and Peter Charles Taylor, Fractals of 'Old' and 'New' Logics: A Post/modern Proposal for Transformative Mathematics Pedagogy
- Peter Collignon and Regina D. Möller, Postmodern Analysis
- Regina D. Möller and Elisabeth Mantel, Postmodern approaches in teacher education
- Allan Tarp, Postmodern Mathematics Education in Practice
- Paul Ernest, The importance of being erroneous in maths: to be wrong or not to be wrong?

Two further papers accepted for the group have been shared online as the presenters were unavoidably detained at the last minute and unable to present in person:

- Mônica Mesquita and Sal Restivo All Human Beings as Mathematical Workers: Sociology of Mathematics as a Voice in Support of the Ethnomathematics Posture and Against Essentialism
- Ilhan M. Izmirli Wittgenstein as a Social Constructivist

An invitation was given to all presenters and participants to comment or publish papers in *The Philosophy of Mathematics Education Journal*. Papers 1, 2, 7 and 8 have been published in *The Philosophy of Mathematics Education Journal* no. 27 (2013), retrievable via http://people.exeter.ac.uk/PErnest/.

The meetings were well attended and there was a lively and controversial set of presentations, questions and responses including extensive audience generated questioning and dialogue. The most frequent question directed to speakers was: "The presentation was interesting, but it could also be given under a cognitivist or constructivist framework! In what way was it postmodern?"

Several presenters answered this from their own perspectives. Paul Ernest's answer is that the social constructivism he has been developing is intended to be a postmodern philosophy of mathematics and mathematics education because:

- It rejects absolutism and accepts multiple perspectives on mathematics and teaching and learning;
- It is grounded in human practices (language games embedded in forms of life— following Wittgenstein)—and this pre-knowable grounding and reality takes precedence over any theorizing (and also installs a deep ethics as the first philosophy of mathematics education);
- It refuses priority to either A the objective or subjective forms of knowledge, B the social or individual forms of being, or C the structural or agentic forms of power—arguing that these are mutually constitutive pairs—two sides of the same coin.

All participants were delighted to be involved with a controversial contemporary topic that is genuinely best approached through discussion.

Open Access This chapter is distributed under the terms of the Creative Commons Attribution Noncommercial License, which permits any noncommercial use, distribution, and reproduction in any medium, provided the original author(s) and source are credited.

Improving Teacher Professional Development Through Lesson Study

Toshiakira Fujii and Akihiko Takahashi

Discussion

The purpose of this proposed Discussion Group is to facilitate discussion and initiate collaborative research with colleagues around the world to seek effective ways to improve teacher professional development through Lesson Study.

Each session, Session 1 (Tuesday, July 10) and Session 2 (Saturday, July 14) was began by two of four key questions addressed by the panel. Then the panelists and the participants had fruitful discussion around the key questions.

Session 1 (Tuesday, July 10).

Key Questions

- What are the key elements of Lesson Study that can help teachers gain mathematical knowledge for teaching?
- What are the key elements of Lesson Study that can help teachers develop expertise in teaching mathematics effectively?

Chair: Toshiakira Fujii (Japan); Discussant: Susie Groves (Australia); Panel: Jennifer Lewis (USA), Yoshinori Shimizu (Japan), Akihiko Takahashi (USA),Tad Watanabe (USA), Nobuki Watanabe (Japan); Reporter: Yo-An Lee (Korea).

Session 2 (Saturday, July 14).

Organizers Co-chairs: Toshiakira Fujii (Japan), Akihiko Takahashi (USA); Team Members: Susie Groves (Australia), Yo-An Lee (Korea); Liaison IPC Member: Mercy Kazima (Malawi).

T. Fujii (✉)
Tokyo Gakugei University, Tokyo, Japan
e-mail: tfujii@u-gakugei.ac.jp

A. Takahashi
DePaul University, Chicago, USA
e-mail: atakahas@depaul.edu

© The Author(s) 2015

609

S.J. Cho (ed.), *The Proceedings of the 12th International Congress on Mathematical Education*, DOI 10.1007/978-3-319-12688-3_69

Key Questions

- How can an established effective professional development model such as Lesson Study be translated for use in different cultures?
- How can a professional development model such as Lesson Study be adapted for use in pre-service teacher education?

Chair: Akihiko Takahashi (USA); Discussant: Lim Chap Sam (Malaysia); Panel: Koichi Nakamura, (Japan), Anika Dreher (Germany), Don Gilmore (USA), Berinderjeet Kaur (Singapore), Thomas E. Ricks (USA); Reporter: Yo-An Lee (Korea).

Results

As the result of two-day discussion the following questions are raised by the discussion group for further discussion:

Question 1: Although we recognize the roles of lesson study facilitators/leaders/outside experts are important, what does expertise mean in conducting lesson study is still not entirely clear.

Outside experts can push LS members in terms of (1) content knowledge (what kinds of knowledge is needed at each particular teaching episode), (2) pedagogical knowledge (how to help students with particular contents), and (3) interactional (whether LS groups are mature enough to take and produce criticism).

In Japan, university in-service training can link to school practices where expertise come into play. In-service training, there are multiple levels. Notable is that LS study groups often have expert among themselves who can handle some aspects of lesson study. In general, it is hard to find a capable expert.

Question 2: Although lesson study is a form of professional development based on collaboration, sometimes teachers feel uncomfortable when criticized by their colleagues.

American teachers are not familiar with or comfortable with criticizing. You want to criticize lesson, not teaching. This could be an effective way of handling the pressure of giver and taker of criticism. One way to handle the pressure is to have groups work on lesson plans and choose one person to teach at the last minute. Lesson study helps teachers work together and to see what kids are doing. Lesson study should try to move their focus from "what I did wrong" to "what the issues were" "what students did (not) learn?"

Question 3: May be good idea to explore more about authentic school based lesson study in Japan.

What students should learn may be clearer in Japan. Developing school research theme might be a good starting point for group of teachers. (Theme means "what do you want to do in the class"). Although novice teachers don't know how to anticipate student responses, they learn how to anticipate student responses through school-based lesson study.

Open Access This chapter is distributed under the terms of the Creative Commons Attribution Noncommercial License, which permits any noncommercial use, distribution, and reproduction in any medium, provided the original author(s) and source are credited.

Theory and Perspective of Mathematics Learning and Teaching from the Asian Regions

Chun Chor Litwin Cheng

Report

The DG has prepared a questionnaire to collect data of teachers' practise in China, Taiwan, Hong Kong, and Korea. The results, together with the literature search in theory and practice in mathematics education was prepared into a booklet of 90 pages for discussion during the ICME-12. There were two sessions of discussion during the ICME-12 and the following is a report of work and discussion of the DG during the ICME-12.

The Chinese Framework and Theories in Mathematics Education

Two practices in China dated back to the 13 century. One is the technique of using analogy by Yang Hui (楊輝) in 1275, which work on two problems which shared the same structure and one can apply the method of the first question to solve the second problems. Another technique is the using of more than one solution to tackle the same problem by Li Zhi (李治, 1248) when he investigates cases of circles inscribed in a right angle triangle.

Organizers Co-chairs: Chun Chor Litwin, Cheng (Hong Kong), Hong Zhang (China); Team Members: Eunmi Choi (Korea), Po-Hung, Liu (Taiwan), Lianghuo Fan (UK); Liaison IPC Member: Shiqi Lee (China).

C.C.L. Cheng (✉)
Hong Kong Institute of Education, Tai Po, Hong Kong
e-mail: cccheng@ied.edu.hk

© The Author(s) 2015
613

S.J. Cho (ed.), *The Proceedings of the 12th International Congress on Mathematical Education*, DOI 10.1007/978-3-319-12688-3_70

Approaches in Understanding and Learning Mathematics in Taiwan

Fou-Lai Lin (National Taiwan Normal University) suggested that the components of being a good teacher include (1) Vary methods, (2) "Skillfully waits to be questioned" and "Hear the questions", (3) Teach students "how to learn", and (4) Know the reasons why teaching is successful or failed. And conjecturing approach is the principle of teaching mathematics.

Different Approaches in Understanding and Learning Mathematics in China

We know that the models and approaches developed in Mainland China these years include:

1. "Four Basic" model (structure approach and heuristic approach)
2. Problem solving model (structure approach)
3. Trial Teaching and Learning approach (heuristic approach),
4. GX experiment and model, (through correspondence, induction and deduction)
5. Teaching through variation approach (structure approach),
6. Demonstration, imitation and practise approach (structure approach) and
7. Dialectic approach for abstraction and internalization (structure approach).

The Characteristics of Chinese Mathematics Education and Four Basics Model

Zhang Dian-zhou (Eastern China Normal University) proposed the 5 aspects of characteristics in China Mathematics Education:

1. good lesson introduction of new topics,
2. technique of interaction among teachers and students in large classes,
3. teaching of mathematical thinking with variation method,
4. variation in teaching and exercises, and
5. fluency in practice-for-sophistication.

Zhang also proposed "Four Basics" model in mathematics teaching. The model has three dimensions and these three dimensions intertwined with each other in the process of learning.

Dimension 1: the accumulation of Basic Mathematics Knowledge (relational and procedural).

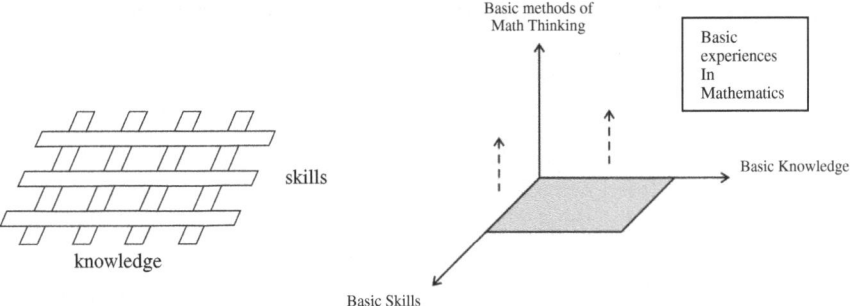

Fig. 1 Dimensional representation of the foundations of mathematics module

Dimension 2: the development of Basic skills (mathematical skill and skills to known procedure).

Dimension 3: the process of Basic Mathematical Thinking (application, formation of method of Mathematical Thinking, and develop new method).

The Basic Experiences in mathematical activities form as glue to connect the three-dimensional module (Fig. 1).

The Korean Framework and Theories in Mathematics Education

A survey conducted by Chung in Korea found that the most important thing that teachers considered in teaching and learning mathematics are (1) understanding 'concepts', (2) 'principles', and (3) 'process'. In Korea, teacher's role is described as "Goon Sa Boo Il Che" (君師父一體), that means King, Teacher and Father are the same one. These circumstances can be explained by culture tradition under Confucian Heritage Culture (CHC) in Korea. Though passive learning in traditional classroom is changing into more active learning in recent reformed classroom. But the zealous of learning under CHC culture still the core of the classroom in Korea.

Kyung Hwa Lee of the Seoul National University indicated that: "Good" Mathematics Teaching and "Good" Teachers usually means typical Korean math teacher have the orchestration of lessons based on the following four areas (a) Systematic instruction, (b) Coherent explanation, (c) Complete practice, and (d) Efficient imprinting.

The Japanese Framework and Theories in Mathematics Education

Masami Isoda (University of Tsukuba) indicated that there are a few traditions in the Japanese teaching of mathematics. The first one is the Japanese Problem Solving Approach for Learning by/for students. The second one is Problematic Situation explained by the Contradiction between Conceptual and Procedural Knowledge Originated from Mathematics Curriculum. And the third one is learning how to develop mathematics.

The aims of the traditions are achieved through the following teaching approaches in classroom:

1. Categorizing students' ideas from Meaning and Procedure.
2. Explaining Contradiction by Meaning/Conceptual and Procedural Framework
3. Procedurization of meaning,

Open Access This chapter is distributed under the terms of the Creative Commons Attribution Noncommercial License, which permits any noncommercial use, distribution, and reproduction in any medium, provided the original author(s) and source are credited.

Using Technology to Integrate Geometry and Algebra in the Study of Functions

Scott Steketee

Report

This group discussed the potential value of using technology-supported geometric transformations to introduce and develop function concepts. This approach (referred to here as *Geometric Functions*) and related representations can be used to help students develop intuitive understandings, avoid and overcome misconceptions, and deepen their understanding of variables and functions.

Why is this approach not more widely used? What are the benefits and obstacles? How can this approach be encouraged and facilitated? The session agendas and notes are available on the DG9 wiki, which also contains links to resources (including movies and existing student activities): http://wiki.geometricfunctions.com/index.php/ICME_12_Discussion_Group_9.

Our DG addressed a number of questions. Why are geometric transformations not more widely integrated into the study of function? What are the benefits, and what are the obstacles? What experiences have discussants had in promoting such an approach? How can we best encourage and facilitate such a change in students' experience of function?

Each session began with a whole-group introduction, broke into small-group discussions addressing particular areas, and concluded with a whole-group summary. The bullet points below are based on the reports from the small groups.

Organizers Co-chairs: Scott Steketee (USA), Cheah Ui Hock (Malaysia); Team Members: Ang Keng Cheng (Singapore), Aleksandra Cizmesija (Croatia), Ali Lelice (Turkey); Liaison IPC Member: Hee-chan Lew (Korea).

S. Steketee (✉)
University of Pennsylvania, Pennsylvania, USA
e-mail: stek@geometricfunctions.org

© The Author(s) 2015
S.J. Cho (ed.), *The Proceedings of the 12th International Congress on Mathematical Education*, DOI 10.1007/978-3-319-12688-3_71

What We Know

- Students need to experience a variety of functions to form a robust conception.
- Though other examples may be given, the conventional approach quickly settles down to $\mathcal{R} \to \mathcal{R}$ functions defined by equations. But many important functions do not merely map real numbers to real numbers.
- The Geometric Functions approach may contribute to a richer concept of function.
- Transforming points versus transforming shapes is an issue; we need to be clear about this distinction.
- We have anecdotal evidence that teachers don't connect algebra and geometry. In most (all?) countries geometric transformations are taught independently of functions. (One participant observed that five mathematics educators from five different countries agreed that geometric transformations are an independent topic from functions in their mathematics curriculum.)
- Students have difficulties with many function-related concepts (variable, function, domain, range, relative rate of change, composition, and inverses).

Research Questions

- How does the Geometric Functions approach differ from current practice? What might it add? Might important elements be lost?
- How can Geometric Functions expand students' understanding of function? How might students' conceptions of variable, function, domain, etc. be strengthened?
- How might this approach help students' concept of function move along the action-process-object (APOS) sequence?
- How does current thinking on embodied cognition support the Geometric Functions approach? How do students experience Geometric Functions as embodied?
- It's important to present functions in a way that does not introduce misconceptions. What impact does this approach have on common misconceptions about function?
- How can physical activities supplement technology-based activities?

Implementation Issues

- When the teacher starts using Geometric Functions activities, how quickly should she go to technology? Can she use some real-world activities before using virtual activities? (See the Function Dance activity, www.geometricfunctions. org/function_dances.html, for one example of this real to virtual transition.)

- Transformations are sometimes taught in the elementary curriculum. How does this affect the use of Geometric Functions in teaching transformations?
- What would teachers need to know about math that may be unfamiliar to them?
- Professional development and support for teachers should be just-in-time.
- Team teaching may be very useful when teaching unfamiliar topics.
- How could teachers get comfortable with the technology? Ideally the technology should be transparent, so that the focus is on the math. Students' experiences should be mathematical rather than magical.
- The goal of experiences with Geometric Functions is to facilitate conversations about what functions really are, and about the connections between Geometric Functions and functions that are normally studied ($\mathcal{R} \rightarrow \mathcal{R}$ functions expressed as equations).
- Assessment is a problem, since students often do not have the opportunity to use technology during tests. How can this situation be corrected?

Conclusion

Discussion Group 9 concluded that teaching geometric transformations as functions has significant potential for improving students' understanding of function concepts and for avoiding common misconceptions, and that dynamic mathematics technology is a promising way for students to experience geometric transformation as a conceptual metaphor on which to ground their conception of function.

DG9 further concluded that more research should be done to establish the benefits of the Geometric Functions approach and to determine effective ways to implement it. Geometric Functions challenge both wide-spread curricular assumptions (that functions belong to algebra, not geometry) and teachers' typical mathematical background and knowledge, and require careful thought and preparation for effective implementation.

Given the poor student understanding of function concepts that results from current practices, and the proven value of incorporating students' sensory-motor systems in the learning process, we encourage mathematics educators and education researchers to take seriously these twin arguments for studying and implementing the Geometric Functions approach.

Open Access This chapter is distributed under the terms of the Creative Commons Attribution Noncommercial License, which permits any noncommercial use, distribution, and reproduction in any medium, provided the original author(s) and source are credited.

New Challenges in Developing Dynamic Software for Teaching Mathematics

Zsolt Lavicza and Balazs Koren

Report

The principal aims of the discussion group were to discuss the development of a variety of mathematical software applications developed during the past decades. Among the most often utilised software types in education are Computer Algebra Systems (e.g. Derive, Mathematica, Maple, Maxima), Dynamic Geometry Systems (e.g. Cabri Geometry, Geometer's Sketchpad, Cinderella, GeoNEXT, GeoGebra), Spreadsheet and Statistics Software (e.g. Excel, SPSS, Fathom, R). Most of this software has been designed by keeping in sight primarily their usability for research purposes while others were predominantly aimed for their use in teaching. In the recent years we could observe, among others, three important trends in the development of these software tools: (1) Designers of research oriented software products started to involve features and support for educational purposes; at the same time teaching oriented software have been becoming increasingly more powerful so their use in some research is increasing; (2) The distinction between different types of software has begun to blur as many products integrate features from other types of software; (3) The computer platforms are diversifying; with the appearance of smart phones, tablets, and Interactive Whiteboards (IWB) in recent years, as well as online services such as Wolfram Alpha, challenging the design and development of mathematics software.

Organizers Co-chairs: Zsolt Lavicza (UK), Markus Hohenwarter (Austria); Andrian Oldknow (UK), Tolga Kabaca (Turkey), Kyeong Choi (Korea); Liasion IPC Member: Frederick Leung (Hong Kong).

Z. Lavicza (✉)
University of Cambridge, Cambridge, UK
e-mail: zl221@cam.ac.uk

B. Koren
Eotvos Lorand University, Budapest, Hungary
e-mail: balazs.koren@gmail.com

© The Author(s) 2015
S.J. Cho (ed.), *The Proceedings of the 12th International Congress on Mathematical Education*, DOI 10.1007/978-3-319-12688-3_72

The Discussion Group aimed to elaborate some of the outlined issues with a mix of short presentations, questions, and reflections from the audience. DG-10 was divided into two sessions both of them were attended by approximately 60–70 participants.

Session I

In the First session Zsolt Lavicza and Balazs Koren outlined the aims of the session and set a schedule for presentations and discussions. The first presentation was given by Balazs Koren, Hungary introducing the different software available and used in mathematics teaching and research. Balazs wanted to have this presentation to challenge thinking of the audience and cluster software into tentative groups that was supposed to be dissected or rearranged into new categories and groups during the sessions. In the past decades, three kinds of tools were mainly used in schools: desktop software, specialised handheld calculators and more recently tablet and mobile phone apps, as well as web-based applications assisting mathematics teaching and learning. However, this grouping is getting outdated and the community should develop new clusters and characterizations to advance software applications and related theories. It seems that the borders between categories are getting more and more overlapped and we are converging towards more complex and adaptive systems in the near future.

The second presentation was given by Tolga Kabaca from Turkey. Dr Kabaca described his experiences with the mathematics community using a variety of software in Turkey. He emphasized that it is necessary to allow teachers to develop their own teaching applications, but at the same time there should be a structured system as well as training that allow them to bring technology into the classroom. In addition, Dr. Kabaca mentioned the importance of getting feedback for both software and material developers directly from the software tools.

Peter Boon from the Freudenthal Institute, Netherlands described the needs for extended environments around mathematical software. It is getting common to embed mathematical software into Learning Management Systems (LMS) or to a so called Digital Mathematical Environment (DME), which are enabling teachers and students manage learning within and outside classrooms, offer assignments of problems and collect data from their solutions. Developing such LMS and DME systems is difficult and may take years of improvements and modifications as well as needs an interaction with the used mathematics software environments.

Chris Sangwin, UK outlined his experiences in developing assessment tools for mathematics. STACK is an assessment environment developed by Dr. Sangwin including a number of own solutions, but at the same time drawing on the resources of other mathematical software. Assessment is one of the most controversial and difficult issues in today's educational debate so that creating an environment is challenging and risky. However, such environments are necessary for the educational system and ample thoughts have been invested into such developments.

The discussions during the sessions reflected the topics of the presentations. At the end there were suggestions for questions to be further investigated through debates and research:

- How do we or should we classify software applications?
- What does Dynamic means in a software environments?
- Do the use technology to enable mathematics learning, if yes, how?
- Do software need to offer and restructure social dynamics in classrooms and on the web?
- We need to emphasize the pedagogical uses of the software and develop them accordingly to enhance further opportunities for learning.

Session II

We consider this session as a historical event as all widely used software creators or leaders of their teams were represented in the room. Jean-Marie Laborde (Cabri), Ulli Kortenkamp (Cindarella), Zsolt Lavicza and Balazs Koren (GeoGebra), and Nicholas Jackiw (Geometer's Sketchpad).

The session also started with the presentation of Ulli Kortenkamp, Germany, who highlighted the difficulties and processes in the development of mathematical software. Dr Kortenkamp emphasized that there are a number of issues could arise when mathematical theories needed to be implemented in a computer environment. For example, matching Euclidian and Hyperbolic geometry into a single software could be challenging thus the community of mathematicians and software developers need to have forums to discuss possibilities for implementation.

Tatsuyoshi Hamada, Japan talked about a wide range of software developed in Japan and the difficulties of their spread across groups and universities. Professor Hamada created a downloadable live Linux application called Math Libre, which is a collection of freely available software tools for mathematics teaching and research. Trough this collection the authors aims that schools, teachers, and students can choose the best applications fitting their needs in education. The project contributed to the involvement of technologies in the curricula in many schools around the world.

Jean-Marie Laborde, France stressed the importance of quality and the mathematical correctness of software development. Professor Laborde described that software development is a costly process and needs to be done in a complex way to ensure the correct mathematical background of the underlining processes within the software. Thus, it is important that while choosing a software must be made based on quality and rather than economy.

Finally Zsolt Lavicza, UK outlined the development of an open source project and the importance of a community surrounding the software. GeoGebra has become a successful mathematical tool, because teachers and students found them on the Internet and started to contribute to both software and material development.

Due to the large user base and the responsiveness of developers to the requests of the users the software was developing quickly and attempt to correct the problems arising during its use.

The debate after the presentations initiated further ideas and questions:

- How we can deal with infrastructure issues in schools, in particular in lesser-developed regions in the world?
- How can we encourage education to use current rather than outdated technologies?
- Do we need to develop specialised or general software? Do we need to connect development with other fields such as with video game?
- How can we learn from the success of long existing and sustained software packages such as R?
- How can we best support LMS and DMS with mathematical software development?
- How can we deal with the complexity of mathematical software development?
- How can we set some guidelines for assessment with computers?
- Possibly we need flexible and customisable tools in the near future
- We need to produce more books and learning materials with different tools

DG-10 offered an inspiring environment to discuss issues for both developers and users of products. The presentations and reflections were fruitful, but because software use in mathematics education is still around the start line with the exception of some larger projects the session ended with more questions than answers. However, the beginning of such discussion is valuable and offered food for thoughts for participants and we believe already impacted the development of software.

Open Access This chapter is distributed under the terms of the Creative Commons Attribution Noncommercial License, which permits any noncommercial use, distribution, and reproduction in any medium, provided the original author(s) and source are credited.

Mathematics Teacher Retention

Axelle Faughn and Barbara Pence

The main question addressed in this Discussion Group is whether professional development can have a positive effect on the retention of mathematics teacher, and if so what is the nature of professional development that leads to teacher retention and how (what are the mechanisms by which) such professional development supports teacher retention.

Background information, literature on Mathematics Teacher Retention, and important guiding questions were listed on the Discussion Group Website at https://sites.google.com/a/cmpso.org/icme2012/ for participants to actively engage in discussions guided through individual contributions by researchers from Israel, New Zealand, Norway, South Africa, India, and the United States. The major theme of *Supporting mathematics teachers: Transition into the workplace and professional development* underlined all discussions. Sub-themes were classified under seven major strands: (1) Mathematics Content and Pedagogy, including Technology; (2) Models of Support; (3) Communities of Practice, including online and lesson study; (4) Teacher Leadership; (5) Research; (6) Policy; (7) Mathematics Teacher Identity. Discussion group organizers were also interested in the magnitude of teacher retention issues in various countries, as well as their local and global impact on mathematics education.

Initial discussions included concerns about what happens at the pre-service level. Choosing teaching training is often a last choice for students, which makes producing qualified teachers a challenge. Many participants indicated trends towards an increasingly diverse range of people choosing teaching as a career choice.

Organizers Co-chairs: Axelle Faughn (USA); Barbara Pence (USA); Team Members: Glenda Anthony (NZ), Mellony Graven (South Africa), Claire V. Berg (Norway); Liasion IPC Member: Oh Nam Kwon (Korea).

A. Faughn (✉)
Western Carolina University, Cullowhee, USA
e-mail: afaughn@email.wcu.edu

B. Pence
San Jose State University, San Jose, USA
e-mail: barbara.pence@sjsu.edu

© The Author(s) 2015 625
S.J. Cho (ed.), *The Proceedings of the 12th International Congress on Mathematical Education*, DOI 10.1007/978-3-319-12688-3_73

In countries such as NZ mathematics teachers recruits included an increasing number of career switchers who did not necessary regard teaching as a life-long career. In other countries such as SA, the critical shortages of mathematics recruits and subsequent high number of out-of-field teachers contributed to high attrition rates. In the United States, 50 % of teachers leave within their first 5 years of teaching, and the mode for teaching expectancy of beginning mathematics teachers is 1 year. Problems contributing to teacher attrition that were highlighted included the inappropriate or challenging placement of new teachers, and the rise in number of unqualified teachers teaching mathematics.

Karsenty described a model of support for new teachers of at-risk students in Israel (SHLAV) based on weekly on-site meetings with a mentor. The support in this model was personalized and included discussions on material/content, teaching strategies, and affective issues of students at risk. Through this support model, mentees felt empowered and gained confidence, but the questions of sustainability of such a model was raised in light of cost-effectiveness and what happens once funds for such programs run out. The idea of fostering deep changes in the school through material sharing and networking to form a community that shares information and resources was put forward as an answer to sustainable change.

Graven noted that support in the form of one day training sessions which negate teacher experiences and communicate a 'fix-it' type of approach based on giving teachers new ways of teaching undermines teacher confidence. In South Africa, 55 % of teachers say they would leave if they could. Emphasizing life-long learning as a continuous professional process and redefining one's identity from hiding shame of not knowing to acknowledging one-selves as life-long learners while knowing where to ask should be embedded in professional development interventions. Other components of successful professional development include providing teacher autonomy, fostering a sense of belonging, empowering through increased confidence, which all bring sustainable changes in practice. The question of sustainability through teacher leadership and promotion was raised, as well as the development of a strong professional identity through leadership. Questions linking teacher retention and identity were also considered.

Common themes across different contexts concerned a lack of qualified mathematics teachers, difficulty with recruitment of quality teachers, and teacher dissatisfaction with the profession. East Asian, where the status of teachers in the society is still high and salaries competitive, were noticeably absent from this discussion. In contrast to interventions that offer 'add on' support, East Asian countries offer systemic opportunities and expectations for teachers to grow and to advance in their professions. China has a well-established and coherent professional development system through the use of teaching researchers who serve as collaborators, facilitators and mentors, and get involved at different levels of teaching research activities by developing research lessons. This is the basis of an intricate ranking and promotion system that includes lesson competitions and ensures that theory is implemented and tested in the classroom. Teachers in East Asia are actively involved at every level of the teaching profession, from training of pre-service teachers, to development of curricular material and delivery of professional development. Madhana Rao described a two-tiered educational system in the district of Warangal of Andhra Pradesh State of India where private institutions are linked to teacher attrition while government schools retain 100 % teachers. Stability in this case is also attributed to a state-led system that provides promotional opportunities and a regular salary to teachers, with professional development interventions.

Participants from other countries reported that teachers who want to become leaders and see their influence increase do not always have the institutional support to do so. When Australian teachers reach the top of the pay scale after 10 years of teaching they can only be promoted as administrators. Anthony examined New Zealand's induction system with time allotted for mentees and mentors to meet regarding concerns pertaining to the teaching profession such as learning about school context, completion of accreditation requirements, and the necessity and tools to become a professional inquirer. The mandated induction program with extra time for planning and support for Year 1 and 2 beginning teachers is a significant factor in low attrition rates. However, the development of a wider community base of support in teacher education programs is necessary to help inform and equip pre-service teachers to proactively counter dissatisfaction and disappointment with the profession and the nature of school culture.

Berg's model is a community of inquiry that aims at replicating material and teaching processes introduced during workshops into the classroom, such as asking good questions during an inquiry activity. Two participating groups qualified as "New Comers" and "Old Timers" showed different levels of appreciation for the professional development, which brings up the question of what is a minimum length for a professional development project to induce sustained change in classroom practices. In this project participants found a community and support outside of their school, a situation echoed by Pence in the California-based Supporting Teachers to Increase Retention project. Pence provided a glimpse on 10 professional development site models aimed at supporting teachers and increase retention. Findings from the project included the emergence of professional communities across all sites and the importance of leadership in keeping teachers motivated and involved. The professional development model for each site was targeted over multiple years, content specific, challenging, and went beyond mathematics content and pedagogy to focus on establishing teaching as a "noble" profession requiring work and preparation, growth that is complex, on-going, and supports the realization that there is a great deal to learn.

Contributions

Contributions to this Discussion group can be found at https://sites.google.com/a/
cmpso.org/icme2012/

Open Access This chapter is distributed under the terms of the Creative Commons Attribution
Noncommercial License, which permits any noncommercial use, distribution, and reproduction in
any medium, provided the original author(s) and source are credited.

Mathematics Teacher Educators' Knowledge for Teaching

Kim Beswick and Olive Chapman

Report

The aims of DG12 were to:

- Facilitate discussion of key issues related to the knowledge required by mathematics teacher educators (MTEs).
- Identify different emergent strands in research that can be related to this area.
- Summarise research and research/theoretical perspectives related to knowledge for mathematics teacher education.
- Identify research directions and potential collaborations that will move the field forward.

Four broad areas were suggested to frame discussions. In summary these were:

- To what extent are the various knowledge types for mathematics teachers described by Shulman (1987), Ball et al. (2008) and others applicable/transferable to MTEs? How does the knowledge needed by MTEs differ from that required by mathematics teachers? Is it a kind of meta-knowledge or something as distinct from the knowledge for teaching mathematics as knowledge for teaching science is?

Organizers Co-chairs: Kim Beswick (Australia), Olive Chapman (Canada); Team members: Merrilyn Goos (Australia), Orit Zaslavsky (USA); Liaison IPC Member: Gail Burill (USA).

K. Beswick (✉)
University of Tasmania, Launceston, Australia
e-mail: kim.beswick@utas.edu.au

O. Chapman
University of Calgary, Calgary, Canada
e-mail: chapman@ucalgary.ca

© The Author(s) 2015
S.J. Cho (ed.), *The Proceedings of the 12th International Congress on Mathematical Education*, DOI 10.1007/978-3-319-12688-3_74

- Who researches MTEs' knowledge? What are the dilemmas and opportunities associated with researching ourselves? What evidence is there of the knowledge required by MTEs? What measures/criteria are there for successful mathematics teacher education and how are they connected to MTEs' knowledge? What methodologies might be effective in building such an evidence base?
- How is knowledge for mathematics teacher education acquired? How is the transition from mathematics teacher to mathematics teacher educator made and what is gained or lost in the transition? To what extent and in what ways is knowledge for teaching mathematics necessary for MTEs? What theories of learning are useful? What models are/should be used?
- Why might it be important to articulate knowledge for MTEs? What contribution can understanding it make to our work and to mathematics education more broadly? Who wants to know about this knowledge and why?

The first session was attended by more than 45 participants from at least 18 different countries. There was a broad range of experience and expertise in relation to the topic with many participants acknowledging that they had not given MTEs' knowledge serious consideration prior to attending the discussion group. Discussion in session 1 focussed on areas 1, 3 and 4 and ended with participants writing down one or more questions that they had about MTEs' knowledge. These were grouped into five themes, summarised below, that formed the basis of discussion in the second session.

Theme 1: The Nature of the Knowledge Needed by MTEs

What knowledge of mathematics is needed by MTEs? What differences are there between teaching at university level and school? What is the distinction between Mathematics Knowledge for Teaching (MKT) and mathematics knowledge for MTEs? Is MKT the 'curriculum' that MTEs teach? How do MTEs' conceptions of teaching and learning develop? How can we research these? How do these conceptions translate into their teaching? Is there a connection to student learning? What aspects of MTEs' knowledge are important? What knowledge do MTEs for in-service teachers need? How is it different from knowledge needed for pre-service teacher education? How can MTEs for in-service MTEs be educated?

Theme 2: Different Types of Mathematics Teacher Educators and Implications for the Knowledge Needed

Who are the MTEs? How does local context impact on MTEs? What kinds of courses would cater for the differences between MTEs (e.g., mathematicians, former mathematics teachers, mathematics education researchers)? Is the same knowledge needed by all MTEs? Is it possible for one person to have/develop all the knowledge necessary? Is it helpful to consider mathematics education as team work?

Theme 3: Research Methodologies/Approaches

In what ways might teacher collaborative inquiry among MTEs provide a methodological framework for research in this area?

Theme 4: Acquisition of Knowledge for Mathematics Teacher Education

How can programs be developed specifically for MTEs of mathematics teachers at different schooling levels? How can professional development for existing MTEs be provided? What is the importance of role models in the development of MTEs? What knowledge is acquired through apprenticeship models? What are the relationships between MTEs' background and the way they acquire knowledge? How can MTEs develop the capacity for inquiry into their own practice? What is the role of collaboration and mentoring?

Theme 5: The Importance of Research in This Area

How can we ensure that the appropriate resources are allocated towards this work?

Future Directions

Many participants indicated their interest in progressing the work through a book or journal publication. There was also interest in international comparative research on MTE backgrounds and the relationship of this to MTE practice and outcomes.

Open Access This chapter is distributed under the terms of the Creative Commons Attribution Noncommercial License, which permits any noncommercial use, distribution, and reproduction in any medium, provided the original author(s) and source are credited.

References

Ball, D. L., Thames, M. H., & Phelps, G. (2008). Content knowledge for teaching: What makes it so special? *Journal of Teacher Education, 59*(5), 389-407.

Shulman, L. S. (1987). Knowledge and teaching: Foundations of the new reform. *Harvard Educational Review, 57*(1), 1-22.

The Role of Mathematics Education in Helping to Produce a Data Literate Society

William Finzer

Report

There were 30 participants from 17 countries. The challenge presented to the group was as follows: "The data revolution is everywhere except the classroom. In general, students finish their schooling seriously under-prepared to participate in the emerging data-driven society. This represents an enormous loss of scientific discovery, solutions to social problems, economic advancement, ..."

Organizers Co-chairs: William Finzer (USA), Cliff Konold (USA); Team Members: Maxine Pfannkuch (New Zealand), Michiko Watanabe (Japan), Yuan Zhiquiang (China); Liasion IPC Member: Yuriko Baldin Yuriko (Brazil).

W. Finzer (✉)
KCP Technologies, Emeryville, USA
e-mail: bfinzer@kcptech.com

© The Author(s) 2015
S.J. Cho (ed.), *The Proceedings of the 12th International Congress on Mathematical Education*, DOI 10.1007/978-3-319-12688-3_75

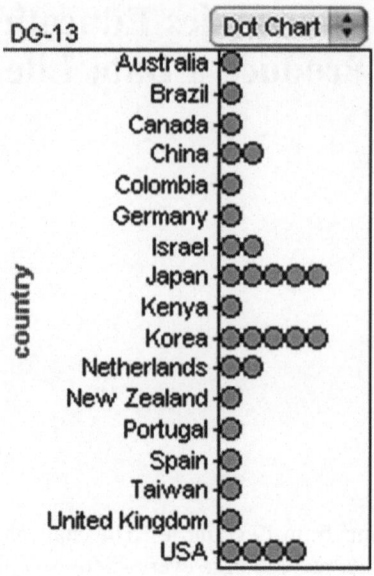

Highlights from Participants' Introductory Comments

- Biggest challenge in Kenya is infrastructure, so there is lack of computers and other technology.
- There is a new program in Portugal to try to emphasize statistical ideas in the early years, however the teachers do not have the knowledge to be able to implement this currently.
- One specific topic in Brazil in general education is data handling, but teachers do not know what it means. There is a new program to introduce data handling to master's students who will be teachers.
- One of the difficulties in Columbia is the different goals among different groups in the country. For example, government versus teacher goals. And the curriculum does not match what is happening in the schools.
- Students are underprepared not just technically, but ideologically. The problem with thinking that if data is on the web, then it's true. The other problem is if you believe in something strongly enough, then you don't need any data.
- Students in Korea explain interpretations very superficially. They are good at computing, but it's difficult for them to do data analysis.
- There is statistical literacy in the curriculum in New Zealand. The students start to talk about their own data, then critique other's data. In high school, it moves to looking at media reports and so on. The challenge is professional development for the teachers because they are not used to having these discussions in the classrooms.

- In China, probability and statistics have been content areas since 1978, but they haven't become the focus area until the new curriculum standards in 2001 and 2003. Teachers will now pay much attention to probability and students might get high scores in examination, but still not be able to solve authentic problems.

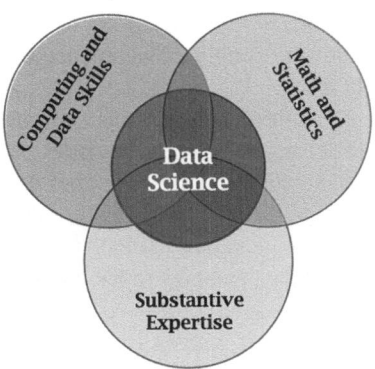

Highlights from Discussion of "Regarding the Problem, to What Extent Is What We Are Talking About Something that Goes Beyond 'the Need for More and Better Statistics Education?'"

- The expansion of statistic education should be manipulation of data, raising critical issues about data, and critiquing data.
- We are looking for boundary crossing between different perspectives including statistical education, heuristics (e.g. representativeness, availability, anchoring, etc.), and responsible critical social perspectives.
- It is not enough to teach more and better statistics in terms of computations. We think students need to be able to communicate, tell a story, and explain their interpretation. There needs to be something else in terms of communication.
- What is taught currently in schools in statistics is too narrow. It is just calculations that even when taught well, is still too narrow. (For example, archival of data is something that statisticians do not think about it, but in economics it is important. Also data cleaning. 90 % of the effort is spent in the data cleaning area, with very little time spent in data analysis.)

Highlights from Discussion of "What Strengths Do Mathematics and Statistics Educators Bring to Bear on the Problem? Conversely, with What Aspects of the Problem Are Mathematics and Statistics Educators Unlikely to Bring Expertise?"

- ability to think quantitatively, defining algebraic expressions, handicap—fear of there not being a right answer
- statistics is one part of mathematics in china, the mathematics teacher is the statistics teacher, statisticians are teaching statistics only, don't consider statistics at school level, good statistical knowledge but don't know how to teach at school level, guide the curriculum design, how to teach have no idea (lecturers)
- know the concepts of scientific enquiry (PPDAC), statisticians know the more concrete of the cycle, huge data and how to handle huge data, important and big role for statistics for big data age
- mathematics doesn't use context, statistics brings context, conflict. Is this content mathematics or social studies? PISA had a strong effect on what to teach in mathematics
- basic concepts in statistics no longer have the same status, huge data sets, different types of data, introduce students to different types of data, how to talk about signal of data, how to present the data, need to work at global level, how do we sample our global data, the basic ideas need to be revisited—data analysis, not statistics as this is an old word
- students learn statistics in the mathematics classroom so always want to know the right answer, "so what is the answer", this is a problem, look at decision making in a comparison situation
- society is not necessarily data driven but there is a flood of data.

Highlights from Discussion of "What Should Be the Role of Mathematics Education in Helping to Produce a Data Literate Society?"

- **classrooms**—Active engagement of students in the investigative process (PPDAC), importance of technology in teaching and learning statistics—creating the displays etc. so can focus on interpretation, also to build conceptual development, exploring outliers as an example.
- **concepts**—Variation and prediction, sample as a starting point, where does data come from, what is data, very important to have simple counting procedures and also by deciding what to measure, defining difficult variables and this is crucial for validity, ordinal and other scales.

- **policy**—What are the key ideas that we need to keep and/or build that we will always need; what do we have to change what we are doing in our classrooms? Ethical issues, big data not always available to all citizens, open data sets being available to all; statistics across the curriculum; look for New Zealand example and follow (blindly) cognizant of the local situation—collaboration amongst statisticians, educators and teachers in leading the reform in the curriculum; may need to collaborate with other disciplines, collaborate with internet experts, data people, computer experts and create discipline of data science.

Open Access This chapter is distributed under the terms of the Creative Commons Attribution Noncommercial License, which permits any noncommercial use, distribution, and reproduction in any medium, provided the original author(s) and source are credited.

Mathematical Modeling in Connecting Concepts to Real World Application

Zhonghe Wu and Lijun Ye

Aim and Rationale

In recent years, achieving mathematics proficiency has received notable attention [RAND 2003; National Research Council (NRC) 2001] What useful, appropriate, practical, and effective strategies can be developed and used to enhance student proficiency in mathematics is still a puzzle to mathematics educators. This urgent need becomes a challenging task for mathematics educators seeking research-based strategies to support classroom teachers to enhance their teaching leading to student proficiency.

The Mathematical Modeling is a research-based teaching model (Lesh and Zawojewski 2007; Niss et al. 1991) that builds conceptual understanding and problem solving skills. The mathematical modeling also reflects the core components of proficiency defined by research studies (Hill and Ball 2004; NRC 2001; RAND 2003)—conceptual understanding, computational skills, problem solving, mathematical reasoning, and mathematical disposition.

Organizers Co-chairs: Zhonghe Wu (USA), Lijun Ye (China); Team Members: Shuhua An (USA), Zhongxiong Fan (China), Ling Wang (China); Liaison IPC Member: Morten Blomhoej (Denmark).

Z. Wu (✉)
National University, California, USA
e-mail: zwu@nu.edu

L. Ye
Hangzhou Normal University, Hangzhou, China
e-mail: yeatsylj@126.com

© The Author(s) 2015

S.J. Cho (ed.), *The Proceedings of the 12th International Congress on Mathematical Education*, DOI 10.1007/978-3-319-12688-3_76

Key Questions

The following five broad areas frame the territory of the discussion.

- What is Mathematics Modeling? Why Mathematics Medeling?
- What is the relationship between mathematical modeling and mathematical proficiency? What does role of Mathematics Modeling play in teaching and learning mathematics for K-12 students?
- How is mathematical modeling used in primary school?
- How is mathematical modeling used in secondary school?
- What are the challenges and issues of mathematical modeling in teacher professional development?

Open Access This chapter is distributed under the terms of the Creative Commons Attribution Noncommercial License, which permits any noncommercial use, distribution, and reproduction in any medium, provided the original author(s) and source are credited.

Mathematics and Culture in Micronesia: An Exploration of the Mathematical Aspects of Indigenous Practices

A.J. (Sandy) Dawson

Aims

The aims of the discussion group are to (1) discuss the findings of the three indigenous authors (Mamangon, Moses and Velasquez) investigations thus far relative to mathematics and culture in Micronesia, (2) explore the challenges and successess achieved in using elders to uncover and validate indigenous knowledge and practices, (3) explore the pedagogical issues of how to translate the findings into materials and approaches suitable for elementary school children, and (4) consider implications for future research in other indigenous cultures. Indigenous mathematics, enthnomathematics, cultural-based mathematics

Key Questions

The discussion will allow an exchange of ideas, successes, and challenges in supporting indigenous activities, capturing the mathematics contained therein, and preserving those activities and the mathematics for future generations. The key questions addressed in the discussion group are:

- What mathematics has been uncovered by examining indigenous practices and activities of Micronesian peoples?
- How can this mathematics and the associated practices be used to teach mathematics to indigenous children?

Organziers Chair: A.J. (Sandy) Dawson (USA); Team Members: Danilo Mamangon (USA), Epi Moses (USA), Rhoda Velasquez (USA); Liaison IPC Member: Bill Barton (New Zealand).

A.J. (Sandy) Dawson (✉)
University of Hawaii-Mãnoa, Honolulu, USA

© The Author(s) 2015
S.J. Cho (ed.), *The Proceedings of the 12th International Congress on Mathematical Education*, DOI 10.1007/978-3-319-12688-3_77

- What are the challenges to conducting such research particularly working with elders and dealing with what, at times, is seen as 'protected' knowledge, and developing approaches to the teaching of mathematics with the focused populations?
- What lessons can be learned from this work with indigenous populations?

Structure of Sessions

Session One: 1.5 h

- 10 min: introductions and setting the scene
- 50 min: discussion of findings of the mathematics found in Micronesian cultural practices
- 25 min: discussion of pedagogical strategies developed for use with indigenous children
- 5 min: summary of session and closing remarks.

Session Two: 1.5 h

- 5 min: overview of previous day's discussion
- 15 min: further discussion of the pedagogical strategies
- 40 min: small group discussions regarding the challenges of conducting the research and devising implementation strategies
- 20 min: large group discussion and consolidation of the major issues, challenge and observations made during the discussion group—lessons learnt, honouring indigenous knowledge and ways of knowing
- 10 min: closing summary and suggestions for furthering the conversation begun here.

Project MACIMISE

Project MACIMISE (Mathematics and Culture in Micronesia: Integrating Societal Experiences) is supported by a National Science Foundation grant (0918309). This material in this paper is based on work supported by that grant. The content does not necessarily reflect the views of the NSF or any other agency of the US government. The Project is a collaborative effort between Pacific Resources for Education and Learning (PREL) and the University of Hawaii-Mānoa (UHM) with PREL as the lead organization.

Open Access This chapter is distributed under the terms of the Creative Commons Attribution Noncommercial License, which permits any noncommercial use, distribution, and reproduction in any medium, provided the original author(s) and source are credited.

Can Art Save Mathematics?

Dirk Huylebrouck

Aim and Rationale

"Can art save the world?" is a well-known catchphrase in art circles. As most participants to the ICME are mathematicians, the title of this DG was reformulated more modestly as: "Can art save mathematics?" Indeed, some call mathematics a supreme art form as it enjoys total freedom, unrestricted by material limitations. An art form with the "collateral advantage" of having many real life applications, sure. However, if it can be considered as art, why don't art and mathematics more often collaborate, for their mutual benefit?

In the past, carpenters or painters sometimes helped mathematicians in the construction of mathematical models which sometimes had artistic ambitions (intarsia, for instance), but today's computers allow mathematicians to express themselves in total freedom, without the help of intermediate persons or tools. However, mathematicians aren't necessarily artists and so this technological improvement does not necessarily guarantee better art. Also, while in the past the lack of mathematical knowledge by artists was a burden for the development of mathematical art, today this should no longer be the case: computer developments make mathematics more accessible to artists, despite their usual aversion for the pure sciences.

Organizers Co-chairs: Dirk Huylebrouck (Belgium), Slavik Jablan (Serbia); Team Members: Jin-Ho Park(Korea), Rinus Roelofs (The Netherlands), Radmila Sazdanovic (USA), J. Scott Carter (USA), Liaison IPC Member: Jaime Carvalho E. Silva (Portugal).

D. Huylebrouck (✉)
Faculty of Architecture, Katholieke Universiteit Leuven,
Paleizenstraat 65, 1030 Brussels, Ghent-Brussels, Belgium
e-mail: huylebrouck@gmail.com

© The Author(s) 2015
S.J. Cho (ed.), *The Proceedings of the 12th International Congress on Mathematical Education*, DOI 10.1007/978-3-319-12688-3_78

Yet how do we bridge the gap between mathematics and art so that mathematical art becomes an equally well "established" art field as, for instance, biological art or kinetic art? It would be beneficial for society because it would help to unite the "two cultures" of J.P. Snow, and because today's society needs designers interested in scientific developments.

Key Questions

- How much art should "artistic mathematicians" know in order to produce more than embellished mathematical results, so that their artistic mathematics are not mere "kitschy attempts"?
- How much mathematics should "mathematical artists" grasp in order to get really involved in the pure sciences, so that their mathematical art is not mere "baby math"?
- Or else, instead of turning mathematicians into 'artists' and artists into 'mathematicians', wouldn't it be better both sides simply cooperate—and if so, what should be the framework for such a collaboration?
- How can mathematics departments take mathematical art achievements into account in their output evaluation? For example, are mathematical art journals included in the journal rankings?
- How should the refereeing process work in this case where "peers" are by definition hard to find since the creative process implies every mathematical artwork should be unique? In the art world, refereeing is seldom done by peers.
- What is the difference between a scientific paper on mathematical art and a poetic artistic portrayal? The objectives of a purely mathematical paper are well known, but what about those of a paper on mathematical art?
- As for its implications in teaching mathematical art to art students, what are their specific needs and aspirations? The scientific "aha-Erlebnis" and "problem solving" are not sufficient, so how do we stimulate the creative mathematical approach?
- Is there a need for teaching mathematical art? The implications could be diverting students' attention from classical mathematics material (leading to "easy credit" courses). However, it could also raise awareness of the usefulness and the beauty of mathematics, inspiring students to continue taking math courses

Open Access This chapter is distributed under the terms of the Creative Commons Attribution Noncommercial License, which permits any noncommercial use, distribution, and reproduction in any medium, provided the original author(s) and source are credited.

Teaching of Problem Solving in School Mathematics Classrooms

Yew Hoong Leong and Rungfa Janjaruporn

Aim and Rationale

The 1980s saw a world-wide push for problem solving to be the central focus of the school mathematics curriculum since the publication of Polya's book about solving mathematics problems in 1954. However, attempts to teach problem solving typically emphasised the learning of heuristics and not the kind of mathematical thinking used by mathematicians. There appears to be a lack of success of any attempt to teach problem solving within school curriculum. Problem solving strategies learned at lower levels tended to be ignored instead of being applied in their mathematical engagements at the higher levels, possibly because of the routine nature of the high-stake national examinations. The era of mathematical problem solving, its research and teaching and learning in schools ended, ambivalent on research findings and imprecise on recommendations for its teaching in schools. Based on the teaching and research experience of the organising team, we think that problem solving should still be the direction for teaching mathematics in schools. As such, this discussion group is proposed to identify the practices in teaching problem solving in school mathematics classrooms across different parts of the world, and how these practices are linked to the success.

Organizers Co-chairs: Yew Hoong (Singapore), Rungfa Janjaruporn (Thailand); Team Members: Tomas Zdrahal (Czech Republic), Khiok Seng Quek (Singapore), Foo Him Ho (Singapore); Liaison IPC Member: Masataka Koyama (Japan).

Y.H. Leong (✉)
National Institute of Education, Singapore, Singapore
e-mail: yewhoong.leong@nie.edu.sg

R. Janjaruporn
Srinakharinwirot University, Bangkok, Thailand
e-mail: rungfajan@yahoo.com

© The Author(s) 2015

S.J. Cho (ed.), *The Proceedings of the 12th International Congress on Mathematical Education*, DOI 10.1007/978-3-319-12688-3_79

What Is the Place of Problem Solving in the School Mathematics Curriculum?

As a result of the publication of Polya's book about solving mathematics problems in 1954, National Council of Teachers of Mathematics and the worldwide educational reforms in school mathematics have recommended the study of problem solving at all levels of the mathematics curriculum. The reform documents indicate that problem solving should be the central focus of these mathematics curricula. It should not be an isolated part of mathematics instruction, it should be an integrated part of mathematics learning. Moreover, there is an expectation in these reform documents that even young students will improve their mathematical knowledge and procedures with understanding of problem solving. As a result of these reforms, problem solving has occupied a major focus in worldwide school mathematics curricula. Problem solving had been identified as both a major goal of instruction and a principal activity in mathematics teaching and learning.

How Much Curriculum Time Is Spent in Problem Solving in Comparison to the Other Components of Mathematics Curriculum?

By learning problem solving in mathematics, students acquire ways of thinking, habits of persistence and curiosity, and confidence in unfamiliar situations that will serve them well outside the mathematics classroom. In addition, problem solving is one of the basic skills that students must take along with them throughout their lives and use long after they have left school. To improve young student's mathematical knowledge and procedures, it is essential that the teacher should know how to teach problem solving and how to approach problem-solving instruction. They should also identify problem solving as a major goal of instruction and a principal activity in mathematics teaching and learning.

What Is the General Perception on the Importance of Mathematical Problem Solving Among the School Teachers?

Although worldwide educational reforms in school mathematics have recommended the study of problem solving at all levels of the mathematics curriculum, problem solving remains an unfamiliar idea for most school mathematics teachers. Many teachers lack the knowledge and confidence to teach mathematical problem solving. They also do not recognize the importance of mathematical problem solving in their classrooms. As such, the question, "How should we go about helping teachers increase their knowledge and confidence to teach problem solving?" has often been asked, and it is timely to ask whether examine teacher professional development programs that are targeted in this area. Apart from workshops on problem solving for teacher—which is the traditional mode of teacher development—there has been a shift towards models that develop teachers as owners and collaborators of innovations in the teaching of problem solving. One example of the latter is the mathematics problem solving for everyone (MProSE) project based in Singapore. Since three members of the organizing committee are investigators of this project, there was substantial capacity to present the details of MProSE in the DG meetings during the conference, which attracted much interest and lively discussion.

How Is Mathematical Problem Solving Assessed?

In order to have experience on problem solving, students should be expected to solve various types of problems in their own way on a regular basis and over a prolonged period of time. Non-routine problems and open-ended problems that provide students with a wide range of possibilities for choosing and making decisions should be used. Students should be asked to show their solutions in writing. A student's written work on a problem can be used to help evaluate progress in problem solving. A rubric scoring scale, both holistic scoring and analytic scoring, are methods for evaluating a student's written work on a problem.

In MProSE, the team introduced the "mathematics practical" into problem solving lessons using a "Practical" Worksheet. The students were encouraged to treat the problem-solving class as a mathematics "practical" lesson. The worksheet contains sections explicitly guiding the students to use Pólya's stages and problem solving heuristics to solve a mathematics problem. The scoring rubric focuses on the problem solving processes highlighted in the Practical Worksheet. The rubric allows the students to score as high as 70 % of the total 20 marks for a correct solution. However, this falls short of obtaining a distinction (75 %) for the problem. The rest would come from the marks in Checking and Extending. The intention is to push students to check and extend the problem (Stage 4 of Pólya's stages), an area of instruction in problem solving that has not been largely successful so far.

Summary

While the participants of the DG are from a wide range of jurisdictions—and hence different social-educational contexts, there is general agreement that mathematics problems solving of the type advocated originally by Polya and subsequently developed by other researchers remain important but elusive. There are many challenges, not least of which is teacher development. It is heartening to note—from the sharing of participants—that innovative projects were conducted, such as MProSE, to address this challenge.

Open Access This chapter is distributed under the terms of the Creative Commons Attribution Noncommercial License, which permits any noncommercial use, distribution, and reproduction in any medium, provided the original author(s) and source are credited.

Erratum to: Congratulatory Remarks: Minister of Education and Science, and Technology

Ju Ho Lee

Erratum to: "Congratulatory Remarks: Minister of Education and Science, and Technology" in: S.J. Cho, *The Proceedings of the 12th International Congress on Mathematical Education*, DOI 10.1007/978-3-319-12688-3_4

In the Opening Chapter 'Congratulatory Remarks: Minister of Education, Science and Technology', the second half of the text is missing. The full text should read as follows:

First of all, congratulations on the opening of the 12th International Congress on Mathematical Education.

I am glad that this important math event is being held in Korea this year.

Also, it is a great pleasure to welcome math education researchers and math teachers from more than 100 countries.

With the aim of transforming Korea into a nation of great science and technology capacity, and a nation of outstanding human talent, the Ministry of Education, Science and Technology of Korea is focusing on three important points in designing and implementing its policies.

The three points are "creativity", "convergence", and "human talent". Creativity enables us to think outside the box, convergence allows us to go beyond the

---

The online version of the original chapter can be found under
DOI 10.1007/978-3-319-12688-3_4

J.H. Lee (✉)
Former Minister of Education, Science and Technology, KDI School
of Public Policy and Management, Seoul, Republic of Korea

© The Author(s) 2015
S.J. Cho (ed.), *The Proceedings of the 12th International Congress on Mathematical Education*, DOI 10.1007/978-3-319-12688-3_80

traditional boundaries between disciplines, and finally human talent builds the very foundation that make all these possible.

Without a doubt, these are the most essential elements in today's knowledge based society. Math is the very subject that can foster much needed creativity and convergence, and is becoming a core factor in raising national competitiveness.

Math is behind everything.

The ICT revolution would have been impossible without the binary system.

The technology behind the CT scans can be traced back to simultaneous equations.

The launching of Korea's first carrier rocket—KSLV-1—will be controlled by a computer program that is based on complicated math equations.

As such, math is behind all technologies we are benefiting from. Anywhere you go in the world, math is considered one of the most important school subjects.

Countries around the world are increasing investment in math education because logical and rational thinking abilities are essential for our students to become creative talents, especially in today's knowledge-based society. Both abilities can be achieved through effective math education. The Korean government is making efforts to improve math education to foster creative talents.

We are planning to revise math textbooks to introduce story-telling methods, so that students can see math principles at work every day in their real lives.

The classroom environment will be changed to help students experience and experiment, rather than simply solving math problems.

The government also stands ready to provide necessary support for math teachers for their professional development.

Once again, congratulations on the opening of the event and I wish all of you a very successful and productive Congress.

Thank you.

Open Access This chapter is distributed under the terms of the Creative Commons Attribution Noncommercial License, which permits any noncommercial use, distribution, and reproduction in any medium, provided the original author(s) and source are credited.